III International Scientific and Technical Conference on Energy Systems (ICES-2018)

IOP Conference Series: Materials Science and Engineering
Volume 552

Belorgod, Russia
29 – 31 November 2018

ISBN: 978-1-5108-9090-9
ISSN: 1757-8981

Printed from e-media with permission by:

Curran Associates, Inc.
57 Morehouse Lane
Red Hook, NY 12571

Some format issues inherent in the e-media version may also appear in this print version.

Printed by Curran Associates, Inc. (2019)

For permission requests, please contact the Institute of Physics at the address below.

Institute of Physics
Dirac House, Temple Back
Bristol BS1 6BE UK

Phone: 44 1 17 929 7481
Fax: 44 1 17 920 0979

techtracking@iop.org

Additional copies of this publication are available from:

Curran Associates, Inc.
57 Morehouse Lane
Red Hook, NY 12571 USA
Phone: 845-758-0400
Fax: 845-758-2633
Email: curran@proceedings.com
Web: www.proceedings.com

Table of contents

Volume 552

III International Scientific and Technical Conference on Energy Systems (ICES-2018)

29–30 November 2018, Belorgod, Russian Federation

Accepted papers received: 13 May 2019

Preface

Papers

Heat and power engineering

Energy and resource saving in technologies

IOP Conf. Series: Materials Science and Engineering **552** (2019) 011001 doi:10.1088/1757-899X/552/1/011001

Preface

Information about the conference
The III international scientific and technical conference " Energy systems" (ICES - 2018) was held on November 29-30, 2018 on the basis of the Federal state budget institution of higher education" Belgorod State Technological University named after V. G. Shukhov" (BGTU named after V. G. Shukhov).

The conference contributed to the solution of fundamental scientific problems in the field of energy. The main point is systematic approach. The studied systems are complex and requires management of large systems and integration of knowledge. Also, one of the objectives of the conference was the wide involvement of young scientists – students, undergraduates and graduate students.

Conference website:

http://energysystem.bstu.ru

The working languages of the conference were English and Russian. The 72 reports were selected for the conference. As a result of the conference 19 Russian-language reports were prepared in the main section of the conference and 7 reports in the section of young scientists, 28 reports were in English. The 18 reports were issued in both Russian and English versions (the articles are not just a translation of the same text into two languages, but have significant differences due to different volumes of publications, requirements for the structure of the material and the number of references in the bibliography). This volume contains reports in English. The reports in Russian can be found on the page http://es.bstu.ru/archive/2018.

IOP Conf. Series: Materials Science and Engineering **552** (2019) 011001 doi:10.1088/1757-899X/552/1/011001

Fundamental and practical issues in the scientific fields of heat technology, energy and resource saving in industrial production are considered.

The conference demonstrated further approval of innovative approaches on the basis of fundamental research. The identified problems and new challenges on the way to strengthening interdisciplinary ties in the energy industry were shown. It was demonstrated an increase in the need to provide the do- mestic industry with modern scientific and technical developments, as well as the strengthening the interaction of Universities and industry.

In addition to the scientific program, the guests of the conference took a tour after BSTU territory, Belgorod city, have visit the objects of alternative energy-biogas station "Luchki" with a capacity of 3.6 MW, solar and wind power plants, have visit the diorama "Fire arc" and the Museum-reserve Prokhorov field – the Third battlefield of Russia.

Belgorod, Russia, 2018
Prof. Pavel Trubaev
E-mail trubaev@gmail.com

RFBR The conference is held with financial support
of the Russian Foundation for Basic Research (RFBR)
project No. 18-08-20138

IOP Conf. Series: Materials Science and Engineering **552** (2019) 011001 doi:10.1088/1757-899X/552/1/011001

Conference sections

The main problems of the research were determined by the selected sections based on the analysis of the received applications:

Topic 1. Heat and power engineering (10 reports)

The fundamental and practical issues in the scientific fields of heat and mass transfer processes, heat generation, heat conversion to oth- er types of energy, providing consumers with energy resources were considered.

Speaker
Alexey Biryukov, DSc

Speaker
Michael Averbukh, DSc

Topic 2. Electric power and electrical engineering (8 reports)

Fundamental and practical issues in the scientific fields of production, distribution, transformation and use of electric energy, development and optimization of electronic components were considered.

Topic 3. Energy saving and energy efficiency, power ecology (13 reports)

Fundamental and practical questions in scientific areas of decrease in an expense of energy resources, increases in energy efficiency in power and heattechnical systems, energy efficiency of buildings, buildings, constructions, impacts of power on the environment, elimination of emissions and pollution, power use of waste were considered.

Speaker Pavel Trubaev, DSc

Speaker Evgeniy Gasho, PhD

Topic 4. Alternative and renewable energy sources (6 reports)

Fundamental and practical issues in the scientific fields of production, distribution, transformation and use of electric energy, development and optimization of electronic components were considered.

Topic 5. Energy and resource saving in technologies (9 reports)

Fundamental and practical issues in the scientific fields of heat tech- nology, energy and resource saving in industrial production are con- sidered.

Speaker
Vladimir Sevostyanov, DSc

IOP Conf. Series: Materials Science and Engineering **552** (2019) 011001 doi:10.1088/1757-899X/552/1/011001

Conferees

The conference was attended by 160 people from 30 organizations (including 26 universities and four enterprises) located in 16 Russian regions and four foreign countries

- Bagdad, University of Technology;
- Barnaul, Altai State Agricultural University;
- Barnaul, Altai State University;
- Belgorod, Belgorod State National Research University;
- Belgorod, Belgorod State Technological University named after V. G. Shukhov;
- Belgorod, LLC "TC "Ekotrans",;
- Bet Yaqub, University of Diala;
- Chita, Transbaikal State University;
- Donetsk, Donetsk National Technical University;
- Donetsk, Donetsk National University of Economics and Trade named after Mikhail Tugan-Baranovsky;
- Ekaterinburg, Ural Federal University named after the first President of Russia B.N. Yeltsin;
- Ivanovo, State Power Engineering University;
- Izhevsk, LLC " Vakuummash Electro";
- Kazan, Kazan National Research Technological University;
- Kazan, Kazan State Power Engineering University;
- Khujand, Polytechnic Institute of Tajik Technical University named after M. S. Osimi;
- Makeyevka, Donbas National Academy of Civil Engineering and architecture;
- Milwaukee, University of Wisconsin-Milwaukee;
- Moscow, LLC "Eurogaz";
- Moscow, Moscow State University of Civil Engineering;
- Moscow, National Research University "MPEI";
- Novosibirsk, Novosibirsk State Technical University;
- Samara, Samara State Technical University;
- Smolensk, Branch of the Moscow Power Engineering Institute;
- Smolensk, LLC "Elektroset-Smolensk";
- St. Petersburg, Peter the Great St. Petersburg Polytechnic University;
- Tver, Tver State technical University;
- Ufa, Ufa State Petroleum Technological University;
- Vladimir, Vladimir State University named after Alexander and Nikolay Stoletov;
- Volzhsky, Branch of the Moscow Power Engineering Institute;

The number of authors is as follows:

- Dr. of tech. Sciences (DSc), professors-28 persons;
- Cand. of tech. Sciences (Phd), associate professors – 56 persons;
- Engineers, assistants, senior teachers – 29 persons;
- Graduate students – 20 persons;
- Graduate and undergraduate students – 27 persons.

The huge demand of this conference was by the young scientists, which were coauthors of the work of their senior colleagues, and the sole authors of the reports.

IOP Conf. Series: Materials Science and Engineering **552** (2019) 011001 doi:10.1088/1757-899X/552/1/011001

PROGRAM COMMITTEE

ORGANIZING COMMITTEE OF THE CONFERENCE

Peer review statement

All papers published in this volume of *IOP Conference Series: Materials Science and Engineering* have been peer reviewed through processes administered by the proceedings Editors. Reviews were conducted by expert referees to the professional and scientific standards expected of a proceedings journal published by IOP Publishing.

IOP Conf. Series: Materials Science and Engineering **552** (2019) 012001 doi:10.1088/1757-899X/552/1/012001

Experimental investigation of hydrodynamics and heat transfer of sphere cooling using air/water mist two phase flow

A H Abed[1, 2], S E Shcheklein[1], V M Pakhaluev[1]

[1]Department of Nuclear Power Plants and Renewable Energy Sources, Ural Federal University named after the first President of Russia B. N. Yeltsin, Mira St., 19, Yekaterinburg 620002, Russia
[2]Department of Electromechanical Engineering, University of Technology, Al-Sina'a Street, Baghdad 10066, Iraq

E-mail: akraaam82@yahoo.com

Abstract. To improve the cooling performance for the future generation of gas-cooled equipment, experimental studies on air/water mist heat transfer of single sphere inside a cylindrical channel have been carried out with an aim to investigate the heat transfer enhancement by suspending tiny water mist into air flow. The effect of the different key factors such the inlet Reynolds number, surface heat flux and water flux density on friction flow and heat transfer characteristics are examined. Experiments were performed in five cases using air as well as air/water mist two phase flow as working coolant for range of water flux density (j = 23.39 - 111.68 kg m-2 hr-1). The results obtained from the related experimental work revealed that the presence of water mist leads to a significant increase in heat transfer over the use of air coolant alone. The Nusselt numbers are respectively 1%, 19.7%, 90.2% and 134% higher than those in single phase - cooling for all cases of water flux density respectively. It was also found that the water flux density has little influence on friction factor. When the surface heat flux is fixed, the heat transfer enhancement factor increases with the increasing of water flux density.

1. Introduction

Heat transfer enhancement of a spherical heated surface or objects that are shaped as spheres is a problem of importance in a set of industrial applications such as chemical process industries and gas-cooled equipments [1]. To solve this problem, many researchers are attempting to find more effective techniques to improve the heat transfer efficiency [2]. One of the new and potentially advantageous of heat transfer enhancement techniques is air/water mist two phase flow [3]. The main features of air/water mist compared to single phase air flow can be outlined as follows: absorbs a large amount of energy in the form of latent heat during the evaporation process of water mist, increase specific heat of mixture, increase the turbulence into the air main flow and inside the boundary layer. [4]. While single-phase heat transfer and aerodynamic phenomena in spherical element has been widely studied, few studies have been found on air/water mist two phase flow. Most studies are found deal with other topics of heat transfer such as flat plate heated wall, gas turbine systems and tubular heat exchangers, including air-fine water droplets and steam-fine water droplets [5-10]. Wang et al. [11] performed a series of experiments to investigate the cooling of hot chrome-plated stainless steel plate using water mist for different surface temperatures and mist characteristics (droplet size and droplet velocity). Also Wang and Li [4, 12, and 13] numerically studied the air/water mist film cooling with 2D and 3D cases in gas turbine system. They found by injecting 2% and 5% water mist with 10 μm droplets diameter into the coolant, the cooling performance can be improved significantly about 50-65%. Allais I and Alvarez G [14] experimentally investigated single-phase as well as two-phase flow of heat transfer rate from packed bed using air/water mist flow. The experiments were performed in low airflow velocity, low water mass flow rate and limited ranges of temperature difference. The airflow velocity and water mass flow rate were in rang (0.35 to 1.4 m s^{-1}) and (0.7 to 1.9 g s^{-1}), respectively with 8μm water droplets diameter. It was found that the total heat flux increased with increasing airflow velocity and the thermal enhancement factor increased with increasing water mass flow rate by 2.8 times under experimental conditions. A theoretical analysis of Heat and mass transfer from/to a water-droplet for the conditions pertaining to the humid air tunnel has been carried out by Barrow and Pope [15]. They reported that the time of droplet evaporation (life time) increase rapidly with increase in droplet diameter. Kim et al. [16] performed computational analysis to investigate the evaporative cooling process using moisture of air and fine water droplets by various heat and mass transfer models. Zhao and Wang [17] presented an experimental investigation to understand the film cooling enhancement by injecting tiny water droplets along with the main coolant. They pointed that the film cooling performance was enhanced significantly by dispersing water droplets with 5 μm diameters in air coolant. They also pointed that the cooling coverage increase by air/water mist coolant compared with air coolant and more uniform surface temperature can be achieved. The present experimental study proposes a new definition of heat transfer enhancement of single sphere inside a cylindrical channel with a constant uniform heat flux using air as well as air/water mist coolant as a test fluids. The influences of the different key factors such water flux density, surface heat flux and inlet Reynolds number on friction flow and heat transfer characteristics were examined.

2. Materials and Methods

An experimental facility was designed and constructed to conduct single phase as well as two phase experiments in the department of nuclear power plants and renewable energy sources / Ural Federal University, Russia. The experimental facility is composed of an air subsystem, mist subsystem, mixing section, test object with heating element and data acquisition system, a schematic layout of the experimental set-up is showed in figure 1. The air subsystem consists of transparent Plexiglas channel, which had 50 mm diameter, 3 mm wall thickness and 940 mm total length. An electrical air vacuum blower of specification (3000RPM, 1000 W) was used with variac voltage regulator to control airflow velocity. On the same axis, a compact pitot tube with digital manometer was used to measure the mean flow velocity. The test object, a 34 mm diameter sphere, was fabricated out of copper metal inserted in flow stream and it is rear supported by textolite rod (k= 0.023 W/m °C, 4mm-diameter) which span across the Plexiglas channel. The test sphere was ohmically heated by using A.C. stainless steel electrical heater type CG- high density cartridge heater whose dimension are 8mm-diameter and 31mm-length. The maximum power of A.C cartridge heater is 100W and the input power has been measured by a voltcraft digital multimeter type APPA 109N with an accuracy (±0.06%). The input power was adjusted using variac voltage regulator so as to achieve the required surface heat flux. To enhance the thermal conductance between the test sphere and A.C cartridge heater, high thermal conductivity material was used. A schematic drawing of the test sphere heat transfer model has been shown in figure 2. To measurements the surface temperature, five calibrated thermocouples type chromel-alumel were tapped in positions which are: two thermocouples placed inside the copper sphere located in positions indicated by the numbers 1, 2 as shown in figure 2, one thermocouple placed in the flow entrance region and the last two placed in the flow outlet region. The thermocouples were inserted in drilling shallow holes (diameter 1 mm) in the sphere, and the copper surface was creased around the holes to install them in place. All thermocouples are connected to data acquisition system that consists from analog signal input module type OWEN MV110-8A with MSD200 data logger. The air/water mist coolant was obtained by blending tiny water droplets (2.73 μm) in diameter into air flow. The mist subsystem was a 1.7 MHz ultrasonic mist generator consists of a piezoelectric transducers, water tank and blower fan. This type of mist generator was chosen due to the low power consumption and a very

IOP Conf. Series: Materials Science and Engineering **552** (2019) 012001 doi:10.1088/1757-899X/552/1/012001

quiet operation comparing with other mist and evaporative generators. The water mass flow rate was measured by the mass of water passing a mist subsystem per unit time. To blend the water mist supplied from mist subsystem with a controlled amount of air, a mixing chamber was installed and used as the blender.

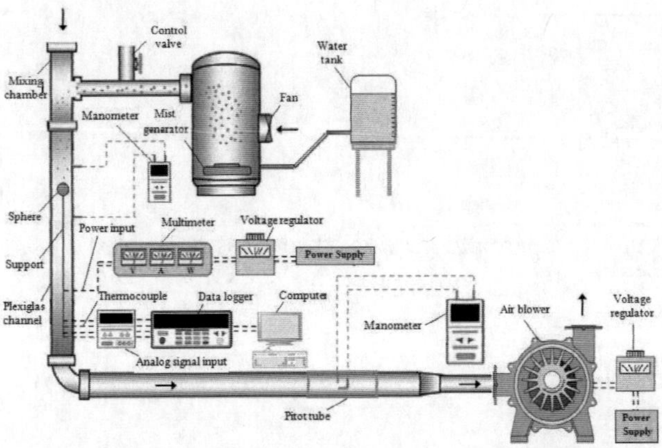

Figure 1. Schematic diagram of the experimental test bench.

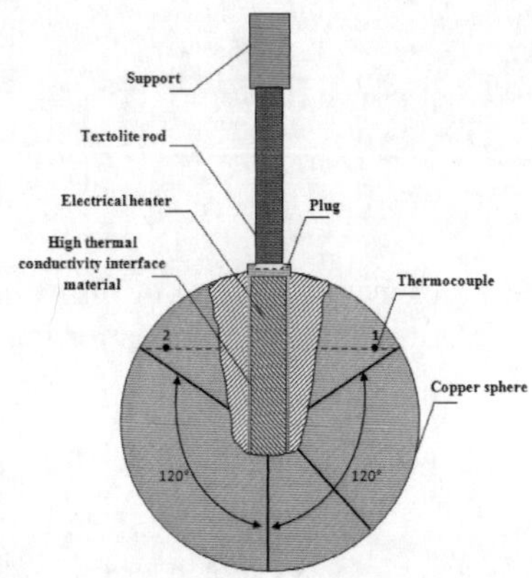

Figure 2. Schematic drawing showing the construction of heat transfer models.

In our study, The heat flux from heating system it was assumed to be uniform. It can be estimated as Eq. (1), [18].

$$q = \frac{Q_{el.} - Q_{los.}}{A_s}, \tag{1}$$

where $Q_{el.}$ - the input electrical power to the test sphere and $Q_{los.}$ - the heat loss. The electrical power of heating system is proportional to the voltage applied to the electrical heater to maintain the required heat flux under a forced convection condition as per Eq. (2), [19].

$$Q_{el.} = \frac{V^2}{R} \tag{2}$$

For a single phase flow, the average heat transfer coefficient can be described as, [20].

$$\alpha = \frac{q}{\left(T_{ave.s} - T_{ia}\right)}, \tag{3}$$

where $T_{ave.s}$ - the average surface temperature, T_{ia} - the main inlet air temperature. The water droplets trajectory can be predicted by applying Newton's second law. The energy balance equation for each droplet can be written as, [21].

$$m_p C_p \frac{dT}{dt} = \pi d^2 \alpha \left(T_{ave.s} - T_{ia}\right) + \frac{dm_p}{dt} h_{fg}, \tag{4}$$

where h_{fg} - the latent heat of evaporation. For the air/water mist flow, the average heat transfer coefficient typically can be calculated by using the wet bulb temperature at the inlet of test section as the reference temperature T_{im} that is, [21].

$$\alpha_{mist} = \frac{q}{\left(T_{ave.s} - T_{im}\right)} \tag{5}$$

$$T_{ave.s} = \sum_{x=1}^{n} \frac{T_{sx}}{n} \tag{6}$$

The water mass flow rate of mist phase is small compared to air flow. The mass flow rate is used to calculate Reynolds number as formula (7), [22].

$$\mathrm{Re} = \frac{\dot{m}\ D}{\mu A_t} \tag{7}$$

The average Nusselt number is obtained by formula (8).

$$Nu = \frac{\alpha d_s}{\lambda_a} \tag{8}$$

where d_s - the diameter of the sphere, λ_a - thermal conductivity of air. Based on the capillary wave mechanism, the droplet diameter can be calculated using Lang equation (9), [23].

$$d_p = 0.34\left(\frac{8\pi\sigma}{\rho F^2}\right)^{1/3} \tag{9}$$

where F- the working frequency of ultrasonic mist generator. The friction factor is obtained by Darcy–Weisbach equation (10), [1].

$$f = \frac{2\Delta P}{N\rho U^2} \tag{10}$$

where N - the number of sphere in channel. For a constant pumping power, the relationship between friction factor and Reynolds number can be expressed as:

$$\left(\dot{V}_a \Delta P_a\right) = \left(\dot{V}_m \Delta P_m\right) \tag{11}$$

$$\left(f\,\mathrm{Re}_a^3\right) = \left(f\,\mathrm{Re}_m^3\right) \tag{12}$$

$$\mathrm{Re}_a = \mathrm{Re}_m \left(f_m / f_a\right)^{1/3} \tag{13}$$

The thermal enhancement factor can be defined as the ratio of the heat transfer coefficient of air/water mist flow to that of a single phase flow at constant pumping power. It can be written as (14), [1].

$$\eta = \frac{\alpha_m}{\alpha_a} = \frac{Nu_m}{Nu_a} = \left(\frac{Nu_m}{Nu_a}\right)\left(\frac{f_m}{f_a}\right)^{-1/3} \tag{14}$$

The reliability of experimental facility has been presented by determining the uncertainties of experimental results [24, 25]. The uncertainty of Nusselt number can be expressed as follows:

$$\left(\frac{\Delta Nu}{Nu}\right)^2 = \frac{1}{Nu}\left[\left(\frac{\partial}{\partial \alpha}(Nu)\Delta\alpha\right)^2 + \left(\frac{\partial}{\partial d_s}(Nu)\Delta d_s\right)^2 + \left(\frac{\partial}{\partial k_s}(Nu)\Delta k_s\right)^2\right], \tag{15}$$

where $\alpha = q/\left(T_{ave.s} - T_{ia}\right)$

$$\left(\frac{\Delta\alpha}{\alpha}\right)^2 = \frac{1}{\alpha}\left[\left(\frac{\partial}{\partial q}(\alpha)\Delta q\right)^2 + \left(\frac{\partial}{\partial T_{ave.s}}(\alpha)\Delta T_{ave.s}\right)^2 + \left(\frac{\partial}{\partial T_{ia}}(\alpha)\Delta T_{ia}\right)^2\right] \tag{16}$$

The uncertainties of Reynolds number and friction factor can be determined as follows:

$$\left(\frac{\Delta \mathrm{Re}}{\mathrm{Re}}\right)^2 = \frac{1}{\mathrm{Re}}\left[\left(\frac{\partial}{\partial u}(\mathrm{Re})\Delta u\right)^2 + \left(\frac{\partial}{\partial \rho}(\mathrm{Re})\Delta\rho\right)^2 + \left(\frac{\partial}{\partial D}(\mathrm{Re})\Delta D\right)^2 + \left(\frac{\partial}{\partial \mu}(\mathrm{Re})\Delta\mu\right)^2\right] \tag{17}$$

$$\left(\frac{\Delta f}{f}\right)^2 = \frac{1}{f}\left[\left(\frac{\partial}{\partial \Delta P}(f)\Delta(\Delta P)\right)^2 + \left(\frac{\partial}{\partial \rho}(f)\Delta\rho\right)^2 + \left(\frac{\partial}{\partial \mathrm{Re}}(f)\Delta \mathrm{Re}\right)^2\right] \tag{18}$$

The uncertainties of non-dimensional parameters were found to be 4%, 1.91% and 1.8%, for Nusselt number, Reynolds number and friction factor respectively.

3. Results and discussion

3.1. Effect of surface temperature

The surface temperature is an important factor that has a great influence on air/water mist heat transfer process. When the surface temperature is very hot, the water mist may be completely evaporated before arriving to the heated surface due to the force of evaporation. On the other hand, at low surface temperature, the water mist may be wet the heated surface and form a liquid film. In this study, several heat fluxes were performed to examine the effect of surface temperature on heat transfer characteristic. Figure 3, shows the average surface temperature under different heat fluxes, Reynolds number and water flux density. It can be seen that the surface temperature in single-phase flow tends to decrease gradually with raising the Reynolds number.

IOP Conf. Series: Materials Science and Engineering **552** (2019) 012001 doi:10.1088/1757-899X/552/1/012001

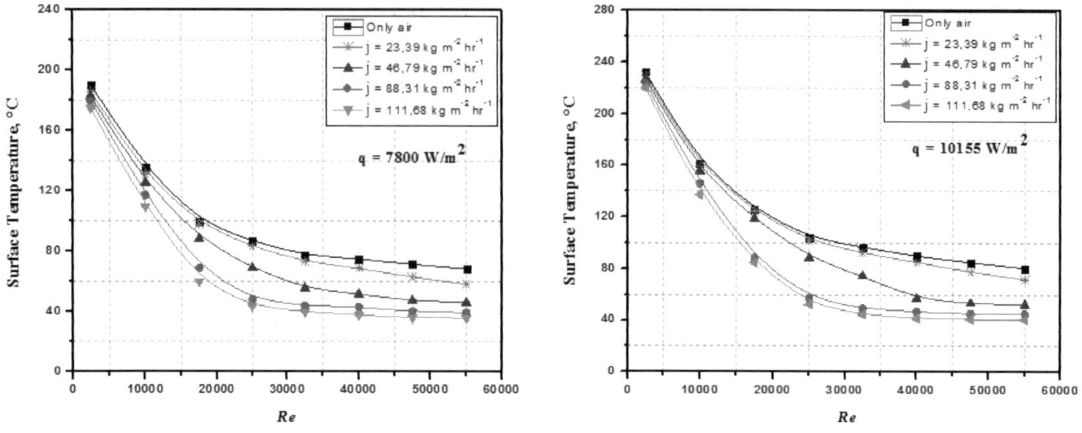

Figure 3. Variations of surface temperature of sphere with Re number under various water flux density.

In air/water mist flow, the influence of water mist evaporation appears and can be observed that the surface temperature decreases as water flux density increases for constant heat flux. The surface temperature decreases about 4%, 17%, 37% and 47% compared with air - cooling under different water flux. By the direct observation method through transparent Plexiglas channel it is observed that the heated surface covered with thin water film at high water flux density and Reynolds number (Re>17500). While the heated surface remains dry even at the lowest heat flux (q = 7800 W /m2) and Reynolds number (Re<17500), which has a surface temperature Ts > 100 °C.

3.2. Effects of Reynolds number and water flux density

The variation of average Nusselt number with Reynolds number under different water flux density is presented in Figure 4 for all cases of heat flux. In this Figures it can be observed that the average Nusselt number significantly increases as the Reynolds number increases under constant water flux density. In single-phase flow, the Nusselt number increases from 66 to 240 at range of Re number (Re = 2500 – 55000) under constant heat flux (q = 10155 W/m^2). For the Reynolds number range considered, the Nusselt numbers of air/water mist flow are respectively 1%, 19.7%, 90.2% and 134% higher than those in the single phase for range of water flux density (j = 23.39 – 111.68 kg m^{-2} hr^{-1}). In current study high Reynolds number means high main flow velocity, which drives more and more of water mist to arrive the heat transfer surface and absorbs a large amount of energy in the form of latent heat during the evaporation process. Besides, the water mist can penetrate the film-layer and enhances the heat transfer process by increasing the surface contact with main flow. As the surface temperature increases (Re<17500) the Nusselt number decrease due to water mist vaporization without wetting the heated surface. The heat transfer enhancement of air/water mist may include four important different parts: the water mist evaporation on the heating surface, sensible heat of impinging water mist, convective heat transfer based on a decline temperature gradient due to water mist evaporation in the air flow near the heated surface and enthalpy transport based on water mist departure which is effectively used for cooling. Under the conditions of q = 7800 and 10155 W/m^2 at different range of Reynolds number the average Nusselt number monotonically increases by increases the water flux density. Also it is important to mention that the maximum enhancement frequently occurs at high water concentration and low temperature surface of sphere.

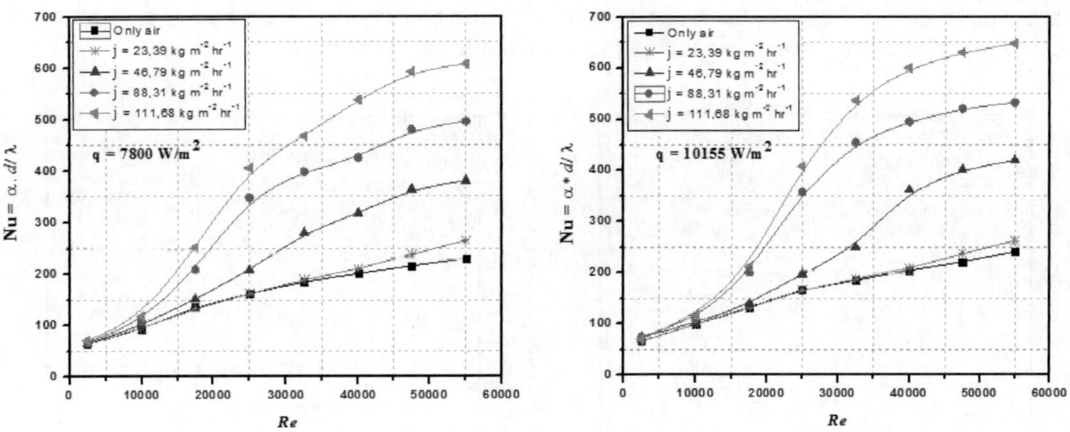

Figure 4. Variations of average Nusselt number with Re number under various water flux density.

Under range of water flux density, j = 23.39 - 46.79 kg/(m^{2}hr), with Re = 47500, the averaged Nusselt number tends to decrease with increasing of heat flux due to the low water content in the air/water mist flow, as is shown in figure 5. As water flux density increases the Nusselt number increase under range of heat flux 7800 to 10155 W/m^2. The maximum change in average Nusselt number is 8 %. The effect of using air/water mist flow on the pressure drop in the terms of friction factor with Reynolds number for all cases of water flux density are summarized in figure 6. Evidently, the friction factor decreases with the increasing of Reynolds number. Under same conditions, the friction factors that generated by air/water mist flow was little more than those generated by single phase - air coolant about 1.8 %, 3.1% 4.9% and 5.9% respectively for all cases of water flux density.

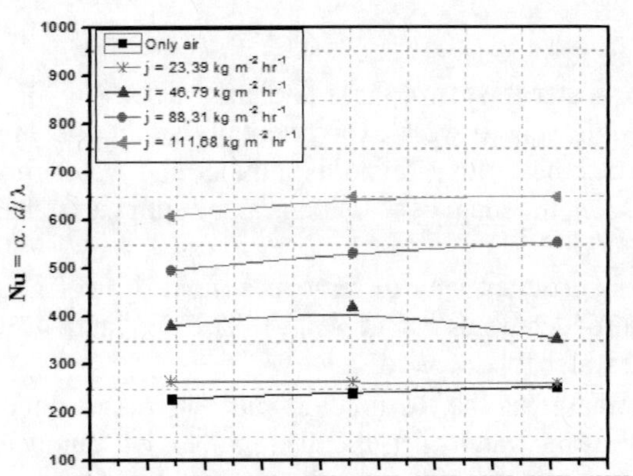

Figure 5. Variations of average Nusselt number with surface heat flux under Re = 47500 and range of water flux density

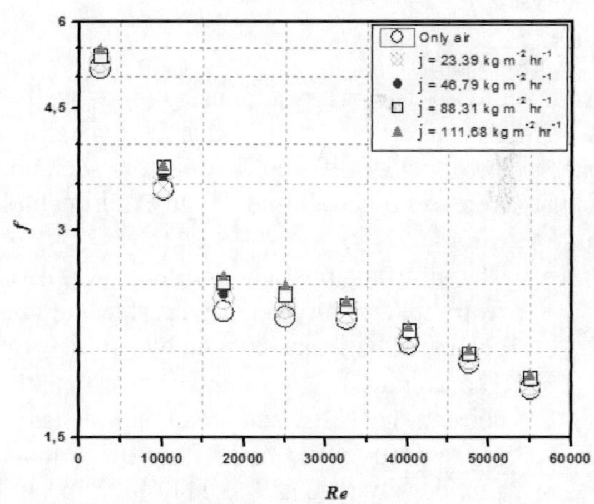

Figure 6. Variations of friction factor with Re number under different of water flux density

3.3. *Thermal Enhancement Factor*

Figure 7 represents the dimensionless thermal enhancement factor with Reynolds number under range of water flux density. It can be obviously seen that the thermal enhancement factor for all cases are generally more than unity and that indicate the influence of dispersing water mist into main flow on heat transfer process. At range of Reynolds number (Re = 2500 - 10000), the thermal enhancement factor were in a range between 1.0 - 1.26 for all water flux cases. This can be expressed that when the surface temperature is very hot, the water mist may be completely evaporated before arriving to the heated surface due to the force of evaporation without wetting the heated surface.

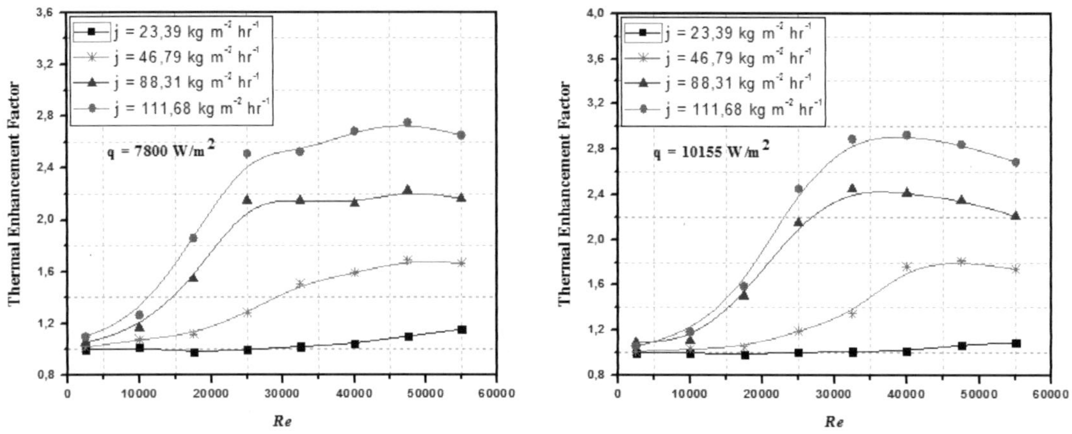

Figure 7. Variations of thermal enhancement factor with Reynolds number under range of water flux density.

The thermal enhancement factor rapidly increases with increasing of Reynolds number values (Re >17500) for all cases. At Reynolds number (Re = 40000), enhancement factor is 3 times of that with only air cooling for water flux density (111.68 kg m^{-2} hr^{-1}) under constant heat flux (q = 10155 W/m^2). The above experimental data indicates that the presence of water mist with different water concentration given more efficient heat transfer enhancement than the application of single phase - air cooling.

4. Conclusions

An open loop - cooling system using air as well as air/water mist two phase coolants was set up. The heat transfer and flow friction characteristics of a single sphere were experimentally studied in this paper. The effect of different key factors such as surface heat flux, Reynolds number and water flux density were examined. Based on the experimental results, the main conclusions can be summarized as follows:

- The air/water mist cooling technique is a very efficient way to remove a great deal of heat from sphere surface. The surface temperature decreases about 4%, 17%, 37% and 47% compared with air - coolant by using different water flux density.
- The average Nusselt number significantly increases as the Reynolds number increases under constant heat flux and water flux density. By using water mist, the average Nusselt numbers are respectively 11%, 19.7%, 90.2% and 134% higher than those in the air - coolant for water flux density (j=23.39 to 111.68 kg m^{-2} hr^{-1}). The average Nusselt number is slowly increased with increasing heat flux when the water flux density and Reynolds number are fixed.
- The friction factor is generally decreases with the increasing of Reynolds number. Under same conditions, the friction factors that generated by air/water mist coolant was little more than those generated by air - coolant about 1.8 %, 3.1%, 4.9% and 5.9% respectively for all cases of water flux density.
- The enhancement factors for all cases are generally more than unity indicating that the presence of water mist with different water concentration given more efficient heat transfer enhancement than the application of single phase - air cooling.

Acknowledgments
The work was supported by Act 211 Government of the Russian Federation, contract № 02.A03.21.0006

References
[1] Jiang G, Shi X, Chen G and Gao J 2018 *Int. J. Heat Mass Transf.* **120** 1101-17
[2] Xishi W, Guangxuan L, Weicheng F and Dobashi R 2004 *J. Fire Sci.* **22(5)** 355-66
[3] Zhao L and Wang T 2014 *J. Turbomach.* **136(7)** 071007
[4] Wang T and Dhanasekaran T 2010 *J. Heat Transf.* **132(12)** 122201
[5] Yang W J and Clark D W 1975 *Int. J. Heat Mass Transf.***18** 311317
[6] Nakayama W, Kuwahara N and Hirasawa S 1988 *Int. J. Heat Mass Transf.* **11(2)** 449-60
[7] Dreyer A, Kriel D and Erens P 1992 *Heat Transfer Engineering* **13(4)** 53-71
[8] Thomas W and Sunderland J 1970 *Ind. Eng. Chem. Fundam.* **9(3)** 368-74
[9] Hwang T 1990 *Int. J. Heat Mass Transf.* **33(5)** 943-52
[10] Hayashi Y, Takimoto A, Matsuda O and Kitagawa T 1990 JSME international journal. Ser. 2, Fluids engineering, heat transfer, power, combustion, thermophysical properties 33(2) 333-39
[11] X S Wang, G X Liao and W C Fan 2004 *J. Fire Sci.* **22** 355–66
[12] X Li and T Wang 2006 *J. Heat Transf.* **128** 509–19
[13] X Li and T Wang 2007 *J. Heat Transf.* **129** 472–82
[14] Allais I and Alvarez G 2001 *Journal of Food Engineering* **49(1)** 37-47
[15] Barrow H and Pope C 2007 *Applied Energy* **84(4)** 404-12
[16] Kim K, Ko H, Kim K and Perez-Blanco H 2012 *Appl. Therm. Eng.* **33-34** 62-69
[17] Zhao L and Wang T 2014 *J. Turbomach.* **136(7)** 071006
[18] Quinn C, Murray D and Persoons T 2017 *Int. J. Heat Mass Transf.* **111** 1234-49
[19] Yoon S and NO H 2017 *Nuclear Engineering and Design* **325** 97-106
[20] Tao Y, Huai X, Wang L and Guo Z 2011 *Appl. Therm. Eng.* **31(10)** 1790-97
[21] Li X and Wang T 2008 *J. Heat Transf.* **130(10)** 102901
[22] Abed A and Shcheklein S 2018 *J. Phys.: Conf. Series* **1015** 032001
[23] R J Lang 1962 *J. Acoust. Soc. Am.* **34** 6–8
[24] S J Kline and F A McClintock 1953 *Mechanical Engineering* **75** 3–8
[25] Kline S J 1953 *Mech. Engineering* **75** 3-8

IOP Conf. Series: Materials Science and Engineering **552** (2019) 012002 doi:10.1088/1757-899X/552/1/012002

Jet-film contact devices for heat and mass transfer processes in heat power engineering

A V Dmitriev[1], A I Khafizova[2] and O S Dmitrieva[3]

[1]Department Teoretical Foundations of Heat Engineering, Kazan State Power Engineering University, Krasnoselskaya St., 51, Kazan, 420066, Russia
[2]Department of Food Production Equipment, Kazan National Research Technological University, Karl Marx St., 68, Kazan, 420015, Russia
[3]Department of Chemical Technology Processes and Apparatus, Kazan National Research Technological University, Karl Marx St., 68, Kazan, 420015, Russia

E-mail: ja_deva@mail.ru

Abstract. One of the main production processes for many branches of industry is cooling of recycled\circulating water, which takes place in the cooling towers. The most important parameter to determine the cooling efficiency of water in the cooling tower is the contact surface of air and water. The existing designs of contact devices are highly effective in cooling the liquid, but they are complex to manufacture and expensive. In this regard, a new design of the jet-film contact device was developed, which allows creating a highly developed contact surface of liquid and gas phases. Mathematical model of the element of jet-film contact device was built by means of specialized software package and numerical calculations were carried out. Paper includes the study of behaviour of liquid and gas phases in the device element, and the mathematical dependences were obtained. Dependency of the liquid entrainment from device when changing the water flow rate was studied. Also, the comparison of designs with different location of holes within the wall elements was also carried out. Modernization of the studied jet-film contact device was carried out in order to identify the direction which to follow for improvement of design properties. This paper proposes an engineering method for calculating the jet-film contact device.

1. Introduction

In the recent years, water reserves have been sharply reduced due to the fact that people cannot rationally use the useful resources. For example, the developed countries spend a huge amount of water for the industrial needs, and the used water from these enterprises is returned back into natural reservoirs, that causes irreparable harm to the environment. In this regard, energy, chemical, petrochemical and many other industries use the recycled water for water cooling. This is due to the fact that the removal of low-temperature industrial waste heat by circulating water allows to reduce the consumption of fresh water and thermal emissions into the environment by several times [1–3].

Circulating water is cooled in cooling towers. The most common type is a fan evaporative cooling tower, which has high cooling efficiency and small dimensions. However, such cooling towers are much more expensive, and the most part of costs is required for the cooling tower fillers. Manufacture of them is very cost-intensive, as well as replacement with new ones. In this regard, there is a need to reconstruct the known devices or develop new designs [4–8].

The most important parameter to determine the water cooling efficiency in the cooling tower is the contact surface of air and water. Designs of contact devices, developed in the recent years, have high efficiency in cooling the liquid, but the manufacture of many of them is expensive and quite complex. Also, many elements have to be replaced with new ones in case of large biological deposits on the device [9–12].

As of today the main task is to develop new high-performance contact devices that will meet all the above requirements, as well as have a large contact area of the liquid and gas phase, good throughput capacity and low hydraulic resistance [13–23].

The authors of this paper created a contact device, which has a high efficiency of heat and mass transfer, and its hydraulic resistance is not inferior to analogues [24–30].

2. Results of studies and discussion of them

The studied device is a stage of a square shape with width of 100 mm and vertical walls of 50 mm high, chequerwise arranged. Distance between the rows of neighboring stages is 25 mm. Geometrical parameters of each element are selected in such a way that the air movement acquires a circular profile. The upper part of the trays is open, there are holes in the lower part. 8 holes with a diameter of 5 mm were executed in the lower part of the walls of each drain cup. Distance from the center of holes to the bottom of drain cup is 2.5 mm. Holes with a diameter of 10 mm were executed within the bottom, which were located in the center of the drain cup.

When conducting experimental studies with two-phase flows, it is almost impossible to observe the processes, taking place in the inner part of the contact element. Reliable data can be obtained only at the input and output of the device, as when introduced into the working area of the measuring devices, they introduce disturbances and distort the flow structure. In this regard, in order to identify the features of interaction between the gas and liquid phases, numerical simulation is used.

In order to calculate the processes, taking place in the element of the device, the ANSYS Fluent software package (figure 1), the liquid volume method (VOF) were used, allowing to take into account the interaction of two immiscible phases. The studied module was divided into 130690 cells. There are 48644 cells at the interface with solid walls. It should be noted that the previous calculations showed satisfactory convergence of the results of modeling and experimental studies. In the course of calculation, the boundary conditions were set by water and air pressures. Air pressure in the lower part of the device was set equal to 101425 Pa, and in the upper part – 101325 Pa. The unsteady issue was being resolved. After 0.9 seconds a quasi-stationary mode was detected, so the calculation was stopped.

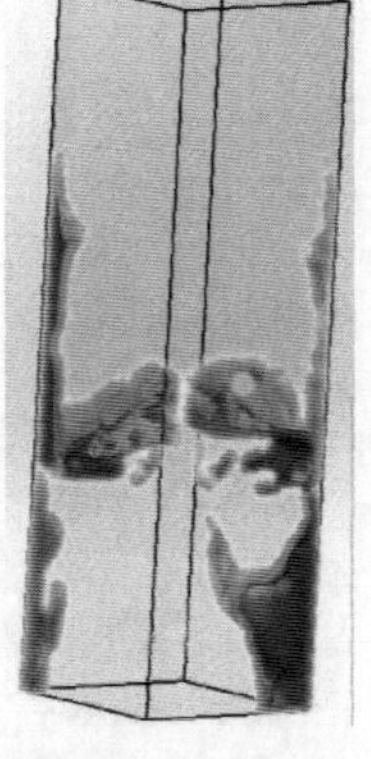

Figure 1. Numerical simulation of liquid and gas phase interaction in the device element.

A recess was executed in the upper drain cup, which had the shape of a circle quarter with a radius of 5 mm and size of 0.1 mm. The water pressure was set equal to 101800 Pa, that corresponded to the level of liquid in the drain cup of 48.57 mm. Calculations were carried out at the temperature of air and water of 20°C.

IOP Conf. Series: Materials Science and Engineering **552** (2019) 012002 doi:10.1088/1757-899X/552/1/012002

In the course of numerical calculation of the contact element, the liquid level in the drain cups was changed. Changing the liquid level, the dependency of the liquid flow rate out of the hole on the maximum average air flow rate (figure 2) was observed.

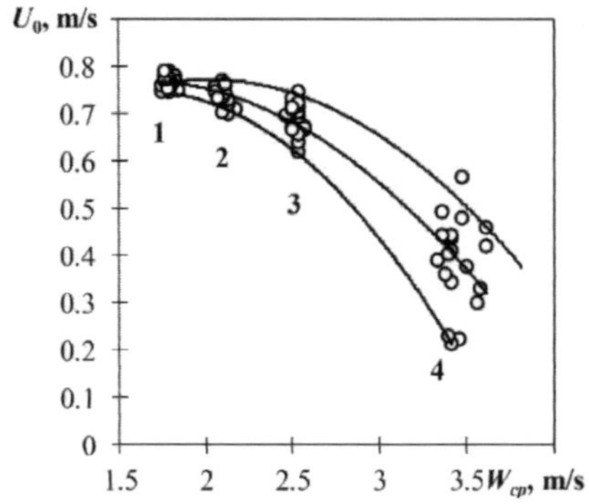

Figure 2. Dependency of the liquid flow rate out of the hole on average air flow rate at different levels of liquid in the trays h, m: 1 – 0.0051; 2–0.0071; 3 – 0.0101; 4 – 0.0204.

The graph shows that with an increase of the liquid level, the limit value of the average air flow rate also increases, which can be set provided that the liquid will not be entrained. It should also be noted that the low liquid flow rates out of the hole correspond to high average air flow rate, otherwise, there is a large percentage of liquid entrainment. Mathematical dependency was obtained by approximation method for this process:

$$U_0 = -0.1126W_{cp}^2 + 0.3647W_{cp} + 0.4753, \tag{1}$$

where U_0 – liquid flow rate out of the hole, m/s; W_{cp} – average air flow rate, m/s.

Equation (1) is an equation of a quadratic polynomial, where the value of the approximation reliability is 0.8954. This equation allows to determine the average air flow rate, at which there is a complete liquid entrainment, its value is equal to 4.2 m/s.

It was also found that the device operates quite effectively at the air flow rate of 3.2 m/s (figure 3), that is a quite high value. This shows a large throughput capacity of the device.

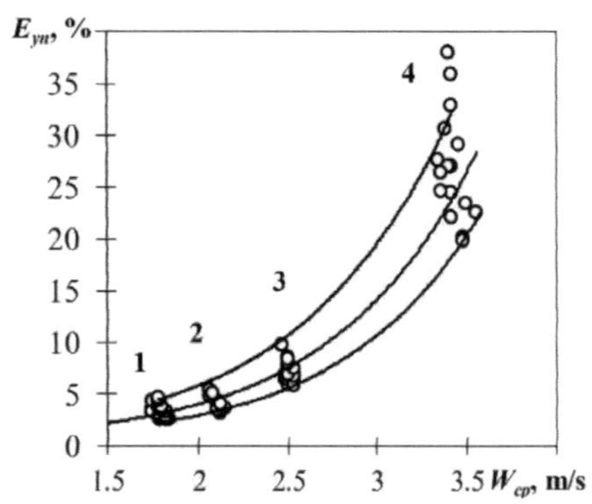

Figure 3. Changing of the entrained liquid amount if the average air flow rate is also changed at different levels of liquid in the trays h, m: 1 - 0.0051; 2 - 0.0071; 3-0.0101; 4 - 0.0204.

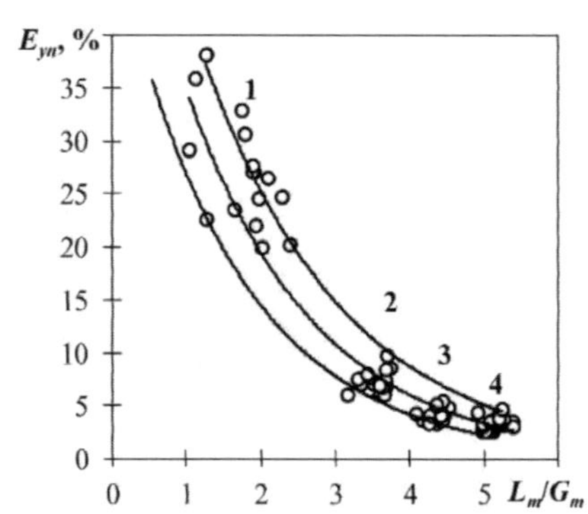

Figure 4. Dependency of the water entrainment on ration of mass water flow rate and mass air flow rate at different levels of liquid in the trays h, m: 1-0.0051; 2 - 0.0071; 3 - 0.0101; 4 - 0.0204.

Dependency of the entrained liquid amount on the average flow rate can be recorded as follows:

$$E_{yn} = 0.3319e^{1.2502W_{cp}}. \tag{2}$$

Then, the ration of mass flow rates of liquid and gas can be set, based on the mathematical balance (figure 4).

The graph in figure 4 shows that with an increase of the mass water flow rate in relation to the mass air flow rate, there is a decrease of the water entrainment.

Water entrainment from the device is not significant, this is due to the fact that the upper stages of the contact device prevent the entrainment of liquid drops, caught into the air stream, that in fact serves as a collector of drops. This is an additional advantage among many other contact devices.

After the ratios of mass flow rates, as well as the maximum average air flow rate are determined, it is possible to determine the cross-sectional area of the apparatus, required not to entrain the liquid by the gained air flow rate and the apparatus will work quite effectively. If the body of the apparatus is already assembled and the cross-sectional area is known, it is possible to determine the maximum air flow rate that can be set.

If we know the liquid flow rate out of the hole (equation (1)), and the mass liquid flow rate is set, it is possible to determine the central diameter of the hole as per the well-known formula:

$$d_0 = \sqrt{\frac{4L_m}{\pi \rho U_0}}. \tag{3}$$

where L_m – mass liquid flow rate, kg/s; ρ – liquid density, kg/m^3.

Thus, this paper introduces an engineering procedure for calculating the jet-film contact device, including the determination of the liquid flow rate out of the hole in the drain cup, average air flow rate, as well as determination of geometrical parameters of the device.

The developed design operates quite effectively, the processes, taking place inside of the device, show that the development of designs of jet-film contact devices is going in a right direction and holes, which had been added within the walls, made a great contribution to operation of device. In this regard, the design can still be finalized.

During the process of design modernization, influence of the holes within the walls of elements on the processes of interaction between two phases, was also studied. These studies were carried out in order to show how the holes affect the movement of flows, as well as to show the direction to follow in order to improve the design.

The next device is a stage of a square shape with width of 100 mm and walls of 50 mm high, chequerwise arranged. Moreover, the walls of the drain cup are arranged so that the bottom of the drain cup passes through the middle of the walls, and the tray has upper and lower walls. Distance between the rows of neighboring stages is 25 mm. Holes with a diameter of 10 mm are executed within the bottoms of the drain cups, and such hole is located directly in the middle of drain cup. One hole with a diameter of 10 mm is also executed within the middle of the wall of each drain cup.

The boundary conditions remained the same: the air pressure in the lower part of device was set equal to 101425 Pa, and in the upper – 101325 Pa. The calculations were carried out at a temperature of the air and water of 20°C. Water flow rate varied from 0.5 m/s to 1.5 m/s. When the liquid flow rate was changed, the percentage of liquid entrainment from the device also changed. We analyzed the obtained data and came to the conclusion that with an increase of the liquid flow rate, the percentage of its entrainment from the device reduces, as well as at the average air flow rates from 1 to 2 m/s, the liquid entrainment is not significant (figure 5). However, the average air flow rate of this contact device is much lower than that of its predecessor, that does not allow to give a high assessment of its properties

IOP Conf. Series: Materials Science and Engineering **552** (2019) 012002 doi:10.1088/1757-899X/552/1/012002

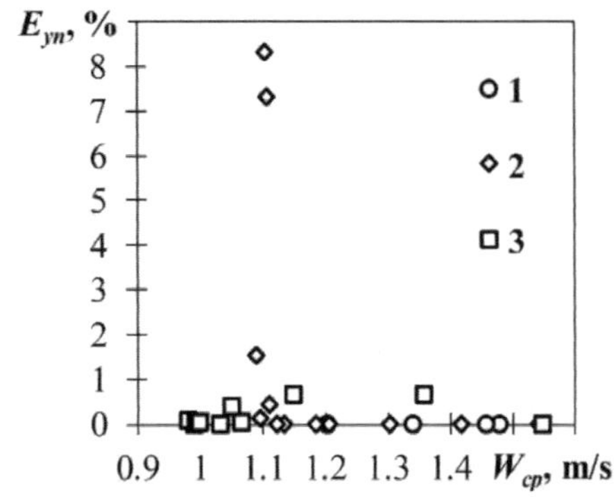

Figure 5. Changing of the entrained liquid amount if the average air flow rate is also changed at different water flow rates U, m/ s: 1 – 0.5; 2 – 1; 3 – 1.5.

The next step to improve the design was the addition of a partition over the hole within the center of the bottom (figure 6). This element was introduced in order to reduce the air carrying capacity, since air bubbles were observed in the previous design, that flew out of the hole.

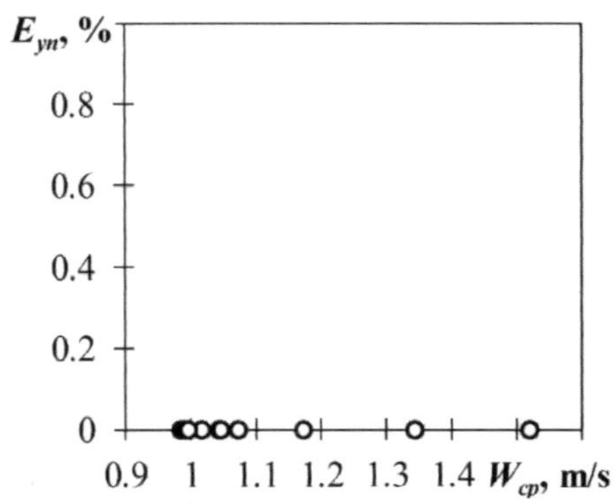

Figure 6. Changing of the entrained liquid amount if the average air flow rate is also changed when the partition over the central hole within the bottom is added into design.

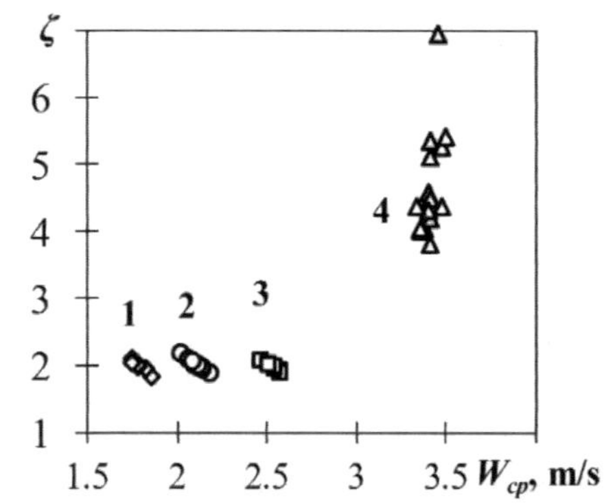

Figure 7. Dependency of the hydraulic resistance coefficient ζ on the average air flow rate at the following pressure drop dp, Pa: 1 – 50; 2 – 70; 3 – 100; 4 – 200.

It was assumed that the partition will hold the air and not allow it to carry out the water up, however, this modernization has not provided a significant effect and we believe that addition of it into design is not worthwhile, due to the fact that it will only bring complexity to the design [31].

Another indicator of the effective operation of contact devices in thermal power plants is the hydraulic resistance that occurs when the air flow passes through the contact device. High values of heat and mass transfer coefficients can be achieved by using the device, having low hydraulic resistance.

Figure 7 shows the dependency of hydraulic resistance on the average air flow rate at different pressure drop between the pressure at the inlet and outlet of the device. As can be seen from the graph, if the average gas flow rate is up to 3 m/s, the hydraulic resistance coefficient remains practically unchanged and fluctuates in the range of values about 2 and the spread of values is minimum. When the rate is more than 3 m/s, the hydraulic resistance coefficient increases sharply, due to the liquid entrainment from the device. In this case, the range of values is between 4 and 6. The proposed device has a relatively low hydraulic resistance at sufficiently high values of the average flow rate and minimum entrainment of liquid drops. All this shows good prospects of its application for heat and mass transfer processes. It should also be noted that at high operational rates of the device there was no distribution between the rows, therefore, there will be no need to install redistributive devices.

For comparison, a device within which the bottom of elements was located in the middle, was studied. As a result of studies, it is found that the hydraulic resistance considerably increases. Moreover, the unacceptable entrainment is observed already at the values of average flow rate of 1.5 m/s. It can be concluded that changing of design in this way is not advisable.

3. Conclusions

Having analysed all three designs and considered their pros and cons, the following conclusions can be drawn:

1) Contact device with holes within the lower part of the wall allows to cool the liquid effectively enough and to gain the air flow rate up to 4.2 m/s, that is a high value among the existing analogues.

2) The average air flow rate within the second variant of design of the contact device reaches 2 m/s.

3) Water entrainment in the devices is observed at an acceptable level, about 15%. In reality, the water entrainment will be much less because of the merging of drops, which is not provided in the model.

4) Water entrainment in the second contact device is lower, however, it should be noted that this data is only at average air flow rates from 1 to 2 m/s.

Acknowledgments
The research was conducted with funding from the RF President`s grant project No. MK-4522.2018.8.

References

[1] Patil M M, Patil S J, Patil P S and Mehta S J 2018 Design and Analysis of Cooling Tower. *IRJET*. **5** pp 2240–45

[2] Chopra M K and Kumar R 2017 An Investigation in Performance Enhancement of Induced Draft Counter Flow Wet Cooling Tower. *IRJET*. **4** pp 2107–11

[3] Ming Gao, Sun F Z H, Shi Y T, Kai Wang and Zhao Y B 2007 Research on the Effect of Cross-wind to Temperature Difference and Efficiency of Natural Draft Counter flow Wet Cooling Tower. *Challenges of Power Engineering and Environment*. pp 513–7

[4] Kryukov O V and Serebryakov A V 2016 Active and adaptive algorithms of autonomous wind power plants control and monitoring. 2nd International Conference on Industrial Engineering, Applications and Manufacturing (*ICIEAM*) pp 1–4

[5] Shibitova N V, Shibitov N S and Kolenchuk S V 2012 Reconstruction of Ventilator Nozzle Cooler. *NEWS of VolgSTU*. **5** pp 22–4

[6] Afanasenko V G, Khafizov F Sh, Khafizov N F, Ivanov S P and Boev E V 2007 Development of designs for polymeric water traps in cooling towers using centrifugal separation forces. *Chem. Pet. Eng*. **43** pp 653–6

[7] Kulkarni S J and Goswami A K 2017 Studies and Experimentation on Cooling Towers: A Review. *IRJET*. **2** pp 279–82

[8] Umacanta B and Nataraj C N 2017 Study on induced draft cooling tower performance analysis in captive power plant. *IRJET*. **4** pp 414–21

[9] Kolev N, Kralev B and Kolev D 2013 Gas side controlled mass transfer in a new packing with stamped horizontal lamellae operating at extremely low liquid loads. *Chem. Eng. Process.: Process Intensification*. **63** pp 44–9

[10] Xiong X, Li L and Zhou X 2017 Numerical Analysis and Optimization Research on Backflow Effect of Cooling Tower. *Procedia Eng*. **205** pp 2003–10

[11] Jia-Gen L I U, Bin Y A N and Yu Q 2009 Study on the mode and strategy of cooling tower fan operation energy saving. *Build. Sci*. **25** pp 35–8

[12] Tsurikova N P, Pushnov A S, Lagutkin M G and Shishov V I 2012 Effect of tier height in a regular-packing block on the effectiveness of evaporative cooling of circulating water. *Chem. Pet. Eng*. **48** pp 3–8

[13] Sánchez F, Kaiser A S, Zamora B, Ruiz J and Lucas M 2015 Prediction of the lifetime of

droplets emitted from mechanical cooling towers by numerical investigation. *Int. J. Heat Mass Transfer*. **89** pp 1190–206

[14] Boev E V, Ivanov S P, Afanasenko V G and Nikolaev E A 2009 Polymeric drop-film sprinklers for cooling towers. *Chem. Pet. Eng*. **45** 454–9

[15] Rahmati M, Alavi S R and Tavakoli M R 2018 Investigation of heat transfer in mechanical draft wet cooling towers using infrared thermal images: An experimental study. *Int. J. Refrig*. **88** pp 229–38

[16] Dmitrieva O S and Dmitriev A V 2013 Cooling of circulating water in a vortex chamber with disk atomizer. *Chem. Pet. Eng*. **49** pp 156–61

[17] Zhou Y, Wang K, Gao M, Dang Z, He S and Sun F 2018 Experimental study on the drag characteristic and thermal performance of non-uniform fillings for wet cooling towers under crosswind conditions. *Appl. Therm. Eng*. **140** 398–405

[18] Hettiarachchi H D M, Golubovic M and Worek W M 2007 The Effect of Longitudinal Heat Conduction in Cross Flow Indirect Evaporative Air Coolers. *Appl. Therm. Eng*. **27** pp 1841–8

[19] Dmitrieva O S, Dmitriev A V and Nikolaev A N 2014 Distribution of Circulating Water in the Work Area of a Vortex Chamber with Disk Atomizer for the Purpose of Increasing the Efficiency of the Cooling Process. *Chem. Pet. Eng*. **50** pp 169–75

[20] Naphon P 2005 Study on the Heat Transfer Characteristics of an Evaporative Cooling Tower. *Int. Commun. Heat Mass Transfer*. **32** pp 1066–74

[21] Golovanchikov A B, Merentsov N A and Balashov V A 2013 Modeling and Analysis of a Mechanical-Draft Cooling Tower with Wire Packing and Drip Irrigation. *Chem. Pet. Eng*. **48** pp 595–601

[22] Laptev A G and Lapteva E A 2017 Mathematical Models and Calculation of the Coefficients of Heat and Mass Transfer in the Packings of Mechanical-Draft Towers. *J. Eng. Phys. Thermophys*. **90** pp 644–50

[23] Dashkov G V, Nogotov E F, Pavlyukevich N V and Tyutyuma V D 2006 Nonstationary Heat and Mass Transfer in Evaporative Cooling of Flowing Liquid Films. *J. Eng. Phys. Thermophys*. **79** pp 1–9

[24] Petruchik A I, Solodukhin A D and Fisenko S P 2001 Simulation of Cooling of Water Droplet and Film Flows in Large Natural Wet Cooling Towers. *J. Eng. Phys. Thermophys*. **74** pp 62–8

[25] Madyshev I N, Dmitrieva O S, Dmitriev A V and Kruglov L V 2017 Impact of the liquid level in the jet-film contact devices on the heat-and-mass transfer process. *MATEC Web Conf.* **129** 06010

[26] Pushnov A S 2013 The effect the pressure drop of regular packings has on the return water evaporation cooling efficiency in cooling towers. *Therm. Eng*. **60** 426–8

[27] Li X, Yang X, Li H, Shi Q, Gao X 2018 Significantly enhanced vapor-liquid mass transfer in distillation process based on carbon foam ring random packing. *Chem Eng Process*. **124** pp 245–54

[28] Gao M, Sun F Z and Turan A 2014 Experimental Study Regarding the Evolution of Temperature Profiles Inside Wet Cooling Tower Under Crosswind Conditions. *Int. J. Therm. Sci.* **86** pp 284–91

[29] Gharagheizi F, Hayati R and Fatemi S 2007 Experimental study on the performance of mechanical cooling tower with two types of film packing. *Energy Convers. Manage*. **48** pp 277–80

[30] Goshayshi H R and Misssenden J F 2000 The investigation of cooling tower packing in various arrangements. *Appl. Therm. Eng*. **20** pp 69–80

[31] Madyshev I N, Dmitrieva O S and Dmitriev A V 2017 Hydraulic resistance of thermal deaerators of thermal power stations (TPS) with jet-film contact devices. *MATEC Web Conf.* **141** 01023

IOP Conf. Series: Materials Science and Engineering **552** (2019) 012003 doi:10.1088/1757-899X/552/1/012003

Mathematical model of fuel heat ignition considering space - time nonlocality

A V Eremin[1], V A Kudinov[2], I V Kudinov[2], V V Zhukov[2] and K V Trubitsyn[3]

[1]Department of Industrial Heat Power Engineering, Samara State Technical University, Molodogvardeyskaya St., 244, Samara, 443100, Russia
[2]Department of Theoretical Foundations of Heat Engineering and Hydromechanics, Samara State Technical University, Molodogvardeyskaya St., 244, Samara, 443100, Russia
[3]Heat Power Engineering Faculty, Samara State Technical University, Molodogvardeyskaya St., 244, Samara, 443100, Russia

E-mail: a.v.eremin@list.ru

Abstract. Using the formula of the Fourier law, which takes into account the relaxation of the heat flux and the temperature gradient, a mathematical model of thermal ignition for systems with an exothermic internal source of heat was developed, taking into account the space-time nonlocality of the heat exchange process. Studies of boundary conditions of the first and third kind showed that taking into account the local nonequilibrium leads to an increase in the delay time of thermal ignition, irrespective of the conditions of heat exchange with the environment. It is shown that if the relaxation properties of materials are taken into account, the boundary conditions (of the first, second, or third kind) cannot be taken instantly, but only in a certain range of initial time. Consequently, the amount of heat supplied to the system has a certain limit, depending on its relaxation properties. Therefore, it is possible to conclude that there is also a limit on the value of heat transfer coefficients, which cannot be exceeded under any conditions of heat exchange with the environment.

1. Introduction

The main content of the theory of thermal ignition (thermal explosion) is a mathematical study of the temperature state of systems during the course of exothermic chemical reactions, which allows finding critical conditions and time of the ignition delay [1 – 3, 12]. Classical models of thermal ignition, based on parabolic equations of heat transfer, are obtained without taking into account the local nonequilibrium. It is known that they are based on the principle of local thermodynamic equilibrium, according to which in any small element of the medium a local equilibrium is assumed, whereas in the system as a whole, the gradients of the potentials of the fields under investigation are observed. Such a state is possible only in the case when the rate of disturbance of the equilibrium caused by the boundary conditions is much less than the relaxation rate of the system to local equilibrium. In this relation, in the differential equations of mathematical models, an infinite velocity of heat propagation is embedded, which is related with the use of the empirical formula of the Fourier law in their derivation for the heat flux

$$q = -\lambda \partial T / \partial x \qquad (1)$$

in which the temperature gradient ($\partial T/\partial x$) and the heat flux q are not separated in time. According to equation (1), any change in the temperature gradient causes an instantaneous change in the heat flux [8, 10, 11, 22]. And, consequently, the rate of heat propagation is assumed to be infinite. Since in real physical processes infinite values of any parameters cannot be observed, the differential equations obtained using the formula (1) can be adequate to these processes only in a certain time range of a non-stationary process. It is known that parabolic equations inadequately describe all fast processes (the time of which is comparable to the relaxation time), as well as the temperature state of the bodies, observed for small and ultra-small values of time and space variables, in any thermal processes [16 – 19]. This is due to the fact that the rate of violation of equilibrium and the system relaxation rate to the equilibrium state in these processes become comparable due to the locality of differential equations describing them [4 – 7]. Taking into account that the processes of thermal ignition are fast, the failure to take account of local non-equilibrium in their mathematical models can lead to a significant difference of the obtained parameters from their real values [3, 9, 23].

2. Formulation of the first problem

To derive a differential equation of a thermal ignition taking into account the local nonequilibrium of the heat exchange process, we represent formula (1) in such a way that it takes into account the acceleration in time of both the heat flux and the scalar quantity of the temperature gradient [5, 12 – 15]

$$q = -\lambda\left(\frac{\partial T}{\partial x} + \tau_1 \frac{\partial^2 T}{\partial x \partial t}\right) - \tau_1 \frac{\partial q}{\partial t} \tag{2}$$

where q – heat flux; T – temperature; x – coordinate; t – time; λ – thermal conductivity coefficient; τ_1 – relaxation time.

Substituting (2) into the heat balance equation

$$c\rho \partial T/\partial t + \partial q/\partial x + \omega(T) = 0 \tag{3}$$

we get

$$c\rho\frac{\partial T}{\partial t} = \lambda\frac{\partial^2 T}{\partial x^2} + \lambda\tau_1\frac{\partial^3 T}{\partial x^2 \partial t} + \tau_1\frac{\partial}{\partial t}\left(\frac{\partial q}{\partial x}\right) + \omega(T), \tag{4}$$

where $\omega(T) = Q\rho k_0 \exp(-E/(RT))$ – power of internal heat source; Q – reaction thermal effect; ρ – density; c – heat capacity; k_0 – preexponential factor; E – activation energy; R – universal gas constant.

Expressing $\partial q/\partial x$ from (3) and substituting into (4), we obtain

$$c\rho\frac{\partial T}{\partial t} = \lambda\frac{\partial^2 T}{\partial x^2} + \lambda\tau_1\frac{\partial^3 T}{\partial x^2 \partial t} - \tau_1\frac{\partial}{\partial t}\left[c\rho\frac{\partial T}{\partial t} + Q\rho k_0 \exp\left(-\frac{E}{RT}\right)\right] + $$

$$+ Q\rho k_0 \exp\left(-\frac{E}{RT}\right). \tag{5}$$

Defining the derivatives in the third term on the right-hand side of equation (5), we have

$$\frac{\partial T}{\partial t} + \tau_1\frac{\partial^2 T}{\partial t^2} = a\left(\frac{\partial^2 T}{\partial x^2} + \tau_1\frac{\partial^3 T}{\partial x^2 \partial t}\right) - \frac{\gamma}{T^2}\exp\left(-\frac{E}{RT}\right)\frac{\partial T}{\partial t} + \frac{Qk_0}{c}\exp\left(-\frac{E}{RT}\right), \tag{6}$$

where $\gamma = \tau_1 Q k_0 E/(cR)$ – complex having dimension K^2; $a = \lambda/(c\rho)$ – temperature conductivity coefficient.

Obviously, for $\tau_1 = 0$, equation (6) is reduced to the classical parabolic heat transfer equation with a nonlinear heat source

$$\frac{\partial T}{\partial t} = a\frac{\partial^2 T}{\partial x^2} + \frac{Qk_0}{c}\exp\left(-\frac{E}{RT}\right). \tag{7}$$

The boundary conditions for equation (6) for an infinite plate under symmetric boundary conditions of the first kind have the form

$$T(x, 0) = T_0; \tag{8}$$

$$\partial T(x, 0)/\partial t = 0; \tag{9}$$

$$\partial T(0, t)/\partial x = 0; \tag{10}$$

$$T(\delta, t) = T_{ст}, \tag{11}$$

where T_0 – initial temperature; $T_{ст}$ – wall temperature ($T_{ст} > T_0$); δ – plate thickness.

We introduce the following dimensionless variables and parameters [20, 21, 24, 25]:

$$\Theta = \frac{T - T_{ст}}{\Delta T};\ \mathrm{Fo} = \frac{at}{\delta^2};\ \xi = \frac{x}{\delta};\ \mathrm{Fo}_1 = \frac{a\tau_1}{\delta^2};\ \tau = \frac{\delta^2 Qk_0}{ca\Delta T};\ \Delta T = T_0 - T_{ст}, \tag{12}$$

where Θ, Fo, ξ – respectively, dimensionless temperature, time, coordinate; τ – dimensionless parameter; Fo_1 – dimensionless relaxation time.

Taking into account the notation (12), problem (6), (8) - (11) will be

$$\frac{\partial\Theta}{\partial \mathrm{Fo}} + \mathrm{Fo}_1\frac{\partial^2\Theta}{\partial \mathrm{Fo}^2} = \frac{\partial^2\Theta}{\partial\xi^2} + \mathrm{Fo}_1\frac{\partial^3\Theta}{\partial\xi^2\partial \mathrm{Fo}} - \frac{\gamma}{(\Theta\Delta T + T_{ст})^2}\exp\left(-\frac{E}{R(\Theta\Delta T + T_{ст})}\right)\frac{\partial\Theta}{\partial \mathrm{Fo}} +$$

$$+\tau\exp\left(-\frac{E}{R(\Theta\Delta T + T_{ст})}\right),\ (\mathrm{Fo} > 0;\quad 0 < \xi < 1); \tag{13}$$

$$\Theta(\xi, 0) = 1; \tag{14}$$

$$\partial\Theta(\xi, 0)/\partial \mathrm{Fo} = 0; \tag{15}$$

$$\partial\Theta(0, \mathrm{Fo})/\xi = 0; \tag{16}$$

$$\Theta(1, \mathrm{Fo}) = 0. \tag{17}$$

Analysis of equation (13) allows us to conclude that it contains a term representing the heat source with a negative sign (the third term on the right). Consequently, in this model the action of a heat source associated with the course of a chemical reaction is weakened by the action of a source taking into account the relaxation properties of the materials.

3. Numerical Solving method

Note that problem (13) - (17) is completely represented in a dimensionless form. To solve it by the method of finite differences in the field under consideration, we introduced a space-time grid with steps $\Delta\xi$, $\Delta\tau$, respectively, for variables ξ and τ

$$\xi_k = k\Delta\xi,\quad (k = \overline{0, K});\quad \mathrm{Fo}_i = i\Delta \mathrm{Fo},\quad (i = \overline{0, I}), \tag{18}$$

where K, I – number of steps by coordinates ξ, Fo.

On the grid (18) grid functions $\Theta_k^i = \Theta(\xi_k, \mathrm{Fo}_i)$ were introduced. Using an explicit scheme for the differential operators approximation, problem (13) - (17) can be written in the form

$$\left[1+\frac{\gamma}{(\Theta_k^i \Delta T + T_{\text{ст}})^2}\exp\left(-\frac{E}{R(\Theta_k^i \Delta T + T_{\text{ст}})}\right)\right]\frac{\Theta_k^i - \Theta_k^{i-1}}{\Delta \text{Fo}} + \text{Fo}_1 \frac{\Theta_k^{i+1} - 2\Theta_k^i + \Theta_k^{i-1}}{\Delta \text{Fo}^2} =$$

$$= \frac{\Theta_{k-1}^i - 2\Theta_k^i + \Theta_{k+1}^i}{\Delta\xi^2} + \text{Fo}_1 \frac{\Theta_{k-1}^i - 2\Theta_k^i + \Theta_{k+1}^i - \Theta_{k-1}^{i-1} + 2\Theta_k^{i-1} - \Theta_{k+1}^{i-1}}{\Delta \text{Fo} \Delta\xi^2} +$$

$$+ \tau \exp\left(-\frac{E}{R(\Theta_k^i \Delta T + T_{\text{ст}})}\right), \quad (\text{Fo} > 0; \quad 0 < \xi < 1); \tag{19}$$

$$\Theta_k^0 = 1; \tag{20}$$

$$\frac{\Theta_k^1 - \Theta_k^0}{\Delta \text{Fo}} = 0; \tag{21}$$

$$\frac{\Theta_1^i - \Theta_0^i}{\Delta\xi} = 0; \tag{22}$$

$$\Theta_K^i = 0. \tag{23}$$

When solving problem (19) - (23), the following initial data were used: $\tau = 3.39 \cdot 10^8$; $E = 127000\ J/mole$; $R = 8.31\ J/(mole \cdot K)$; $\Delta T = 10\ K$; $T_{\text{ст}} = 850\ K$; $k_0 = 1.925 \cdot 10^6\ 1/s$.

4. Descussion of the results

The results of calculations are shown in figures 1 – 3. Their analysis allows us to conclude that the magnitude of the relaxation coefficients has a significant effect on the delay time of thermal ignition. Thus, for $\text{Fo}_1 = 0.15$ in case of heating a body $(T_{\text{ст}} > T_0)$, the ignition delay time increases (in comparison with the case when the relaxation properties of the material are not taken into account $(\text{Fo}_1 = 0)$ (see Fig. 1).

The fact of delay of thermal ignition is explained by the relaxation properties of materials. The results of investigations showed that taking into account the relaxation properties of materials leads to the impossibility of instantaneous acceptance of boundary conditions (first, second or third kind) on the wall $(\xi = 1)$ - the process of their establishment takes some initial time range $0 \le \text{Fo} \le \text{Fo}^*$. And, in particular, for $\text{Fo}_1 = 0.15$ $\text{Fo}^* = 0.5$ (figure 2). It follows that the boundary conditions of the first kind (thermal shock) cannot be established instantaneously under any conditions of heat exchange with the environment. Consequently, in real physical processes the heat transfer coefficient is limited with a certain constant determined by the relaxation properties of the material.

IOP Conf. Series: Materials Science and Engineering **552** (2019) 012003 doi:10.1088/1757-899X/552/1/012003

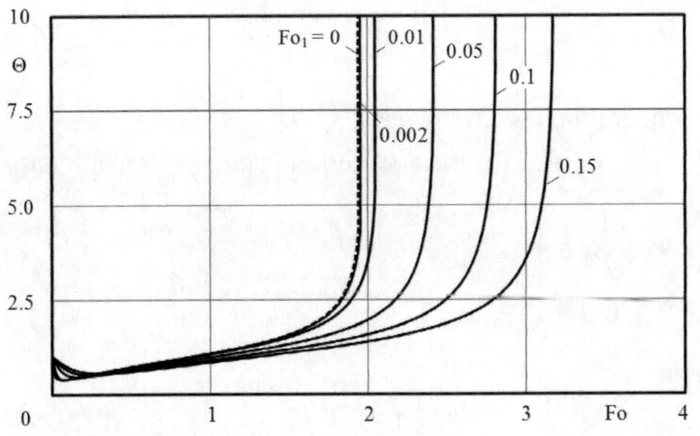

Figure 1. Temperature change in time ($\xi = 0.9$; $\gamma = 0$)

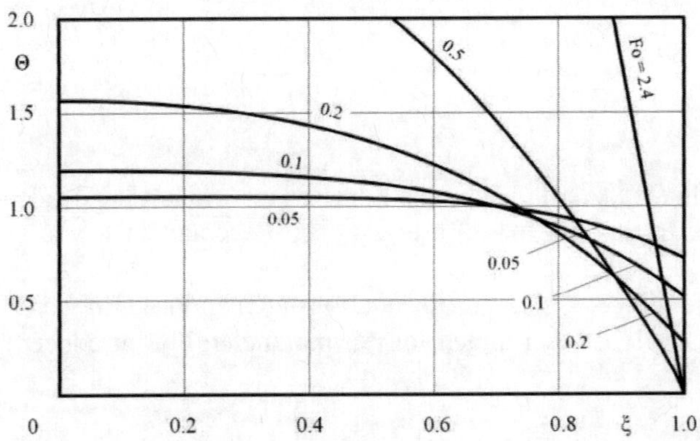

Figure 2. Temperature change with respect to the ξ coordinate in time ($Fo_1 = 0.15$; $\gamma = 0$)

Figure 3 shows the results of calculations in the case when $\gamma \neq 0$. Their analysis allows us to conclude that for small values of Fo_1, the value of γ has a minor effect on the ignition delay time. With an increase of the relaxation time Fo_1, its influence substantially increases.

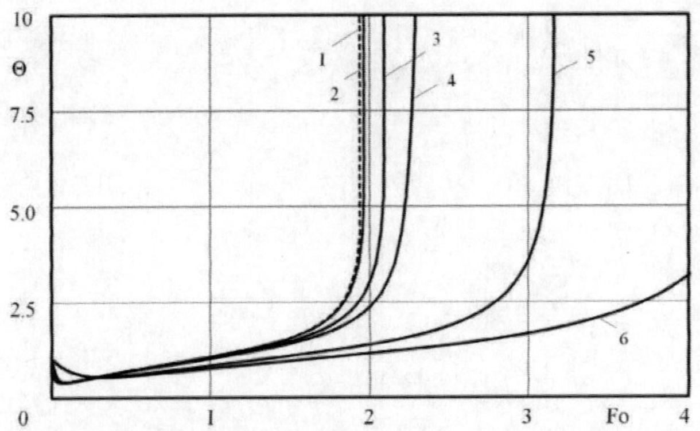

Figure 3. Temperature change in time ($\xi = 0.9$):

1 – $Fo_1 = 0.0015$; $\gamma = 0$

2 – $Fo_1 = 0.0015$; $\gamma = 10^{11}$

3 – $Fo_1 = 0.015$; $\gamma = 0$

4 – $Fo_1 = 0.015$; $\gamma = 10^{12}$

5 – $Fo_1 = 0.15$; $\gamma = 0$

6 – $Fo_1 = 0.15$; $\gamma = 10^{13}$

5. Formulation of the second problem

Let us find a solution of equation (6) for inhomogeneous boundary conditions of the third kind, which for the infinite plate have the form

$$\lambda \frac{\partial T(0,t)}{\partial x} = \alpha_1 [T(0,t) - T_{c1}]; \tag{24}$$

$$\lambda \frac{\partial T(\delta,t)}{\partial x} = \alpha_2 [T_{c2} - T(\delta,t)], \tag{25}$$

where T_{c1}, T_{c2} – ambient temperature; α_1, α_2 – heat transfer coefficients.

The boundary conditions for equation (6) with allowance for local nonequilibrium in the boundary conditions (24), (25) are written in the form

$$T(x, 0) = T_0; \tag{26}$$

$$\partial T(x, 0)/\partial t = 0; \tag{27}$$

$$\lambda \frac{\partial T(0,t)}{\partial x} + \lambda \tau_1 \frac{\partial^2 T(0,t)}{\partial x \partial t} = \alpha_1 [T(0,t) - T_{c1}] + \alpha_1 \tau_1 \frac{\partial T(0,t)}{\partial t}; \tag{28}$$

$$\lambda \frac{\partial T(\delta,t)}{\partial x} + \lambda \tau_1 \frac{\partial^2 T(\delta,t)}{\partial x \partial t} = \alpha_2 [T_{c2} - T(\delta,t)] - \alpha_2 \tau_1 \frac{\partial T(\delta,t)}{\partial t}, \tag{29}$$

where T_0 – initial temperature.

Note that the accelerations in time of heat fluxes, determined by the Fourier law (the left-hand sides of these equations) and the Newton law (their right-hand parts) are taken into account in formulas (28), (29).

To reduce the problem (6), (26) - (29) to the dimensionless form, equations (12) are used, in which only the dimensionless temperature changes, as well as other dimensionless parameters are introduced

$$\Theta = \frac{T - T_0}{\Delta T_1}; \ \mathrm{Bi}_1 = \frac{\alpha_1 \delta}{\lambda}; \ \mathrm{Bi}_2 = \frac{\alpha_2 \delta}{\lambda}; \ \Delta T_1 = T_{c1} - T_0 .$$

The problem (6), (26) - (29), taking into account the introduced notation, takes the form

$$\frac{\partial \Theta}{\partial \mathrm{Fo}} + \mathrm{Fo}_1 \frac{\partial^2 \Theta}{\partial \mathrm{Fo}^2} = \frac{\partial^2 \Theta}{\partial \xi^2} + \mathrm{Fo}_1 \frac{\partial^3 \Theta}{\partial \xi^2 \partial \mathrm{Fo}} - \frac{\gamma}{(\Theta \Delta T_1 + T_0)^2} \exp\left(-\frac{E}{R(\Theta \Delta T_1 + T_0)} \right) \frac{\partial \Theta}{\partial \mathrm{Fo}} +$$

$$+ \tau \exp\left(-\frac{E}{R(\Theta \Delta T_1 + T_0)} \right), \quad (\mathrm{Fo} > 0; \quad 0 < \xi < 1); \tag{30}$$

$$\Theta(\xi, 0) = 0; \tag{31}$$

$$\partial \Theta(\xi, 0)/\partial \mathrm{Fo} = 0; \tag{32}$$

$$\left[\frac{\partial \Theta}{\partial \xi} + \mathrm{Fo}_2 \frac{\partial^2 \Theta}{\partial \xi \partial \mathrm{Fo}} - \mathrm{Bi}_1 \left(\Theta + \mathrm{Fo}_1 \frac{\partial \Theta}{\partial \mathrm{Fo}} + \Delta T_2 \right) \right]_{\xi=0} = 0; \tag{33}$$

$$\left[\frac{\partial \Theta}{\partial \xi} + \mathrm{Fo}_2 \frac{\partial^2 \Theta}{\partial \xi \partial \mathrm{Fo}} - \mathrm{Bi}_2 \left(\Delta T_3 - \Theta - \mathrm{Fo}_1 \frac{\partial \Theta}{\partial \mathrm{Fo}} \right) \right]_{\xi=1} = 0, \tag{34}$$

where $\Delta T_2 = (T_0 - T_{c1})/\Delta T_1$; $\Delta T_3 = (T_{c2} - T_0)/\Delta T_1$.

6. Numerical Solving method

Using the explicit scheme of differential operators approximation, problem (30) - (34) can be represented in the form

$$\left[1+\frac{\gamma}{(\Theta_k^i \Delta T_1+T_0)^2}\exp\left(-\frac{E}{R(\Theta_k^i \Delta T_1+T_0)}\right)\right]\frac{\Theta_k^i-\Theta_k^{i-1}}{\Delta \text{Fo}}+\text{Fo}_1\frac{\Theta_k^{i+1}-2\Theta_k^i+\Theta_k^{i-1}}{\Delta \text{Fo}^2}=$$

$$=\frac{\Theta_{k-1}^i-2\Theta_k^i+\Theta_{k+1}^i}{\Delta\xi^2}+\text{Fo}_1\frac{\Theta_{k-1}^i-2\Theta_k^i+\Theta_{k+1}^i-\Theta_{k-1}^{i-1}+2\Theta_k^{i-1}-\Theta_{k+1}^{i-1}}{\Delta \text{Fo}\Delta\xi^2}+$$

$$+\tau\exp\left(-\frac{E}{R(\Theta_k^i \Delta T_1+T_0)}\right), \quad (\text{Fo}>0;\quad 0<\xi<1); \tag{35}$$

$$\Theta_k^0=0; \tag{36}$$

$$\frac{\Theta_k^1-\Theta_k^0}{\Delta \text{Fo}}=0; \tag{37}$$

$$\frac{\Theta_1^i-\Theta_0^i}{\Delta\xi}+\text{Fo}_1\frac{\Theta_1^{i+1}-\Theta_0^{i+1}-\Theta_1^i+\Theta_0^i}{\Delta \text{Fo}\Delta\xi}-\text{Bi}_1\left(\Theta_0^i+\text{Fo}_1\frac{\Theta_0^{i+1}-\Theta_0^i}{\Delta \text{Fo}}+\Delta T_2\right)=0; \tag{38}$$

$$\frac{\Theta_K^i-\Theta_{K-1}^i}{\Delta\xi}+\text{Fo}_1\frac{\Theta_K^{i+1}-\Theta_{K-1}^{i+1}-\Theta_K^i+\Theta_{K-1}^i}{\Delta \text{Fo}\Delta\xi}-$$

$$-\text{Bi}_2\left(\Delta T_3-\Theta_{K-1}^i-\text{Fo}_1\frac{\Theta_{K-1}^{i+1}-\Theta_{K-1}^i}{\Delta \text{Fo}}\right)=0. \tag{39}$$

7. Descussion of the results

The results of calculations of the problem (35) - (39) are shown in Fig. 4 - 6. Their analysis allows us to conclude that, with $\text{Bi}_1=5, \text{Bi}_2=10$ an unlimited increase in temperature (thermal explosion) does not take place, and with $\text{Fo}\geq 1$there is an established (steady) state is observed, in which the heat input from the heat source is compensated with the heat removal to the environment (the dashed line in Fig. 4, see also Fig. 6).

When $\text{Bi}_1=5, \text{Bi}_2=1$ the thermal ignition occurs at any Fo_1, and the time of its delay essentially depends on their value (see Fig. 4, 5). Thus, with an increase of Fo_1 from its value equal to $2\cdot 10^{-3}$ to the value 0.15, the delay time for thermal ignition at the point $\xi=0.9$ increases more than twice (see Fig. 4). The cause for its delay is the fact that when the relaxation properties of the material are taken into account, the thermal fluxes at the boundaries in a certain time range increase from zero to values determined by the boundary conditions of the third kind. In view of the relaxation properties of the material, the boundary conditions (regardless of their type - the first, the second, the third kind) can be satisfied only after a certain initial time range $0\leq \text{Fo}\leq \text{Fo}^*$, where Fo^* is the time at which the specified boundary condition of the boundary-value problem begins to be satisfied. Consequently, in a certain range of the initial time interval, plate heating takes place under conditions of limited heat exchange with the environment involved in the heating of the structure (for $T_{c1}>T_0$; $T_{c2}>T_0$ which along with the negative heat source in equation (13) is also a constraint to the thermal ignition process. With an increase of the relaxation time, regardless of the heating or cooling of the structure, the delay time for thermal ignition increases, including the case of setting the boundary conditions for its total thermal insulation.

IOP Conf. Series: Materials Science and Engineering **552** (2019) 012003 doi:10.1088/1757-899X/552/1/012003

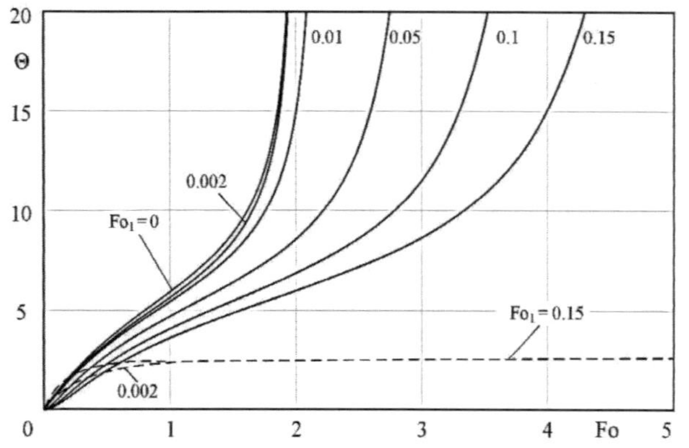

Figure 4. Temperature change in time ($\xi = 0.9$): solid line $\mathrm{Bi}_1 = 5; \mathrm{Bi}_2 = 1$; dotted line – $\mathrm{Bi}_1 = 5; \mathrm{Bi}_2 = 10$

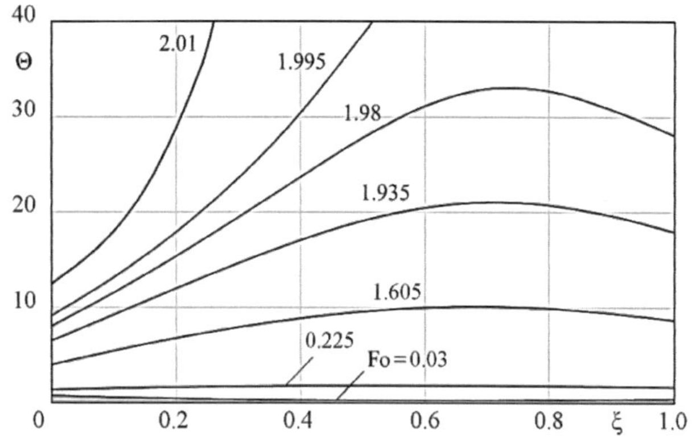

Figure 5. Temperature change with respect to the coordinate ξ in time $\mathrm{Fo}_1 = 0.002$; $\mathrm{Bi}_1 = 5$; $\mathrm{Bi}_2 = 1$; $\gamma = 9.3 \cdot 10^{10}$

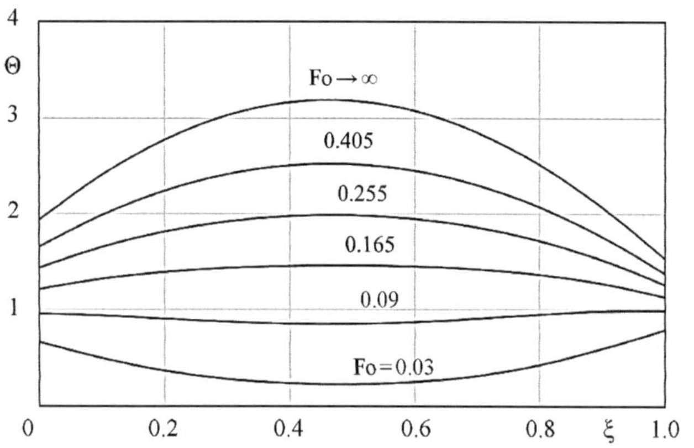

Figure 6. Temperature change with respect to the coordinate ξ in time $\mathrm{Fo}_1 = 0.002$; $\mathrm{Bi}_1 = 5$; $\mathrm{Bi}_2 = 10$; $\gamma = 9.3 \cdot 10^{10}$

As a particular example, a solution of the thermal ignition problem was also obtained with the following initial data: $\mathrm{Bi}_1 = 5$; $\mathrm{Bi}_2 = 1$; $\delta = 0.00876$ m; $a = 0.8 \times 10^{-6}$ m^2/s; $\tau_1 = 0.014$. Analysis of the obtained results allows us to conclude that the delay in the onset of thermal ignition is 0.33 s, in comparison with the calculation option at $\tau_1 = 0$.

8. Conclusions

1. Based on the modified formula of the Fourier law for a heat flux, a mathematical model of thermal ignition in a plate with an exponentially temperature dependent source of heat under boundary conditions of the first and third kind was developed with allowance for space-time nonlocality.

2. Studies of the obtained model, performed by the finite difference method, showed an essential dependence of the delay time of thermal ignition on the value of the relaxation coefficient, which is related to the thermal inertia of materials caused by the nonequilibrium thermal conductivity process, owing to which the transition of heat into the form of the thermal motion of particles in a nonstationary process occurs not instantaneously, but only after a certain time (time of relaxation). Consequently, the heat transfer coefficients cannot exceed certain limiting values depending on the thermophysical properties of the body, under no conditions of heat exchange with the external environment, including under boundary conditions of the first kind, for which in classical models infinite quantities are assumed and, consequently, an infinite value of the heat flux, which leads to an instantaneous establishment on the surface of the body of the temperature given by the boundary condition of the first kind.

3. It is shown that irrespective of body heating or cooling, when the relaxation properties of materials are taken into account, the delay time of thermal ignition increases. It also increases in the absence of heat exchange at the boundaries, that is, when the temperature field is determined only by the action of the internal heat source. The increase in the relaxation coefficients leads to an increase in the delay time for thermal ignition.

Acknowledgment
The reported study was funded by Russian Science Foundation (RSF) according to the research project № 18-79-00171.

References

[1] Frank–Kamenetsky D A 1967 *Diffuziya i teploobmen v khimicheskoy kinetike* [Diffusion and heat transfer in chemical kinetics] (Moscow: Nauka) p 492 [In Russian]

[2] Merzhanov A G and Dubovitsky F I 1966 Uspekhi Khimii Tekushchee sostoyanie teplovogo vzryva [Current state of the theory of thermal explosion] *Uspexi ximii* **35(4)** pp 656 – 82 [In Russian]

[3] Kudinov V A, Eremin A V and Stefanyuk E V 2016 Critical conditions for thermal explosion in a plate with a nonlinear heat source *Journal Of Machinery Manufacture And Reliability* **45** pp 38-43

[4] Sobolev S L 1991 *Peredayushchiye protsessy i dvizhushchiyesya volny v lokalno neravnovesnykh sistemakh* [Transmitting and moving processes are free in locally nonequilibrium systems] (Moscow: Nauka) [In Russian]

[5] Sobolev S L 1997 *Uspekhi fizicheskikh nauk Lokalno neravnovesnyye modeli protsessov perenosa* [Locally nonequilibrium models of transfer processes] **167(10)** pp 1096 – 106 [In Russian]

[6] Zhou D, Casas–Basques J and Lebon J 1999 Extended irreversible thermodynamics *Reports on Progress in Physics* **51(8)** p 528

[7] Day U A 1974 *Termodinamika prostyh sred s pamyat'yu* [Thermodynamics of simple environments with memory] (Moscow: Mir) [In Russian]

[8] Eremin A V, Kudinov V A and Kudinov I V 2017 Mathematical model of heat transfer in a fluid with account for its relaxation properties *Fluid Dynamics* **51** pp 33-44

[9] Kudinov V A, Eremin A V and Zhukov V V 2017 Mathematical models of heat ignition and explosion considering local non–equilibrium of processes **24(6)** *Journal of Physics Conf. Series*

[10] Petrov N and Brankov I 1986 *Sovremennyye problemy teromidinamiki* [Current problems of trotidynamics] (Moskva: Nauka) [In Russian]

[11] Bubnov V A 1975 Zamechaniya k volnovym uravneniyam teorii teploprovodnosti [Remarks on the wave equations of the theory of heat conductivity] *Engineering physics journal* **28** pp 670 – 6 [In Russian]

[12] Vilyunov V S 1984 *Teoriya vosplameneniya kondensirovannyh sred* [Condensed matter ignition theory] (Moscow: Nauka) p 190 [In Russian]

IOP Conf. Series: Materials Science and Engineering **552** (2019) 012003 doi:10.1088/1757-899X/552/1/012003

[13] Vernotte P 1985 Les paradoxes de la theorie continue de l'equation de la chaleur *Compte Rendus* **246(22)** pp 3154-3155.
[14] Lykov A V 1967 *Teoriya teploprovodnosti* [Heat conduction theory] (Moscow: Vysshaya shkola) [In Russian]
[15] Cattaneo C A 1958 Form of heat conduction equation which eliminates the paradox of instantaneous propagation *Compte Rendus* **247(4)** pp 431-3
[16] Lykov A V 1965 Primenenie metodov termodinamiki neobratimyh processov k issledovaniyu teplo-i massoobmena [Application of the methods of thermodynamics of irreversible processes to the investigation of heat and mass transfer] *Inzhenerno-fizichesky zhurnal* **9(3)** pp 287-304 [In Russian]
[17] Lykov A V and Mikhailov Y A 1965 *Teoriya perenosa energii i veshchestva* [Theory of energy and matter transfer] (Moscow: Izdatel'stvo akademii nauk SSSR) [In Russian]
[18] Baumeister K and Khamill T 1969 *Teploperedacha* **4** p 112 [In Russian]
[19] Shashkov A G, Bubnov V A and Yanovskii S Y 2004 *Volnovye yavleniya teploprovodnosti: Sistemno-strukturnyi podkhod* [Heat conduction wave phenomena: A system-structural approach] (Moscow: Editorial URSS) [In Russian]
[20] Brovkin L A 1984 *Izvestiya vysshih uchebnyh zavedeny SSSR, Energiya* **10** p 111 [In Russian]
[21] Kartashov E M 2008 *Vestnik moskovskogo instituta tonkoi khimicheskoy tekhnoljgii im. M.V. Lomonosova* **3(3)** p 20 [In Russian]
[22] Kirsanov Y A 2007 *Tsiklicheskie teplovye protsessy i teoriya teploprovodnosti v regenerativnykh vozdukhopodogrevatelyakh* [Cyclic thermal processes and the theory of thermal conductivity in regenerative air heaters] (Moscow: Fizmatlit) [In Russian]
[23] Tsoi P V 2005 *Sistemnye metody rascheta kraevykh zadach teplomassoperenosa* [System methods for calculating boundary problems of heat and mass transfer] (Moscow: Izdatel'stvo MEI) [In Russian]
[24] Tsirelman N M 2005 *Pryamye i obratnye zadachi teplomassoperenosa* [Direct and inverse problems of heat and mass transfer] (Moscow: Energoatomizdat) [In Russian]
[25] Nikitenko N I 1971 *Issledovanie nestatsionarnykh protsessov teplo- i massoobmena metodom setok* [Study of non-stationary processes of heat and mass transfer by the method of grids] (Kiev: Naukova dumka) [In Russian]

IOP Conf. Series: Materials Science and Engineering **552** (2019) 012004 doi:10.1088/1757-899X/552/1/012004

Construction and variants of the modernization of the condensing hot water boiler

A V Gubarev[1] and N M Lozovoj[2]

[1]Department of Energy Engineering of Heat Technologie, Belgorod State Technological University of V.G. Shukhov, Kostyukov St., 46, Belgorod, 308012, Russia
[2]Department of Urban Cadastre and Engineering Surveys, Belgorod State Technological University of V.G. Shukhov, Kostyukov St., 46, Belgorod, 308012, Russia

E-mail: artwo0248@mail.ru

Abstract. In this article are considered the main problems facing mankind, which are related with production, transformation and use of energy and fuel resources. For purposes of fuel saving and improve environmental safety of the heat supply systems of various consumers the use of decentralized heat supply systems with the installation of a condensing hot water boilers in heat source, is proposed. The substantiation of application of these heat supply systems is made. There are considered appointment, relevance of use, design features and principle of action of the fuel-saving condensing hot water boiler, developed by the employees of the Belgorod State Technological University named after V.G. Shukhov, proposed for installation as a heat generator in the heat sources of autonomous heating systems. There are considered the variants of construction of various units of the proposed boiler and the conditions of use of a particular design variant. Also, the advantages and disadvantages of various versions of the boiler design and its main components were critically evaluated, the comparison was made and the advantages of these or other design variants in front of others for different conditions were indicated.
Keywords.
Decentralized heat supply systems, condensing hot water boiler, radiation part, contact-recuperative part.

1. Introduction

Nowadays, among the main problems facing mankind, the following can be singled out: the gradual depletion of hydrocarbon fuel in the bowels of the planet (this is especially true for those deposits from which fuel is relatively inexpensive) [1], environmental pollution (thermal, toxic compounds, greenhouse gases) and climate change associated with the latter [2–4]. In this regard, in the energy and industry there are tendencies to reduce the consumption of traditional hydrocarbon fuels for energy generation and production work due to the use of unconventional and alternative sources and methods of transformation energy [5–7], secondary energy resources [8, 9], as well as improving energy efficiency and environmental friendliness of units with hydrocarbon fuels and systems in which such units are used [10–13].

The most important life support systems are heat supply systems. In Russian Federation, the production of heat energy for these systems consumes about 1/3 of all used fuel and energy resources [14]. The purpose of this work is fuel saving and improve environmental safety of heat supply systems.

To achieve this purpose, the following tasks are set:

- the justification of higher efficiency of decentralized heat supply systems in comparison with centralized systems;
- the justification of advantages of installation of condensing hot-water boilers and water heaters in a heat source of the decentralized heat supply system in comparison with boilers and water heaters, in which there is no condensation of the water vapor, which is a component of organic fuel's combustion products;
- comprehensive consideration of the design of the condensing hot water boiler, which is proposed for installation in heat sources of decentralized heat supply systems;
- consideration and comparison of various design options of the proposed condensing hot water boiler and its main elements, their advantages and disadvantages, in relation to the different working conditions.

2. Materials and Methods

As one of the ways to reduce the consumption of hydrocarbon fuels and improve environmental safety in such an important area of human activity as heat supply for various consumers, the use of decentralized heat supply systems with the installation of a fuel-saving condensing water boiler, designed by employees of Belgorod State Technological University named after V.G. Shukhov can be proposed [15].

High energy and environmental performance of such systems are due to the following features. In decentralized systems, there are no large-scale heat networks, in which very significant heat losses occur, reaching 20% of the output, resulting both from the losses to the environment, through the walls of heat pipes and thermal insulation, and losses with coolant leakage. In addition, load control in decentralized heating systems is much simpler than regulation in centralized systems. Thus, the heat supply by the heat source of decentralized systems most closely corresponds to its consumption by subscribers [16], which also reduces heat losses by an amount reaching 15%. In addition, heat networks require maintenance and repair, which increases operating costs [17, 18].

It should be noted, that this article discusses the advantages of decentralized heat supply systems over centralized systems, the heat source in which is a large or medium-capacity boiler plant [19]. The centralized systems, in which heat supply of consumers is carried out from the thermal power plants, working on a heating cycle, when the generated thermal energy is an additional product in the production of electrical energy, is not considered in this article.

The installation of a decentralized heat supply system for condensing-type heat generators in a heat source also contributes to significant fuel savings [20, 21]. As is known, the products of combustion of hydrocarbon fuels contain a significant amount of water vapor. For example, the share of water vapor in the composition of natural gas combustion products is 10...12%. In condensing boilers, the combustion products are cooled to temperatures that are below the dew point for gases of this composition. As a result, the water vapor contained in the flue gases is partially condensed, and their heat of condensation can be useful. Thus, in condensation-type heat generators, heat losses with flue gases can be significantly reduced (for example, when natural gas is burned, by up to 15%) by lowering the temperature of gases and due to the beneficial use of the latent heat of condensation of water vapor contained in them [22].

Improving the environmental performance of decentralized heating systems with a condensing hot water boiler, developed by employees of Belgorod State Technological University named after V.G. Shukhov is achieved thanks to the following merits. First, in such systems, the specific consumption of combustible fuel is two times lower than in centralized systems, because of their higher energy efficiency. Secondly, the condensation part of the specified heat generator is an effective particulate trap [23]. Thirdly, in the case of organizing the injection of water into the furnace of such a boiler, the formation of thermal nitrogen oxides is significantly reduced, and the heat losses with flue gases do not increase [24].

3. Results

The design of the condensing hot water boiler under consideration can be adapted to the specific conditions of its operation, the requirements for the layout of equipment in the boiler room, as well as the characteristics and parameters of working environments.

The main units of the condensing hot water boiler are the radiation and contact-recuperative parts, as well as the adiabatic part connecting them, which includes a condensate tank [25]. In the radiation section, which is carried out mainly in the same way as the traditional designs of modern gas-tube boilers, fuel is burned and the water for heating is heated. In the adiabatic part, the combustion products are moistened, passing through the condensate plumes, and adiabatically cooled. In the contact-recuperative part, there is a deep cooling of the combustion products (to a temperature of about 35° C), the condensation of most of the water vapor contained in them, as well as the heating of water supplied to meet the needs of hot water.

4. Discussion

For technological reasons the contact-recuperative part is always positioned vertically, while the radiation part can be placed either horizontally or vertically (figure 1).

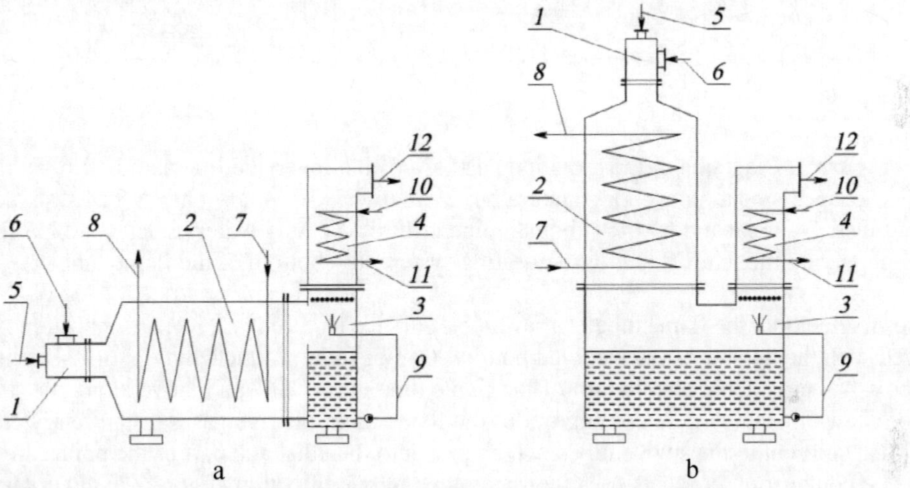

Figure 1. Options for the location of the radiation part: a – horizontal; b – vertical; 1 – burner; 2 – radiation part; 3 – adiabatic part; 4 – contact-recuperative part; 5 – natural gas; 6 – air 7 – back water heating system; 8 – direct water heating system; 9 – condensate; 10 – cold water; 11 – heated water to hot water needs; 12 – flue gases

As an advantage of the horizontal arrangement of the radiation part can highlight the possibility of using such structures in any industrial gas burners. In the embodiment of the vertical furnace with a burner cannot be used as it does not allow the working position "burner on the torch" [26]. A significant drawback of the horizontal layout of the radiation part is the uneven heating of the flame tube over its cross section [27], since there is an elevation of the torch as it is removed from the mouth of the burner. The result of this phenomenon will be an uneven temperature lengthening of the forming flame tube and an increase in temperature stresses in it, which can lead to serious accidents. Another advantage of boilers with a vertical layout of the radiation part is their smaller dimensions in the plan compared to boilers equipped with a horizontal firebox. At the same time, the height of the room must ensure compliance with the requirements of regulatory documents for boiler rooms, and, in addition, ensure the possibility of mounting and dismounting the burner and the radiation unit.

For technological reasons, it is necessary to provide a three-way scheme for the movement of combustion products in the radiation part of the boiler. There are two most obvious ways to organize such a gas flow pattern (figure 2).

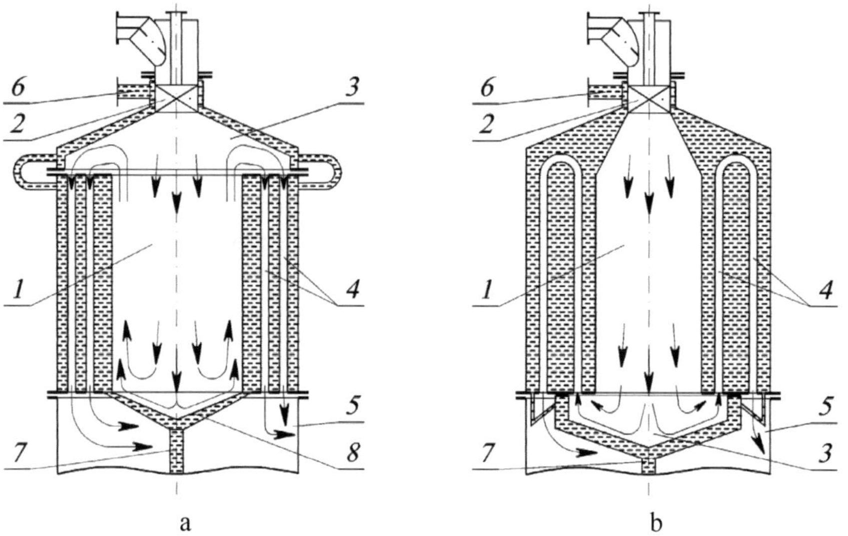

Figure 2. Diagrams of the radiation part: a – with a reversive fire chamber; b – with U-shaped smoke tubes; 1 – flame tube; 2 – burner; 3 – rotary chamber; 4 – smoke tubes; 5 – adiabatic part; 6 – branch pipe of direct heating water outlet; 7 – pipe for supplying the return water to the heating system; 8 – bottom of the flame tube

In the first method, the flame tube is a reversible furnace (figure 2, a), its part opposite to the burner is plugged with the help of a water-cooled bottom. Combustion products moving in the central part of the firebox, having reached the bottom of the flame tube, turn 180 ° and move along the periphery of the firebox towards the burner. Coming out of the flame tube, the flue gases in the rotary chamber are turned 180 ° and sent to the smoke tubes, which pass into the adiabatic part of the boiler. In the second method, the combustion products pass the flame tube in one direction (figure 2, b). A rotary chamber is located at the exit of the flame tube. In this rotary chamber, the flue gases are rotated 180 ° and then flow into the smoke tubes. Smoke tubes are U-shaped, so the products of combustion in them make two moves.

A significant drawback of the design of the radiation part with reversing furnace is the need to use compensating devices in order to avoid the destruction of elements of this part due to stresses arising due to the uneven thermal lengthening of the flame tube and fire tubes, as well as the body. One option to compensate for temperature lengthening in this design is the use of a corrugated flame tube. In the design of the radiation section with U-shaped pipes, the fire tubes can freely lengthen in the longitudinal direction, and the probability of occurrence of metal damage or welded joints caused by thermal stresses is significantly reduced [28]. However, in this design of the radiation part, access to the welds in the attachment points of the flame tube and U-shaped tubes to the tube plate is very difficult.

In order to provide access to the specified welds without destroying the details of the radiating part of the boiler, it can be offered as shown in figure 3 variant of the design of the considered node, when the rotary chamber is made in the form of a water-cooled truncated cone, and the diameter of the smaller base of the cone provides access to the welds at the attachment points of the flame tube and fire tubes to the tube plate [29].

IOP Conf. Series: Materials Science and Engineering **552** (2019) 012004 doi:10.1088/1757-899X/552/1/012004

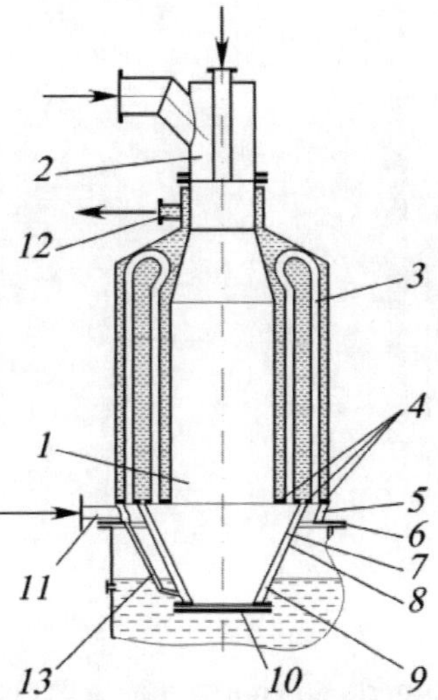

Figure 3. Diagram of the radiation part with increased maintainability: 1 – flame tube; 2 – burner; 3 – smoke tubes; 4 – tube sheet; 5 – ring water jacket; 6 – mounting flange of the radiation part; 7 – inner bottom of the flame tube; 8 – the outer bottom of the flame tube; 9 – water jacket; 10 – a cover; 11 – pipe for supplying return water to the heating system; 12 – branch pipe of direct heating water outlet

After leaving the radiation part, the combustion products of the fuel pass through the adiabatic part, the main elements of which are the condensate tank, nozzles and the reference-distribution grid. The nozzles spray condensate from the tank into the flue gas stream, the temperature of which exceeds 250 °C. In this case, the condensate partially evaporates, and the gases are respectively moistened and adiabatically cooled to a temperature approaching the temperature of the wet thermometer. With the passage of the resulting two-phase gas-liquid flow through the reference distribution grid, phase inversion occurs. Thus, above the grate - in the contact-recuperative part of the boiler - a foamy two-phase flow is formed, which is characterized by high intensity of heat and mass transfer between the phases. In the contact-recuperative part, a tube bundle is installed, into which cold water of drinking quality is supplied. Water, heated in a tube bundle, to temperatures reaching 50...60 °C is supplied to the needs of hot water supply. The gas-liquid stream, passing through the inter-tubular space of the tube bundle, is cooled. When this occurs, the condensation of the majority (up to 80%) of water vapor contained in the gas. Above the tube bundle, the velocity of the gas-liquid flow drops sharply and gravitational separation of condensate droplets from the gas flow occurs. Condensate is discharged to the walls of the casing and discharged into the condensate tank. For a better separation of gas and liquid, an inertial droplet separator should be installed in front of the flue gas pipe.

The heat exchange surface of the tube bundle is formed by straight tubes. The use of round tubes in the contact-recuperative part of the condensing hot water boiler under consideration is not recommended, since in this case the increase in hydraulic resistance is much higher than the increase in heat transfer efficiency. In addition, in this case, the dimensions of the contact-recuperative part and the whole boiler increase. To increase the intensity of heat transfer, and both the compactness of the contact-recuperative part, the use of staggered flat-oval tubes arranged in a checkerboard pattern, which can be made by squeezing standard round tubes, seems most appropriate. It should be noted, that the pipes must be oriented with a larger axis along the flow in the inter-tubular space of the contact-recuperative part.

In this case, the ends of the pipes can be additionally crimped in the horizontal direction (figure 4). This additional compression of the ends of the pipes allows you to meet the technological requirements for fastening pipes in tube plates.

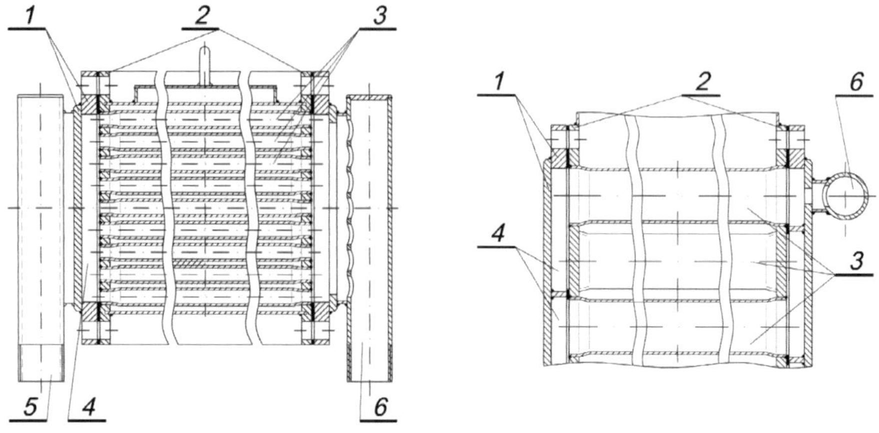

a b

Figure 4. Pipe bundle of the contact-recuperative part: a – horizontal section; b – vertical section; 1 – tube cover; 2 – tube sheet; 3 – flat oval tube; 4 – bypass water chamber; 5 – outlet pipe for heated water; 6 – cold water inlet

The use of flat-oval tubes with an additional crimping of their ends in the tube bundle of the contact-recuperative part of the hot water boiler allows reducing horizontal gaps between the tubes. This, respectively, reduces the flow area in the inter-tubular space of the heat exchanger. In this case, the speed of the two-phase gas-liquid flow in the inter-tubular space of the contact-recuperative part of the boiler increases, and the intensity of heat transfer from the condensate to the outer surface of the heat exchange tubes increases too. In addition, the use of flat-oval tubes with additional crimping of their ends in the tubular heat exchanger of the contact-recuperative part of the condensing hot water boiler allows reducing the relative horizontal pitch of the tubes. This, in turn, allows reducing the width of the casing of the contact-recuperative part of the boiler, to reduce its mass and material consumption while maintaining the required heat exchange surface [30].

5. Conclusion

The use of decentralized systems with the installation of condensing-type hot-water boilers in their heat source allows a 2-fold reduction in the unit cost of fuel for the needs of heat supply to consumers. As such heat generators can be very effectively used condensing boiler, developed by employees of Belgorod State Technological University named after V.G. Shukhov.

When using this hot water boiler in addition to increasing the energy efficiency of the heat supply system, its negative environmental impact can be significantly reduced. This can be achieved both by reducing the specific fuel consumption for the production of a unit of thermal energy, and due to the fact that the proposed apparatus can be an effective wet particulate trap, and in addition, in the case of organizing the injection of water into the furnace chamber of the heat generator, the formation of thermal nitrogen oxides can be reduced. Since the hot water boiler is condensing, the injection of water into the combustion zone will not have a negative impact on its technical and economic indicators.

The design of this boiler allows you to adapt it to different conditions of use. It is also distinctive features of this heat generator are its compact size and low metal consumption.

Acknowledgments

The article was prepared within development program of the Flagship Regional University on the basis of Belgorod State Technological University named after V.G. Shukhov.

References

[1] Muradov N, Choi P, Smith F and Bokerman G 2010 Integration of direct carbon and hydrogen fuel cells for highly efficient power generation from hydrocarbon fuels *J. Power Sources* **195** 1112-21

[2] Abdel-Shafy H I and Mansour M S M 2016 A review on polycyclic aromatic hydrocarbons: Source, environmental impact, effect on human health and remediation *Egyptian J. Of Petroleum* **25** 107-23

[3] Nordell B 2003 Thermal pollution causes global warning *Global and Planetary Change* **38** 305-12

[4] Zhu L, Chen L, Wu X and Ding X 2018 Developing a greenhouse gas management evaluation system for Chinese textile enterprises *Ecological Indicators* **91** 470-7

[5] Salameh M G 2003 Can renewable and unconventional energy sources bridge the global energy gap in the 21st century? *Applied Energy* **75** 33-42

[6] Zerrahn A, Schill W-P and Kemfert C 2018 On the economics of electrical storage for variable renewable energy sources *European Economic Review* **108** 259-79

[7] Müller-Steinhagen H and Nitsch J 2005 The Contribution of Renewable Energies to a Sustainable Energy Economy *Process Safety and Environmental Protection* **83** 285-97

[8] Richterová V 1984 Utilization of secondary energy resources in chemical industry *Energy Conversion and Management* **24** 27-32

[9] Datsuk T, Pozin G, Ulyasheva V and Kanev M 2015 Energy-saving Technologies in Transportation of Natural Gas Facilities *Procedia Engineering* **117** 874-82

[10] Delpech B, Milani M, Montorsi L, Boscardin D, Chauhan A, Almahmoud S, Axcell B and Jouhara H 2018 Energy-efficiency enhancement and waste heat recovery in industrial processes by means of the heat pipe technology: Case of the ceramic idustry *Energy* **158** 656-65

[11] Jouhara H, Khordehgah N, Almahmoud S, Delpech B, Chauhan A and Tassou S A 2018 Waste heat recovery technologies and applications *Thermal science and Engineering Progress* **6** 268-89

[12] Woolley E, Luo Y and Simeone A 2018 Industrial waste heat recovery: A systematic approach *Sustainable Energy Technologies and Assessments* **29** 50-9

[13] Trubaev P A, Zajtsev E A and Besedin P V 2016 Exergic analysis of technological heat devices *Inter. J. of Pharm. and Technology* **8** 24968-75

[14] Bashmakov I A 2010 Povy'shenie e'nergoe'ffektivnosti v sistemah teplosnabzheniya Chast' 1. Problemy' Rossijskih sistem teplosnabzheniya [Increase energy efficiency in heat supply systems Part 1. Problems of Russian heat supply systems] *E'nergosberezhenie* **2** 46-54 [In Russian]

[15] Kozhevnikov V P, Kuleshov M I, Gubarev A V, Trubaev P A and Andreev A B 2014 Condensing Hot Water Boiler: Applicability, Design, Research *Advances in Environmental Biol* **8** 122-7

[16] Koshlich Y, Trubaev P and Belousov A 2018 An approach of energy resources control system design *International Multidisciplinary Scientific GeoConference Surveying Geology and Mining Ecology Management (SGEM) Vol 18(2.1)* 63-9

[17] Lichter J M and Konstantinov V A 1995 Avtonomnoe teplosnabzhenie zdanij [Independent heating of buildings] *Energetik* **4** 9-10 [In Russian]

[18] Kozhevnikov V P, Kuznetsov V A, Mochalin A A, Titarenko R Y and Sokolova L V 2014 Independent heating modules with condensing Hot Water Boilers as heat generators as an effective alternative to centralized heat supply *Advances in Environmental Biol* **8** 89-93

[19] Terés-Zubiaga J, Campos-Celador A, González-Pino I and Diarce G 2016 The role of the design and operation of individual heating systems for the energy retrofits of residential buildings *Energy Conversion and Management* **126** 736-47

[20] Trubaev P A and Tarasyuk P N 2017 Evaluation of energy-saving projects for generation of heat and heat supply by prime cost forecasting method *International Journal of Energy*

Economics and Policy **7(5)** 201-8

[21] Varlamov G, Romanova K, Daschenko O, Ocheretyanko M and Kasyanchuk S 2016 The use of contact heat generators of the new generation for heat production *EasternEuropean J. Of Enterprise Technologies* **1** 52-8

[22] Grinenko G P, Kozhevnikov V P, Kuleshov M I and Pogonin A A 2012 Perspektivy' razvitiya ry'nka kondensacionny'h kotlov v Rossii [Prospects of the market development of condensation boilers in Russia] *Vestnik Belgorodskogo gosudarstvennogo tehnologicheskogo universiteta im. V.G. Shuhova* **3** 145-9 [In Russian]

[23] Kuleshov M I, Belyaeva V I, Kozhevnikov V P, Pogonin A A and Mochalin A A 2012 Povy'shenie e'kologicheskoj bezopasnosti sistem teplosnabzheniya [Increase of ecological safety of heat supply systems] *Ekologija i promyshlennost' Rossii* **7** 12-3 [In Russian]

[24] Kuleshov M I and Gubarev A V 2015 On Effectiveness And Ways Of Ecological Characteristics Increase Of The Condensation Boiler For Heat Supply Independent Systems *Inter. J. of App. Engin. Res* **10** 44872-8

[25] Gorshkolepov G M, Kuleshov M I and Belyaeva V I 2002 Reshenie problemy' e'nergosberezheniya i povy'sheniya bezopasnosti v kommunal'noj sfere [Solving the problem of energy saving and improving safety in the municipal sphere] *Bezopasnost' truda v promy'shlennosti* **7** 33-5 [In Russian]

[26] Kuleshov M I, Gerasimov M D and Gerasimov D M 2011 Kondensacionny`j vodogrejny`j kotel [Condensation hot water boiler] Patent RU 2411420

[27] Kuznetsov V A and Trubaev P A 2018 Resources and Problems of the Mathematical Simulating Thermo-Technological Processes *J. Phys.: Conf. Ser.* **1066** 012024

[28] Gubarev A V, Kuleshov M I, Vasilev B P and Gubareva V V 2010 Komponovka trubnogo puchka radiacionnoj chasti toplivosberegayushhego gazovogo vodonagrevatelya [The arrangement of the tube bundle assembly of the radiation part of the fuel-efficient gas water heater] *Promy'shlennaya e'nergetika* **2** 37-9 [In Russian]

[29] Kuleshov M I and Gubarev A V 2015 Modernizaciya konstrukcii vy'sokotemperaturnoj chasti kondensacionnogo vodogrejnogo kotla, napravlennaya na povy'shenie ego remontoprigodnosti [Upgrading the construction of high-temperature part of the condensing hot water boiler, aimed to increasing its maintainability] *Vestnik Belgorodskogo gosudarstvennogo tehnologicheskogo universiteta im. V.G. Shuhova* **2** 196-8 [In Russian]

[30] Gubarev A V and Lozovoi N M 2017 Intensifikaciya teploobmena v kontaktno-rekuperativnoj chasti kondensacionnogo vodogrejnogo kotla [Heat exchange intensification in a recuperator of a condensing water boiler] *E'nergobezopasnost' i e'nergosberejenie* **5** 15-8 [In Russian]

IOP Conf. Series: Materials Science and Engineering **552** (2019) 012005 doi:10.1088/1757-899X/552/1/012005

Convective heat transfer in the near-wall turbulent gas stratum

V A Kuznetsov and P A Trubaev

Department of Energy Engineering of Heat Technologie, Belgorod State Technological University of V.G. Shuchov, Belgorod, 308012, Russia

E-mail: v.kousnezov@mail.ru

Abstract. In order to diminish the computational cost of heat transfer simulations, some wall functions had been often introduced in numerical algorithms. More universal and robust dimensionless velocity functions that can be employed both in buffer and equilibrium parts of the near-wall turbulent gas stratum are offered here for consideration. They have been obtained on the basis of Prandtl's semiempirical hypothesis of turbulence. A mode, how to determine the one-sided derivatives of the gas velocity and temperature at the wall surface on a coarse grid, has been substantiated. That allows applying the boundary conditions of Neumann and Robin type for differential equations of the convective heat transfer to the confining walls from the turbulent gas flow. A semiempirical formula for the eddy viscosity in the wall proximity has been deduced, and a valid boundary value has been determined also for the rate of turbulence-energy dissipation. It is demonstrated that the simulation results coincide sufficiently with experimental data.

Key words: numerical simulation, coarse-mesh grid, convective heat transfer, turbulent boundary layer, boundary conditions, eddy viscosity.

1. Introduction

Many power engineering problems that occur in gas flows can be settled today with the help of numerical simulations. Mathematical models include, as a rule, partial-differential transport equations with their discrete algebraic counterparts [1], which base themselves on the universal conservation laws stated for the momentum, heat and mass transfer. In order to obtain numerical solution of the discrete equations with required exactness, multiple iterations have to be performed through means of difference grids, formed sometimes by hundreds of thousands of small enough meshes. When mathematic simulation relates to the convective heat transfer in turbulent flows, the very fine grid ought to be employed near flow-confining walls, in consequence of strong nonlinear alteration of the fluid velocity, eddy viscosity and temperature in the immediate wall proximity [2].

Most of industrial problems with turbulent gas velocities might be decided by means of a Reynolds averaged Navier–Stokes (RANS) simulation together with the standard dissipative turbulence model, which is often adopted for evaluating the eddy viscosity. In order to decrease in a way a computational cost of the numerical solution of the differential equations, some approximate wall functions can be deployed while setting the boundary conditions to the differential equations. For such a purpose, some peculiarities of the turbulent gas movement beside the wall surface are usually utilized.

The near-wall stratum of the turbulent boundary layer can be divided theoretically in three sublayers. A thin viscous sublayer, in which molecular viscosity prevails over eddy one, adjoins to the wall surface. An equilibrium sublayer, in which the rate of turbulent-energy generation approximately equals that of its dissipation, is allocated at some distance from the wall [3]. A buffer sublayer would be disposed between them. The gas velocity, tangent to the wall, increases almost linearly in the viscous sublayer, and it alters logarithmically in the equilibrium sublayer. But mathematical statements for the velocity and fluid temperature in the buffer sublayer remain uncertain. As a rule, this crucial region comes out of consideration.

For instance, a logarithmic wall function, taken from [4], had been adopted in [5] for computing a boundary value of the tangential velocity at first nodes off the wall. Also several wall functions have been obtained in [6] by means of analytical solutions of some differential equations in the viscous and equilibrium sublayers. Since the computational logarithmic-velocity region is strictly limited, the simulation results became very sensitive in these cases to the size of grid meshes adjacent the wall. Moreover, the used wall functions seem to be incapable to predict a value of the wall-normal derivatives of the velocity and temperature in the boundary conditions of Neumann or Robin type.

Similar drawbacks are also inherent to dissipative models of turbulence, because a boundary value of dissipation rate of turbulent energy had be defined analytically only in the equilibrium sublayer. Such an impediment was tried to be removed in [7] by empirically choosing the near-wall viscosity profile and in [8] by modifying diffusion terms of turbulent scalar equations.

The indicated imperfections can be attenuated by applying some complicated numerical wall functions instead of the analytical ones [9–11]. The boundary conditions would be specified more accurately with their help, but only due to an additional set of partial differential equations, which should be solved on a sub-grid, created close to the wall. Moreover, this approach needs some empirical functions for defining the eddy viscosity in the low-Reynolds-number wall-adjacent stratum.

With such a background, it is worthwhile to develop further the turbulent boundary-layer theory, in order to substantiate more perfect wall functions with its help. The near-wall turbulence, being anisotropic, yields to the theoretical analysis with difficulty. In these circumstances, Prandtl's semiempirical hypothesis, which remained unreclaimed for a long time, seems very inviting. It defines near-wall eddy stresses as follows [12]:

$$\sigma = \rho\left(l\,\mathrm{d}u/\mathrm{d}y\right)^2,$$

where ρ is the fluid density; l is the mixing length; u is a fluid velocity tangent to wall; y is a distance along the normal to the wall.

Using this formula, a balance equation had been composed for shearing stress in isothermal fluid flows [13]. Then a dimensionless expression for the near-wall velocity derivative has been deduced from it, though the mixing lengths beside the wall remained uncertain.

2. Methods

The eddy viscosity in the nonisothermal wall proximity

Sum of the molecular and eddy stresses in a thin nonisothermal gas stratum that is adjacent to the wall would be approximately equal to the shear stress σ_w at the wall surface:

$$\sigma_w \equiv \rho_w u_*^2 \approx \rho\nu\,\mathrm{d}u/\mathrm{d}y + \rho l^2\left(\mathrm{d}u/\mathrm{d}y\right)^2,$$

where ν is the local kinematic viscosity; ρ_w is the fluid density at the wall temperature; u_* is the skin-friction velocity, which replaces the shear stress in formulae.

This equality can be reduced to a dimensionless form,

$$l_+^2 (du_+/dy_+)^2 \nu_w/\nu + du_+/dy_+ - \rho_w \nu_w/(\rho\nu) = 0,$$

where ν_w is the kinematic viscosity at the wall temperature; u_+, y_+, l_+ are dimensionless variables which are determined as follows:

$$u_+ \equiv u/u_*, \qquad y_+ \equiv y u_*/\nu_w, \qquad l_+ \equiv l u_*/\nu_w .$$

Solution of the quadratic equation for the dimensionless velocity derivative can be written in the form of its inverse proportionality to the mixing length:

$$du_+/dy_+ = f_u (\rho_w/\rho)^{0.5}/l_+, \tag{1}$$

where f_u is a theoretical function, which approaches to one at large mixing lengths in the far distances from the wall:

$$f_u = \left(1 + \frac{1}{4l_+^2}\frac{\nu^2}{\nu_w^2}\frac{\rho}{\rho_w}\right)^{0.5} - \frac{1}{2l_+}\frac{\nu}{\nu_w}\left(\frac{\rho}{\rho_w}\right)^{0.5}.$$

In the special case of the isothermal boundary layer, these formulae become simplified as follows:

$$(du_+/dy_+)^{iso} = f_l/l_+, \tag{2}$$

$$f_l = \left(1 + \frac{1}{4l_+^2}\right)^{0.5} - \frac{1}{2l_+}.$$

Superscript 'iso' marks here a ratio of quantities in the isothermal medium.

When equality (1) would be taken into account, a dimensionless expression for the ratio of the eddy viscosity ν_e to the kinematic one ν_w follows from Prandtl's hypothesis for the near-wall gas stratum:

$$\nu_e/\nu_w = l_+ f_u (\rho_w/\rho)^{0.5}. \tag{3}$$

Formulae (1)–(3) can be useful for mathematic simulating the turbulent gas flow, when the dimensionless mixing length l_+ would be determined.

Isothermal boundary conditions for momentum transport equations

The mixing-length alteration is described by van Driest's exponential law [14], which takes into account the turbulence damping beside walls,

$$l_+ = [1 - \exp(-y_+/A_+)]\kappa y_+, \tag{4}$$

where κ is the universal constant, A_+ is a configurational factor.

This factor had been evaluated in [14] at the condition that computed near-wall velocities would conform on average to experiments. However, the result turned out to be too inaccurate for deploying the formula (4) in mathematical models. More effective mode of its determination had been realized in [15], where the factor A_+ was computed from each experimental velocity value. Mathematical treat-

IOP Conf. Series: Materials Science and Engineering **552** (2019) 012005 doi:10.1088/1757-899X/552/1/012005

ment of the derived data revealed that the configurational factor changes in accordance with the following quadratic function:

$$A_+ \approx 30\left[1-\left(y_+/y_+^{\log}\right)^2\right].$$

Constants $\kappa = 0.41$ и $y_+^{\log} = 50$ were found from experimental velocity distributions for the developed turbulence in the immediate wall proximity.

As figure 1 shows, curve *2* that corresponds to the exponential function (4) proceeds smoothly into Prandtl's linear law *1* at a dimensionless distance from the wall $y_+ = y_+^{\log}$ and establishes as such a formal edge of the equilibrium sublayer, which can be distinguished by the velocity logarithmic law in it. In fact, the logarithmic-velocity law begins at $y_+ = 0.7y_+^{\log}$. Hence, a substantial part of the formally buffer sublayer with dimensionless coordinates $(0.7\ldots1)y_+^{\log}$ reveals some peculiarity that is more proper to the equilibrium sublayer of the turbulent boundary layer.

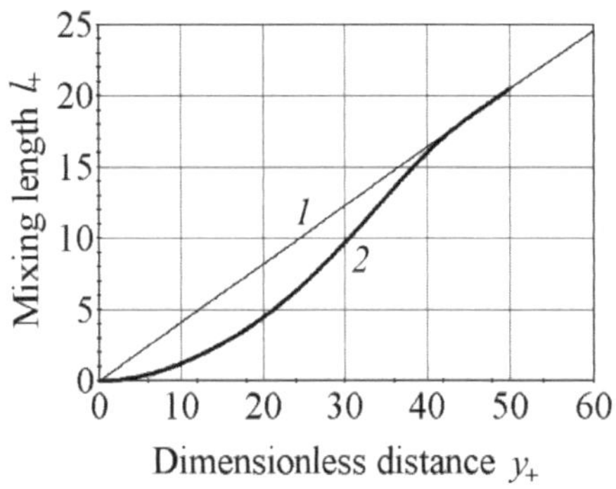

Figure 1. Functional dependence of the dimensionless mixing length on wall-normal coordinate: *1* – Prandtl's linear law; *2* – exponential law (4)

When the factor A_+ in formula (4) had become defined, opportunity appeared for numerical integration of the differential equality (2). The right side of it had been approximated by means of a sufficiently precise polynomial expression, which had been then integrated. As result, a polynomial of eighth power had been derived for dimensionless-velocity alteration in the viscous and buffer sublayers [16]:

$$u_+ = a_0 + a_1 y_+ + a_2 y_+^2 + a_3 y_+^3 + a_4 y_+^4 + a_5 y_+^5 + a_6 y_+^6 + a_7 y_+^7 + a_8 y_+^8. \quad (5)$$

Its coefficients, related to the fully developed turbulence, are presented in the following table.

Table 1. Coefficients of the polynomial for computing dimensionless velocities

y+	a0	a1	104a2	104a3	104a4	106a5	108a6	1010a7	1012a8
0…11	0	1	1	–10.6	7.18	–248	2990	–15870	32000
11…50	–2.58	1.986	–1296	55.4	–1.57	2.9135	–3.395	2.255	–0.6528

In the equilibrium sublayer, the dimensionless velocity is recommended to be determined with the help of the logarithmic law:

$$u_+ = \frac{1}{0.41}\ln y_+ + 5.1.$$

IOP Conf. Series: Materials Science and Engineering **552** (2019) 012005 doi:10.1088/1757-899X/552/1/012005

The polynomial (5) permits finding a ratio $(u_+/y_+)^{\text{iso}}$ of the dimensionless velocity to its wall-normal coordinate at a wall-neighbour grid node P in the isothermal turbulent fluid. That was employed in [16] for computing one-sided gas-velocity derivatives at the wall surface and for improving consequently boundary conditions to the differential and discrete equations of turbulent gas flow:

$$(\mathrm{d}u/\mathrm{d}y)_{\mathrm{w}}^{\text{iso}} = \frac{u_P}{y_P}(y_+/u_+)_P^{\text{iso}}. \tag{6}$$

Since the dimensionless derivative (2) equals one at the wall, the formula (6) might be considered as an application of the rule for calculating the whole of some quantity when a part of its value is known. In this case, the known part of the whole quantity would be a discrete value u_P/y_P of the velocity derivative.

Nonisothermal boundary conditions for momentum transport equations

It is more difficult to realize the same approach to setting boundary conditions for the nonisothermal gas flow because of uncertain altering of the gas density and viscosity near the wall in the general case. In this connection, some probability might be considered how the isothermal-stratum velocity derivative, taken at the wall, could be used in nonisothermal circumstances of turbulent-gas cooling.

The derivative of dimensionless isothermal-gas velocity has to be inserted into the right-side part of expression (1) with the help of equality (2):

$$\frac{\mathrm{d}u_+}{\mathrm{d}y_+} = \frac{\mathrm{d}u_+^{\text{iso}}}{\mathrm{d}y_+}\left(\frac{\rho_{\mathrm{w}}}{\rho}\right)^{0.5}\frac{f_u}{f_l}.$$

This equation can be integrated formally:

$$u_+ = u_+^{\text{iso}}\langle\rho_{\mathrm{w}}/\rho\rangle^{0.5}\langle f_u\rangle/\langle f_l\rangle.$$

Angle brackets here make note of quantities that have been averaged over the integration area from the wall to the neighbouring grid node P.

The result of such a manipulation tenders an expression that would allow computing correctly the one-sided velocity derivative at the wall in the nonisothermal medium:

$$(\mathrm{d}u/\mathrm{d}y)_{\mathrm{w}} = m_u\, u_P/y_P, \tag{7}$$

where m_u is a correction factor at discrete value of the velocity derivative,

$$m_u = \left(\frac{y_+}{u_+}\right)_P^{\text{iso}}\left\langle\frac{\rho}{\rho_{\mathrm{w}}}\right\rangle^{0.5}\frac{\langle f_l\rangle}{\langle f_u\rangle}. \tag{8}$$

In order to simplify equations and computational algorithm of the mathematical model, it would be advisable to introduce in the computational formulae a by-wall transfer factor, which in fact determines Neumann type boundary conditions at the wall surface for the differential momentum-transport equations:

$$\zeta \equiv \frac{\nu_{\rm w}}{u_P}\left(\frac{\partial u}{\partial y}\right)_{\rm w},$$

where u_P is the velocity at an wall-adjacent grid node P.

It is evident that the by-wall transfer factor is proportional to the squared skin-friction velocity:

$$\zeta = u_*^2 / u_P\,. \tag{9}$$

If the expression (7) of the one-sided velocity derivative would be taken into consideration, a discrete form of the by-wall transfer factor can be found:

$$\zeta = m_{u,P} \nu_{\rm w} / y_P\,. \tag{10}$$

While numerical iterations are being fulfilled at numerical solving of discrete counterparts of the differential equations, the skin-friction velocity u_* might be computed from a rough value of the factor ζ in formula (9). After that, the dimensionless coordinate y_+ should be calculated at the wall-adjacent node P. Then the correction factor (8) can be found, whereupon the previous value of the by-wall transfer factor becomes defined more exactly by means of ratio (10).

A boundary condition for the turbulence-energy dissipation model

In order to find the eddy viscosity for its utilizing in the momentum equations, two differential transport equations of the standard dissipative k-ε turbulence model have to be numerically solved for quantities k and ε [17]. The boundary condition for turbulence kinetic energy k had been approximately spread over the whole wall proximity. On the contrary, the boundary condition for the dissipation rate of the turbulence energy remains valid only in equilibrium sublayer.

When this turbulence model would be applied to the low-Reynolds-number turbulent fluid flow in the wall proximity, the following formula had been proposed to find the eddy viscosity [18]:

$$\nu_{\rm e} = C_\mu f_\mu k^2 / \varepsilon\,. \tag{11}$$

where C_μ is an empirical coefficient; k is the kinetic turbulence energy; ε is the rate of the turbulence-energy dissipation; f_μ is an empirical function that equals one at far distances from the wall.

The empirical function equals, as a matter of fact, a ratio of values of the eddy viscosity that have to be calculated with and without the turbulence damping, correspondingly:

$$f_\mu \equiv \nu_{\rm e} / \upsilon_{\rm e}\,,$$

where $\upsilon_{\rm e}$ is a conventional eddy viscosity that can be found by means of formula (3) with using in it Prandtl's linear mixing-length law,

$$\upsilon_{\rm e} / \nu_{\rm w} = \sqrt{(\kappa y_+)^2 \rho_w/\rho + (\nu/\nu_w)^2/4} - (\nu/\nu_w)/2\,.$$

Inserting this ratio into the formula (11) reduces it to a new form:

$$\upsilon_{\rm e} = C_\mu k^2 / \varepsilon\,.$$

IOP Conf. Series: Materials Science and Engineering **552** (2019) 012005 doi:10.1088/1757-899X/552/1/012005

Since the eddy viscosity ν_e already had been determined with the formula (3), the last equality might be used now as the boundary condition for the transport equation of the rate of the turbulence-energy dissipation at the wall-adjacent grid nodes *P*:

$$\varepsilon_P = C_\mu k_P^2 / \upsilon_{eP} .$$

This expression can serve as the boundary condition in all the cases when the node *P* falls into equilibrium, buffer or viscous sublayers.

Boundary conditions for the heat-transfer equation

When considering the heat transfer in the turbulent near-wall stratum, a notion of dimensionless temperature of medium can be applied:

$$T_+ \equiv \rho_w c_w u_* (T - T_w) / q_w ,$$

where ρ_w, c_w are the gas density and thermal capacity at the wall temperature; T_w, T are temperatures of the wall and gases beside it, respectively; q_w is the heat-flux density on the wall surface.

It should be taken into consideration, that the advective heat transfer to the wall is usually neglected in the boundary layer. As result, a differential expression of the conductive heat-flux density in the near-wall gas stratum would be equalized to its value on the wall surface:

$$q_w = \rho c \left(\frac{\nu}{\mathrm{Pr}} + \frac{\nu_e}{\mathrm{Pr_t}} \right) \frac{dT}{dy},$$

where Pr is the gas Prandtl number; $\mathrm{Pr_t}$ is the turbulent Prandtl number.

If the gas thermal capacity can be supposed constant, a dimensionless equality follows from the previous equation:

$$1 = \frac{\rho}{\rho_w} \left(\frac{1}{\mathrm{Pr}} \frac{\nu}{\nu_w} + \frac{1}{\mathrm{Pr_t}} \frac{\nu_e}{\nu_w} \right) \frac{dT_+}{dy_+} .$$

The first item in its right-side part may be cast aside, because it is too small at the equilibrium sublayer in comparison with the second one. Then, applying formula (3) of the eddy viscosity, we can obtain an approximate expression for the dimensionless temperature derivative:

$$dT_+ / dy_+ \approx \mathrm{Pr_t} (\rho_w / \rho)^{0.5} / (f_u l_+) .$$

Comparing this expression with formula (1), we ascertain that an approximate proportionality between the dimensionless derivatives of the temperature and velocity exists in nonisothermal gases at some distance from the wall:

$$dT_+ / dy_+ \approx (du_+ / dy_+) \mathrm{Pr_t} / f_u^2 .$$

It might be expected that such proportionality will remain in the equilibrium sublayer and outer part of the buffer sublayer after integrating the previous approximate equality:

IOP Conf. Series: Materials Science and Engineering **552** (2019) 012005 doi:10.1088/1757-899X/552/1/012005

$$T_+ \approx \mathrm{Pr_t}\, u_+ / \langle f_u \rangle^2 .$$

The derivative of the function T_+/Pr equals one at the coordinate origin [13]. Hence, in order to find the one-sided temperature derivative on the wall surface, its discrete value has to be divided by the quantity $T_+/(y_+\mathrm{Pr})$ at the wall-adjacent grid node P with taking into account formula (8):

$$\left(\frac{\mathrm{d}T}{\mathrm{d}y}\right)_\mathrm{w} \approx \frac{\mathrm{Pr}}{\mathrm{Pr_t}} \frac{T_P - T_\mathrm{w}}{y_P} m_u \langle f_u \rangle^2 . \tag{12}$$

The convective heat-flux density from the turbulent gases to the wall would be determined on the bases of the generally accepted assumption that heat transfer in the thin gas stratum adjacent to a solid surface is realized only by conduction in accordance with Fourier's law:

$$q_\mathrm{w} = \rho_\mathrm{w} c_\mathrm{w} \nu_\mathrm{w} \left(\mathrm{d}T/\mathrm{d}y\right)_\mathrm{w} / \mathrm{Pr} .$$

If the one-sided derivative in this equation is replaced with its expression (12), the following formula for the convective heat transfer in the nonisothermal turbulent-gas boundary layer would be deduced:

$$q_\mathrm{w} = \frac{\rho_\mathrm{w} \nu_\mathrm{w} c_\mathrm{w}}{\mathrm{Pr_t}\, y_P} (T_P - T_\mathrm{w}) m_u \langle f_u \rangle^2 .$$

The correction factor m_u can be replaced here by its expression that follows from of equality (10). Then the last formula would be reshaped into a more appropriate computational form:

$$q_\mathrm{w} = \zeta \frac{\rho_\mathrm{w} c_\mathrm{w}}{\mathrm{Pr_t}} (T_P - T_\mathrm{w}) \langle f_{uP} \rangle^2 , \tag{13}$$

The obtained expressions can be used for setting boundary conditions of Neumann or Robin type to the convective heat-transfer equations.

3. Discussion

Adequacy of the mathematical model of the convective heat transfer was proved by comparing its numerical results with experimental data of Tamonis [19], who had measured the heat transfer from a turbulent hot-air flow to cooled walls of a tube which diameter was 150 mm. Dots on figure 2 demonstrate ratios of the experimental Nusselt ($\mathrm{Nu_{ex}}$) and Reynolds ($\mathrm{Re_{ex}}$) numbers, which had been calculated in [19] with using values of the physical quantities on the tube axis at distance 3.52 m from the experimental-section inlet.

In the mathematical model, a cooled tube of 2 m length was taken into account before the experimental section. The air temperature at its inlet was 480 K. Better accordance with the experiment data has been found at the turbulent Prandtl number $\mathrm{Pr_t} = 0.85$. When radial grid-mesh sizes equaled 2.5 mm, the wall-adjacent grid nodes were disposed in the dimensionless interval $y_+ = 25...84$ within limits of the buffer and equilibrium sublayers. Quadratic roots of the density ratio ρ_w/ρ and the functions f_u and f_l were defined in the wall-adjacent grid nodes, what has rather raised their values in compare with the averaged ones, especially when the grid nodes happened to fall into the inner part of the buffer sublayer.

The results of the numerical simulation are represented at figure 2 in the form of correlation *1* between the Nusselt and Reynolds numbers. They accord satisfactory with the experimental data on the whole. Nevertheless, as it could be expected, a part of the curve *1* that corresponds to grid nodes in the buffer sublayer disposes somewhat lower than the respective group of experimental dots.

In order to eliminate such a discrepancy, an empirical function has been found:

$$f_m = \left[1 - \exp\left(-y_+ / A_m\right)\right]^{-0.5},$$

where factor A_m has been determined as follows:

$$A_m = 18\left[1 - \left(y_+ / 100\right)^2\right].$$

The ratio of Nusselt and Reynolds numbers, computed after multiplying the right-side parts of the formula (13) by the function f_m, is presented by means of line *2* at figure 2.

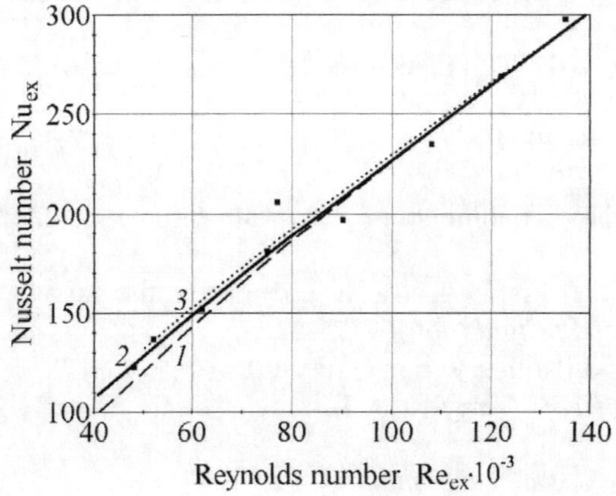

Figure 2. Numerical simulating the convective heat transfer: *1* – theoretical equations; *2* – their empirical correction; *3* – power function; dots are experimental data [19]

This curve does not differ much from the power function that is showed there by dotted line *3*:

$$Nu_{ex} = 0{,}023\,Re_{ex}^{0.8}.$$

Hence, the results of the numerical simulation on the coarse grid almost coincide with the function of power 0.8, which is specific to the empirical heat-exchange equations at the turbulent fluid flow in tubes [20]. Such a coincidence can serve as a proof for adequacy of the applied empirical correction.

4. Conclusion

In the whole, the adaptation of Prandtl's semiempirical hypothesis to nonisothermal turbulent gas flow allowed us to employ the coarse-mesh grid near the wall while simulating the convective heat transfer to the walls. The boundary conditions to the differential and discrete equations that are usually used for computing the gas velocity, eddy viscosity, and gas temperature have been correctly formulated at the walls confining the turbulent gas flow. A new semiempirical formula has been deduced for eddy viscosity of the turbulent gas flow in the immediate wall proximity. Adequacy of the mathematical model has been proved by means of comparison of the obtained simulation results with experimental data.

Acknowledgments
The article was prepared within development program of the Flagship Regional University on the basis of Belgorod State Technological University named after V.G. Shukhov.

References
[1] Pletcher R H, Tannehill J C and Anderson D 2012 *Computational Fluid Mechanics and Heat Transfer* (Hoboken, NJ: CRC Press)
[2] Aleksin V A 2015 Method of near-wall conditions for investigating flows and heat transfer at high turbulence intensity *Fluid Dynamics* **50** 412–29
[3] Townsend A A 1961 Equilibrium layers and wall turbulence *J. Fluid Mech.* **11** 97–120
[4] Kays W M and Crawford M E. 1993 *Convective Heat and Mass Transfer* (McGraw-Hill Sci.)
[5] Lacasse D, Turgeon E and Pelletier D 2004 On the judicious use of the k-ε model, wall functions and adaptivity *Int. J. Therm. Sci.* **43** 925–38
[6] Kalitzin G, Medic G, Iaccarino G and Durbin P 2005 Near wall behavior of RANS turbulence models and implications for wall functions *J Comput. Phys.* **204** 265–91
[7] Nazif H R, Tabrizi H B and Farhadpour F A 2014 Comparative analysis of the boundary transfer method with other near-wall treatments based on the k-ε turbulence model *Europ. J. Mech. - B/Fluids* **44** 22–31
[8] Otero G J, Patel A, Dies R and Pecnik R 2018 Turbulence modelling for flows with strong variations in thermophysical properties *Int. J Heat Fluid Flow* **73** 114–23
[9] Craft T, Gant S, Iacovides H and Launder B. 2004 A new wall function strategy for complex turbulent flows *Numer. Heat Transf.* **45** 301–18
[10] Utyuzhnikov S V 2012 Interface boundary conditions in near-wall turbulence modeling *Computers Fluids* **68** 186–91
[11] Bäckar J-A and Davidson L 2017 Evaluation of numerical wall functions on the axisymmetric impinging jet using OpenFOAM *Int. J. Heat Fluid Flow* **67** 27–42
[12] Schlichting H 1958 *Grenzschicht-Theorie* (Karlsruhe: Verlag G. Braun)
[13] Zukauskas A and Slanciauscas A 1987 *Heat Transfer in Turbulent Fluid Flows* (Berlin: Springer)
[14] Van Driest E R 1956 On turbulent flow near a wall *J. Aeronaut. Sci.* **23** 1007–11
[15] Kuznetsov V A 1986 Refinement of wall-turbulence hypotheses *J. Eng. Phys.* **50** 640–4
[16] Kuznetsov V A 2017 Computational simulation of convective heat transfer of turbulent gas flows *Theoret. Found. Chem. Engrg* **51** 1063–9
[17] Lander B E and Spalding D B 1974 The numerical computation of turbulent flows *Comput. Methods Appl. Mech. Engrg* **3** 269–89
[18] Patel V C, Rodi W and Scheuerer G Turbulence models for near-wall and low Reynolds 1985 number flows *AIAA Journal* **23** 1308–19
[19] Tamonis M 1987 *Radiation and Combined Heat Transfer in Channels* (New York: Hemisphere Publ. Corp.)
[20] Mikheyev M 1956 *Fundamentals of Heat Transfer* (Moscow: Peace Publishers)

IOP Conf. Series: Materials Science and Engineering **552** (2019) 012006 doi:10.1088/1757-899X/552/1/012006

Parametric model for predicting the reliability of heat power equipment of TPP

M M Sultanov[1], V M Truhanov[1] and Yu A Gorban[2]

[1]Department of Thermal Engineering and Heat Engineering, Branch Moscow Power Engineering Institute Volzhsky, Lenina Ave., 69, Volzhskij, 404110, Russia
[2]Department of Automation of Technological Processes and Production, Branch Moscow Power Engineering Institute Volzhsky, Lenina Ave., 69, Volzhskij, 404110, Russia

E-mail: vfmei@vfmei.ru

Abstract. Forecasting and ensuring the required high reliability level of heat power plants (thermal power plants) at a stage of their creation and operation is an important problem of economic value. For the solution of this problem in the present article the task is to develop such mathematical models of forecasting of reliability at all stages of life cycle of a thermal power plant from designing before removal from operation. It would be possible to reduce costs of realization of concrete actions. Parametrical mathematical models of growth of reliability have been developed for the solution of an objective. In particular two-parametrical, three-parametrical and multiple parameter models of growth of reliability have been developed and presented. For calculation of estimates of parameters of distribution standard programs are offered. All developed mathematical models of forecasting of reliability level are based on use of the statistical data obtained in the course of tests or operation of products analogs. Change of reliability level on five stages of life cycle of the power equipment of a thermal power plant is graphically shown in the article.

Keywords. Heat power equipment, reliability level, methods, parametrical mathematical models, laws of distribution.

1. Introduction

Mathematical models of growth of reliability allow to predict change of reliability of the power equipment of a thermal power plant in the course of its creation and operation and also to give an assessment to reliability at any moment of prototypes' tests and in use at the moments of carrying out completions [1-4]. It is known that in the course of carrying out tests or operation there are refusals which reason can be not always unambiguously established [5-7]. Owing to this fact the carried-out completions also ambiguously influence change of reliability, nevertheless the tendency of growth of reliability from one completion to another is observed. Thus, process of change of reliability due to completions has a casual reason. Graphically it is shown in figure 1.

From the schedule is visible that at a design stage thanks to introduction of new constructive decisions and other actions reliability increases to the required value.

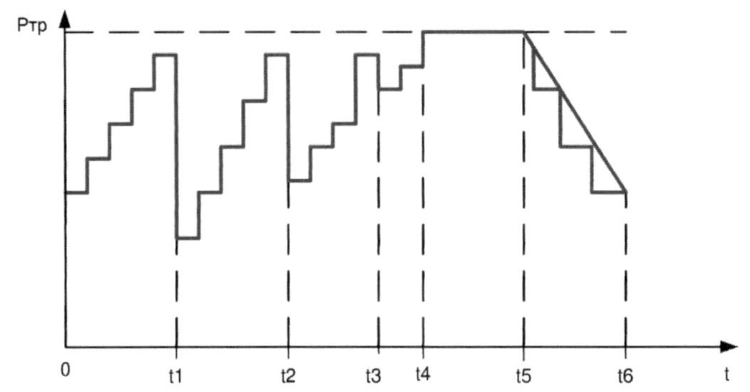

Figure 1. Change of reliability level of the power equipment of a thermal power plant in the course of creation and operation

2. Results

Upon transition to prototypes' tests there is a sharp falling of function of reliability at the moment t_1. It is caused by the fact that at a design stage it is almost impossible to consider all features of the created prototype and therefore the high design reliability coinciding with the required reliability, as a rule, isn't maintained.

At a stage of prototypes' tests ($t_1 - t_2$) at the expense of the carried-out completions there is an increase in a reliability level to the size close to the required value. On an initial stage of operation ($t_2 - t_3$) at the beginning there is some falling of reliability level due to constructive and technological refusals. On the site of operation ($t_4 - t_5$) within a warranty period at the expense of the held events reliability to the required level increases. Outside a warranty period of operation, the site ($t_5 - t_6$), there is a sharp falling of reliability level because of material aging and wear of the rubbing surfaces.

Thus, for quality control and reliability of the power equipment of a thermal power plant is necessary to find analytical model of change of reliability at all stages of its creation and operation. The function of reliability presented in fig. 1 includes sites on which reliability passes with jump from one stage to another, remains to a constant or grows at the expense of the carried-out completions. Therefore, for the description of all function of reliability is necessary to receive such models which would reflect three characteristic manifestations of reliability.

Creation of the model of reliability's change by results of tests is carried out at rather large number of tests and a large number of completions. In the case under consideration carrying out completions only after establishment of causes of refusals is provided in mathematical model and provided that these completions don't reduce reliability level. In the course of creation of model results of tests are presented in the form of separate series with a certain number of completions in each of them. For the description of an increment of reliability of a product from one series of tests to another the linear dependence can be as a first approximation used (1):

$$\Delta P_i = a_i(P_\text{r} - P_{i-1}), \tag{1}$$

where a_i – the coefficient characterizing efficiency of i-oh completion which is in an interval $(0{,}5 \le a_i \le 1{,}0)$;

P_r– the required value of probability of no-refusal operation;

P_{i-1} – probability of successful test after $(i$-1) - completions.

As in the course of tests completion is carried out in case of obvious clarification of a cause of refusal, reliability of a product changes only after completion performance.

Change of function of reliability in the course of working off has spasmodic character (figure 2).

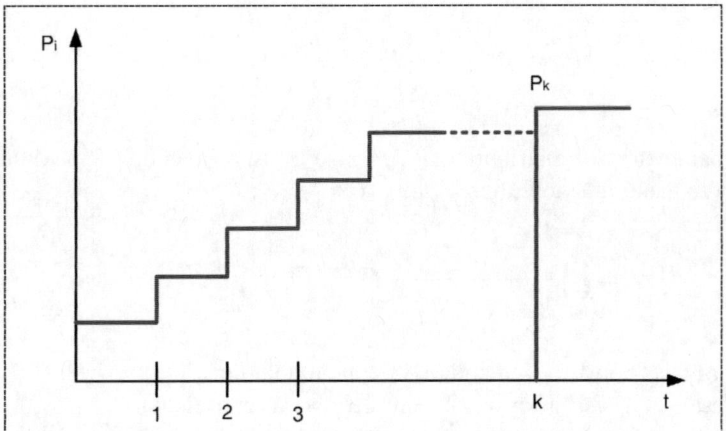

Figure 2. Change of Function of Reliability (probability of no-refusal operation) at working off of a product

Function P_i has jumps only in those points of i after which completions have been carried out, and these jumps depend on the reached reliability P_{i-1} and also the coefficient a_i characterizing success of elimination of a cause of refusal i.e. it is possible to write down (2):

$$P_i = P_{i-1} + a_i(P_r - P_{i-1}). \tag{2}$$

Then required function of reliability (probability of no-refusal operation) describing all process of working off has an appearance (3):

$$P_i = P_0 + \sum_{i=1}^{k} a_i(P_r - P_{i-1}), \tag{3}$$

where P_0 – initial reliability (probability of no-refusal operation) of a product received before carrying out the first completion;

k – number of the carried-out completions.

Thus, the probability of no-refusal operation depends on two parameters – P_0 and a_i, so-called two-parametrical model. If these parameters are determined by results of tests, then it is possible to calculate values of function of reliability in points of carrying out completions and to predict change of reliability level and costs of carrying out completions [8-12].

The similar formula can be written down for function of reliability expressed as average value of a time between refusals (4):

$$T_i = T_0 + \sum_{i=1}^{k} a_i(T_r - T_{i-1}) \tag{4}$$

where T_0 – the average value of a time between refusals received before carrying out the first completion;

ΔT_i – an increment of average value of a time between refusals from (i-1) - oh before i-oh completion, $\Delta T_i = T_i - T_{i-1}$.

On the basis of the assumptions accepted above for working off of products the three-parametrical model is received (5) [13]:

$$P_i = P_r - (P_r - P_0)\left(1 - \frac{a_i}{P_r}\right)^i \tag{5}$$

For finding of parameters of distribution P_0, P_r also a_i use a method of maximum likelihood for what it is necessary to make function of credibility (6):

$$L = \prod_{i=0}^{k} \frac{n_i!}{m_i!(n_i - m_i)!} P_i^{n_i - m_i}(1 - P_i)^{m_i} \tag{6}$$

where n_i – number of tests between i-oh and (i+1) - oh completions;
m_i – number of refusals in n_i tests after which completion was carried out.

Substituting in a formula (6) instead P_i of its value from a formula (5), we will receive the required function of credibility (7) depending on three parameters – P, P_r and a_i:

$$L = \prod_{i=0}^{k} C_{n_i}^{m_i} \left[P_r - (P_r - P_0)\left(1 - \frac{a_i}{P_r}\right)^i \right]^{n_i - m_i} \times \left[P_r - (P_r - P_0)\left(1 - \frac{a_i}{P_r}\right)^i \right]^{m_i} \tag{7}$$

The equation (7) includes the known skilled data m_i and n_i also unknown parameters P_o, P_r and a_i. Estimates of maximum likelihood are estimates P_o, P_r and a_i at which function of credibility (7) addresses in a maximum. For the solution of this task standard programs of a minimum of linear function are used.

For the description of change of function of reliability P_i the method based on representation P_i in the form of function of casual arguments which values can be determined by results of tests of this type of products or products analogs can be used. Such approach allows to establish an analytical form of dependence which is function of reliability [14-19].

In a general view the function (7) can be written down as follows (8):

$$P_i = P(\alpha_1, \alpha_2, ..., \alpha_s, t) \tag{8}$$

where α_i – unknown parameters.

Having chosen several various forms of function P_i, it is possible to estimate which of them in the best way describes an experimental curve.

We will consider hyperbolic model of growth of function of reliability and other mathematical models.

The hyperbolic model of change of reliability is based on the assumption that all program of tests is presented in the form of k-stages, on each of which make a certain number of tests, build a curve of growth of reliability which has an type of a hyperbole and is expressed by dependence (9):

$$P_i = P_r - \frac{\alpha}{i} \tag{9}$$

where α – the parameter characterizing reliability growth rate;
i = 1, 2, 3, … – stages of tests.

IOP Conf. Series: Materials Science and Engineering **552** (2019) 012006 doi:10.1088/1757-899X/552/1/012006

The described model depends on two parameters α and P_r.
Then for i-oh of a stage of working off we will have (10):

$$L_i = P_i^{n_i - m_i}(1 - P_i)^{m_i} \tag{10}$$

Believing that the results of all stages are statistically independent, we will receive (11):

$$L = \prod_{i=1}^{k} L_i = \prod_{i=1}^{k} P_i^{n_i - m_i}(1 - P_i)^{m_i} \tag{11}$$

or after logarithming:

$$\ln L = \sum_{i=1}^{k}(n_i - m_i)\ln\left(P_r - \frac{\alpha}{i}\right) + \sum_{i=1}^{k} m_i \ln\left(1 - P_r + \frac{\alpha}{i}\right)$$

The equations of credibility can be written down in the form of (12):

$$\frac{d\ln L}{dP_r} = \sum_{i=1}^{k} \frac{n_i - m_i}{P_r - \frac{\alpha}{i}} + \sum_{i=1}^{k} \frac{m_i}{1 - P_r + \frac{\alpha}{i}} = 0;$$

$$\frac{d\ln L}{d\alpha} = -\sum_{i=1}^{k} \frac{(n_i - m_i)/i}{P_r - \frac{\alpha}{i}} + \sum_{i=1}^{k} \frac{m_i / i}{1 - P_r + \frac{\alpha}{i}} = 0. \tag{12}$$

As a result of the solution of the equation (12) we will receive estimates of parameters (13), (14):

$$\hat{\alpha} = \frac{\sum_{i=1}^{k} n_i \left[\sum_{i=1}^{k} i(n_i - m_i) - \frac{k+1}{2}\sum_{i=1}^{k}(n_i - m_i)\right]}{\frac{k+1}{2}C - k}, \tag{13}$$

$$\hat{P}_r = \frac{\sum_{i=1}^{k} n_i \left[\frac{C}{k}\sum_{i=1}^{k} i(n_i - m_i) - \frac{k+1}{2}\sum_{i=1}^{k}(n_i - m_i)\right]}{\frac{k+1}{2}C - k}, \tag{14}$$

$$C = \sum_{i=1}^{k} \frac{1}{i} \approx \ln\left(k + \frac{1}{2}\right) + E$$

where E = 0,57721– constant.
As the approximating dependences it is possible to use models of a look (15):

$$P_i = 1 - (1 - P_0)\exp(-\alpha_i) \tag{15}$$

Or model (16):

$$P_i = \frac{P_0}{P_0 + (1 - P_0)\exp(-\alpha_i)}. \tag{16}$$

The most widespread is the model described by the indicative law (17):

$$P_i = 1 - A\exp[-c(i-1)], \tag{17}$$

where $A = \beta \cdot P$; $c = ln\frac{1}{1-\alpha \cdot P} > 0$.

This function depends on three parameters β, P and α which can be determined by method of maximum likelihood or method of the smallest squares by standard programs. More general model of growth of reliability [20-22] has a type of (18):

$$P_i = P_r - \alpha \cdot f(i) \tag{18}$$

where $f(i)$– the positive monotonously decreasing function.

For finding of estimates of parameters α and P_r method of the smallest squares calculate the sums:

$$C_1 = \sum_{i=1}^{k} f(i),\ C_2 = \sum_{i=1}^{k} f^2(i).$$

For some types of the $f(i)$ functions approximate formulas for calculations of C_1 and C_2 (table 1) are given.

Table 1. Approximate formulas for calculations of C_1 and и C_2

f(i)	C_1	C_2	Note
$i^{-\frac{1}{2}}$		$\ln\left(k + \frac{1}{2}\right) + 0{,}577$	
i^{-1}	$\ln\left(k + \frac{1}{2}\right) + 0{,}577$	$\frac{\pi^2}{6} - \left(k + \frac{1}{2}\right)^{-1}$	$\frac{\pi^2}{6} = 1{,}645$
i^{-2}	$\frac{\pi^2}{6} - \left(k + \frac{1}{2}\right)^{-1}$	$\frac{\pi^4}{90} - \frac{1}{3}\left(k + \frac{1}{2}\right)^{-3}$	$\frac{\pi^4}{90} = 1{,}082$
i^{-3}	$1{,}2 - \frac{1}{2\left(k + \frac{1}{2}\right)^{-2}}$	$\frac{\pi^6}{945} - \frac{1}{5}\left(k + \frac{1}{2}\right)^{-5}$	$\frac{\pi^6}{945} = 1{,}017$

Assessment of parameter of distribution is found, solving the equations (19) and (20):

$$Q = \sum_{i=1}^{k}\left[\frac{n_i - m_i}{n_i} - P_r + \alpha \cdot f(i)\right]^2;$$

$$\frac{dQ}{dP_r} = -2\sum_{i=1}^{k}\left[\frac{n_i - m_i}{n_i} - P_r + \alpha\cdot f(i)\right] = 0; \tag{19}$$

$$\frac{dQ}{d\alpha} = 2\sum_{i=1}^{k}\left[\frac{n_i - m_i}{n_i} - P_r + \alpha\cdot f(i)\right]\cdot f(i). \tag{20}$$

As a result of the solution of the equations (19) and (20) we will receive estimates of parameters (21) and (22):

$$\hat{P}_r = \frac{C_2\sum_{i=1}^{k}\left(\frac{n_i - m_i}{n_i}\right) - C_1\sum_{i=1}^{k}\left(\frac{n_i - m_i}{n_i}\right)\cdot f(i)}{k\cdot C_2 - C_1^2} \tag{21}$$

$$\hat{\alpha} = \frac{C_1\sum_{i=1}^{k}\left(\frac{n_i - m_i}{n_i}\right) - k\sum_{i=1}^{k}\left(\frac{n_i - m_i}{n_i}\right)\cdot f(i)}{k\cdot C_2 - C_1^2} \tag{22}$$

Change of reliability of a product at its stage-by-stage test can also be estimated statistically. Reliability assessment after i-oh series of tests is determined by dependence of a type (23):

$$\hat{P}_i = \alpha\cdot \hat{P}_i' + (1-\alpha)\hat{P}_{i-1} \tag{23}$$

where $\hat{P}_i'$ and $\hat{P}_{i-1}$ – reliability estimates by results of respectively only i-oh series of the tests and all tests which are carried out to i-oh series;
α – the coefficient characterizing the weight of each assessment.

Let's say that reliability of the tested samples in each series is identical. Features of this model are that that at α =1 in its all tests to *i*-oh series aren't considered, and at α=0 – results of the last series. The value of assessment $\hat{P}_i'$ is defined by results of tests on a formula (24):

$$\hat{P}_i' = 1 - \frac{m_i}{n_i} \tag{24}$$

It is possible to be set by value of parameter, proceeding from practical reasons that it reduces efficiency of use of this model.

The obtained results can be used in the development of new turbines.

The developed parametric models of reliability forecasting were used at the stage of creation of complex technical systems such as mobile units for special purposes and found practical confirmation in published scientific publications [23-25].

3. Conclusions

1. For the purpose of a solution of the problem of providing the power equipment of a thermal power plant required the reliability level in the present article methods and mathematical models of forecasting of reliability with use of estimates of parameters of distribution of function of reliability are developed.

2. The method of maximum likelihood and a method of the smallest squares are applied to finding of estimates of parameters of distribution.
3. Mathematical models of forecasting are developed and presented in the form of function of reliability with two, three and large number of parameters.

4. Acknowledgments

The research is executed within the state task No. 13.9602.2017/BCh on a subject 1041170.
The results of researches presented in article are received with financial support of the Russian Federal Property Fund within the scientific project No. 17-01-00018.

References

[1] Leojd D and Lipov M 1964 *Nadezhnost'. Organizatsiya issledovaniya, metody, matematicheskiy apparat* [Reliability. Organization of research, methods, mathematical apparatus] (Moscow: Sovetskoe radio) [In Russian]
[2] Truhanov V M 2013 2013 *Nadezhnost' mashinostroitel'noy produktsii. Teoriya i praktika* [Reliability of engineering products. Theory and practice] (Moscow: OOO «Izdatel'stvo Spektr») [In Russian]
[3] Chervonyj A A, Luk'yashchenko V I and Kotin L V 1972 *Nadezhnost' slozhnykh sistem* [Reliability of complex systems] (Moscow: Mashinostroyeniye) [In Russian]
[4] Barlou R and Proshan F 1969 *Matematicheskaya teoriya nadezhnosti* [Mathematical theory of reliability] (Moscow: Sovetskoe radio) [In Russian]
[5] CHuev YU V and Spekhova G P 1971 *Tekhnicheskie zadachi issledovaniya operacij* [Technical tasks of operations research] (Moscow: Sovetskoe radio) [In Russian]
[6] Truhanov V M 1993 *Slozhnye tekhnicheskie sistemy tipa podvizhnyh ustanovok. Razrabotka i organizaciya proizvodstva* [Complex technical systems such as mobile units. Development and organization of production] (Moscow: Mashinostroyeniye) [In Russian]
[7] RD 50-650-87. *Metodicheskie ukazaniya. Nadezhnost' v tekhnike. Sostav i obshchie pravila zadaniya trebovanij k nadezhnosti* [Methodical instructions. Reliability in technology. Composition and general rules for setting reliability requirements] (Moscow: Izdatel'stvo standartov) [In Russian]
[8] Rembez A I 1986 *Nadezhnost' i ehffektivnost' v tekhnike. Metodologiya. Organizaciya. Terminologiya* [Reliability and efficiency in technology. Methodology. Organization. Terminology] (Moscow: Engineering) [In Russian]
[9] Salomon D and Motta G 2010 *Handbook of data compression* (Springer Science & Business Media)
[10] Mashkov A V 2012 *Ekonomika kombinirovannogo proizvodstva teplovoj i ehlektricheskoj ehnergii: uchebnoe posobie* [Economics of Combined Heat and Power Generation: A Tutorial] (Volzhskij : Fil. MEHI v g. Volzhskom) [In Russian]
[11] Sovremennaya rynochnaya ehnergetika Rossijskoj Federacii [Modern market energy of the Russian Federation] (Moscow: ANO «Uchebnyj centr NP «Sovet Rynka») [In Russian]
[12] Sultanov M M and Truhanov V M 2017 Matematicheskaya model' izmeneniya vyhodnyh ehkspluatacionnyh harakteristik parovyh turbin TEHS [Mathematical model for changing the output performance of steam turbines TEHS] *Nadyozhnost' i bezopasnost' ehnergetiki* **1** pp 42-7 [In Russian]
[13] Gartung YU A 1984 *Issledovanie razvitiya dinamicheskih sistem, obuslovlennogo nekotorymi differencial'nymi programmami.* [Investigation of the development of dynamic systems, due to some differential programs] (Moscow: Science) [In Russian]
[14] Sultanov M M and Truhanov V M 2017 Matematicheskaya model' izmeneniya vyhodnyh ehkspluatacionnyh harakteristik parovyh turbin TEHS [Mathematical model of changes in the output performance of steam turbines of thermal power plants] *Reliability and safety of energy* pp 81-6 [In Russian]

[15] Truhanov V M 2017 *Nadezhnost' v tekhnike* [Reliability in technology] (Moscow: Izdatel'skij dom «Spektr») [In Russian]

[16] SHor YA B 1962 *Statisticheskie metody analiza i kontrolya kachestva i nadezhnosti* [Statistical methods of analysis and quality control and reliability] (Moscow: Sovetskoe radio) [In Russian]

[17] Hal'd A 1956 *Matematicheskaya statistika s tekhnicheskimi prilozheniyami* [Mathematical statistics with technical applications] (Moscow: Inostrannaya literatura) [In Russian]

[18] Smirnov N V and Dunin-Barkovskij I V 1969 *Kurs teorii veroyatnostej i matematicheskoj statistiki dlya tekhnicheskih prilozhenij* [The course of the theory of reliability and mathematical statistics for technical applications] (Moscow: Nauka) [In Russian]

[19] Mitropol'skij A K 1971 *Tekhnika statisticheskih vychislenij* [Statistical Computation Technique] (Moscow: Nauka) [In Russian]

[20] GOST 15.309-98. *Sistema razrabotki i postanovki produkcii na proizvodstvo. Ispytaniya i priemka vypuskaemoj produkcii* [System development and production of products. Testing and acceptance of products] (Moscow: Standartinform) [In Russian]

[21] GOST 27.202-83. *Nadezhnost' v tekhnike. Tekhnologicheskie sistemy. Metody ocenki nadezhnosti po parametram kachestva izgotovlyaemoj produkcii* [Reliability in technology. Technological systems. Methods for assessing the propensity to the quality parameters of the displayed products] (Moscow: Izdatel'stvo standartov) [In Russian]

[22] RD 50-109-89. *Rekomendacii. Nadezhnost' v tekhnike. Obespechenie nadezhnosti izdelij. Obshchie trebovaniya* [Recommendations. Reliability in technology. Ensuring the reliability of products. General requirements] (Moscow: Izdatel'stvo standartov) [In Russian]

[23] Sultanov M M, Truhanov V M, Arakelyan EH K and Kulikova M A 2018 Metody dostizheniya i obespecheniya vysokogo urovnya nadezhnosti i bezopasnosti ehnergeticheskogo oborudovaniya TEHS, GEHS, AEHS na vsekh ehtapah zhiznennogo cikla [Methods of achieving and ensuring a high level of reliability and safety of the energy equipment of TPP, HPP, and NPP at all stages of the life cycle] *Novoye v rossiyskoy elektroenergetike* pp 06-14 [In Russian]

[24] Truhanov V N and Sultanov M M 2017 Estimation technique of corrective effects for forecasting of reliability of the designed and operated objects of the generating systems *Journal of Physics: Conf. Series* pp 01-13

[25] Sultanov M M 2016 *Ocenka nadezhnosti, prodlenie resursa i optimizaciya remonta oborudovaniya TEHS i ehnergeticheskih sistem* [Reliability assessment, life extension and optimization of TEHS equipment and energy systems] (Volzhskij: Filial FGBOU VO «NIU «MEHI» v g. Volzhskom) [In Russian]

Research of dynamic characteristics of signals for calculation of technical and economic indicators and increase in power efficiency of the equipment of thermal power plant

M M Sultanov[1], I A Boldyrev[2] and Yu A Gorban[2]

[1]Department of Thermal Engineering and Heat Engineering, Branch Moscow Power Engineering Institute Volzhsky, Lenina Ave., 69, Volzhskij, 404110, Russia
[2]Department of Automation of Technological Processes and Production, Branch Moscow Power Engineering Institute Volzhsky, Lenina Ave., 69, Volzhskij, 404110, Russia

E-mail: vf_mei@mail.ru

Abstract. The article is devoted to questions of effective management of the power equipment of a thermal power plant in the conditions of quickly changing operating modes of power systems. The results of a research of dynamic characteristics of the signals used for calculation of technical and economic indicators of the generating sources are given. The analysis is carried out and an assessment of a possibility of increase in efficiency of updating of the current values of technical and economic indicators on the example of package boilers of block and not block thermal power plants is given. The practical recommendations of use of a new approach of measurement of dynamic characteristics of signals for calculation of efficiency of the boiler and change of parameters of operation of the power equipment are formulated.

Keywords: power efficiency, technical and economic indicators, the package boiler, averaging, a mismatch, dynamic characteristics, thermal power plant

1. Introduction

The calculation of various parameters, including technological ones, related to the operation of the equipment, is an important task in the industry and, in particular, in the heat and power industry, where it is necessary to determine the technical and economic indicators (TEI), to assess the quality of the start/stop mode of the equipment, analysis of the water-chemical regime, etc. Performance of certain calculation tasks allows to estimate not only quality of work of the equipment, but also object as a whole, and also to give an assessment to actions of operational personnel that finally affects service life and reliability of work of the equipment [1].

For example, a comprehensive analysis of the start/stop of heat power equipment should contribute to the improvement of the operation technology and improve the technical level of personnel [2]. High-quality water treatment and compliance with the water-chemical regime is a guarantee of reliability and accident-free heat power equipment and heating networks, prevents corrosion damage to the equipment and elements of the network water route, as well as helps to save fuel. TEI calculation provides information to ensure the most economical operation of the equipment, allows to predict its repair, as well as to assess the quality of the operating personnel, to assess the performance of commissioning and operational tests [3, 4].

On the basis of the corresponding array (operational, shift, daily, monthly) of accumulated values and an array of regulatory and reference information, an array of output indicators (operational, shift, daily and monthly intervals) is formed, which is the result of the calculation of TEI.

In the considered cases, the input data are the values of technological parameters of the equipment obtained from a variety of different sensors (temperature, pressure, etc.). The set of sensors is part of the automatic control system (ACS), and the values of the process parameters that they receive and convert, allow you to create a comprehensive picture of the operating mode of the object under study [5].

The purpose of the APCS is the purposeful management of the technological process, and it includes the following components: operational personnel, information, organizational, software and technical support (see figure 1) [6].

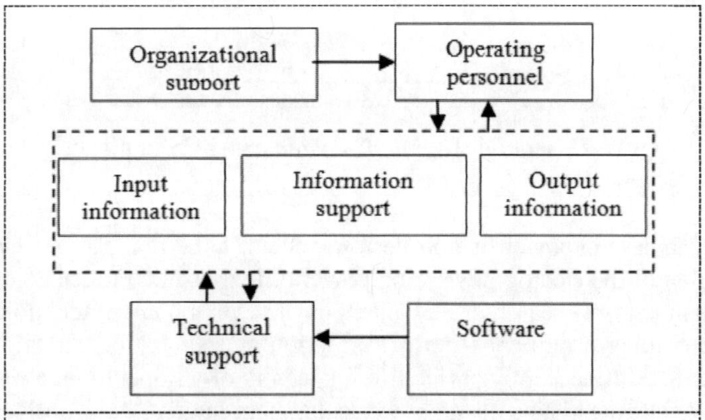

Figure 1. Simplified scheme of interaction of the main elements of ACS TP

There are also software and hardware systems (PTC), which are a comprehensive solution for the creation and interaction of systems of each level of ACS TP (see figure 2) therefore, as a rule, they embrace the entire system, organizing the process of removing and converting analog signals, their transmission, storage, processing and accessing them, control. Server equipment and automated workstations make up the upper level of PTC, which uses its own SCADA-system, but it should be noted that many PTC have the ability to integrate third-party [7].

For the calculation and evaluation of various parameters of the object or the enterprise as a whole at the upper level of the PTC used calculation stations with access to the archive station, recording the values of technological parameters during the operation of the object. As well as SCADA-system, the calculation station can be made in the form of a separate module, independent software (SOFTWARE), and work together with PTC; in this case, it needs to have access to the values of the process parameters of the APCS, which can be implemented in different ways depending on the permissions and functions of the PTC integration:

- reading the archive station directly if it is accessed or not encrypted;
- receiving data from PTC by sending it through the server;
- importing the data.

The need to use a third-party calculation station can be caused by the presence of the necessary functions, user-friendliness of the interface or a more advanced calculation algorithm.

Analysis of the averaging period of initial parameters for calculation of technical and economic indicators

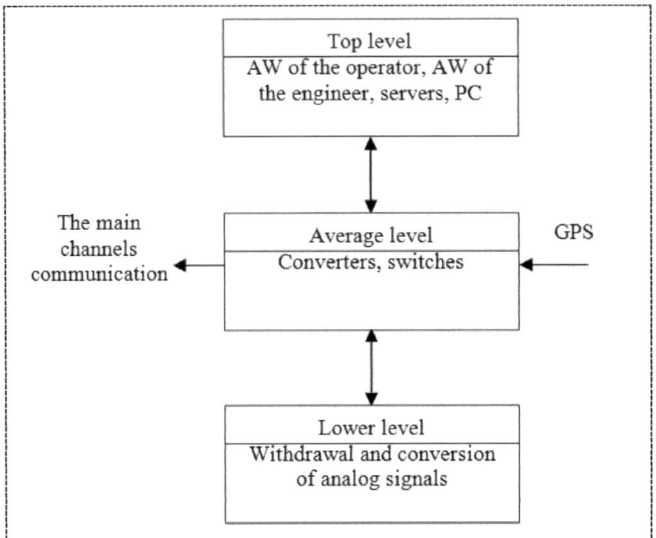

Figure 2. General structural scheme of APCS in power engineering

The questions of increase in power overall performance of a thermal power plant in the conditions of the wholesale market of the electric power and power (OREM) and a fierce competition of the generating sources demand not only search of new methods of development of technological and economic advantages, but providing a rational fuel consumption and an ecological safety of the operating thermal power plants [8]. Influence of service life, indicators of an operation, a repair and, in this regard, change of power characteristics of the power equipment have to be coordinated by questions of ensuring high of reliability level [9] and use of calculation procedures and optimization of an operation of the capital equipment of a thermal power plant. Sufficient attention is paid to researches of mathematical modeling of output characteristics of the equipment of the generating systems [10, 11], questions of calculations and optimization of parameters of management and their errors [12, 13, 14].

In the automated control system (ACS) for power units of a thermal power plant the calculation of the technical and economic indicators (TEI) takes an important place. The results of calculations of TEI allow to create information base for automation of the solution of problems of optimum control of technological process of the equipment of the power unit, including for operational impact on adjustable parameters. It is known that calculation of TEI with use of instant values of the parameters of technological process entering them is incorrect. Between the warmth introduced in a copper fire chamber with the burned fuel, and thermal power of primary and secondary steam there is a time shift because of thermal inertia of the boiler. Averaging of controlled parameters which enter settlement formulas of assessment of TEI is made for compensation of the specified shift. According to [15] calculation of TEI it is made on the values of quick and adjustable parameters average in 15 minutes.

In the analysis of a mismatch the value of the period of averaging two minutes has been determined by time of the current values of technological parameters of the power unit of Konakovo state district power plant for power units with direct-flow coppers optimum [13]. The research of values of parameters of work of a not block thermal power plant with drum coppers conducted on the example of the Volzhskiy TPP shows (see figure 3) existence of a mismatch on time bigger, than for direct-flow package boilers.

Apparently from figure 3, account characteristics of fuel and steam at change of loading of the boiler have a mismatch (see the period 504,5 – 507,8 minutes). The analysis of signals shows that time of a mismatch of input and output parameters of a copper is 2,3 minutes.

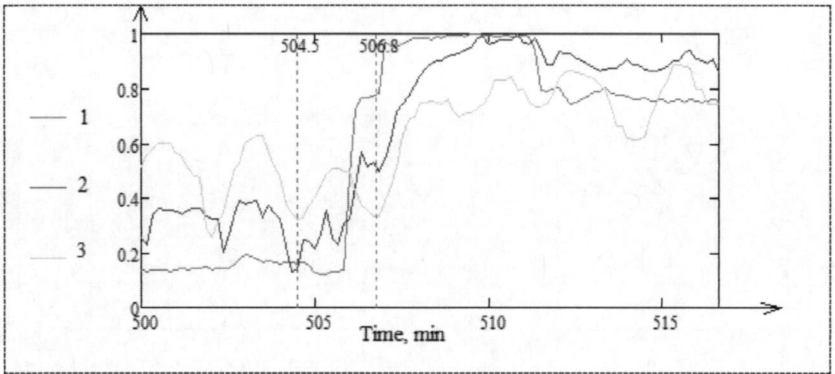

Figure 3. Account characteristics of the boiler, Art. No. 7 of the Volzhskiy TPP: 1 – gas consumption, 2 – the consumption of superheated steam, 3 – a feed water expense

2. Materials and Methods

2.1. Research of algorithms efficiency of averaging of values

The basic data for calculation of TEI are average values of technological parameters. A number of software and hardware complexes, including Kvint-SI, contains the built-in functions of averaging by an arithmetic average, however application of such methods as the sliding average, the digital exponential filter of the first order, etc. is possible [16-19].

The arithmetic average value is often used as an averaging algorithm. However, this way of averaging is subject to strong influence of big deviations. The received average value can not correspond to real "average". When calculating of the sliding average the value of function is calculated every time anew, at the same time the final significant set of the previous values is considered. The exponential filter – the simplest and widespread filter, but having a big delay [20].

The research of a mean square error of a deviation of average values shows that, in relation to values of technological parameters of a thermal power plant, the algorithm of the sliding average has the largest accuracy. Values of errors are presented in table 1.

Table 1. The error of the deviation average from reference values

	Arithmetic average algorithm	**Algorithm of the sliding average**	**Algorithm of the exponential filter**
mean square deviation, %	7,8	37,1	9,0

In figure 4 the results of use of various algorithms of averaging of technological parameters at change of an operating mode of the power unit are given.

The further analysis of the calculation results of TEI was carried out for the values average by an arithmetic average algorithm by criterion of a minimum of a mean square deviation.

IOP Conf. Series: Materials Science and Engineering **552** (2019) 012007 doi:10.1088/1757-899X/552/1/012007

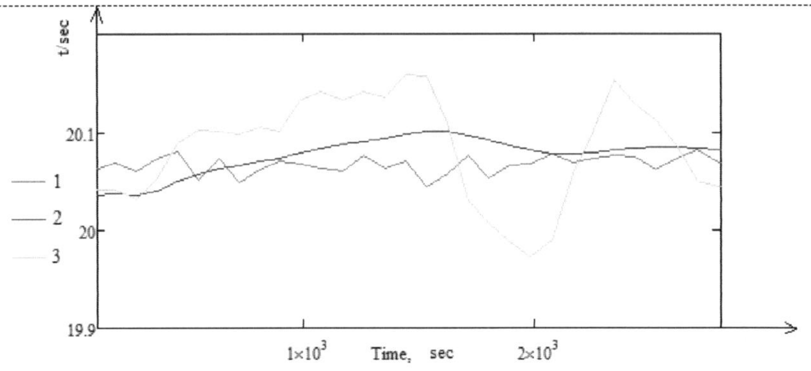

Figure 4. Average Signal of the Gas Consumption: 1 – The gas consumption average by an arithmetic average algorithm, 2 – the gas consumption average by an algorithm of the sliding average, 3 – the gas consumption by an average algorithm of the exponential digital filter

2.2. CalculationerrorofTEI atchange research

Assessment of influence of the period of averaging of initial signals on the result of calculation of TEI has been carried out for the power unit of Konakovo state district power plant [21]. As a key indicator efficiency gross of a copper is used (further ηбр). Calculation ηбр is made with use of functions of determination of specific volume of superheated steam and feedwater, enthalpy of superheated steam, steam of hot and cold intermediate overheating and feedwater, expenses of gas, superheated steam, feedwater, network water behind the reducing cooling installation (RCI). Also in calculation ηбр interpolation of the two-dimensional massif is used.

The calculation algorithm ηбр when determining TEI of power units of Konakovo state district power plant is presented in [22-25].

In figure 5 the comparison of the results of calculation ηбр is shown at averaging intervals of 2 minutes and 15 minutes.

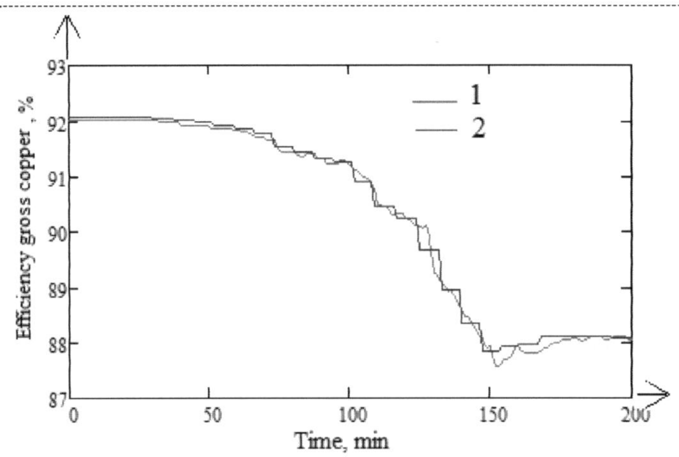

Figure 5. Efficiency gross copper: 1 – an averaging period 2 minutes, 2 - a period of averaging of 15 minutes

It is possible to estimate change of accuracy of calculation of TEI at variation of an interval of averaging a relative dynamic error δд. The mathematical model of calculation of a relative dynamic error is presented in [13]. The results of calculation of a relative dynamic error ηбр on various periods of averaging are presented in figure 6.

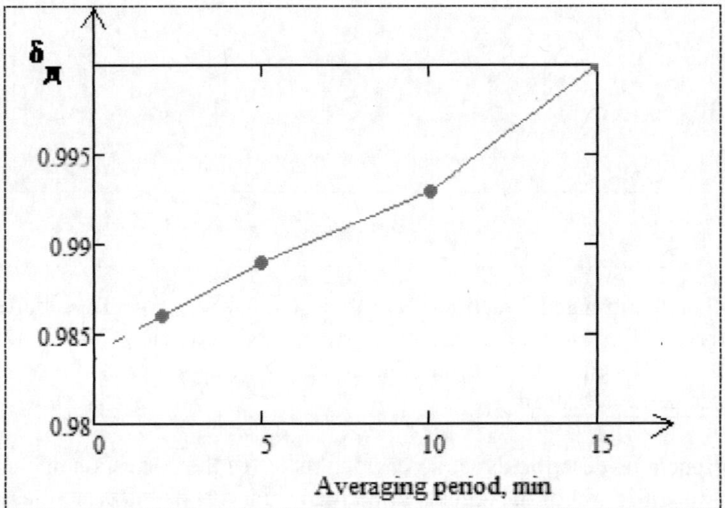

Figure 6. Dependence of a Relative Dynamic Error on the Averaging Period

2.3. Dynamic characteristics of initial signals for calculation of TEIresearch

Measuring information on the basis of which calculation of TEI is carried out represents the values of technological parameters transferred on measuring channels. Owing to influence of production factors and also hindrances, measuring information is stochastic therefore for the analysis and creation of metrological characteristics of system for measurement of TEI it is expedient to use statistical methods. The exponential model of correlation functions of signals and hindrances of a look has been applied to the description of properties of measuring information:

$$K_x(t) = D_x \cdot \exp(-\alpha t),$$

where D_x - dispersion of a signal; α – the speed of change of correlation function of a signal.

3. Results

A type of autocorrelated functions of signals of key parameters of the drum package boiler on the example of a drum copper of the station No. 7 of the Volzhskiy TPP are presented in figure 7. Characteristics of signals are specified in table 2.

Table 2. Characteristics of correlation functions of signals of key parameters of the package boiler

	Feedwater expense	**Gas consumption**	**Superheated steam consumption**
Dx, $(t/h)^2$	125,8	0,5	92,4
α, t/h·s	0,2	10,0	1,0

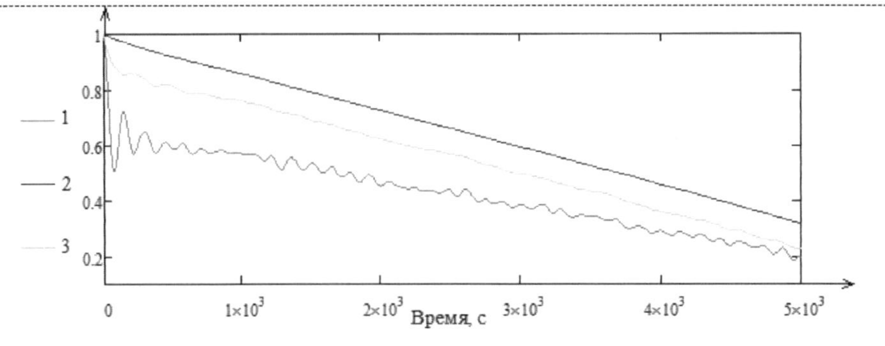

Figure 7. Autocorrelated functions of initial signals of the package boiler: 1 – Autocorrelated function of a signal of an expense of feedwater, 2 – Autocorrelated function of a signal of a consumption of gas, 3 – Autocorrelated function of a signal of superheated steam

The correlation functions describe dynamic characteristics of the corresponding signals and can be used for formation of realization of the signals imitating initial for the purpose of assessment of accuracy of calculation of TEI in various modes.

4. Discussion

Reduction of an interval of averaging will allow to increase the accuracy and efficiency of calculation of TEI for more optimum entering of the correcting influences into an operating mode of the equipment of power units. It will allow to improve, both ecological, and economic indicators of a power plant.

5. Conclusion

1. Providing a rational fuel consumption and technical and ecological safety of the operating thermal power plants is important. It can be reached, including increase in accuracy and efficiency of calculation of technical and economic indicators of a thermal power plant. The conducted researches show a possibility of a reduction period of calculation of TEI from current 15 minutes to 3 minutes.

2. .An assessment of accuracy of various algorithms of averaging in relation to values of initial parameters for calculation of TEI shows optimality of use of an algorithm of average algorithm.

3. The models of the description of properties of measuring information - values of the technological parameters used for calculation of TEI – in the form of correlation functions of an exponential look which can be used for imitation of signals in various rates are received.

References

[1] Pletnev G P 2007 *Avtomatizaciya tekhnologicheskih processov i proizvodstv v teploehnergetike* (Moscow: MPEI) [In Russian]

[2] Pletnev G P 1994 *Avtomatizirovannye sistemy upravleniya ob"ektami teplovyh ehlektrostancij* [Automated control systems of thermal power plants] (Moscow: MPEI) [In Russian]

[3] SHul'ginov N G, ZHukov A V, Demchuk A T and Koshcheev L A 2009 *Energetic Koncepciya protivoavrijnogo avtomaticheskogo upravleniya EEHS Rossii v sovremennyh usloviyah* [The concept of emergency control of UES of Russia in modern conditions] **10** pp 2-4 [In Russian]

[4] Pletnev G P 2001 Metody modelirovaniya raspredelennyh sistem upravleniya tekhnologicheskim processom ehnergoblokov TEHS [Methods for modeling distributed process control systems of power units of thermal power plants] *Teploehnergetika* **10** pp 12-9 [In Russian]

[5] Ivanov V A 1982 *Regulirovanie ehnergoblokov* [Regulation of power units] (Leningrad: Engineering) [In Russian]

[6] Denisenko V V 2008 *Komp'yuternoe upravlenie tekhnologicheskim processom, ehksperimentov, oborudovaniem* [Computer control of technological process, experiments, equipment] (Moscow: Goryachaya liniya – Telekom) [In Russian]

[7] Voropaj N I, Efimov D N, Karataev B N, Novikov E A, Osak A B and Panaseckij D A 2016 Adaptivnye algoritmy avtomatiki raspredelennogo otklyucheniya nagruzki [Adaptive algorithms for distributed load shedding automation] *Elektrostantsii* **11** pp 27-35 [In Russian]

[8] Sultanov M M, Konstantinov A A and Ivanitckii M S 2017 *Environmental Aspects of Thermal Power Equipment Operation Modes Optimization* pp 13300-6

[9] Sultanov M M, Truhanov V M, Arakelyan EH K and Kulikova M A 2018 Metody dostizheniya i obespecheniya vysokogo urovnya nadezhnosti i bezopasnosti ehnergeticheskogo oborudovaniya TEHS, GEHS, AEHS na vsekh ehtapah zhiznennogo cikla [Methods of achieving and ensuring a high level of reliability and safety of the energy equipment of TPP, HPP, and NPP at all stages of the life cycle] *Novoye v rossiyskoy elektroenergetike* pp 06-14 [In Russian]

[10] Truhanov V N and Sultanov M M 2017 Estimation technique of corrective effects for forecasting of reliability of the designed and operated objects of the generating systems *Journal of Physics: Conf. Series* pp 01-13

[11] Sultanov M M and Truhanov V M 2017 Matematicheskaya model' izmeneniya vyhodnyh ehkspluatacionnyh harakteristik parovyh turbin TEHS [Mathematical model of changes in the output performance of steam turbines of thermal power plants] *Nadezhnost' i bezopasnost' energii* pp 81-6 [In Russian]

[12] Arakelyan EH K, Koryagin A V and Solov'ev I A 1995 Optimizaciya raspredeleniya tochnosti izmerenij parametrov turboustanovok [Optimization of accuracy distribution] *Vestnik Moskovskogo energeticheskogo instituta* pp 01-16 [In Russian]

[13] Boldyrev I A and Struzhinskaya YU A 2016 Optimizaciya perioda rascheta koehfficienta poleznogo dejstviya brutto kotla ehnergobloka [Optimization of the period for calculating the efficiency of the gross boiler power unit] *Al'ternativnaâ ènergetika i èkologiâ* pp 76-9 [In Russian]

[14] Solov'ev I A, Zuev A V, Kirillov V A and Samsonova E A 1992 Obrabotka dannyh stacionarnyh teplofizicheskih ehksperimentov s uchetom pogreshnostej vsekh izmeryaemyh velichin [Processing of data from stationary thermophysical experiments taking into account the errors of all measured values] *Fizicheskiy Inzhenernyy Zhurnal* **2** pp 294-300 [In Russian]

[15] Ivanova G M, Kuznecov N D and CHistyakov V S 2005 *Teplotekhnicheskie izmereniya i pribory* [Physical Engineering Journal] (Moscow: MPEI) [In Russian]

[16] Semenov A D, Avdeeva O V and Nikitin A S 2012 Algoritm ehkstremal'nogo regulirovaniya na osnove rekurrentnoj procedury metoda naimen'shih kvadratov [Algorithm of extremal regulation based on the recurrent least squares procedure] *Izvestiya vysshikh uchebnykh zavedeniy. Povolzh'ye. Tekhnicheskiye nauki* **1(21)** pp 03-11 [In Russian]

[17] Smirnov A and Mihajlov S 2003 Valyutnyj spekulyant Vybor tipa skol'zyashchih srednih [The choice of the type of moving averages] pp 50-5 [In Russian]

[18] Denisenko V V 2011 Dinamicheskaya pogreshnost' izmeritel'nyh kanalov ASU TP [Dynamic error of measuring channels of industrial control system] *V zapisnuyu knizhku inzhenera* **2** pp 93-101 [In Russian]

[19] Demidovich B P and Maron I A 1960 *Osnovy vychislitel'noj matematiki* [Basics of Computational Mathematics] (Moscow: Fizmatgiz) [In Russian]

[20] RD 34.11.321-96 Normy pogreshnosti izmerenij tekhnologicheskih parametrov teplovyh ehlektrostancij i podstancij [Norms of measurement error of technological parameters of

thermal power plants and substations] [In Russian]
[21] Actez 2004 Konakovskaya teplovaya ehlektrostanciya [Konakovo Thermal Power Plant] (Moscow: Vneshtorgizdat) [In Russian]
[22] RD 34.09.453 Tipovoj algoritm rascheta TEHP moshchnyh otopitel'nyh TEHC [Typical algorithm for calculating TEC of powerful heating CHP] (Moscow: Soyuztechenergo)
[23] Komarov N F et al. 1990 *Tipovoj algoritm rascheta TEHP kondensacionnyh ehnergoblokov moshchnost'yu 300, 500, 800 i 1200 MVt* [Typical algorithm for calculating the TEC of condensing units with a capacity of 300, 500, 800 and 1200 MW] (Moscow: Soyuztechenergo) [In Russian]
[24] Zak I D, Hant N G and SHul'man A E 1989 Trudy CNIIKA Metody vychisleniya tekhniko-ehkonomicheskih pokazatelej [Methods for calculating technical and economic indicators]
[25] Aleksandrov A A and Grigor'ev B 1999 *Tablicy teplofizicheskih svojstv vody i vodyanogo para* [Tables of thermophysical properties of water and water steam] (Moscow: MPEI) [In Russian]

Efficiency of application of various layout arrangements of oil-gas burners in thermal power plant boilers

M A Taymarov[1], R V Akhmetova[2], E A Akhmetov[3]

[1]Department of Power Engineering, Kazan State Power Engineering University, Krasnoselskaya St., 51, Kazan, 420066, Russia
[2]Department of Electric Stations, Kazan State Power Engineering University, Krasnoselskaya St., 51, Kazan, 420066, Russia
[3]Department of Industrial Heat and Power Engineering and Heat Supply Systems, Kazan State Power Engineering University, Krasnoselskaya St., 51, Kazan, 420066, Russia

E-mail: ahmetova_rv@bk.ru

Abstract. The necessary intensity of burning and completeness of fuel burning out in furnace volume are reached by the correct supply organization and the subsequent mixture of fuel with secondary air that is provided with burner devices. Experimental research of fuel oil and gas combustion efficiency in boiler furnaces of thermal power plant has shown that, the peripheral axial twist executed in the form of two channels with axial scapular air rotator is preferable. The combined gas supply is perspective: tubular peripheral and pilot central. For fuel oil combustion at large expenses the most effective burning is reached due to application of steam-mechanical nozzles which main lack is the considerable length of a torch. The two-tiered arrangement of burners in boiler at combustion of gas and fuel oil is more effective, in comparison with single-tier, both at the counter direction, and at unilateral frontal. For increasing the efficiency of combustion of fuel at some burners types in a boiler furnace the ascending vertical rotation of a torch can be created by the asymmetrical counter arrangement of burners of the first and second tiers. Use of the smoke exhauster of gases recirculation for reduction of harmful emissions of nitrogen oxides with boilers flue gases gives bigger effect in comparison with optimization of the choice of a layout arrangement of burners and parameter of a twist.

Keywords. Boiler efficiency, fuel oil, burner, flame temperature, air swing, burning, natural gas, torch, tier, radiation.

1. Introduction

The efficiency of combustion of fuel oil in power coppers depends on a number of factors which configuration and design features of torches is among. The general is fall of temperature of products of combustion of fuel oil at the exit from a fire chamber and decrease in efficiency of coppers upon transition from combustion of gas to combustion of fuel oil. However by application of optimal technical solutions when designing new torches, modernization of the existing torches and the choice of parameters of process of combustion of fuel oil can increase the efficiency of coppers at combustion of fuel oil. In the real work the efficiency of combustion of fuel oil in coppers with various configuration of torches is experimentally investigated. Experimental research of fuel oil and gas combustion efficiency in boiler furnaces of thermal power plant has shown that, the peripheral axial twist executed in

IOP Conf. Series: Materials Science and Engineering **552** (2019) 012008 doi:10.1088/1757-899X/552/1/012008

the form of two channels with axial scapular air rotator is preferable. The combined gas supply is perspective: tubular peripheral and pilot central.

2. Materials and Methods

Objects of a research are steam power boilers of the thermal power plants using for gas and fuel oil combustion oil-gas burners with a mechanical and steam-mechanical fuel oil atomization. Experiments were made on drumbeat boilers of TGM-84A, TGM-84B, TGM-96B, TG-104, PK-47, TPE-429, TGME-464 and direct-flow PK-41boilers. PK-47 and PK-41 double-hulled boilers , and PK-41 is a supercritical pressure boiler. TGME-464 boiler with natural circulation is drumbeat gas-tight under pressurization with balanced thrust and also has the smoke exhauster of gases recirculation . TG-104 boiler has the smoke exhauster of flue gases recirculation which is switched-off at big steam loadings.

Schemes of burners arrangement on the researched boilers are given on fig. 1. TGM-84A and TGM-84B boilers have the two-light screen in a boiler furnace, and TGM-96 boiler at the same sizes of a boiler furnace and bigger thermal power doesn't contain the screen. On TGM-84A TGM-84B, TGM-96, TG-104 boilers burners are located unilaterally on frontal walls, and on PK-47 and PK-41boilers – opposite on side walls. PK-47 boiler has an asymmetrical arrangement of burners which provides at the top view the vortex ascending in height current of combustion products with right-hand rotation.

On a TPE-429 boiler burners are located opposite in one tier on frontal and back walls of a boiler furnace. On a TGME-464 boiler burners are located in two tiers on a back wall of a boiler.

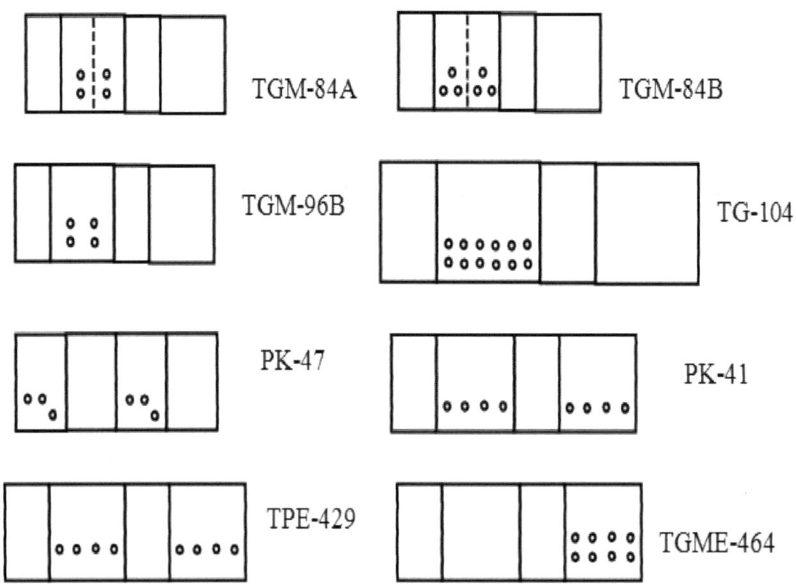

Figure 1. Schemes of burners arrangement on the researched boilers

Rating technical characteristics of the researched boilers are given in table 1.

Width of a boiler furnace is measured on a frontal wall, height – from the middle of a cold funnel to the middle of the horizontal gas flue. On the Surgut state district power plant reserve fuel oil isn't used. The technical characteristics of burners brands, given in tab. 1, can differ considerably in comparison with factory owing to their modernization in use. Such modernization is executed on TGM-96B boiler by application of nozzle gas-distribution, instead of conic nozzle with openings, either on peripheral, or on the central gas supply.

Table 1. Rating technical characteristics of the researched boilers.

Brand of a boiler and installation place	Width, depth and height of a boiler furnace, m	Productivity, t/h	Pressure and steam temperature	Brand of a burner and power, MW	Brand of a nozzle
TGM-84A, Nizhnekamsk TPP-1	16×7,5×23	420	14 MPa, 560°C	HF-CKB-VTI-TKZ, 77	FUZ-7500
TGM-84B, Nizhnekamsk TPP-1, Naberezhnye Chelny TPP	16×7,5×23	420	14 MPa, 560°C	TKZ, 53	FUZ-5000
TGM-96B, Nizhnekamsk TPP-1	16×7,5×23	480	14 MPa, 560°C	HF CKB-VTI-TKZ, 89	FUZ-9000
TG-104,Surgut SDPP-1	18,6×7,2×29	670	14 MPa, 560°C	GMU-45m, 45	-
PK-47, Zainsk SDPP	9,5×9,5×28	640 (on 2 housings)	14 MPa, 545 °C	ZIO, 50	FM-5000
PK-41, Karmanovo SDPP	17,3×8,7×32	950 (on 2 housings)	25 MPa, 545 °C	ZIO, 50	FM-5000
TPE-429, Kazan TPP-3	12×8×30	400	14 MPa, 560°C	GMU-45-12, 45	FUZ - 5000
TGME-464, Hizhnekamsk TPP-2	12×8×30	500	14 MPa, 560°C	GMU-45-12, 45	FUZ - 5000

Technical characteristics of burners applied on the researched boilers are given in table 2.

Table 2. Technical characteristics of burners applied on the researched boilers.

Brand of a boiler	Gas supply	Gas-distribution	Air twist	Parameter of a twist, n
TGM-84A	Peripheral	Tubular	Peripheral tangential	1,1
TGM-84B	Central	Conic nozzle with openings	Peripheral tangential	1,05
TGM-96B	Peripheral and central	Nozzle	Peripheral tangential	0,94
TG-104	Peripheral	Tubular	Peripheral axial	1,8
	Central	Conic nozzle with openings		
PK-47	Peripheral	Tubular	Peripheral tangential	0,92
PK-41	Peripheral	Tubular	Peripheral tangential	0,92
TPE-429	Central	Conic nozzle with openings	Peripheral tangential	1,5
TGME-464	Central	Conic nozzle with openings	Peripheral tangential	1,5

On TG-104 boiler (see table 2) burners with the combined gas supply are applied: the main peripheral tubular and additional pilot central with conic nozzle with openings.

IOP Conf. Series: Materials Science and Engineering **552** (2019) 012008 doi:10.1088/1757-899X/552/1/012008

Apparently from table 2, the largest parameter of a twist is reached with application of peripheral axial scapular air stream vortex generator that provides the best mixing him with fuel.

During the experiments mode parameters of boilers operation and heat fluxes falling from a torch on boiler furnace screens were measured. During the measurements of operational mode parameters tenured instruments were used. Measurements of nitrogen oxides content and oxygen content in the leaving combustion products in mode cross section were carried out using DAG-500 gas analyzer. Temperature of the leaving gases in balance cross section was measured by a method of two thermocouples [1-5] along with the tenured station temperature sensor. Measurement of the heat fluxes falling from a torch on boiler furnace screens is executed by a noncontact method using the radiometer TERA-50 with RK-15 graduation via hatches within the walls of boiler furnaces. The error of experiments during the measurement of heat fluxes from a torch for determination of temperature reached ±2,76 %. The error of determination of boilers efficiency on the reverse balance calculated by a technique [10-13,15] reached ±3,51 %.

3. The received results and their discussion

Comparative experimental data in parameters of boilers operation at various steam loadings at gas combustion in burners of various designs and layout in a boiler furnace are given in table 3.

Table 3. Experimental data in parameters of boilers operation at various steam loadings at gas combustion in burners of various designs.

Boiler	Loading D, t/h	Air excess coefficient, α	Leaving gases temperature t_{yx}, °C	The NO_x content in flue gases calculated as α=1,4, mg/m^3	Efficiency gross η, %
TGM-84A	210	1,07	109	92	94,96
	300	1,06	116	114	95,04
	350	1,06	120	124	94,97
	420	1,05	125	132	94,83
TGM-84B	210	1,09	105	105	94,65
	260	1,07	113	125	94,77
	320	1,06	123	150	94,75
	420	1,05	136	190	94,39
TGM-96B	240	1,07	114	128	94,23
	300	1,06	118	160	94,34
	420	1,05	125	220	94,34
	480	1,05	130	235	94,17
TG-104 (with SEGR)	335	1,1	112	30	94,54
	440	1,05	116	40	95,11
	510	1,045	118	50	95,22
	670	1,04	124	160	95,12
PK-47, housing A	180	1,12	120	210	92,8
	255	1,06	125	250	93,6
	320	1,05	130	288	93,4
PK-41, housing A	205	1,09	124	190	91,7
	265	1,07	128	200	92,2
	365	1,06	130	300	91,8

	480	1,05	135	380	91,2
TPE-429	250	1,1	120	240	92,5
	300	1,09	125	250	92,3
	400	1,07	130	255	92,1
TGME-464 (under pressurization with SEGR)	300	1,04	113	75	94,54
	400	1,03	129	101	94,02
	500	1,03	141	126	93,59

Note: SEGR – smoke exhauster of flue gases recirculation

Apparently from comparison of data in tables 2 and 3 at peripheral tubular gas-distribution and a peripheral axial twist in the applied burners of TG-104 boiler it is reached the high efficiency values and low values of harmful NO_x nitrogen oxides emissions concentrations in all range of steam loadings. Air twist parameter is high at the same time and is also equal to 1,8. Except the applied burners small NO_x nitrogen oxides emissions concentrations at low steam loadings in a TG-104 boiler were reached due to recirculation of flue gases in a boiler furnace. Technically recirculation was carried out by selection from the lowering gas flue up to 10 % of all volume of flue gases in the field of the economizer by means of the recirculation smoke exhauster and supplying them through the mixer in an air duct in front of burners [6-9]. Increase of NO_x nitrogen oxides concentration at loadings of 670 t/h is explained by the termination of recirculation in connection with technical need of turning on of the intermediate steam reheater in work. At the same time temperature in a boiler furnace increases that leads to an intensification of thermal NO_x nitrogen oxides formation.

The increased efficiency values of TGM-84A boiler are explained with a good mixing of air with natural gas in burners at peripheral tubular gas-distribution and a peripheral tangential twist and reached parameter of twist of 1,1. However at such combination of gas-distribution and a twist and existence of four burners it is created raised throw of a torch on the back screen in a boiler furnace and the chamotte brickwork envelope thermally collapses. The increased temperature of an external steel covering of a boiler furnace was demonstrated to it during the experiments. The same negative result of heating the steel boiler furnace covering took place during the experiments on TGM-96B boiler.

The counter arrangement of burners in PK-47 and PK-41 boiler furnaces doesn't increase efficiency of natural gas combustion. Values of parameter of a twist are low. PK-47 boiler, in comparison with PK-41 boiler, provides the best mixing of air and fuel at the expense of the ascending vertical twist of a torch that allows to increase efficiency. However in both cases in PK-47 and PK-41 boiler furnaces due to counter collision of torches there are sites with the increased temperature which is the reason of increase of harmful NO_x nitrogen oxides generation [14,16-20].

Experimental data in parameters of boilers' work at various steam loadings at M100 fuel oil combustion are given in tab. 4.

Experiments on TG-104 and TPE-429 boilers at combustion of fuel oil weren't conducted as they weren't exploited in this operating mode.

From tab. 4 it is visible that pressurization in combination with recirculation of flue gases in a TGME-464 boiler furnace allows to receive at M100 fuel oil combustion quite high efficiency which exceeds value of this efficiency during the work on gas. It is connected with the fact that temperature of the leaving gases during the smoke exhauster operation of flue gases recirculation in a boiler furnace decreases. At the same time during the work of TGME-464 boiler on fuel oil it is equal to 106…118 °C, and at gas combustion to 115…141 °C.

The two-tiered unilateral arrangement of burners in TGM-84B and TGME-464 boilers at their increased number gives the best mixing of the dispersed fuel oil drops with air and, as a result increase in efficiency (see tab. 4) in comparison with an arrangement of burners and their number in amount of four on TGM-84A, TGM-96B boilers and a single-tier counter arrangement on PK-41 boiler or with the reduced number of burners on the first tier in PK-47 boiler (see figure 1).

IOP Conf. Series: Materials Science and Engineering **552** (2019) 012008 doi:10.1088/1757-899X/552/1/012008

Table 4. Experimental data in parameters of boilers' work at various steam loadings at M100 fuel oil combustion

Boiler	Loading D, t/h	Air excess coefficient, α	Leaving gases temperature t_{yx}, °C	The NO_x content in flue gases calculated as α=1,4, mg/m^3	Efficiency gross η, %
TGM-84A	220	1,1	150	339	90,85
	300	1,08	157	351	91,29
	360	1,07	163	360	91,25
	420	1,06	167	365	91,17
TGM-84B	300	1,14	144	520	93,42
	360	1,13	150	545	93,34
	400	1,12	152	555	93,32
	420	1,11	155	565	93,25
TGM-96B	340	1,11	148	333	92,23
	400	1,08	153	331	92,1
PK-47, housing A	186	1,15	154	349	92,6
	255	1,09	155	360	92,8
	313	1,08	159	370	92,7
PK-41, housing A	265	1,12	150	320	92,9
	365	1,09	155	350	92,8
	480	1,08	160	450	92,6
TGME-464 (under pressurization with SEGR)	300	1,06	106	240	94,95
	400	1,05	113	300	94,71
	500	1,05	118	370	94,35

The important characteristic of efficiency of fuel combustion in a boiler furnace is the uniformity and value of density of the thermal radiation falling on screens from a torch. As process of burning happens in diffusive area, density of thermal radiation is connected with intensity of fuel and air mixture.

Experimental values of streams of a torch $q_ф$ thermal radiation at height h of boiler furnaces of the researched boilers at gas combustion at the largest steam loadings specified in tab. 3 are given in figure 2.

Apparently from figure 2, TG-104 boiler has the largest thermal streams from torch $q_ф$ at the level of burners equal to 480 kW/sq.m, that is connected with its steam loading of 670 t/h and applied peripheral axial twisting of air stream with the parameter of a twist of 1,8 (see table 2). TGM-84A boiler has the smallest streams of radiation from torch $q_ф$ at the level of burners equal to 220 kW/sq.m that is connected with the applied peripheral tubular gas supply and a peripheral tangential air twist with the parameter of a twist of 1,1 (see table 2).

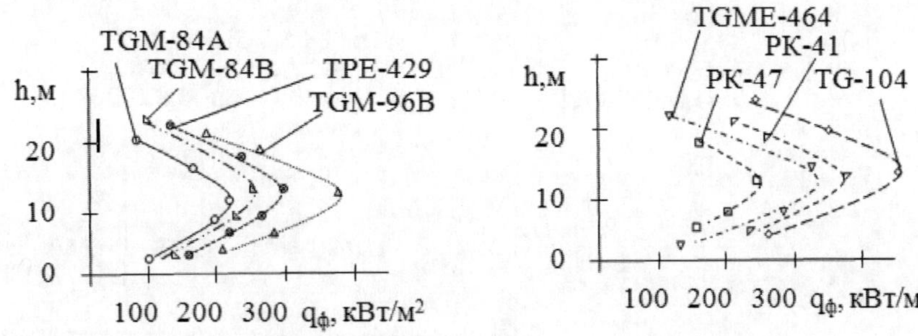

Figure 2. Streams of a torch $q_ф$ thermal radiation at height h of boiler furnaces of the researched boilers at gas combustion at the largest steam loadings specified in table 3.

Experimental values of temperature of torch $t_ф$ at height h of boiler furnaces of the researched boilers at gas combustion at the largest steam loadings specified in table 3 are given in figure 3.

From figure 3 it is visible that at the level of burners the highest values of a torch temperature at gas combustion are received in TGM-96B and TG-104 boiler furnaces respectively for steam loadings of 480 t/h and 670 t/h.

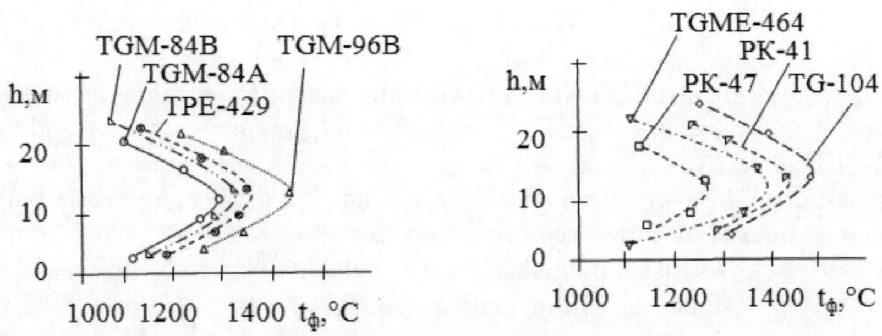

Figure 3. Torch $t_ф$ temperature at height h of boiler furnaces of the researched boilers at gas combustion at the largest steam loadings specified in table 3.

High values of temperature of a torch for TGM-96B boiler are connected with the nozzle gas-distribution applied in burners (see table 2) at which the stream of gas leaves perpendicular to the twisted air stream. At the same time length of a torch becomes shorter in comparison with gas-distribution with conic nozzle with openings and concentration of combustion products at small value of parameter of a twist equal to 0,94 leads to growth of temperature. Values of nitrogen oxides content in combustion products increase and are equal to 235 mg/m^3 for TGM-96B boiler (see table 3).

In TG-104 boiler at an axial peripheral twist of air with high value of parameter of a twist equal to 1,8 time of stay of burning products in a zone of high temperatures is reduced and generation of nitrogen oxides decreases that is characterized by their contents in the leaving combustion products by value of 160 mg/m^3 (see table 3).

Efficiency values and temperature t_{yx} of the leaving gases depending on the parameter of a peripheral tangential twist and steam loadings of boilers are given in figure 4-a for gas combustion during gas-distribution with conic nozzle with openings and in figure 4-b for M100 fuel oil combustion with application of steam-mechanical nozzles.

Apparently from figure 4, efficiency η considerably depends on temperature t_{yx} of the leaving gases which takes place at this value of parameter of a twist n and arrangement of burners in boiler furnaces.

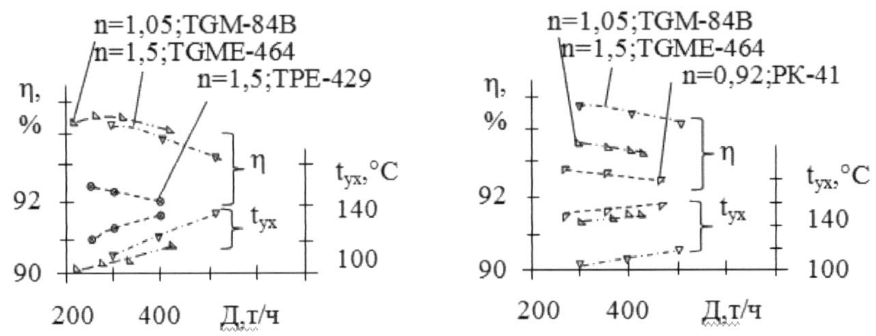

a) combustion of gas b) combustion of fuel oil

Figure 4. Efficiency η values and temperature t_{yx} of the leaving gases depending on the parameter of a peripheral tangential twist n and steam loadings of boilers D.

At gas combustion at identical parameters of the twist n=1,5 a two-tiered arrangement of burners in TGME-464 boiler (see figure 1) and smaller values of α air excess coefficient at the expense of a gas-dense furnace efficiency are higher in comparison with TPE-429 boiler.

Reduction of burners number to two on the second tier in a TGM-84B boiler furnace lowers temperature at the exit from furnace in comparison with a TGME-464 boiler furnace having four burners on the second tier. At TGM-84B boiler furnace exit it leads to decrease temperature t_{yx} of the leaving gases and to increase in efficiency (see figure 4-a).

At M100 fuel oil combustion in TGME-464 boiler the high efficiency η values are connected with the lowered temperatures on the exit from a boiler furnace at four burners on the second tier in comparison with TGM-84B boiler.

The counter single-tier arrangement of burners on frontal and back walls of a PK-41 boiler furnace at fuel oil combustion as a result of torches collision increases temperature at the exit from boiler furnace, leads to growth of temperature t_{yx} of the leaving gases and is followed by decrease in efficiency η in comparison with TGM-84B and TGME-464 boilers (see figure 4-b).

Values of NO_x nitrogen oxides (mg/m^3) concentration in flue gases and α air excess coefficient depending on the parameter of a peripheral tangential twist and steam loadings of boilers are given in figure 5-a for gas combustion during gas-distribution with conic nozzle with openings and in figure 5-b for M100 fuel oil combustion with application of steam-mechanical nozzles.

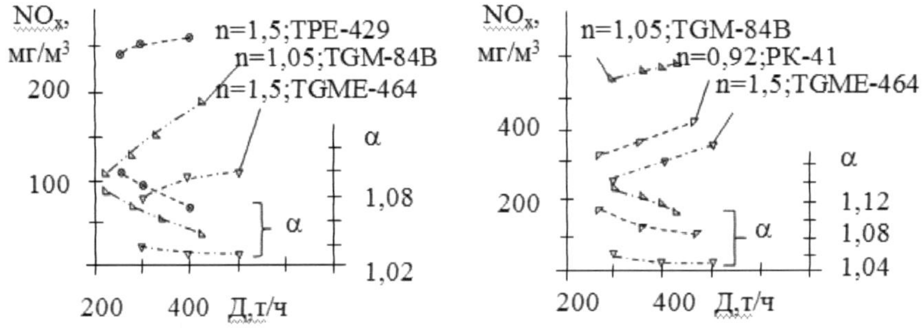

a) gas combustion b) fuel oil combustion

Figure 5. Nitrogen oxides NO_x (mg/m^3) concentration in flue gases and α air excess coefficient depending on the parameter of a peripheral tangential twist n and steam loadings boilers D.

Apparently from figure 5-a, during gas combustion the smallest values of harmful nitrogen oxides NO_x concentration are observed for TGME-464 boiler that is explained by recirculation by means of SEGR of flue gases in boiler furnace from the lowering gas flue and smaller, in comparison with TGM-84B boiler, values of α air excess coefficient. For TPE-429 boiler increase in parameter of a twist up to n=1,5, in comparison with n=1,05 for TGM-84B boiler, doesn't lead to reduction of nitrogen oxides NO_x concentration . The frontal two-tiered arrangement of burners in a TGM-84B boiler furnace is constructively more effective in comparison with a counter single-tier arrangement of burners in TPE-429 boiler furnace for elimination of zones in a torch with the increased temperatures at which there is an intensive thermal nitrogen oxides generation [21-30].

As it is already noted above (see table 3), during gas combustion TG-104 boiler has the smallest emissions of nitrogen oxides (NO_x=50 mg/m^3 at D=510t/h) with recirculation of flue gases and a frontal two-tiered arrangement of burners with an axial peripheral twist and parameter of a twist n=1,8. The gas-dense TGME-464 boiler under pressurization and with recirculation has nitrogen oxides concentration in flue gases higher and equal to 100 mg/m^3 at the same loadings. The shutdown of the smoke exhauster of gases recirculation in TG-104 boiler at loading of 670 t/h connected with steam reheaters operation has led up to growth of temperature of a torch and intensification of nitrogen oxides emission to 160 mg/m^3.

During M100 fuel oil combustion (see figure 5-a) gas-dense TGME-464 boiler under pressurization and with gases recirculation in a boiler furnace gives the smaller nitrogen oxides content in flue gases in comparison with PK-41 and TGM-84B boilers. Experiments on M-100 fuel oil combustion in TPE-429 and TG-104 boilers weren't conducted in this work. As well as during gas combustion , for fuel oil combustion essential decrease in nitrogen oxides content in flue gases with reduction of α air excess coefficient during the work of various boilers with various layout arrangement of burners in a boiler furnace is characteristic. With growth of steam loading D there is a decrease in α air excess coefficient in flue gases [31]. However at the same time there is a growth of temperature of a torch that causes increase in nitrogen oxides concentration in flue gases. Lower values of air excess coefficient for PK-41 boiler and counter layout arrangement of burners, in comparison with TGM-84B boiler, have reduced nitrogen oxides concentration in flue gases at fuel oil combustion.

4. Conclusions

1. The efficiency of fuel combustion at a frontal two-tiered arrangement of oil-gas burners in a boiler furnace very strongly depends on the applied way of a peripheral air twist and the most optimum is using of axial peripheral turning scapular air rotator.
2. For gas combustion the most effective is the combined peripheral tubular and central with conic nozzle gas supply that allows at rotary heads of the gas-distributing pipes to eliminate torch throw on the back screen at a frontal arrangement of burners. 3. Using of the smoke exhauster of gases recirculation for reduction of harmful nitrogen oxides emissions with flue gases of boilers gives bigger effect in comparison with optimization of the choice of a layout arrangement of burners and parameter of a twist.

References

[1] Taymarov M A, Sungatullin R G, Stepanova T O and Almuhametov R S 2016 Optimization of management of physical and chemical process of gas fuel burning in a radiant boiler furnace *Bulletin of Kazan technological university* **19(13)** pp 142-4

[2] Taymarov M A, Akhmetova R V, Chiklyayev D Ye, Chiklyayev Ye G and Sungatullin R G 2016 Education and ways of decrease in nitrogen oxides in TG-104 boilers with direct-flow and vortex burners and peripheral supply of gas *News of HEI. Power problems* vol 9-10 pp 83-90

[3] Taymarov M A, Akhmetova R V, Safin R G and Lavirko Y V 2016 Calculation of Fuel Oil Drop Burnup Time Dependence on Intensity of Flame Radiation *Research Journal of Applied Sciences Year* vol 11 pp 1660-5

[4] Taymarov M A, Akhmetova R V, Chiklyayev D Ye, Chiklyayev Ye G and Sungatullin R G

2016 Burning of methane-hydrogen fraction and fuel oil in TGM-84A boilers *KSPEU bulletin* vol 4 pp 83-95

[5] Taymarov M A and Lavirko Y V 2018 Reduction in emissions of nitrogen oxides energy boilers *European Journal of Technical and Natural Sciences* vol 1 pp 27-30

[6] Taymarov M A, Lavirko Y V and Chiklyayev D G 2017 *Burner for combustion of gas and fuel oil Patent for useful model № 170609*

[7] Taymarov M A, Lavirko Y V and Akhmetova R V 2017 *The nozzle Patent for useful model № 174497*

[8] Taymarov M A, Akhmetova R V and Sungatullin R G 2017 *Boiler installation Patent for useful model № 169930*

[9] Taymarov M A and Lavirko Y V 2015 *Boiler installation Patent for useful model № 154647*

[10] Akhmetova R V, Taymarov M A, Chiklyayev Ye G and Sungatullin R G 2017 Indicators of mode parameters of TGM-84B boilers when burning methane-hydrogen fraction in them *KSPEU bulletin* vol 3-4 pp 58-63

[11] Taymarov M A, Akhmetova R V, Sungatullin R G, Saltanayeva Ye A and Khusainov D G 2017 Research of mode parameters of boilers work at combustion of fuel oil with the increased content of water *KSPEU bulletin* vol 2 (34) pp 68-75

[12] Taymarov M A, Akhmetova R V, Margulis S M and Saltanayeva Ye A 2017 *Thermal radiation in boiler furnaces* (Kazan: KSPEU) p 124

[13] Taymarov M A, Akhmetova R V, Lavirko Y V, Sungatullin R G and Zheltuhina Ye S 2017 Decrease in harmful emissions in the atmosphere of nitrogen oxides by boilers of thermal power plant *KGACU news* **1(39)** pp 180-8

[14] Roslyakov P V and Yegorova L Ye 1996 Influence of the main characteristics of a zone of active burning on nitrogen oxides output *Power system* vol 9 pp 22-6

[15] Taymarov M A 2013 Development of methods of decrease in emissions of oxides of nitrogen by boilers of thermal power plant (Kazan: KSPEU) p 69

[16] Shimizu K, Hibi A, Koshi M, Morii Y and Tsuboi N 2011 *Updated Kinetic Mechanism for High-Pressure Hydrogen Combustion* Journal of Propulsion and Power **vol 27** pp 383–95

[17] John C C 2004 Transient radiative transfer in 2D cylindrical medium with collimated pulse irradiation *Journal of Quantitative spectroscopy and Radiative* pp 299-313

[18] Mishra S C and Rath P 2003 Transient radiative heat transfer in participating media of Numerical heat transfer *Int Journal of Heat and Mass* pp 746-52

[19] Abdallah P B 2000 Thermal Emission of Semitransparent slab with variable spatial refractive index *Journal of Quantitative spectroscopy and Radiative* pp 185-98

[20] Krishna N A and Subhash C 2006 Discrete transfer method applied to radiative transfer in a variable refractive Index semi-transparent medium *Journal of Quantitative Spectroscopy and Radiative* pp 432-40

[21] Burke M P, Chaos M, Ju Y, Dryer F L and Klippenstein S J 2012 Comprehensive H_2/O_2 kinetic model for high-pressure combustion *International Journal of Chemical Kinetics* **44(7)** pp 444-74

[22] Lu T J and Jin J M 2017 Electrical-thermal co-simulation for analysis of high-power rf/microwave components *IEEE Trans. Electromagn. Compat.* pp 93–102

[23] Skoplaki E and Palyvos J A 2009 On the temperature dependence of photovoltaic module electrical performance *Solar Energy* vol 63 pp 614-24

[24] Sánchez A L and Williams F A 2014 Recent advances in understanding of flammability characteristics of hydrogen *Progress in Energy and Combustion Science* vol 41 pp 1-55

[25] Rui Lou and Shu Bin Wu 2011 Products properties from fast pyrolysis of enzymatic/mild acidolysis lignin *Applied Energy* vol **88** pp 316-22

[26] Shen Y and Yoshikawa K 2013 Recent progresses in catalytic tar elimination during biomass gasification or pyrolysis *Renewable and Sustainable Energy Reviews* vol 21 pp 371-92

[27] Irannejad A and Jaberi F 2014 Large eddy simulation of turbulent spray breakup and

IOP Conf. Series: Materials Science and Engineering **552** (2019) 012008 doi:10.1088/1757-899X/552/1/012008

evaporation *International Journal of Multiphase Flow* vol 61 pp108-28

[28] Taymarov M A, Sadykov R A and Chaykovsky V G 2016 Measurement of temperature of the falling streams in TG-104 boiler furnace of Surgut SDPP-1 *Messenger of mechanical engineering* vol 2 pp. 36-8

[29] Ahrenfeldt J, Thomsen T P, Henriksen U and Clausen L R 2013 Biomass gasification cogeneration- A review of state of the art technology and near future perspectives *Applied Thermal Engineering* vol 50 pp 1407-17

[30] Gelderen L, Malmquist L M V and Linus M V 2017 Vaporization order and burning efficiency of crude oils during in-situ burning on water *Fuel* vol 191 pp 528-37

IOP Conf. Series: Materials Science and Engineering **552** (2019) 012009 doi:10.1088/1757-899X/552/1/012009

Approximate analytical solution to the stationary two-dimensional heat conduction problem on infinite bar with the source of heat

V K Tkachev

Department of Industrial Heat and Power, Samara State Technical University, Molodogvardeyskaya St., 244, Samara, 443100, Russia

E-mail: Totig@yandex.ru

Abstract. An approximate analytical solution to the boundary- value heat conduction problem for an infinite bar with a heat source was obtained with the use of the integral method of heat balance, by introducing a complementary required function and complementary boundary conditions. The boundary - value problem for a partial differential equation is reduced to an ordinary differential equation with respect to this function due to the complementary required function that characterizes the change in temperature along the axis of symmetry in the cross-section of the bar. The complementary boundary conditions determined by the initial differential equation and the given boundary conditions are found so that their satisfaction is equivalent to the solution of the initial equation of the boundary value problem at the boundary points. The fulfilment of the equation at the boundary points as well as the heat balance integral results in the fulfilment of the initial equation inside the domain. The approximate analytical solution obtained can be used to identify the amount of internal heat generated by various production processes (vibration and deformation loads, electromagnetic fields effects, etc.) in thermal and nuclear power plants, in the rocket and space industry and other industrial facilities.

1. Introduction

Among all approximate analytical solutions for boundary value problems of heat conduction, the integral heat balance method related to the group of orthogonal weighted residual methods, has been widely used [1–4]. It helps to obtain approximate analytical solutions for boundary value problems, the accurate solution to which can not be found (nonlinearity of the problem, variable physical properties of the medium, etc.) [5–9].

However, the main disadvantage of these methods is low accuracy as these methods require solving the averaged initial equation (heat balance integral). The application of the complementary desired function and complementary boundary conditions provides satisfying the initial differential equation with a specified degree of accuracy which depends on the number of approximations of the initial solution. Note that the methods based on the solving the equation at the boundary points have been also considered in papers [10–13].

2. Theoretical justification and application of the method

Consider the use of the integral heat balance method with complementary boundary conditions and a complementary required function for solving a stationary 2D heat conduction problem in an infinite bar of square cross section with a constant heat source (figure 1) [14–17].

IOP Conf. Series: Materials Science and Engineering **552** (2019) 012009 doi:10.1088/1757-899X/552/1/012009

Assume that the temperature on the lateral faces of the bar is known and is equal to T_w. It is necessary to determine the temperature distribution in the cross section of the bar. Due to the symmetry of the boundary value problem, only a quarter of the cross section can be considered fixing the boundary conditions of the first kind along EF and AF curves, and the conditions of the absence of heat transfer along OE and AO curves. The mathematical setting of the problem in this case is as follows (figure 1)

$$\frac{\partial^2 T(x,y)}{\partial x^2}+\frac{\partial^2 T(x,y)}{\partial y^2}=-\frac{v}{\lambda} \quad (0<x<\delta; 0<y<\delta); \tag{1}$$

$$T(\delta, y)=T_w;\ \frac{\partial T(0,y)}{\partial x}=0;\ \frac{\partial T(x,0)}{\partial y}=0;\ T(x,\delta)=T_w \tag{2}$$

where T - temperature, K; x, y - coordinates, m; v — power of the internal heat source, W / m^3; λ — thermal conductivity coefficient, W / (mK); T_w - wall temperature, K; δ - half width of the bar face, m. Finding a simple type of analytical solutions for 2D boundary-value problems based on the Poisson equation (1) is of considerable practical interest because of its wide use when analyzing many physical processes (heat conduction, fluid flow, theory of elasticity, thermoelasticity, torsion of prismatic objects, etc.).

The difficulty in constructing a solution is caused by the two-dimensionality of the problem and the non-homogeneity of the initial differential equation. Exact analytical solutions to such problems with complex infinite functional series were obtained only in individual particular cases and, moreover, with substantial assumptions (only under boundary conditions of the first kind and with a heat source to be constant in space). Note that regardless of the simplified mathematical problem setting (1), (2) the method described below provides obtaining solutions with variable heat sources, without regard to the type of boundary conditions and the cross section shape (rectangle, square) [18–21].

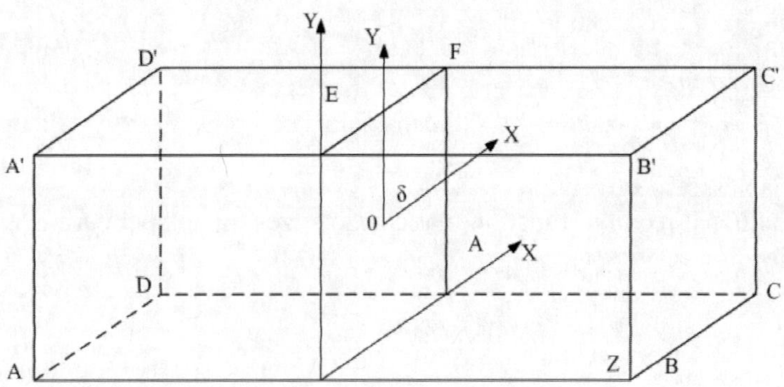

Figure 1. The computational domain for the cross-section of the infinite rectangular bar.

Introduce the following dimensionless variables and parameters:

$$\Theta=\frac{T-T_w}{T_w};\ \xi=\frac{x}{\delta};\ \eta=\frac{y}{\delta};\ B=\frac{v\delta^2}{\lambda T_w},$$

where Θ - dimensionless temperature; ξ, η — dimensionless coordinates; B –dimensionless parameter.

Taking into account notations introduced, the heat exchange diagram in the problem (1), (2) will be (figure 2)

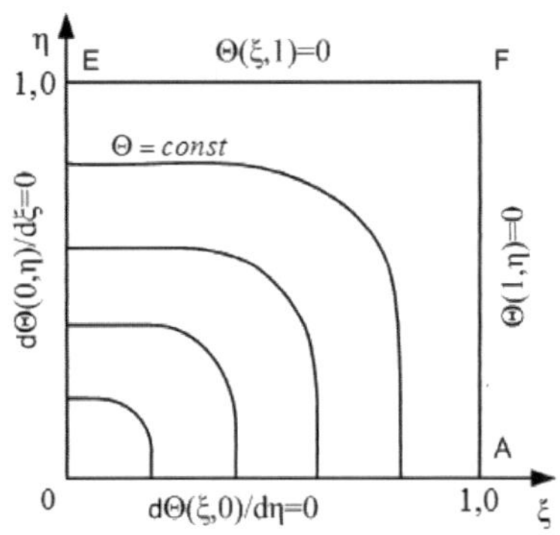

Figure 2. Heat exchange diagram for the cross section of an infinite rectangular bar.

The approximate analytical solution of the stationary 2D heat conduction problem for an infinite bar with a heat source is described by the following relations

$$\frac{\partial^2\Theta(\xi,\eta)}{\partial\xi^2}+\frac{\partial^2\Theta(\xi,\eta)}{\partial\eta^2}+B=0 \quad (0<\xi<1;\ 0<\eta<1)\,; \tag{3}$$

$$\Theta(1,\eta)=0\,; \tag{4}$$

$$\frac{\partial\Theta(0,\eta)}{\partial\xi}=0\,; \tag{5}$$

$$\frac{\partial\Theta(\xi,0)}{\partial\eta}=0\,; \tag{6}$$

$$\Theta(\xi,1)=0\,; \tag{7}$$

where $\Theta=\frac{T-T_\alpha}{T_\alpha}$; $\xi=\frac{x}{\delta}$; $\eta=\frac{y}{\delta}$; $B=\frac{v\delta^2}{\lambda T_\alpha}$ - dimensionless complex; Θ, ξ, η - dimensionless temperature and spatial variables, respectively.

Introduce the additional required function that characterizes the temperature distribution along the OA curve (Fig. 2) by

$$q(\xi)=\Theta(\xi,0)\,. \tag{8}$$

An approximate solution to the problem (3) - (7) may be found in the form

$$\Theta(\xi,\eta)=\sum_{k=1}^{n}b_k(q)\varphi_k(\eta)\,, \tag{9}$$

where $b_k(q)$ – required coefficients; $\varphi_k(\eta)=\cos\left(\frac{r\pi\eta}{2}\right)$ – coordinate functions $(r=2k-1,\ k=\overline{1,n})$.

Relation (7) due to the adopted system of coordinate functions satisfies the boundary conditions (6), (7). To obtain a solution to the problem (3) - (7) in the first approximation we substitute (9) (taking into account only one term of the series with k = 1) to the relation (8)

$$b_1(q)\cos\left(\frac{\pi\eta}{2}\right)\Bigg|_{\eta=0}=q(\xi)\,. \tag{10}$$

Relation (10) with respect to the required coefficient $b_1(q)$ offers an algebraic linear equation, solving which one can find $b_1(q)=q(\xi)$. By taking into account the value found of the required coefficient $b_1(q)$, the relation (9) is written as

$$\Theta(\xi,\eta)=q(\xi)\cos\left(\frac{\pi\eta}{2}\right). \tag{11}$$

We will demand that relation (11) to be fulfilled the averaged equation (3) with respect to the coordinate η (heat balance integral).

$$\int_0^1\left[\frac{\partial^2\Theta(\xi,\eta)}{\partial\xi^2}+\frac{\partial^2\Theta(\xi,\eta)}{\partial\eta^2}+B\right]d\eta=0. \tag{12}$$

Substituting (11) into (12), after determining the integrals with respect to the required function $q(\xi)$, we obtain the following ordinary differential equation

$$\frac{d^2q(\xi)}{d\xi^2}-\frac{\pi^2q(\xi)}{4}+\frac{\pi B}{2}=0. \tag{13}$$

The general integral of the equation (13) is as follows

$$q(\xi)=C_1\exp\left(\frac{\pi\xi}{2}\right)+C_2\exp\left(\frac{-\pi\xi}{2}\right)+\frac{2B}{\pi}, \tag{14}$$

where C_1, C_2 - constants of integration. Substituting (14) into (11) we have

$$\Theta(\xi,\eta)=\left(C_1e^{\frac{\pi\xi}{2}}+C_2e^{\frac{-\pi\xi}{2}}+\frac{2B}{\pi}\right)\cos\left(\frac{\pi\eta}{2}\right). \tag{15}$$

The constants of integration C_1 and C_2 are found from boundary conditions (4), (5). Substituting (15) into (4), (5) with respect to C_1 and C_2 we will have a system of two algebraic linear equations, and solving them we obtain

$$C_1=C_2=\frac{-2B\exp\left(\frac{\pi}{2}\right)}{\pi+\pi\exp\pi}.$$

Taking into account values found C_1 and C_2 the relation (15) is as follows

$$\Theta(\xi,\eta)=\frac{2B}{\pi}\left[\frac{\exp\frac{\pi}{2}}{1+\exp\pi}\exp\frac{\pi\xi}{2}+\exp\frac{-\pi\xi}{2}+1\right]\cos\frac{\pi\eta}{2}. \tag{16}$$

The solution of the problem (3) - (7) in the first approximation is given by relation (16). It exactly satisfies the boundary conditions (4) - (7), the heat balance integral (12) and approximately satisfies the equation (3) (in the first approximation). The standard deviation of the solution (16) compared to the exact one [22] according to Gauss form is 0.7%.

To increase the accuracy of the solution, it is necessary to enlarge the number of terms of series (9). To determine the required coefficients it is necessary to apply complementary boundary conditions (CBC). These conditions are found in such a form that their fulfillment by the desired solution is equivalent to the fulfillment of equation (3) at the borders of OE and OA of the bar section (Fig. 2). To obtain the first condition, let us differentiate the boundary condition (6) twice with respect to the variable ξ

$$\frac{\partial}{\partial\eta}\left(\frac{\partial^2\Theta(\xi,0)}{\partial\xi^2}\right)=0. \tag{17}$$

The relation (17) taking into account equation (3) is reduced to the following CBC along OA curve

$$\frac{\partial^3\Theta(\xi,0)}{\partial\eta^3}=0. \tag{18}$$

To obtain the second CBC (for the border set OA), let us differentiate (18) twice with respect to ξ

$$\frac{\partial^3}{\partial\eta^3}\left(\frac{\partial^2\Theta(\xi,0)}{\partial\xi^2}\right)=0. \tag{19}$$

Relation (19) taking into account equation (3) is reduced to CBC in the form

$$\frac{\partial^5\Theta(\xi,0)}{\partial\eta^5}=0 \tag{20}$$

Similarly, we can get any arbitrary number of CBCs on the axis of symmetry (along the OA curve). The general formula for them will be

$$\frac{\partial^{2i-1}\Theta(\xi,0)}{\partial\eta^{2i-1}}=0\ (i=2,3,4\ldots) \tag{21}$$

Note that due to the adopted system of coordinate functions $\varphi_k(\eta)$ $k=\overline{1,n}$, all CBCs determined by formula (19) by relation (9) are fulfilled.

To determine the CBC along the axis of symmetry η (along OE curve), we differentiate the boundary condition (5) twice with respect to the variable η

$$\frac{\partial}{\partial\xi}\left(\frac{\partial^2\Theta(0,\eta)}{\partial\eta^2}\right)=0. \tag{22}$$

Relation (22) taking into account equation (3) is reduced to the following CBC

$$\frac{\partial^3\Theta(0,\eta)}{\partial\xi^3}=0. \tag{23}$$

If we differentiate (23) twice with respect to variable η, we can find

$$\frac{\partial^3}{\partial \xi^3}\left(\frac{\partial^2 \Theta(0,\eta)}{\partial \eta^2}\right)=0. \tag{24}$$

Relation (24) taking into account equation (3) is reduced to the following CBC

$$\frac{\partial^5 \Theta(0,\eta)}{\partial \xi^5}=0. \tag{25}$$

The general formula of CBC along OE curve will be

$$\frac{\partial^{2i-1} \Theta(\xi,0)}{\partial \xi^{2i-1}}=0 \ (i=2,3,4...) \tag{26}$$

To determine the CBC along AF curve, we differentiate the boundary condition (3) twice with respect to the variable η

$$\frac{\partial^2 \Theta(1,\eta)}{\partial \eta^2}=0. \tag{27}$$

Equation (3) taking into account (27) is reduced to the following CBC

$$\frac{\partial^2 \Theta(1,\eta)}{\partial \xi^2}+B=0. \tag{28}$$

If we differentiate (28) twice with respect to the variable η, we find

$$\frac{\partial^2}{\partial \xi^2}\left(\frac{\partial^2 \Theta(1,\eta)}{\partial \eta^2}\right)=0. \tag{29}$$

Relation (29) taking into account equation (3) is reduced to the CBC in the form

$$\frac{\partial^4 \Theta(1,\eta)}{\partial \xi^4}=0. \tag{30}$$

The general formula for all subsequent CBCs along the AF curve will be

$$\frac{\partial^{2i} \Theta(1,\eta)}{\partial \xi^{2i}}=0 \ (i=2,3,4...) \tag{31}$$

Let us differentiate relation (28) twice with respect to the variable η

$$\frac{\partial^2 q(\xi)}{\partial \xi^2}=\frac{\partial^2 \Theta(\xi,0)}{\partial \xi^2}. \tag{32}$$

Relation (32) taking into account equation (3) is reduced to the following CBC

$$\frac{\partial^2 q(\xi)}{\partial \xi^2}+\frac{\partial^2 \Theta(\xi,0)}{\partial \eta^2}+B=0. \tag{33}$$

IOP Conf. Series: Materials Science and Engineering **552** (2019) 012009 doi:10.1088/1757-899X/552/1/012009

Let us differentiate (33) twice with respect to the variable ξ

$$\frac{\partial^4 q(\xi)}{\partial \xi^4} + \frac{\partial^2}{\partial \eta^2}\left(\frac{\partial^2 \Theta(\xi,0)}{\partial \xi^2}\right) = 0. \tag{34}$$

Relation (34) taking into account equation (3) is reduced to the CBC in the form

$$\frac{\partial^4 q(\xi)}{\partial \xi^4} + \frac{\partial^4 \Theta(\xi,0)}{\partial \eta^4} = 0. \tag{35}$$

The general formula for all subsequent CBCs, starting with CBC (35), will be

$$\frac{\partial^{2i} q(\xi)}{\partial \xi^{2i}} \pm \frac{\partial^{2i} \Theta(\xi,0)}{\partial \eta^{2i}} = 0 \quad (i = 2,3,4,\ldots) \tag{36}$$

where «+» sign is for even i, and «–» sign is for odd i.

To obtain a solution to the problem (3) - (7) in the second approximation, substitute (9) (taking into account two terms of the series) to relations (8) and (33). With respect to $b_1(q)$ and $b_2(q)$we will have a system of two algebraic linear equations, solving which we can find

$$b_1(q) = -\frac{4q' - 9\pi^2 q + 4}{8\pi^2}; \; b_2(q) = \frac{4q' - \pi^2 q + 4}{8\pi^2}; \tag{37}$$

Relation (9) taking into account (37) is as follows

$$\Theta(\xi,\mathrm{Fo}) = \frac{\left(-q' + 2{,}25\pi^2 q - 1\right)\cos\left(\frac{\pi\eta}{2}\right) + \left(q' - 0{,}25\pi^2 q + 1\right)\cos\left(\frac{3\pi\eta}{2}\right)}{\pi^2} \tag{38}$$

Substituting (38) into the heat balance integral (12), with respect to the required function $q(\mathrm{Fo})$ we will have the following ordinary differential equation

$$\frac{4}{3\pi^3} q^{IV} - \frac{10}{3\pi} q'' + \frac{3\pi}{4} q = 0, \tag{39}$$

where $q^{IV} = \frac{d^4 q(\xi)}{d\xi^4}$; $q'' = \frac{d^2 q(\xi)}{d\xi^2}$

The general integral of equation (39) is as follows

$$q(\xi) = \frac{4B(\pi + 1)}{3\pi^2} + C_1 e^{\frac{\pi\xi}{2}} + C_2 e^{\frac{3\pi\xi}{2}} + C_3 e^{\frac{-\pi\xi}{2}} + C_4 e^{\frac{-3\pi\xi}{2}}, \tag{40}$$

where C_1, C_2, C_3, C_4 - constants of integration determined from the basic (4), (5) and complementary (23), (28) boundary conditions, which for the function $q(\xi) = \Theta(\xi,0)$ are written as $q(1) = 0$; $\frac{dq(0)}{d\xi} = 0$; $\frac{d^3 q(0)}{d\xi^3} = 0$; $\frac{d^2 q(1)}{d\xi^2} + B = 0$. Substituting (40) into all these boundary conditions with respect to C_1, C_2, C_3, C_4 we will have a system of four algebraic equations, solving which we can find

$$C_1 = C_3 = -\frac{(3\pi + 2)e^{\frac{\pi}{2}}}{2\pi^2(e^{\pi} + 0{,}5)};\ C_2 = C_4 = \frac{(\pi + 2)e^{\frac{3\pi}{2}}}{6\pi^2(e^{3\pi} + 1)}. \tag{41}$$

Relation (33) taking into account (40), (41) offers the solution to the problem (3) - (7) in the second approximation. It exactly satisfies the boundary conditions (4) - (7), the heat balance integral (12) and approximately (in the second approximation) the equation (3). The results of calculations by the formula (38) are shown in figure 3

Compared to the exact solution [8], it follows that the standard deviation according to the Gauss formula is 1.3%.

To increase the accuracy of the solution, it is necessary to enlarge the number of terms of the series (9). To find the required coefficients $b_i(q)$, we use the condition (8) and complementary boundary conditions obtained by the general formula (36). To find the constants of integration C_k, ($k = 2n$) the initial boundary conditions (2), (3), CBC (21) and CBCs obtained by the general formulas (26) and (31) are used.

It should be noted that the adopted solution method has rather rapid convergence. So, already in the second approximation, the maximum ratio error of the solution obtained (Chebyshev norm) compared to the exact solution [8] decreases from 33% to 10%.

Note also that the use of an exact analytical solution (which is an infinite series) is difficult in the cases where the solution of the problem (3) - (7) is an intermediate stage of the researches for other problems, such as thermoelasticity, inverse boundary value problems, control problems, etc [23-24]. In these cases, a solution that contains a limited number of terms of the series and has sufficient accuracy for engineering applications will be more effective.

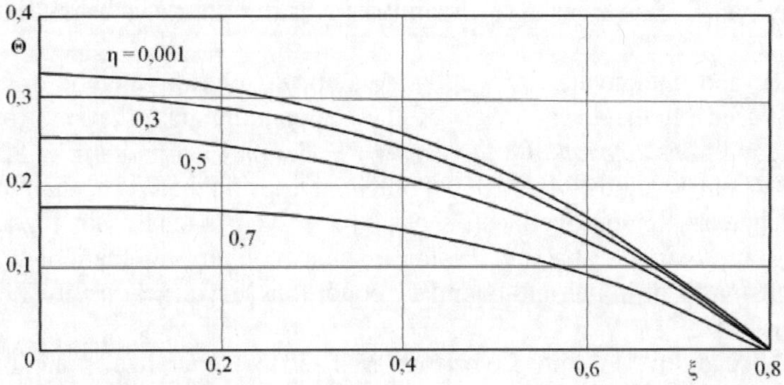

Figure 3. Temperature distribution in the bar section in the second approximation.

3. Conclusions.

1. Based on the use of a complementary required function and complementary boundary conditions in the integral heat balance method, an approximate analytical solution to the stationary two-dimensional heat conduction problem for an infinite bar with a uniformly distributed heat source is obtained. The use of the complementary required function provided reducing the solution of the partial differential equation to the integration of an ordinary differential equation.

2. To find the required coefficients and the constants of integration we used complementary boundary conditions determined in such a way that their fulfilling by the required solution is equivalent to the fulfillment of the initial differential equation at domain borders. It is shown that the fulfillment of the equation at the boundary points also leads to its fulfillment within the considered domain, with the accuracy depending on the number of approximations.

3. The use of the heat balance integral provides applying this method to solving boundary value problems for the Poisson equation with space-wise variables by physical properties of the medium and internal heat source.

4. Using the analytical solution (15) found, by solving the inverse heat conduction problem, one can fulfill quantitative identification of the internal heat sources generated in solids due to the processes of different nature (chemical reactions, heating by electromagnetic fields, vibration loads, deformation, friction, etc.) when included an experimental temperature value at any point in the structure.

4. Acknowledgment

The reported study was funded by Russian Science Foundation (RSF) according to the research project № 18-79-00171.

References

[1] Kudinov V A and Kudinov I V 2013 *Analiticheskiye resheniya dlya parabolicheskikh i giperbolicheskikh uravneniy teplomassoperenosa* [Analytic solutions for parabolic and hyperbolic equations of heat and mass transfer] (Moscow: Infra – M) p 391 [In Russian]

[2] Kudinov V A and Kudinov I V 2017 *Metody resheniya parabolicheskikh i giperbolicheskikh uravneniy teplomassoperenosa i peredachi impul'sa* [Methods for solving parabolic and hyperbolic equations of heat, mass and momentum transfer] (Moscow: Lenand) p 336 [In Russian]

[3] Timoshpolsky V I, Postolnik Y S and Andrianov D N 2005 *Teoreticheskiye osnovy teplofiziki i termomekhaniki v metallurgii* [Theoretical Foundations of Thermophysics and Thermomechanics in Metallurgy] (Minsk: Bel. navuka) p 560 [In Russian]

[4] Glazunov Y T 2006 *Variatsionnyye metody* [Variational methods] (Moscow-Izhevsk: NITS "Regulyarnaya i khaoticheskaya dinamika"; Institut komp'yuternykh nauk) p 470 [In Russian]

[5] Kudinov V A and Stefanyuk E V 2009 Analytical solution method for heat conduction problems based on the introduction of the temperature perturbation front and additional boundary conditions *Journal of Engineering Physics and Thermophysics* **82(3)** pp 537-55.

[6] Stefanyuk E V and Kudinov V A 2010 Polucheniye priblizhennykh analiticheskikh resheniy dlya zadach teorii teploprovodnosti s nesovpadeniyem nachal'nykh i granichnykh usloviy [Obtaining approximate analytical solutions for problems in thermal-conductivity theory with the mismatch of initial and boundary conditions] *Izvestiya vuzov. Matematika* **4** pp 63-71 [In Russian]

[7] Kudinov V A, Kudinov I V and Skvortsova M P 2015 Generalized functions and additional boundary conditions in heat conduction problems for multilayered bodies *Zhurnal Vychislitelnoi Matematiki i Matematicheskoi Fiziki* **55(4)** pp 129-40

[8] Kantorovich L V and Krylov V I 1962 *Priblizhennyye metody vysshego analiza* [Approximate methods of higher analysis] (Leningrad: Fizmatgiz) p 708 [In Russian]

[9] Fedorov F M 2000 *Granichnyy metod resheniya prikladnykh zadach matematicheskoy fiziki* [Boundary Method of Solving Applied Problems of Math-ematical Physics] (Novosibirsk: Nauka) p 220 [In Russian]

[10] Kudryashov L I and Menshih N L 1979 *Priblizhennyye resheniya nelineynykh zadach teploprovodnosti* [Approximate solutions to nonlinear heat conduction problems] (Moscow: Mashinostroyeniye) p 232 [In Russian]

[11] Eremin A V, Kudinov I V, Dovgyallo A I and Kudinov V A 2017 Heat Exchange in a Liquid with Energy Dissipation *Journal of Engineering Physics and Thermophysics* **90(5)** 234-42

[12] Averin B V, Kolotilkin D I and Kudinov V A 2000 Sturm-Liouville problem for a differential equation of second order with discontinuous coefficients *J. Eng. Phys. Thermophys* **73 (4)** pp 735-40

[13] Vladimirov V S 1979 *Obobshchennye funktsii v matematicheskoi fizike* [Generalized Functions

in Mathematical Physics] (Moscow: Nauka) [In Russian]

[14] Kolyano Y M 1978 Primeneniye obobshchennykh funktsiy v termomekhanike kusochno-odnorodnykh tel, v Matematicheskiye metody i fiziko-mekhanicheskiye polya [Application of generalized functions in thermal mechanics of piecewise homogeneous bodies, in Matematicheskie metody i fiziko-mekhanicheskie polya] *Matematicheskiye metody i fiziko-mekhanicheskiye polya* (Kiev: Naukova Dumka) **7** pp 7-11 [In Russian]

[15] Tikhonov A N and Samarskii A A 1999 *Uravneniya matematicheskoy fiziki* [Equations of mathematical physics] (Moscow: Izdatelstvo MGU) p 798 [In Russian]

[16] Bollati J, Semitiel J and Tarzia D A 2018 Heat balance integral methods applied to the onephase Stefan problem with a convective boundary condition at the fixed face *Applied mathematics and computation* **331** pp 1–19

[17] Fedorov F M 2000 *Granichnyy metod resheniya prikladnykh zadach matematicheskoy fiziki* [The boundary method for solving applied problems of mathematical physics] (Novosibirsk: Nauka) p 220 [In Russian]

[18] Hristov J 2017 Multiple integral-balance method basic idea and an example with Mullin's model of thermal grooving *Thermal science* **21(3)** pp 1555–60

[19] Tsoi P V 2005 *Sistemnyye metody rascheta krayevykh zadach teplomassoperenosa* [Systemic methods for calculating boundary-value problems of heat and mass transfer] (Moscow: Izdatelstvo MEI) p 568 [In Russian]

[20] Kudinov V A, Averin B V and Stefanyuk E V 2008 *Teploprovodnost' i termouprugost' v mnogosloinykh konstruktsiyakh* [Heat Conductivity and Thermoelasticity and Multilayered Constructions] (Moscow: Vysshaya Shkola) [In Russian]

[21] Kantorovich L V 1934 Ob odnom metode priblizhennogo resheniya differentsial'nykh uravneniy v chastnykh proizvodnykh [On one method for the approximate solution of partial differential equations] *Izvestiya Akademii nauk USSR* **2(9)** pp 532-4 [In Russian]

[22] Kudinov V A and Kudinov I V 2010 About one method of obtaining an exact analytical solution for the hyperbolic heat equation using orthogonal methods *Vestnik SamGTU. Seriya Fiziko-Matematicheskie Nauki* **5 (21)** pp 159-69

[23] Kudinov V A, Dikop V V, Gabdushev R Z, Levin D V and Stefanyuk S A 2002 On one method of determining eigenvalues in nonstationary heat conduction problems. *Izv. Ross. Akad. Nauk, Energetika* **4** p 112

[24] Wang G T, Pei K, Agarwal R P, Zhang L H and Ahmad B 2018 Nonlocal Hadamard fractional boundary value problem with Hadamard integral and discrete boundary conditions on a half-line *Journal of computational and applied mathematics* **343** pp 230–9

[25] Chen C S, Muleshkov A S, Golberg M A, Mattheij RM M 2005 A mash-free approach to solving the axisymmetric Poisson's equation *Numerical methods for pertial differential equations* **21(2)** pp 349–67

IOP Conf. Series: Materials Science and Engineering **552** (2019) 012010 doi:10.1088/1757-899X/552/1/012010

Problems of energy supply of the main consumers of distributive networks of Iraq

M A Averbukh[1], E V Zhilin[1] and M W Abdulwahhab[1,2]

[1]Department of Power Engineering and Automation, Belgorod State Technological University of V.G. Shukhov, Kostyukov St., 46, Belgorod, 308012, Russia
[2]Daiyla University, Baquba City, Daiyla Governorate, 32001 ZIP, Iraq

E-mail: zhilinevg@mail.ru

Abstract. The main purpose in this paper to study the main features of the distribution network for Iraq, which showed that the distribution network is physically and morally obsolete. Analyzing of electrical distribution networks is considered of highest priority at the present time in Iraq. This paper presents efficient practical methods for calculating the main parameters of the distribution network based on scientific and practical fundamentals. In this work the analysis is based on the advanced RasterWin3 software as a tool for the calculation of a basic parameters for distribution network performing the required analysis. RasterWin3 software is practical and efficient analysis software used by many electrical companies worldwide. The distribution network calculation was implemented (11/33) kV. The calculation results showed that the voltage deviation exceeds the maximum permissible value and is 12% in normal mode and 40% in emergency mode. Power losses in the branches of the power system of Iraq reach 20% of the consumed electricity. This is justified by the low quality of electricity and overload of the power supply system.

1. Introduction

The state of Iraq is located in the Mesopotamian lowland, in the valley of the Tigris and Euphrates rivers in the Middle East, and covers an area of 444,000 km2. In the south-east it lies near the Persian Gulf. Most of the territory is occupied by steppe and desert, however, irrigated lands are located in the valleys of the Tigris and Euphrates rivers. The climate is continental, with dry and exceptionally hot summers and relatively rainy, cool winters. Average temperature in winter ranges from 7 to 12 ° C, and in summer - 34 ° C (on some days they can reach 50 ° C). The annual precipitation is 50–150 mm in the plains and up to 1,500 mm in the mountains [1-5].

The history of Iraq's electric networks dates back to 1918, when the first Alaboukhana power plant built began generating electricity for the needs of the population. Thanks to the energy resources located in the republic, by 1990 the total generation capacity was 12,000 MW.

Iraq's distributive networks consist of twenty-four 400/132 kV substations with a total transmission capacity of about 17,000 MVA. They are generally located near cities. Electric power networks of class 132 kV are represented by twenty-eight power lines, with a length of about 4,500 km. The system also includes two hundred and eight 132/33 kV substations with a total capacity of 27,000 MVA. The total length of distributive networks of 33 kV voltage class is about 12,000 km. All these substations are interconnected through the central control station. In addition, many 33/11 kV substations are connected to the distributive networks of Iraq. The largest share of distributive networks falls on the line voltage of 11 kV with a total length of 52,000 km [6-10].

However, after two wars in the Gulf and an economic blockade, the conditions and technical completeness of the networks deteriorated sharply, which led to a significant overload of the power system of Iraq. For example, the electricity demand in Iraq for 2018 is about 18,000 MW, while the Iraqi power system generates about 1,5500 MW. Overloading Iraq's distributive networks has many negative consequences, the main of which are [11-13]:

- voltage deviation;
- increase in power losses;
- reducing the life of electrical equipment;
- reduction of power quality.

Reducing the cost of electricity, as well as reducing losses in the distributive networks of Iraq is the main problem today.

2. Materials and Methods

To analyze the operation of the Iraqi energy system, a part of the distribution network was considered, containing 33 and 11 kV lines, shown in figure 1.

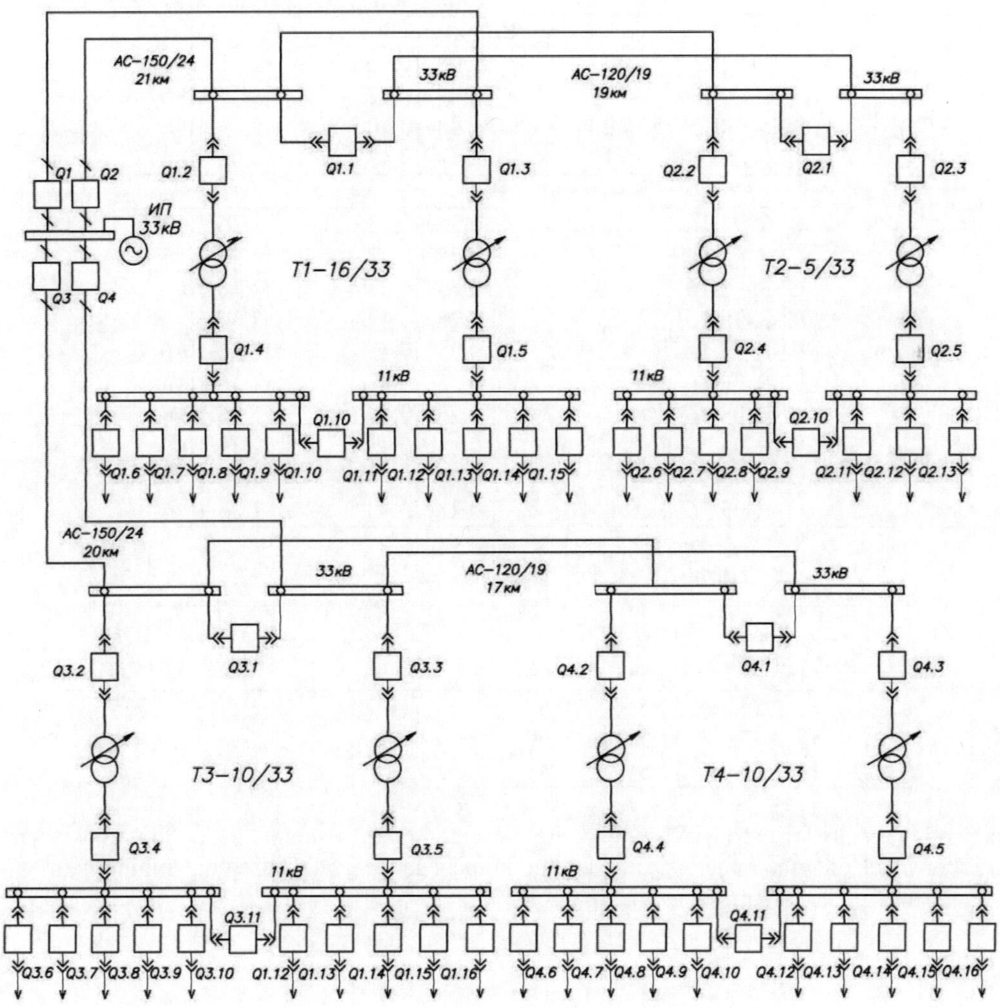

Figure 1. Fragment of the power supply circuit of Iraq

IOP Conf. Series: Materials Science and Engineering **552** (2019) 012010 doi:10.1088/1757-899X/552/1/012010

The calculation of the mode of the considered network is performed in the RastrWin3 software package, for steady-state and emergency operation.

RastrWin allows you to solve various problems, such as calculating, analyzing and optimizing the modes of electrical systems. A special feature of the software complex is a large number of design modules: starting with calculations of steady-state modes of electrical networks of any topology and voltage and ending with structural optimization in terms of voltage level, power loss and distribution of reactive power. The user interface should be mentioned because it is a tabular processor that allows individual or group correction and parameter input, the ability to sort the results of calculations by one or several columns, as well as the ability to "highlight" the data, depending on its value (for example, in the case of its output from normal values) [14-16].

3. Results

The initial data for the calculation of the regime are the topology and parameters of the equivalent circuit, which are presented in figure 2 and in table 1, 2.

Table 1. Initial data for branches

Branch	Wire type	Length (km)	Specific resistance (ohm / km)	Conductivity (cm / km)
		33 кВ		
0-1	ACSR-150/24	21	0,198+j0,406	j2,75·10-6
1-13	ACSR-120/19	19	0,249+j0,427	j2,66·10-6
0-22	ACSR-150/24	20	0,198+j0,406	j2,75·10-6
22-34	ACSR-120/19	17	0,249+j0,427	j2,66·10-6
		11 кВ		
2-3	ACSR-95/16	5	0,25+j0,43	j2,36·10-6
2-4	ACSR-95/16	4	0,25+j0,43	j2,36·10-6
2-5	ACSR-95/16	4	0,25+j0,43	j2,36·10-6
2-6	ACSR-95/16	3	0,25+j0,43	j2,36·10-6
2-7	ACSR-95/16	3	0,25+j0,43	j2,36·10-6
2-8	ACSR-95/16	4	0,25+j0,43	j2,36·10-6
2-9	ACSR-95/16	4	0,25+j0,43	j2,36·10-6
2-10	ACSR-95/16	5	0,25+j0,43	j2,36·10-6
2-11	ACSR-95/16	3	0,25+j0,43	j2,36·10-6
2-12	ACSR-95/16	4	0,25+j0,43	j2,36·10-6

Table 2. Passport data of transformers

	Power (MVA)	BH (KV)	HH (KV)	RT (Ohм)	XT (Ohм)	ΔPxx (кВт)	ΔQxx (Квар)
T1	16	33	11.5	0,52	7,4	9,2	56,7
T3, T4	10	33	11.5	0,96	11,1	14,5	80
T2	5	33	11.5	1,48	15,2	60	96

The calculation results for the 11 kV network are presented on the example of the first transformer substation. The values of the parameters of nodes and branches when calculating the minimum, maximum and emergency modes are presented in table 3, 4.

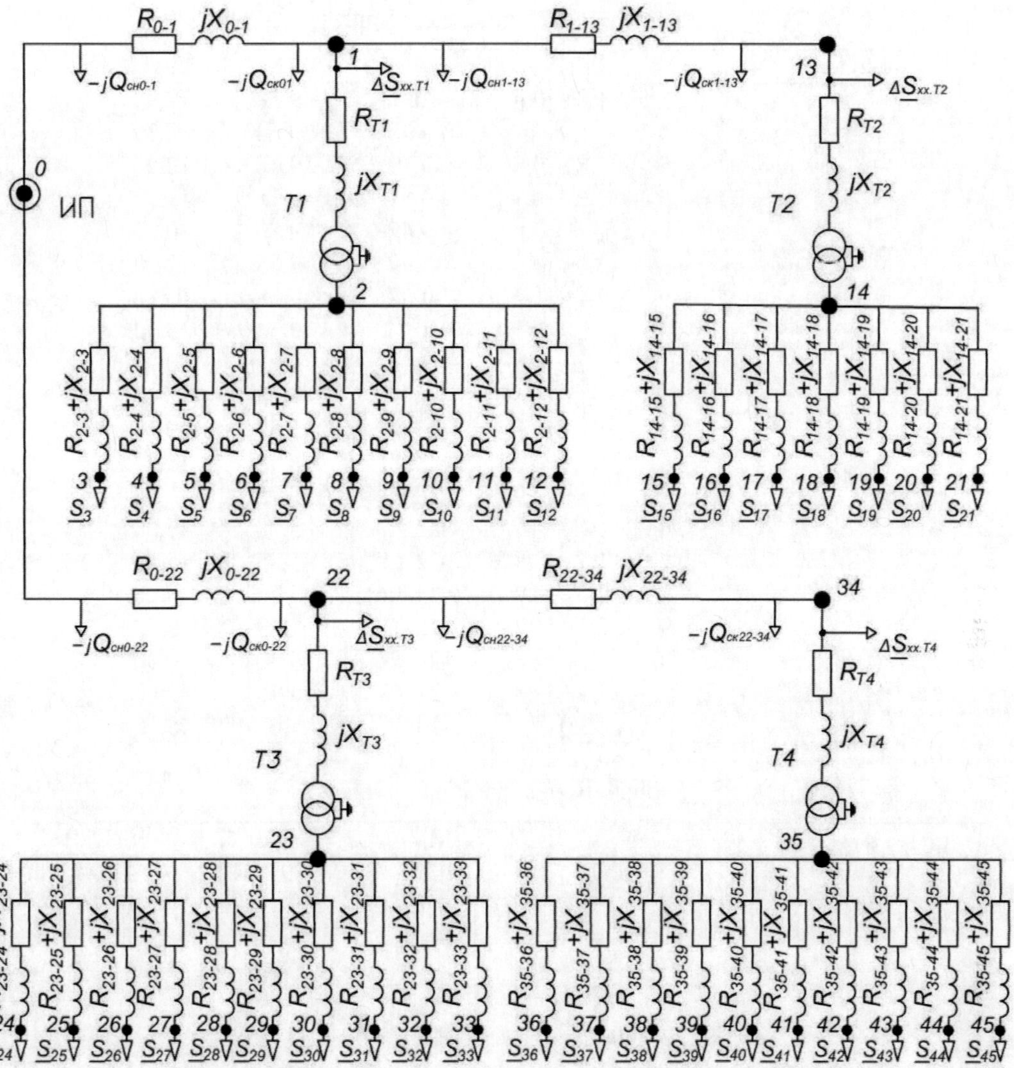

Figure 2. The replacement scheme of the electrical network of Iraq

From the results of the calculation of the steady-state mode, it can be seen that in the nodes of the power system of Iraq, the voltage deviation from the nominal is about 12%. In emergency situations, the voltage deviation can reach 40%. Normally, the allowable voltage deviation from the nominal value according to IEEE Std 1366-2012 should be no more than ± 5%, and the maximum allowable no more than ± 10%. Non-compliance with the requirements of the standard indicates the unreliability of the electricity grid and the low quality of electricity.

Electricity losses in the branches of the electricity system of Iraq reach about 20% of the electricity consumed. This is due to the low power factor and significant overload of the power supply system.

4. Discussion

To improve the quality of electricity and reduce power losses in the power system of Iraq, you can use various methods such as: voltage regulation with the help of on-load tap-changers, installation of a battery of static capacitors (BSC), installation of longitudinal and transverse compensation devices, installation of static thyristor compensators (STC) or static compensator (STATCOM).

Table 3. Node Parameters Values

Mode	№	S (MVA)	Angle (^{O})	V (KV)	ΔV, %
Steady	3	1,8+ j0,9	-10,08	9,58	12,95
	4	1,9+ j	-9,78	9,64	12,39
	5	1,8+ j0,9	-9,79	9,68	12,01
	6	1,9+ j	-9,51	9,75	11,32
	7	1,7+ j0,8	-9,47	9,80	10,90
	8	2,0+ j0,9	-9,99	9,65	12,27
	9	2,1+ j1,2	-9,87	9,58	12,88
	10	1,9+ j	-10,12	9,54	13,32
	11	1,7+ j0,8	-9,47	9,80	10,90
	12	2,0+ j	-9,92	9,63	12,43
Emergency	3	1,8+ j0,9	-23,63	5,57	49,36
	4	1,9+ j	-22,90	5,68	48,37
	5	1,8+ j0,9	-22,90	5,75	47,74
	6	1,9+ j	-22,26	5,88	46,58
	7	1,7+ j0,8	-22,15	5,95	45,91
	8	2,0+ j0,9	-23,38	5,70	48,18
	9	2,1+ j1,2	-23,14	5,59	49,22
	10	1,9+ j	-23,75	5,50	50,00
	11	1,7+ j0,8	-22,15	5,95	45,90
	12	2,0+ j	-23,22	5,67	48,45

Table 4. Branch parameters values

Mode	№	№	S (MVA)	Angle (^{O})	V (KV)
Steady	3	1,8+ j0,9	-10,08	9,58	12,95
	4	1,9+ j	-9,78	9,64	12,39
	5	1,8+ j0,9	-9,79	9,68	12,01
	6	1,9+ j	-9,51	9,75	11,32
	7	1,7+ j0,8	-9,47	9,80	10,90
	8	2,0+ j0,9	-9,99	9,65	12,27
	9	2,1+ j1,2	-9,87	9,58	12,88
	10	1,9+ j	-10,12	9,54	13,32
	11	1,7+ j0,8	-9,47	9,80	10,90
	12	2,0+ j	-9,92	9,63	12,43
Emergency	3	1,8+ j0,9	-23,63	5,57	49,36
	4	1,9+ j	-22,90	5,68	48,37
	5	1,8+ j0,9	-22,90	5,75	47,74
	6	1,9+ j	-22,26	5,88	46,58
	7	1,7+ j0,8	-22,15	5,95	45,91
	8	2,0+ j0,9	-23,38	5,70	48,18
	9	2,1+ j1,2	-23,14	5,59	49,22
	10	1,9+ j	-23,75	5,50	50,00
	11	1,7+ j0,8	-22,15	5,95	45,90
	12	2,0+ j	-23,22	5,67	48,45

Different sections of the power system are characterized by different values of voltages, which are determined primarily by the load and the network diagram. The on-load tap-changer provides for voltage regulation in various limits depending on the power and voltage of transformers (from ± 10 to ± 16% in steps of approximately 1.5%). The adjustment steps are carried out on the high voltage side, since the lower value of the current makes it easier to switch the device. To extend the control range without increasing the number of branches, coarse and fine adjustment steps are used.

The use of BSC allows reducing the reactive power flows in the network, which leads to a significant reduction in active energy losses in the networks of 11-132 kV, and this, in turn, reduces the load on power lines and network transformers. Increasing the power ratio in the trunk and distribution power facilities makes it possible to increase their throughput without increasing the power of transformers and building or upgrading power lines. The main functions of BSC are voltage regulation and reduction of energy losses [17-18].

Longitudinal compensation devices (LCD) (fiure. 3) contribute to increasing the capacity of overhead lines and ensuring more efficient operation of existing power lines. The composition of the LCD in the form of batteries includes longitudinal compensation capacitors, which are included in the electric transmission lines in series to compensate for some part of the inductive longitudinal resistance. Due to this, the capacity of overhead lines increases significantly [19-20].

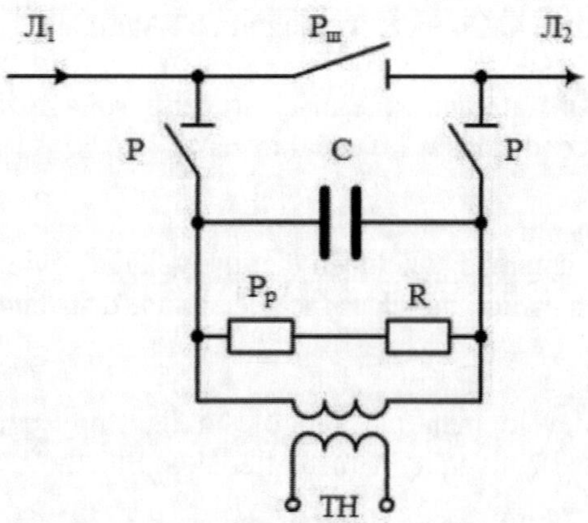

Figure 3. Longitudinal compensation device

For longitudinal reactive power compensation, the capacitors are connected in series with the load via isolating or booster transformers. Longitudinal compensation provides automatic regulation of the voltage depending on the load current. However, with longitudinal compensation, emergency modes occur. Their causes may be Ferro resonance oscillations, over voltages during capacitor bypassing, internal damage to capacitors.

Static thyristor compensator (STC) is a device for smooth control of reactive power and keeping its value within the specified limits. The principle of operation of the STC is based on the parallel inclusion in the network of filter-compensating circuits that perform the function of high-harmonic filters and thyristor-reactor reactors switched by thyristors.

STC functions when it is installed on power lines [21-24]:

- voltage stabilization;
- lower harmonic levels;
- reduction of voltage fluctuations in the supply network;
- load balancing.

STATCOM is a controlled static device, connected between two AC systems, consisting of a connecting transformer connecting the power supply system with a synchronous controller that generates voltage pulses, comparing them with one of the electrical systems to realize the exchange of reactive power (figure 4) [25].

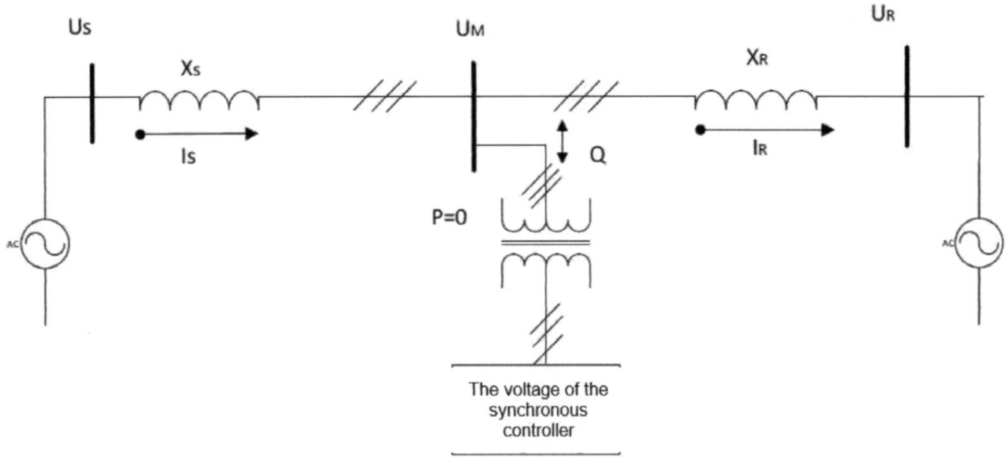

Figure 4. Connecting diagram STATKOM

The control system STATCOM at each time of the reverse voltage regulates so that the power supplied to the network, in these conditions were equal to zero.

5. Conclusion

Based on the analysis performed, it can be concluded that the voltage deviation in the networks of Iraq exceeds the permissible maximum, which indicates the need to use distributed generation.

Acknowledgments
The article was prepared within development program of the Flagship Regional University on the basis of Belgorod State Technological University named after V.G. Shukhov.

References

[1] Averbukh M A and Abdulwahhab M W 2018 Influence of Non-Sinusoidality of Voltage on Electricity Loss in Distribution Networks *Journal of Physics: Conference Series IOP Publishing* **1066(1)** pp 12-4
[2] Saeed I M, Ramli A T and Saleh M A 2016 Assessment of sustainability in energy of Iraq, and achievable opportunities in the long run *Renewable and Sustainable Energy Reviews* **58** pp 1207-15
[3] Ter-Martirosyan A and Kwoka J 2010 Incentive regulation, service quality, and standards in US electricity distribution *Journal of Regulatory Economics* **38(3)** pp 258-73
[4] Rao L, Liu X, Xie L and Liu W 2010 Minimizing electricity cost: optimization of distributed internet data centers in a multi-electricity-market environment *INFOCOM Proceedings IEEE.* pp 1-9
[5] Branker K, Pathak M J M and Pearce J M 2011 A review of solar photovoltaic levelized cost of electricity *Renewable and sustainable energy reviews* **15(9)** pp 4470-82
[6] Saeed I M, Ramli A T and Saleh M A 2016 Assessment of sustainability in energy of Iraq, and achievable opportunities in the long run *Renewable and Sustainable Energy Reviews* **58** pp 1207-15
[7] Hu J, Zhu J and Platt G 2011 Smart grid-the next generation electricity grid with power flow optimization and high power quality *Electrical Machines and Systems (ICEMS) International Conference on IEEE* pp 1-6

[8] Sadorsky P 2009 Renewable energy consumption and income in emerging economies *Energy policy* **37(10)** pp 4021-8.

[9] Dihrab S S and Sopian K 2010 Electricity generation of hybrid PV/wind systems in Iraq *Renewable Energy*. **35(6)** pp 1303-7.

[10] Payne J E 2010 A survey of the electricity consumption-growth literature *Applied energy* **7(3)** pp 723-31.

[11] Averbukh V A, Zhilin V E and Roschubkin P V 2017 Experimental Analysis of Electrical Modes in a Residential Estate Electrical Power Supply System *Journal of Engineering and Applied Sciences* **12** pp 3446-51.

[12] AL-TAHIR A L I A A L R 2008 Methods of improving perfomance for electrical power systems in Iraq *Journal of kerbala university* **6(4)** pp 14-32.

[13] Obais A M and Pasupuleti J 2011 Harmonics reduction of thyristor controlled reactor with minimal no load operating losses *Canadian Journal on Electrical and Electronics Engineering* **2(7)** pp 261-9

[14] Bliznyuk D I, Bannykh P Y and Khalyasmaa A I 2015 Power Flow Calculation Realization in Software for Grids with Four-Phase Power Lines *Applied Mechanics and Materials. – Trans Tech Publications* **698** pp 694-8

[15] Averbukh M, Prasol D and Khvorostenko S 2017 E`ksperimental`naya ocenka parametrov rezhimov v vy`sokovol`tny`x shaxtny`x setyax s moshhny`m nelinejny`m e`lektricheskim priemnikom [Experimental estimation of parameters of regimes in high-voltage mine networks with powerful non-linear electric receiversvol] *Vestnik Irkutskogo gosudarstvennogo texnicheskogo universiteta* vol 2 pp 75-84 [In Russian]

[16] Dolinger S, Lyutarevich A and Goryunov V 2013 Ocenka dopolnitel`ny`x poter` moshhnosti iz-za snizheniya kachestva e`lektricheskoj e`nergii v e`lementax sistemy` e`lektrosnabzheniya [Estimation of additional power losses due to a decrease in the quality of electrical energy in the elements of the power supply system] *Omskij nauchny`j vestnik* vol 2 pp 178-83 [In Russian]

[17] Kartashev I, Tulsky V and Chamonov R *Upravlenie kachestvom e`lektroe`nergii* [Electricity quality management] (Izdatel`stvo ME`I) vol 3 [In Russian]

[18] Averbukh M, Kuznetsov V and Korzhov D 2013 Problemy` obespecheniya e`lektromagnitnoj sovmestimosti v e`lektroustanovkax promy`shlenny`x predpriyatij [Problems of ensuring electromagnetic compatibility in electrical installations of industrial enterprises] *Bulletin of Belgorod State Technological University named after V G Shukhov* vol 5 pp 203-7 [In Russian]

[19] Demirci T et al 2011 Nationwide real-time monitoring system for electrical quantities and power quality of the electricity transmission system *IET Generation, Transmission & Distribution* **5(5)** pp 540-50.

[20] Dihrab S S and Sopian K 2010 Electricity generation of hybrid PV wind systems in Iraq *Renewable Energy* **35(6)** pp 1303-7

[21] Hazaa H K et al. 2011 Evaluation of electric energy losses in kirkuk distribution electric system area *Iraqi Journal for Electrical And Electronic Engineering*. **7(2)** pp 144-50

[22] Ostovar F and Legha M M 2013 An imperialist competitive algorithm for sitting and sizing of distributed generation in radial distribution network to improve reliability and losses reduction *Iraqi Journal for Electrical And Electronic Engineering* **9(2)** pp 59-66

[23] Zehra E J A, Moghavvemi M, Hashim M M I and Muttaqi K 2010 Network reconfiguration using PSAT for loss reduction in distribution *systems Energy, Power and Control (EPC-IQ), 2010 1st International Conference on IEEE* pp 62-6

[24] Kazem H A and Chaichan M T 2012 Status and future prospects of renewable energy in Iraq *Renewable and Sustainable Energy Reviews* **16(8)** pp 6007-12

[25] Rao P, Crow M L and Yang Z 2000 STATCOM control for power system voltage control applications *IEEE Transactions on power delivery* **15(4)** pp 1311-7

The device for monitoring the LED display high-voltage insulators state

R N Balobanov, D K Zaripov and AA Akhmadeev

Department of Electric Stations, Kazan State Power Engineering University, Krasnoselskaya St., 51, Kazan, 420066, Russia

E-mail: rassel_ipek@mail.ru

Abstract. The article discusses the possibility of using devices for monitoring the state of high-voltage insulators with LED indication, mounted on insulating structures of high-voltage lines and substations. The scheme of the device is given and its principle of operation is described. The feasibility of indicating the early stage of the an insulation defect development in a laboratory and full-scale experiments on real power system facilities using devices with LED indication are shown. A new method of insulation control using the developed device will allow, when visually inspecting an overhead line, to additionally provide information about the change in the electric field and the appearance of discharges in the insulation. In addition to the early warning of the birth of a defect, this information will be useful for routine maintenance (cleaning) of insulators.

Keywords: polymer insulator, indicator, defects, LEDs, power line, insulation condition, assessment, outdoor switchgear, contactless method, dynistor.

1. Introduction

The object of study is high-voltage insulators. According to the brochure "The main results of the electric power facilities operation in 2016. The results of the 2016–2017 autumn-winter period (OZP), issued by the Ministry of Energy of the Russian Federation in 2017, the number of accidents that led to the cessation of electricity supply to consumers with a capacity of 10 MW or more due to the damage of overhead line insulators (OHL) amounted to 12% of total number of accidents with such damage on overhead transmission lines (HVTL), substations (PS) and open switchgear (ORU). If we consider only VL, then it is 22.6%. These indicators are inferior only to those due to extreme external influences and significantly exceed these indicators due to, for example, thunderstorms or ice [1]. The damage to insulators in most cases occurs as a result of aging, surface contamination, manufacturing defects and accidental damage during repairing work.

A special place among the types of damage to the overhead lines insulation is taken by the damage to the suspension polymer insulators associated with their breakdown or overlap. According to the Research Institute of Electric Power Industry (EPRI) in the USA, where the operating experience of polymer insulators on overhead lines is more than 30 years, over the past 10 years, the US electrical grid companies have begun to experience an increasing number of polymer insulators failures on transmission lines 115 and 138 kV. EPRI studies have shown that these failures are usually associated with high electric field strengths occurring near or on high-voltage insulators. The decrease of overhead lines reliability is due to the constant activity of electrical discharges on metal fittings. Constant impact on the insulation of these corona discharges leads to the formation of cracks in the polymer shell and destruction of the mechanical seal, therefore, constant monitoring of this type of insulators condition is necessary [2].

At present, radiowave [3-6], acoustic [7-10], optica [11-14] and thermal imaging [15-18] methods of insulator control are used in electric power industry. The use of these methods is convenient and effective in diagnosing the supporting and suspension insulators of substations, but it is difficult in examining extended objects such as overhead lines. Instrumental examination of overhead lines from aircraft also did not find wide application. Therefore, according to researchers, more than 40% of companies still use mostly visual inspection of equipment. Infrared diagnostics and night vision devices are used by approximately 15% of enterprises, and only a few enterprises use measurements of radio emission and ultrasound monitoring [19-20].

Remote control methods of isolators have not been able to supplant a simple visual inspection of overhead lines and are only an addition to it. At the same time, during visually inspecting polymer insulators, it is not always possible to identify defects, since in most cases they are not visible. Consequently, there is an actual scientific problem associated with the lack of effectiveness of the existing methods for controlling insulators of overhead lines. Its analysis revealed the need for the development and research of new methods for controlling insulation using devices installed on insulators, which would allow for visual inspection of overhead lines to signal their condition directly.

2. Materials and Methods

One of these devices is our proposed indicator. The scheme of the device is made on the basis of dynistors [21-24] figure 1. In Figure 1, the dotted lines indicate the connection of the indicator with the high-voltage and grounded part of the structure. A feature of this scheme is the ability to display discharges that occur in the insulation during damage and / or contamination. The circuit consists of two loops. The first one includes the following elements: capacitance C1, resistances R1 and R2, dinistor VD6 and green LED HL1. The second one consists of the capacitance C2, the resistance R3, the dynistor VD7 and the red LED HL2.

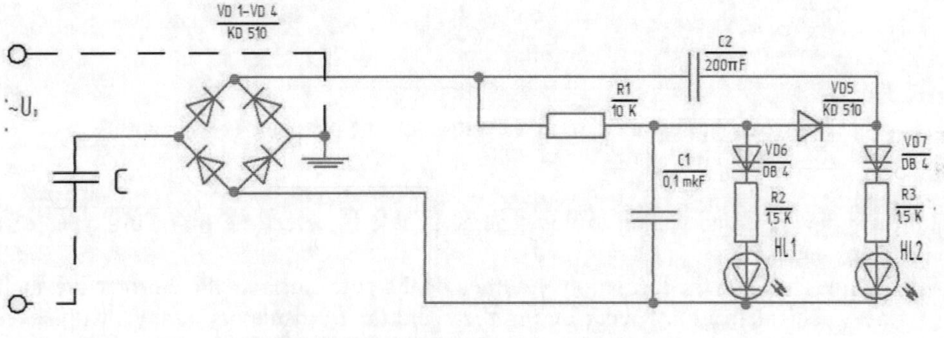

Figure1. Diagram based on dynistors

Let's consider the principle of the indicator. The distribution of potentials along an insulating structure changes in violation of the integrity of its individual parts. The potential difference in the damaged area decreases, which causes an increase in voltage on the intact part of the structure. The defective condition of the insulator can be detected, for example, by increasing the intensity or frequency of the pulses of the light emitted by the indicator mounted on the portion of the insulating part of the structure or support, as this increases the voltage on it. One of the electrodes of the radiator can be earthed through the grounded part of the insulator or support, the second electrode can be positioned freely [25-26].

To ensure reliable operation of the radiator with a constant installation, it is necessary to connect a discharger in parallel to it. In the event of a surge overvoltage, the discharger shunts the radiator, protecting it from failure.

The scheme works as follows. The current from the diode bridge VD1-VD4 charges the capacitance C1. As charging C1 voltage on it increases. The dynistor VD6 does not pass the current until the voltage on its terminals (on C1) reaches 40 V.As soon as the voltage reaches the desired value, the dynistor opens and the charge from the capacitor C1 is discharged to the LED HL1, as a result of which the LED emits a flash of green light. Simultaneously with the flash, the voltage on the dynistor VD6 drops, and it closes. Next, the capacitance C1 is again charged until the opening voltage of the dynistor and the cycle repeats. Thus, the green LED is pulsed.

If there are no discharges on the insulation, only the green HL1 LED will always be triggered, because due to the voltage drop across the VD5 diode, the dynistor VD7 will always be closed.

At a certain intensity of the discharges on the insulator, the pulses, passing through the second (high-frequency) circuit with a capacitance C2, will increase the voltage on the VD7 dynistor to a level higher than that on the VD6 dynistor. Then VD7 will open first and the red LED HL2 will light up. The value of capacitance C2 is selected in such a way as to react to the appearance of short pulses (≤ 1 μs) caused by discharges on the insulator. More details ot the work of the LED indicator are considered in the thesis.

3. Results.

According to the scheme described above, experimental samples of the indicator were fabricated, which were tested in the high-voltage laboratory of Kazan State Power Engineering University and at existing OHL and open switchgear. The appearance of the indicator and the location on the insulator are shown in figure 2

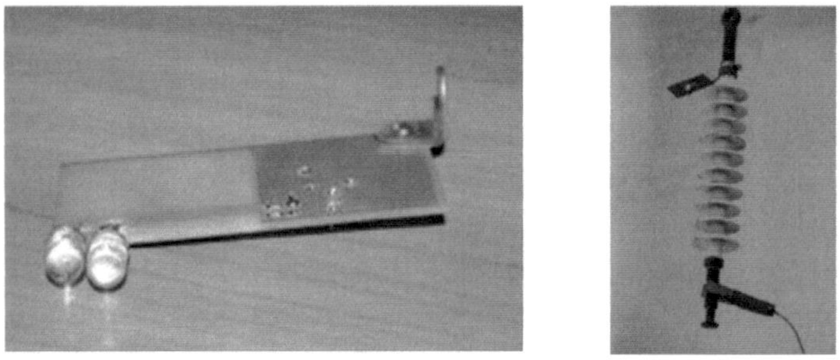

a b

Figure 2. Appearance (a) and location on the insulator (b) variants of the LED indicator

The indicator developed according to this scheme is able to detect the following types of defects and damage to insulators:

1. Erosion of the shell and the formation of a track on the outer surface of the protective shell of the insulator figure 3. This defect may occur during the operation of insulators in areas with severe contamination due to the impact of surface discharges in wetted insulators.

Figure. 3. Erosion of the containment shell

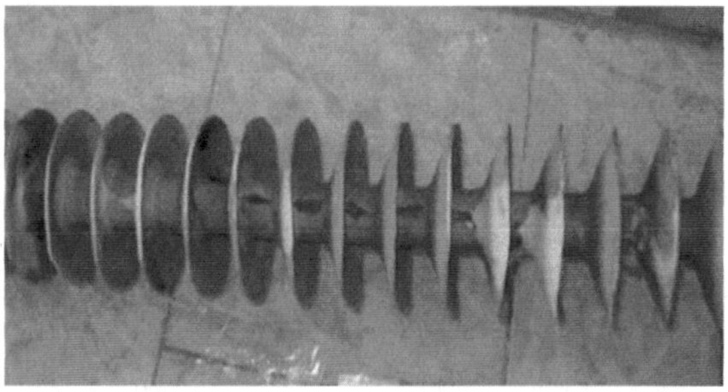

Figure. 4. Breakdown of a linear polymer insulator on HVL 330 kV

2. Formation of dendrites in a fiberglass rod and tracks at the interface of the "core-shell" section in a polymer insulator figure 4.

Often this is accompanied by an internal erosion of the containment up to the formation of through holes. This damage is also associated with the penetration of moisture into the insulator. The root cause of damage is most often a defect in the design of the insulator, namely:

- insufficient tightness of the interface between the shell and metal reinforcement;
- poor adhesion of the shell to the fiberglass rod.

In addition, the leakage may be due to damage (breaks) of the shell during transportation of insulators or installation.

According to the operating experience, this type of damage to polymer insulators is the most common. The breakdown of the insulator is accompanied by the occurrence of an electric arc with the destruction of part of the protective sheath.

It is almost impossible to identify the location of the indicated damage while visually inspecting insulators on an overhead line because of the large distances to them and the need to choose an observation angle each time.

Similar defects were simulated in the laboratory. In the picture (Fig. 5), two polymer insulators LC 70/35 are shown, to which AC voltage was applied from 0 to 25 kV. One of the insulators (right) has a defect in the form of a longitudinal conducting channel that shunts a part of the structure (~ 30%), and the second (left) is serviceable. LED indicators are attached to the upper electrodes of both insulators.

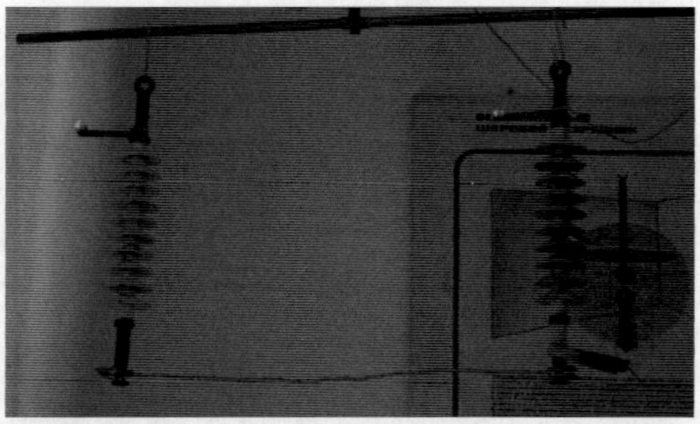

Figure. 5. Image of defective and serviceable polymer insulators with LED indicators mounted on them

For the experimental part of the work, an electrical circuit was assembled (Fig. 6), designed to measure the leakage currents of the insulator, as well as the parameters of the LED indicator of the defect.

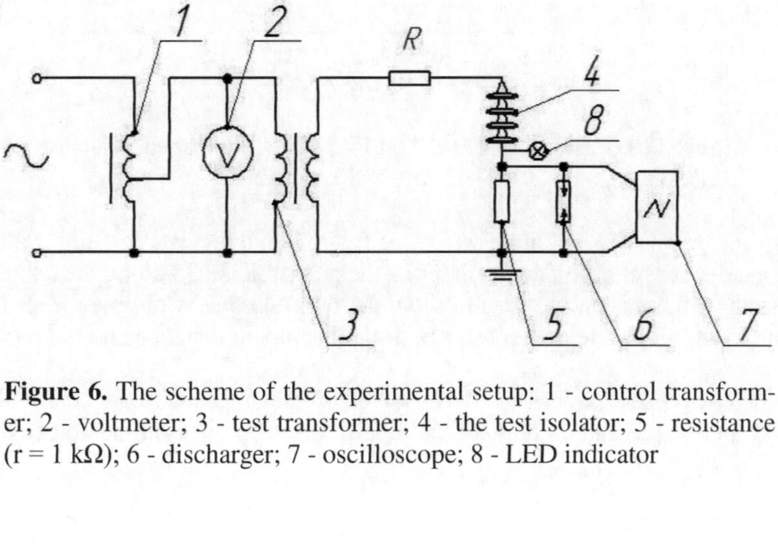

Figure 6. The scheme of the experimental setup: 1 - control transformer; 2 - voltmeter; 3 - test transformer; 4 - the test isolator; 5 - resistance (r = 1 kΩ); 6 - discharger; 7 - oscilloscope; 8 - LED indicator

The source of voltage of industrial frequency is the test transformer IOM - 110 kV (3). The voltage on it is regulated by the autotransformer RNO-10/250 (1). Resistor R is used to protect the transformer from large currents and a large slope of the voltage at the overlap of the test insulator (4). The measuring circuit consists of a resistor r = 1 kΩ (5), a spark gap (6), designed to protect the oscilloscope (7) in the event of an overvoltage.

4. Discussion.

The experiments were carried out on suspension polymer insulators brand LK 70/35 for voltage of 35 kV. A working voltage of 20 kV was applied to an insulator of brand LK 70/35, and a defect was artificially created by short-circuiting an insulator from 0 to 50%. The dependence of the pulse repetition rate of the green LED on the degree of damage to the insulator Fv = f (n) was determined. The results of the experiment are listed in Table 1 and displayed on the graphs (Fig. 7)

Table 1 The results of experiment No.

U, кВ	20	20	20	20	20	20
The proportion of defect in the insulator*n*, %	0	10	20	30	40	50
LED pulse repetition rateF_v, 1/c	1,23	1,42	1,69	1,88	-	-
$K(n)=F_v(n)/F_v(0)$	1	1,15	1,37	1,52	-	-
LED color*	G	G	G	G	R	R

*G- green; R-red

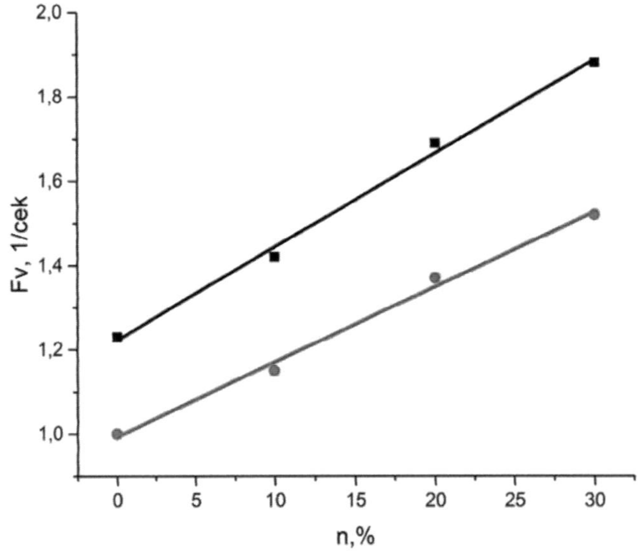

Figure 7. Dependences K (n) and Fv (n) on a damaged insulator LC 70/35

An analysis of the graphs of K (n) and Fv (n) (see figure7) and the data in Table 1 allows us to conclude that when a defect develops on an insulator of the brand LK 70/35 along the length up to 30% of the insulating distance, there is an increase in the repetition rate pulses of green LED indicator. With an increase in the length of the defect up to 40% of the insulation distance and the red LED lights up more.

When the voltage was raised, it was clearly observed on the one hand, the increase in the pulse repetition rate of both indicators, and on the other, the pulse repetition rate on the defective insulator was noticeably higher. When a voltage of 20 kV was reached, pulses of a red LED appeared on the indicator mounted on the defective insulator. On a good insulator, the green LED pulses continued.

The results obtained in the laboratory, allowed to go to full-scale experiments on real objects of the power system. The tests of the indicators took place in the 110 kV outdoor switchgear of the Kazan CHP-3 (OJSC TGK-16). The defect indicator was mounted on an insulating rod and was brought into contact with a grounded part of the supporting insulators by a single electrode (Fig. 8). On the inspected support insulators, as can be seen in Figure 9, the indicator glow was sufficiently bright for visual detection in sunlight.

Figure 8. Order of work

Figure 9. Defect indicator at the support insulator in the open switchgear 110 kV

As a result of the survey conducted on 07/11/2017, the 110 kV outdoor switchgear of Kazan CHP-3 using an LED indicator of high-voltage insulation attached to an isolation rod, insulators were found, on which increased intensity of discharges was observed (red indicators glowed): cell 1 reference insulators of phase A , B, C; cell 2 is a phase A supporting insulator; cell 4, phase A supporting insulator;cell 6 supporting insulators of phase A, B, C; cell 7 supporting insulators of phase A, B, C; cell 8 supporting insulators of phase A, B, C; cell 10 supporting insulators of phase A, B, C; cell 11 supporting insulator phase C; cell 14 is a phase C supporting insulator, cell 17 is a phase A, B, C reference insulators; cell 18, phase A, B, C reference insulators. This assumption was confirmed by partial discharges fixed on the surface of these insulators by the UD-8V ultrasonic flaw detector.

The likely reason for the increased intensity of discharges is the non-standard mounting of the flexible bus to the insulator using a standard corner, as well as the pollution of the insulators.

It was recommended to replace the corners with standard clips and to maintain the insulators.

The experimental work was continued in Almetyevsk electrical networks of JSC "Grid Company". During the experiments, about 50 defect indicators were installed on 35 kV and 110 kV lines.

The purpose of the work in Almetyevsk electrical networks, where the indicators were introduced, is to study the possibility of large-scale use of indicators to monitor the condition of overhead polymer insulation of overhead lines, as well as the possibility of excluding from the regulations periodic monitoring using thermal imagers and electron-optical flaw detectors.

The prototypes of the indicators worked on the 35 and 110 kV lines for more than six months and showed good reliability. Of the 39 pieces of checked bypass indicators after six months of operation did not work 5 pieces.

As for the performance of indicators according to their intended purpose, today only about 50 indicators have been installed on the lines and it is still too early to carry out a full-scale assessment of the effectiveness.However, when going round together with the head of the VL service of a section of the 110 kV line "Abdrakhmanovo-Uzlovaya" with indicators installed on all phases on the polymer insulators of the individual phases of the three pillars, an increased blinking of the green LED was detected. This anomaly was taken by the VL service under control. The indicator operation on the 35 kV "Wastewater treatment plant - Substation 126" line is shown in figure 10.

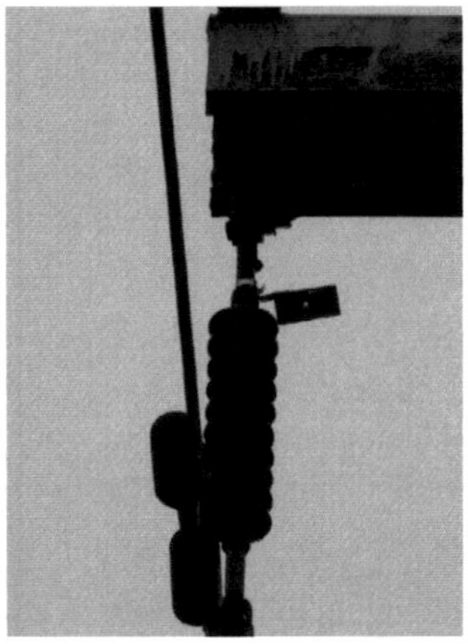

Figure 10. Indicator operation on the 35 kV “Wastewater treatment plant - Substation 126” line

The ability to use indicators on various types of insulators and voltage classes. One of the main advantages of the developed indicator is its versatility.As laboratory and field experiments have shown, it can be used without any changes in the design on various types of insulators and on various classes of voltages, both on high-voltage lines and on the open switchgear.The indicators were tested and tested on garlands of suspended glass insulators, suspended polymer insulators of 35 and 110 kV lines. The same indicators installed on the isolation rod were inspected on the open switchgear, supporting porcelain and polymer insulators with a voltage of 110-220 kV. We consider the most promising use of indicators to control polymer insulators of voltage class from 35 kV to 500 kV. However, additional evidence is needed to confirm this finding.

5. Conclusion.

A device for monitoring the state of high-voltage insulators (LED) on power lines and switchgears, based on the registration of changes in the electric field in the area of the insulator when a defect appears, was developed.

The developed LED indicator of the state of high-voltage insulation does not require an autonomous power source and allows to record both the breakdowns of the insulator part and the occurrence of electrical discharges on it.

Laboratory and field experiments showed the possibility of using the developed indicator to monitor the insulation status of suspension and support insulation of overhead lines and open switchgear.

Indicators can be used as stationary devices mounted directly on the insulator, and as a portable device mounted on an insulating rod.

These indicators will help the electrical network personnel to assess the condition of the insulators during planned inspections of the switchyard.

References

[1] 2018 «Osnovnye rezul'taty funktsionirovaniya ob"ektov elektroenergetiki v 2016 godu. Itogi prokhozhdeniya OZP 2016–2017 godov» URL https://minenergo.gov.ru/node/6224

[2] 2018 Learning from Service Experience with Composite Line Insulators URL http://www.inmr.com/learning-from-service-experience-with-composite-line-insulators/

[3] James R E and Condition Q Su 2008 *Assessment of High Voltage Insulation in Power System Equipment*(Great Britain:Institute of Educational Technology) p 276

[4] Arbuzov R S , Ovsyannikov A G, Tolchin V M, Kolesnikov A A and Nelaev A A 2003 Akusticheskaya diagnostika energeticheskogo oborudovaniya [Acoustic diagnostics of power equipment] *Sbornik trudov 7-go simpoziuma «Elektrotekhnika 2010 god»: sbornik Moskovskaya oblast', may 27-29* vol 3 pp 186-8 [In Russian].

[5] Khusnutdinov R A and Mardanov G D Monitoring sostoyaniya vysokovol'tnoy izolyatsii [Monitoring the state of high-voltage insulation] *Radioelektronika, elektrotekhnika i energetika: tezisy dokladov Dvadtsat' tret'yey Mezhdunarodnoy nauchno-tekhnicheskoy*

konferentsii studentov i aspirantov vol 3 p 352 [In Russian].

[6] SviP M 1992 Metody i sredstva diagnostiki oborudovaniya vysokogo napryazheniya: proizvodstvenno-prakticheskoe izdanie [Methods and means of diagnostics of high voltage equipment: production and practical edition](Moscow: Energoatomizdat) p 240 [In Russian].

[7] Reykh E N, Sychenko V Gand and Kim E D 2012 Analiz metodov i sredstv diagnostirovaniya dlya otsenki sostoyaniya izolyatorov kontaktnoy seti [Analysis of diagnostic methods and tools for assessing the condition of contact network insulators] *Elektrifikatsiya transportu* **4** pp 54-62 [In Russian].

[8] Gaivoronsky A S 2010 Povrezhdeniya polimernykh izolyatorov i ikh diagnostika pri ekspluatatsii [Damage of composite insulators and their diagnosis during operation] *Glavnyy energetik* **2** pp 23-7 [In Russian].

[9] BazanovV P, SpirinM V and Turaev V A 2000 Ul'trazvukovoy metod kontrolya farforovoy izolyatsii vozdushnykh liniy elektroperedachi 35 -220 kV [Ultrasonic method of control of porcelain insulation of overhead power lines 35 -220 kV] *Energetik* **4** pp 16-7 [In Russian].

[10] Zavidey V I 2010 Distantsionnye metody i sistemy defektoskopii vysokovol'tnoy izolyatsii elektrooborudovaniya po opticheskomu izlucheniyu [Remote methods and systems of flaw detection of high-voltage insulation of electrical equipment by optical radiation] *Novoe v rossiyskoy elektroenergetike* **9** pp 5-12 [In Russian].

[11] Ma Bin, Zhou Wenjun and Wang Tao 2006 Study on corona discharge test under the power frequency voltage of the severe non-uniform electric field based on the UV-light imaging technology CEEM 2006

[12] Stolper R, J Hart J and Mahatho N 2006 *The design and évaluation of a Multi-Spectral Imaging Caméra for the inspection of transmission lines and substation equipment* (South Africa: Applied Optics)

[13] 2018 Metodicheskie ukazaniya po distantsionnomu opticheskomu kontrolyu izolyatsii vozdushnykh liniy elektroperedachi i raspredelitel'nykh ustroystv peremennogo toka napryazheniem 35–1150 kV URL http://www.fsk-ees.ru

[14] GigreWorkingGroup 22.03 (convener C. de Tourreil) 2001 *Composite insulators handling guide* **195** pp 50-3

[15] 2018 Teplovizionnaya diagnostika kak sovremennoe sredstvo monitoring URL http://www.centert.ru/ articles/22/

[16] Krishna N A and Subhash C 2006 *Discrete transfer method applied to radiative transfer in a variable refractive Index semi-transparent medium Journal of Quantitative Spectroscopy and Radiative Transfer* **102** pp 432-40

[17] Zaripov D K 2006 Metod distantsionnoy diagnostiki izoliruyushchey konstruktsii vysokogo napryazheniya [Method of remote diagnostics of high voltage insulating structure] *Problemy Energetiki* **3-4** pp 56-60 [In Russian].

[18] Salustiano R, Capelini R M, S R de Abreu, M L B Martinez, I C Tavares, G M F Ferraz and M A A Romano 2014 Development of new methodology for insulators inspections on aerial distribution lines based on partial discharge detection tools:High Voltage Engineering and Application (ICHVE) *International Conference* pp 1-4

[19] VdovikoVP, Ovsyannikov AG and Pospelov AI 1995 Diagnostika elektricheskoy izolyatsii vysokovol'tnogo oborudovaniya pod rabochim napryazheniem [Diagnostics of electrical isolation of high-voltage equipment under operating voltage] *Energetik* **10** pp 16-8 [In Russian].

[20] Balobanov R N, Zaripov D K and Margulis S M 2015 Novye vozmozhnosti kontrolya sostoyaniya izoliruyushchikh konstruktsiy vysokovol'tnykh podstantsiy [New possibilities of monitoring the state of insulating structures of high-voltage substations] *Energetika Tatarstana* **4** pp 48-52 [In Russian].

[21] Rui Lou and Shu Bin Wu 2011 *Products properties from fast pyrolysis of enzymatic/mild acidolysis ligninApplied Energy* **88** pp 316-22

[22] ZaripovDK and Balobanov RN 2016 Indikator defekta vysokovol'tnoy izoliruyushchey konstruktsii [The indicator of the defect of high voltage insulation design] *Elektrotekhnika* **6**

pp 16-21 [In Russian].

[23] BalobanovRN, ZaripovDK, NasibullinR A and MargulisSM 2017 Ustroystvo opticheskoy indikatsii defekta vysokovol'tnoy izoliruyushchey konstruktsii [The optical display device of the defect of high voltage insulation design] *Izvestiya vysshikh uchebnykh zavedeniy. Problemy energetiki* **3-4** pp119-25 [In Russian].

[24] Zaripov D K and Balobanov R N 2016 *An Indicator of a Defect in a High-Voltage Insulating Structure -Russian Electrical Engineering* vol 87 **6** pp 311-15

[25] Balobanov R N and Zatsarinnaya Yu N 2015 Osobennosti diagnostirovaniya vysokovol'tnogo oborudovaniya s elegazovoy izolyatsiey [Features of diagnosing high-voltage equipment with gas-insulated] *Vestnik tekhnologicheskogo universiteta* vol 18 **2** pp 257-8 [In Russian].

IOP Conf. Series: Materials Science and Engineering **552** (2019) 012012 doi:10.1088/1757-899X/552/1/012012

Composite structural elements of overhead lines as a means of increasing technical efficiency

F Byk[1], V Kityshin[2], L Myshkina[1] and E Kaminskaya[3]

[1]Automated Electrical Power Systems Department, Novosibirsk State Technical University, Karl Marks Ave., 20, Novosibirsk, 630073, Russia
[2]Management and Economic Systems in Power Engineering Department, Novosibirsk State Technical University, Karl Marks Ave., 20, Novosibirsk, 630073, Russia
[3]Philology Department, Novosibirsk State Technical University, Karl Marks Ave., 20, Novosibirsk, 630073, Russia

E-mail: L-zakirova@yandex.ru

Abstract. Innovations in engineering and technologies depend not so much on emergence of new properties of the equipment, production process, system of management of the enterprise, but to a greater extent on their occurrence in the necessary set of combination. Technical and economic indicators reflecting technical and economic efficiency are usually used as characteristics. The degree of technical efficiency is determined by a combination of three properties – safety, reliability and productivity. These basic properties determine the economic effect of an innovative product performing certain functions, enabling to estimate their usefulness and significance. One of the means of increasing the overhead lines technical efficiency in the electric grid is the introduction of composite structural elements. The overhead line model is proposed for assessing the impact of elements on reliability as a component of technical efficiency. Changes in the reliability of the line when introducing innovative elements are calculated and priorities are defined.

1. Introduction

The terms "innovative products", "innovation activity", "innovative development" are used privately. However, it remains unclear how the term "innovative" differs from the others. It is not clear what criteria can be used to distinguish innovations. This determines the relevance of the work. The issues of the feasibility study of innovative techniques effectiveness and technologies are extremely important. There is no unified understanding of terminology, in most cases there has been no analysis of technical efficiency.

Taking into consideration only economic effects in the analysis of innovative products has been a priority. It has lead to creating barriers to the innovation application. It has caused a slowdown of the innovative development of the enterprise, industry and the country. These factors determine the feasibility of the study.

Often, a qualitative feature enabling to allocate innovative products is the availability of an appropriate technical characteristic, which essentially distinguishes it from analogues. It is vital to determine quantitative technical indicators the achievement of those will enable to distinguish innovation.

2. Problem definition

IOP Conf. Series: Materials Science and Engineering **552** (2019) 012012 doi:10.1088/1757-899X/552/1/012012

The objective of the work is to define objectives of innovation, as the basis for innovative development. Presenting a change in the innovation activity is supposed to be vital and understood as the process of innovation commercialization.

From the authors' view, innovation concept can be considered as the improvement of the technical efficiency of products. If this goal is achieved and the particular tasks are solved, a basis for obtaining the economic effect from the introduction and use of innovative products appears. The economic efficiency is a necessary condition for the commercialization of innovations. The activity of entities forming innovative infrastructure is aimed at that resulting in the innovative development of the enterprise, industry, country.

Tasks:

- Systematization, analysis and terminology arrangement in the field of innovation and innovative development.
- Criteria determination for comparison of technologies, their classification as "innovative".
- Analysis of the technical efficiency of innovative products by the example of composite structural elements of overhead lines.

Scientific novelty of the study. It is shown that the subject of innovation activity should increase the technical efficiency of innovative products but not as the commercialization of scientific and technological achievements. The definition of innovative development of an enterprise, industry, and country, as revolutionary and intensive changes in the technological structure in combination with the reduction of man-caused risks is proposed.

The practical and theoretical significance of the results obtained is the increase of scientific knowledge in the subject area of innovative development. Results can be used to estimate the introduction of innovations in solving practical tasks of increasing reliability, safety and productivity.

3. Materials and methods

The research is based on the application of the methodology of the systematic approach to the process of development and implementation of innovations. Authors use methods of the system analysis allowing to reveal system of interrelations of the basic concepts and to define the term "innovative development".

There is a great number of definitions of the term "innovation" [1 – 8, etc.]. Denoting "upgrade" or "improvement" the term itself came from the Latin word "innovato". In scientific research it appeared in the XIX century [1]. At the same time, some authors tend to understand innovation as a process which is considered as an invention or an idea acquiring economic content [1,2], while others treat innovation as a result. This implies the understanding of innovation activity and innovative products.

So, Th. Schumpeter identified innovation as a new scientific and organizational combination of production factors motivated by the entrepreneurial spirit [3].

Within the framework of the "Oslo Manual" (the international standard adopted by the OECD in 1992) innovation is defined as the final result of innovative activity, has been implemented as a new or improved product launched on the market, new or improved technological process used in practical activities or a new approach to social services [6].

This definition allows us to consider the innovation of any new benefit having use-value, reflecting the combination of its useful properties. Being distinguished by its inherent combination of useful properties, determines the innovation importance allowing us to judge the value of innovation. The utility should be distinguished from significance.

Values are traditionally classified as material, economic, political, social and spiritual changing the development of society, economy and technology. The importance of innovation can be characterized by appropriate efficiency. It is generally accepted that the availability of economic and technical efficiency determines the production efficiency, the increase of that indicates the development of the enterprise, industry and the country.

Economic efficiency enables to estimate the possibilities of obtaining high profits with the minimum possible cost reflecting the significance of the innovations. Allowing us to estimate the useful-

IOP Conf. Series: Materials Science and Engineering **552** (2019) 012012 doi:10.1088/1757-899X/552/1/012012

ness technical efficiency indicates the ability to produce a certain amount of products or services using a given amount of resources. The concept of innovation is considered to include the process of creation, improvement, dissemination and application of innovations, which including identification of the usefulness and significance of innovations, the implementation of a measures set aimed at innovation commercialization.

Naturally innovation infrastructure is required for innovation activity requires necessary for the organization and arrangement of this activity. The presence of this infrastructure determing, first of all, economic relations and interconnections allows us to talk about the transition to innovative development in technology and other areas that ultimately determine the level and quality of life of people.

The very understanding and defining of the "development" concept can be multisided, and the lack of clear criteria for distinguishing development from other changes is crucial to practice. It is now clear that not every change, even if it is aimed at achieving a certain goal, can be considered as development [9 – 12].

In literature and practice, the combining the noun "development" with various adjectives does not make the definition of this concept clear and certain as it allows to identify differences in order to change the objects and the methods and means chosen for this. Types of development opposed and different from each other towards the growth of the number and its transition to a new quality are so widely known:

- intensive and extensive;
- evolutionary and revolutionary.

Often, the concept of development is used in relation to an individual enterprise and entire sectors of the economy, particular countries and continents in general representing a particular system. By carrying out certain targeted changes, condition of achieving the goals set, the results of these activities are often related to the socio – economic, scientific, technical and other development of systems. A certain interpretation of the term "innovative development" is presently widespread. In economic science and practice, the issues of developing a new theory of economic development taking into account the innovative orientation of technological, managerial, organizational, institutional and other systemic transformations in the economy are being actively discussed [13]. However, there are no generally accepted criteria to distinguish innovative development of the enterprise from other types of development. Therefore, innovative development is widely believed to be the process of introducing effective innovations into production and/or its management systems.

The authors propose to consider that the distinctive feature of innovative development is the approach to environmental management. Innovative development is aimed at keeping environment safe, as a basis to keep human life sustaining activity. The emergence of environmental safety standards stems from an understanding of the value of nature. Therefore, it is possible to estimate the innovative development by reducing land pollution by industrial and household waste, reducing harmful emissions into the atmosphere and other indicators characterizing changes in the ecosystem as human habitat.

Innovative development is proposed to be related to the enterprise in a combination of intensive and revolutionary development of the business entity providing a substantial, multiple increase in its production efficiency due to radical changes in both the technology and technology of production, as well as organization and management of the enterprise. At the same time, environmental measures, reduction of consumption of limited natural resources, reduction of risks of man-made accidents and catastrophes are necessarily taken.

Actually, the innovative development of the enterprise is based on activities aimed at the production of new or improved consumer goods, which have a new quality and useful and significant properties for the consumer.

Due to their usefulness and significance, innovative products have high use value and value. Sometimes this enables the company to have a short-term monopoly on the market and this provides a significant increase in economic efficiency in a market relationships.

The above stated enables to note that the increase in efficiency achieved through the use of scientific achievements and advanced technologies in production and the development of modern innovative management can serve as a characteristic of innovation in improving engineering activity and technology. However, if this leads to an increase in the risks of environmental degradation, then this cannot be considered as an innovative development, and if the task is to achieve it, then it requires determining what innovations can provide this.

The development of science and technology is obvious to provide more opportunities for creating new technology with the desired combination of properties. It can be achieved by applying knowledge about methods and methods of creating technical means for the effective performance of specified functions under certain conditions. To match the technical means one can use the same technical indicators characterizing their properties. The choice is made according to the technical efficiency, reflecting the importance of certain properties, i.e. their value for the specified conditions. And only if there is multiple similar technical efficiency of the options, one should compare their economic efficiency. Therefore, any innovative technical solution, first of all, should have increased technical efficiency. This is a crucial condition for the carrying out innovative activity.

It is important to distinguish between increasing the technical efficiency of performing certain functions and the economic effect derived from this. Naturally, the presence and size of the effect depends on improving the quality, increasing the number and volume of functions performed. However, experience shows that often an innovative solution at the stage of its appearance is inferior in economic efficiency to existing, outdated analogs. This can be explained by the need for the transition of economic entities to a new order of relationships and the abandonment of established relationships. Innovations require a reassessment of established values, leading to certain difficulties in society due to its conservatism. There is a well-known, rule "the best is enemy of good".

One can point to a sufficient condition allowing to rely on success of innovation. As it is known the main problem is to arouse the interest of investors seeking to get revenues from the commercialization of innovations.

But often to achieve this it is necessary to create a market demand for an innovative technical solution, demonstrating and proving the usefulness and significance of innovation. The emergence of demand is a sufficient condition for the commercialization of new technical products, considering it an innovation.

The availability of necessary (utility) and sufficient (significancy) conditions enables individual enterprises, entire countries and industries to change their technological structure. This ensures the growth of their productivity and competitiveness. The innovative infrastructure should create favorable conditions. Often speaking about innovative development, one means the development of this infrastructure, within which there is a quantitative growth of qualitatively new use values, primarily those created in the service sector.

Basing on the notion that innovative technology and technology have a certain combination of properties, determining their technical effectiveness, you can point out three main properties. These properties include - safety, reliability and productivity, figure 1. Their combination enables to obtain an economic effect, but the value and significance of each of these basic properties depends on many factors.

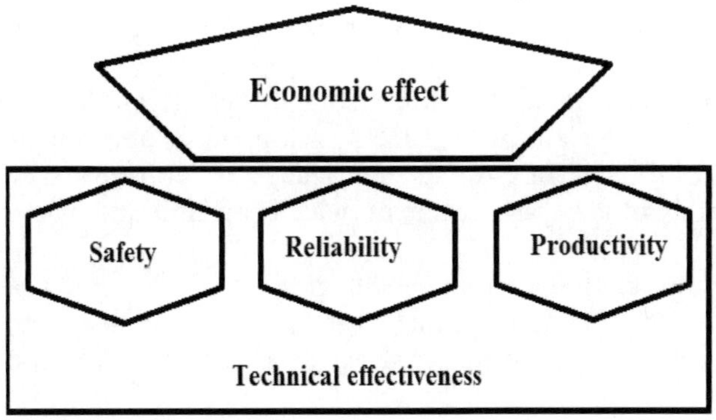

Figure 1. Basic properties of technical efficiency.

Being complex each of these basic properties includes certain properties. Technical security is a property characterizing the ability to prevent internal and external threats when using new equipment and technology and indicates the degree of protection of a person, society, objects and the environment from the threats connected with their introduction into the production process. Maintenance of technical safety means the creation of appropriate technical products and technologies. Providing technical safety means the creation of appropriate technical products and technologies. These include process equipment that provides control and management of emissions of hazardous substances, energies and information. This requires the introduction of modern diagnostic and monitoring tools, various types of automation, designed to prevent and protect against dangerous failures, accidents, and disasters. Therefore, the probability and risks of technical accidents and man-made disasters, as defined in [14], can be classified as complex indicators of technical safety. Innovative products as a result of innovation should ensure the reduction of these indicators.

Technical reliability is a complex property reflecting the ability of an object to be in working condition under certain internal and external influences. The main components of this property are reliability, maintainability and durability. The main indicators are the ratio of readiness and conservation of efficiency, allowing us to estimate the structural reliability of the system, i.e. on the existence of a sufficient number of interrelations between the elements forming the system that are necessary for the reliable performance of its functions. Shaping of this property is quite a difficult task, often its solution affects the economic efficiency, because requires not only redundancy of elements and connections, but also functional, temporal and information redundancy. However, if we seek a solution based on the principle of "improving reliability without reducing productivity", then this provides innovation activity aimed at improving technical efficiency to be considered Pareto optimal.

Technical productivity is a property indicating the ability of the system to perform a certain number of functions to provide services and the maximum amount output, using the specified size and type of resources. The technical performance of the system is determined by the basic production parameters of its elements and links, and is calculated per unit of time in relation to the operating conditions of the object when it is fully loaded.

Technical productivity depends on the influence of internal and external factors, reflecting the nature and conditions of the facility. In essence, the technical performance indicators reflect the production potential of the enterprise. Naturally, it is often used not to the full extent, but only in the demand for products and services provided. Therefore, coefficients of utilization of installed capacity and certain raw materials and other resources, such as electricity are used to characterize it.

Innovations aimed at productivity growth are usually implemented by science and technology achievements aimed at reducing the size and changing the type of resources used by the system, especially limited and non-renewable in nature. Usually productivity growth does not change the range of products and a service provided but affects the economic performance of their production. Therefore, they usually do not require radical changes in the system of economic relations and connections, therefore, the role of innovation infrastructure is not so important for the commercialization of these innovations.

Innovations aimed at increasing safety and reliability often depend not only the replacement of morally and physically obsolete elements and links with more reliable and safe ones, but also with changes in the structure of the system. With the appearance of new or exclusion of existing elements and connections, the quality of the system changes, which is manifested by a change in these properties.

Below, as an example of the proposed methodological approach to the analysis of technical efficiency, one of the options for solving this problem is shown for an electric grid enterprise by using composite materials for manufacturing the basic structural elements of overhead transmission lines.

4. Results

Since the 60s of the 20th century, various composite materials have been developed. Today in the world, as in Russia, these materials are experiencing their second wave. Presently the most significant reasons for their appearance include the development of the material itself (improvement of the properties due to the introduction of various additives) and the emergence of technologies for manufacturing products from composite materials.

Modern manufacturers offer a great number of products for the power grid complex. Mainly, composite materials are introduced in overhead transmission lines (OL). The properties of overhead lines are largely determined by the choice of their structural elements: wires, insulators, traverses, supports and other additional equipment. All the elements are closely related to each other and the low quality of one of them can eliminate all the advantages of the others, which can ultimately lead to a breakdown in the electricity supply to consumers.

There is a great number of composite insulators, wires, poles, insulating traverse of domestic and foreign production on the market. Composite structural elements (CSE) are widely used abroad. Russia is lagging behind. The creation of demand for these products is facilitated by the policies of technical and innovative development of enterprises based on foreign practices and on the observed, including in the field of experimental equipment in Russia, the effects of increasing safety, reliability and productivity.

Insulators were the first to appear in the Russian market of composite materials and presently they are massively used.

The composite wire was also in demand. Due to the increased conductivity and the ability to withstand high operating temperatures and the uniqueness of the core, it enables to double the transmission capacity of the line. The smooth surface of the composite wire with eco-friendly coating, its compact internal structure, the strength of the core and low corrosion resistance ensure an increase in trouble-free operation and long service life of the overhead line. The research of the authors show that the use of a composite wire makes it possible to reduce the frequency of wire failures and the recovery time of OL by 15% and 23%, respectively [15].

From the of innovative development point of view, the reduction of the negative impact on the environment is of greater importance. Minimizing the slack of the wire enables to significantly reduce the elimination of land under the line, and consequently, avoid deforestation when overhead lines go through resort or protected areas. The use of conductive material in the conductor makes it possible to reduce transmission losses, in comparison with aluminum wires of the AC type, by almost 20%, and correspondingly, associated emissions to the atmosphere. Together, this indicates a significant reduction in environmental impact [15, 16].

Composite power line pylons are being most often considered as quick-mounted emergency reserve ones presently in Russia. This enables us to identify the features of design and operation of composite power line pylons and determine the opportunities for their application for the construction and reconstruction of lines of various voltage classes. Such an attitude is due to the simplicity and short installation time, small weight of the support and the convenience of transportation and storage (it is a set of cone-shaped parts) [17 - 22].

However, their technical effectiveness is obvious and caused by the properties of the material itself. Power line pylons can withstand severe weather, they are invulnerable to woodpeckers, termites and, of course, for decay and corrosion, because they are waterproof. Composite power line pylons are also fire-resistant and can become completely incombustible when coated with fire-resistant solutions. The increased elasticity of such supports made it possible to make a more secure installation of the pylons along highways, since when colliding with a composite support, the harm to human health is minimal if compared with a collision with a metal or reinforced concrete pylon [20]. From the environmental point of view, the use of composite pylons is not contaminating during disposal, similar to those caused by creosote coated wooden pylons. Composite pylons are 100% recyclable, non-toxic, contain no styrene and dangerous air pollutants or volatile organic compounds. The possibility of painting the support in any color enables you to fit them into any terrain and make the line almost invisible [18].

As for strength, these pylons are comparable to steel, and their service life is about 70 years [17, 19]. In addition to increasing the durability, it should be noted that the composite pylon has a reliability almost 40% higher compared with the metal and 30% - compared with the reinforced concrete one. And due to the speed of installation, the recovery time of this pylon is only 4 hours instead of 32 for the classical version [18, 22].

Crossbeams made of composite materials have proved to be more attractive economically due to their operational, technological and environmental parameters abroad. Like other CSE, isolating composite crossbeams (ICC) are used in the reconstruction and construction of high-voltage lines of various voltage classes [19].

Reliability and environmental friendliness makes ICC highly effective. No linear insulation has made the reliability of the overhead line already increased by 30%, and the recoverability by 23%.

As for the impact on the environment, several factors can be pointed out. Firstly, the required width of the clearing, under equal operating conditions, when using ICC s is less by 35% (for example, in plantations of Group I forests, parks and orchards, for 110 kV it is 21-31 m compared to 42-51 m for metal crossbeams). For a voltage level of 6-10 kV, the use of ICC solves the problem of mass death of birds as a result of electric shock. Also these crossbeams are impenetrable for insects and woodpeckers. Do not contain harmful chemicals [23 - 25].

The insulator extraction, eliminates the appearance of a high potential near the reinforced concrete pylons. The maximum possible convergence of the overhead lines during the use of ICC makes enables to reduce the intensity of electric and magnetic fields under the overhead lines. These factors are relevant today in terms of security.

5. Evaluation of the technical effectiveness

Peculiarities of the introduction of CSE during the reconstruction and construction of overhead lines are the set of their various combinations and the number of lines to be replaced (from 1 span to the entire length of the overhead line). In this regard and the technical efficiency of the use of the CSE can be different depending on the action being implemented.

Obviously, the reliability of the overhead power lines is determined by the reliability of its

constituent elements. Therefore, a mathematical model of overhead line for analyzing the technical efficiency of CSE has been developed. The model reflects the influence of the main structural elements on the failure-free and recoverability of the line [26]. The OL is modeled as spans having a serial connection. The failure of one span leads to the failure of the line. Therefore, the failure rate is determined by the number of spans, their type and reliability:

$$\omega_l = \sum_{z=1}^{Z} (\omega_{span_z} \cdot r_z) \quad (1)$$

where ω_{span_z} - failure rate of z-type span, Z – the number of spans types in the lines, r_Z - the number of z-type spans in the line.

The failure rate of the span is dependent on the reliability of its structural elements: pole, cross-arms, insulators and wires.

$$\omega_{span} = \omega_p + \gamma(\omega_{is} + \omega_c + \omega_w) \tag{2}$$

where $\omega_p, \omega_{is}, \omega_c, \omega_w$ - the failure rate of pole, insulator, cross-arm and wires, respectively, γ=3 for earthed neutral, γ=3 for OL with low-grounded neutral, γ=2 for OL with isolated neutral.

If the influence of structural elements on the OL failure rate and recovery time is known, then the OL recovery time can be determined:

$$t = \sum_1^4 d_e \cdot \tau_e \tag{3}$$

where d_e and τ_e - the percentage and recovery time of OL due to elements failure.

In most cases, these indicators are known from the statistics of observations over the lines. Using the OL model [26] and information about the properties and features of CSE [15 - 25 and etc.], the influence of CSE on the OL reliability was calculated.

In table 1, a possible change in failure-free and reducibility when introducing composite wires, supports and crossbeams on the 110 kV line is shown. The selection of this or that option, with a certain effect on technical efficiency, is an optimization task, depending on the objectives of the introduction of the CSE.

Table 1. Influence of CSE on 110 kV overhead line reliability.

Option	Classical	ICC	Pylon	Wire	ICC and Pylon	ICC, Wire and Pylon
Failure rate per 100 km, times / 100 km	9.4	6.49	9.01	8.65	6.09	5.34
Recovery time of the line 110 kV, h	16.7	12.70	12.84	15.47	8.87	4.98
Change in failure rate, %	-	31	4,2	8	35.2	43.2
Change in recovery time for a complete reconstruction, %	-	23.8	23	23.3	46.8	70.1

The composite structural elements of overhead lines are obvious to have an increase in all three basic properties: safety, reliability and productivity. The necessary condition for innovation has been fulfilled. The infrastructure of the power industry, the policy being implemented, are aimed at shaping the demand for CSE in the electric grid complex - a sufficient condition for innovation activity. Therefore, there is reason to expect the economic effect. Orientation towards the environmental friendliness of this innovative product, the implementation of which is accompanied by environmental measures, makes us speak about innovative development.

6. Discussion

As a result of the system analysis the following system of concepts is offered: "innovative activity", "innovative infrastructure", and "innovative development". This differs from existing views.

In accordance with the interrelationships in this system, the result of innovation is innovative products having high technical efficiency. The objectives and tasks of innovation are reduced to increase the safety, reliability and productivity of products. Manifestation of these properties is crucial for the comparison of innovations. Technical indicators known as technical properties enable to quantify the efficiency. Only an increase of some indicators without a decrease in others let us consider improve-

IOP Conf. Series: Materials Science and Engineering **552** (2019) 012012 doi:10.1088/1757-899X/552/1/012012

ments. Technical efficiency of new products must at least meet the technical requirements of consumers.

The presence of high technical efficiency can be a necessary condition. The emergence of demand is considered to be a sufficient condition for the commercialization of new technical products. Entities forming an innovative infrastructure should eliminate various barriers preventing commercialization of innovative products. Having the purpose to launch mechanisms of innovative development it is the object of their activities.

The analysis of technical efficiency of using composite materials in the electric power industry has been made. The appearance of composite structural elements of overhead lines increases the technical efficiency of lines and networks.

7. Conclusion

Crucial and sufficient conditions for the implementation of innovative activities have been determined. An explanation of the term innovative development has been proposed. Innovative development is a combination of revolutionary and intensive changes combined with environmental measures, reduction of consumption of limited natural resources, reduction of risks of man-caused accidents, i.e. a set of factors indicating the environmental friendliness of innovations. Analysis of composite structural elements of air lines as innovative products has showed that their use in electrical networks leads to a significant increase in technical efficiency. But the shortcomings of existing techniques of feasibility study prevent this innovative product from being widely used in practice. Therefore, it is required to improve the methodological support for the innovation infrastructure. This can be the scientific task of further research.

References

[1] Twiss B 1989 *Upravlenie nauchno-tekhnicheskimi novovvedeniyami* [Management of scientific and technological innovations] (Moscow: Ekonomika) p 271 [In Russian]

[2] Santo B 1990 *Innovaciya kak sredstvo ehkonomicheskogo razvitiya* [Innovation as a means of economic development] (Moscow: Progress) p 296 [In Russian]

[3] Schumpeter J 1982 *Teoriya ehkonomicheskogo razvitiya* [The theory of economic development] (Moscow: Progress) p 455 [In Russian]

[4] Yakovets Yu 2004 *EHpohal'nye innovacii XXI veka* [The epochal innovations of the XXI century] (Moscow: Izdatel'stvo Ekonomiki) p 444 [In Russian]

[5] Milner B 2003 Koncepciya upravleniya znaniyami v sovremennyh organizaciyah [The concept of knowledge management in modern organizations] *Rossiyskiy zhurnal menedzhmenta* **1** pp 57-76 [In Russian]

[6] Balakina Iu 2016 Teoreticheskie aspekty innovacionnoj deyatel'nosti i puti vnedreniya innovacij v organizacii [Theoretical aspects of innovative activity and the ways of implementation of innovations in organizations] *Upravleniye ekonomicheskimi sistemami: nauchnyy elektronnyy zhurnal* **12** [In Russian]

[7] Balabanov I 2000 *Innovacionnyj menedzhment* [Innovation management] (St Petersburg: Izdatel'skiy Dom Piter) p 208

[8] Anshina V and Dagaeva A 2003 *Innovacionnyj menedzhment* [Innovation management] (Moscow: Biznes) p 528 [In Russian]

[9] Byk F and Kitushin V 2008 Konceptual'naya model' razvitiya i zadachi menedzhmenta [The conceptual model of development and the task of management] *Menedzhment v Rossii i za rubezhom* **6** pp 3-8 [In Russian]

[10] Byk F and Kitushin V 2009 Konceptual'naya model' upravleniya razvitiem [The conceptual model of development management] *Menedzhment v Rossii i za rubezhom* **4** pp 112-8 [In Russian]

[11] Byk F and Kitushin V 2008 Mekhanizmy razvitiya i upravlenie im [Mechanisms of development and management of it] *Menedzhment v Rossii i za rubezhom* **4** pp 3-8 [In

Russian]

[12] Byk F and Kitushin V 2013 Monitoring izmenenij i diagnostika razvitiya proizvodstvennoj organizacii [Monitoring of changes and diagnostics of the development of the production organization] *Menedzhment v Rossii i za rubezhom* **4** pp 92-102 [In Russian]

[13] Kuralenko O 2011 Metodologicheskie voprosy innovacionnogo razvitiya ehkonomicheskih sistem [Methodological issues of economic systems innovative development] *Molodoy Uchenyy* **10** pp 127-30 [In Russian]

[14] Makhutov N 2005 *Konstrukcionnaya prochnost', resursnaya i tekhnogennaya bezopasnost'* [Structural strength, resource and technogenic safety] (Novosibirsk: Nauka) p 610 [In Russian]

[15] Rashmi B, Shivashankar G and Poornima 2017 Overview of different overhead transmission line conductors *Materials to day: Proceedings* **4** pp 11318-24

[16] Nepomnyashchij V 2011 Ocenka ehffektivnosti ispol'zovaniya v ehlektricheskih setyah provodov s povyshennoj propusknoj sposobnost'yu [Evaluation of the effectiveness of use electrical wire networks with increased bandwidth] *Ekspert po energetike* **3** pp 38-44 [In Russian]

[17] Bocharov Yu and Zhuk V 2013 Kompozitnye opory. Perspektivy primeneniya dlya VL 110 – 750 kV [Composite supports. Prospects for the use of 110 – 750 kV overhead lines] *News Elektrotekhnika* **10** [In Russian]

[18] Rodin M 2008 Composite Poles Support the Circuit of the Future *Transmission and Distribution World*

[19] Byk F, Goldobin D and Levin V 2013 Perspektivy primeneniya vozdushnyh linij ehlektroperedachi na kompozitnyh oporah v ehlektrosetevom komplekse Rossii [Prospects for the use of overhead transmission lines on composite poles in the power grid of Russia] *Glavnyy Energetik* **10** pp 52-9 [In Russian]

[20] Dutt V 2006 Composite Utility Poles: Advances in Design, Materials & Manufacturing 2005/2006 IEEE/PES Transmission and Distribution Conf. and Exhibition (IEEE) pp 1243

[21] Donnelly R 2017 The conductivity of composite fibre glass poles *52nd Int. Universities Power Engineering Conf.* (IEEE) pp 1-6

[22] Fan W, Wang G, Liu H, Xie Z, Chen Y, Yang Z and Wang H 2014 Study on 110kV Composite Material Transmission Poles and Towers *Advanced Materials Research* **986-987** pp 967-70

[23] Rowland S, Cotton I, Zachariades S, Allison F, Peesapatti V and Chambers D 2014 *Developing Composite Insulating Cross-Arms for 400 kV Lattice Towers* (INMR) pp 86-90

[24] Zhao X, Fu Z, Zhao M, Wang L and Wu Y 2015 Mechanics Analysis of 110kV Suspension-Type Composite Insulators in Ice Coating and Wind Loads Conditions *Int. Symp. on Material, Energy and Environment Engineering* (IEEE) pp 216-19

[25] Afiqah Nadhirah et al 2017 Properties of Fiberglass Crossarm in Transmission Tower - A Review *International Journal of Applied Engineering Research* **12** pp 15228-33

[26] Byk F and Myshkina L 2016 Comparison and choice of measures to improve the reliability of distribution grid companies *2nd Int. Conf. on industrial engineering, applications and manufacturing* (IEEE) art 3.23

On the forecasting of trends of specific mixed loads of electric consumers of microdistricts and cities change in the conditions of inadequate data

E G Gasho[1], S V Guzhov[2] and A A Krolin[3]

[1]Department of Industrial Heat and Power Systems, National Research University MPEI, Krasnokazarmennaya St., 14, Moscow, 111250, Russia
[2]Department of Heat and Mass Transfer Processes and Installations, National Research University MPEI, Krasnokazarmennaya St., 14, Moscow, 111250, Russia
[3]Energy Management Department, National Research University MPEI, Krasnokazarmennaya St., 14, Moscow, 111250, Russia

E-mail: GuzhovSV@yandex.ru

Abstract. The paper gives a brief analysis of modern methods of analysis, confirmation and prediction of specific mixed loads of electric consumers in housing estates and cities and formulates the main problems afflicting the reorganization of the industry, taking into account the requirements of digitalization of the economy. In order to identify the advantages and disadvantages of various solutions for optimizing the compilation of energy balances, a mathematical model is constructed, which allows analyzing the needs of microdistricts and cities in electrical energy and realizes the forecast of the development of urban energy systems in different socio-economic and climatic conditions.

Keywords: forecasting, approksimation, mixed electric loads, predicting FEB, quality life of the population, autoregressive model, correlation, microdistrict.

1. Introduction

Planning of operational parameters and technical and economic indicators is one of the most important tasks of the industrial and municipal complexes of the Russian Federation. One of the main indicators in planning the operation of the power system is the forecast of the expected demand for secondary energy resources in the whole system, groups and individual consumers, hubs of the power system. The need for accurate forecasting is due to technological and economic reasons. Accurate calculation provides reliability and reserve system capacity, which should be optimal from technical point of view. In the economic aspect, a correct forecast allows consumers to better plan the purchase of energy resources, taking into account changes in the tariff policy, climatic and meteorological factors, technological, social and other features.

Reducing non-production costs and specific energy consumption by 40% by 2020, as well as increasing energy efficiency by industrial and municipal systems are among the main tasks set for the energy sector and fixed in Presidential Decree of June 4, 2008 N 889 "On some measures to improve the energy and environmental efficiency of the Russian economy" and the requirements for sectoral and regional energy and resource saving programs.

IOP Conf. Series: Materials Science and Engineering **552** (2019) 012013 doi:10.1088/1757-899X/552/1/012013

The achievement of the task set by industrial and municipal systems is impossible without digitizing the tools for analyzing and planning consumption of fuel and energy resources by objects of municipal and industrial spheres of various scales, which corresponds to the "Digital Economy Development Program in the Russian Federation till 2035".

Conversion of fuel and energy balances to a digital format will allow improving the quality and reliability of energy declarations introduced instead of mandatory energy audit in accordance with the Federal Law No. 261-ФЗ "On energy saving and energy efficiency increase" dated December 23, 2009. At the same time, a significant national economic effect is to reduce the volume and cost of work both on filling energy declarations with data on fuel and energy resources in physical terms, and on aggregating their economic equivalents when making a forecast of demand for energy resources at the level of consumers and their associations, as well as for the entire power system.

2. The status of the task of forecasting the loads of mixed electricity consumers in Russia

The solution of the actual for the power industry problem of compiling forecast balances of fuel and energy resources is based on the use of complete and reliable information about the processes in equipment and the energy system as a whole. In view of the well-known specifics of the power systems, the main method of obtaining such information is the mathematical modeling [1]. Dominant today, purely numerical simulation and various software and computing systems for its implementation do not always provide the necessary completeness and reliability of the information indicated [2]. In this regard, inevitably there is a need to apply very significant simplifications and limitations, which leads to a decrease in the accuracy of the forecast [3].

As a result of insufficient theoretical development of mathematical models for forecasting fuel and energy balances (FEB) for heat supply systems of industrial and municipal facilities, there is a contradiction between the need for high-precision mathematical models and the results of calculating and predicting FEB on one hand, and the existing level of accumulation and analysis of statistical data on the another hand.

The aim of the work is to develop methods, algorithms and software tools for analyzing and planning the balances of energy resources consumed by industrial and municipal systems under conditions of insufficient data.

Achievement of the goal has been provided by consistently solving the following groups of tasks:

1. Preparation of a brief overview of the methods of energy consumption planning recommended by the current legislative and regulatory framework in order to ensure a balanced industrial and regional development of the country, as well as a description of the methodology and sufficiency of the application of specialized geo-information systems and engineering software compiling forecast balances of fuel and energy resources.

2. Assessment of the limits of applicability of the most used deterministic and stochastic information methods for planning balances of fuel and energy resources consumed by industrial and communal energy systems and forming requirements for the completeness of the array of source data sufficient to compile the predicted function of energy consumption by an object.

3. Development of principles, models and methods for improving the procedure for formulating forecast balances of fuel and energy resources for groups and individual industrial and communal systems, both using statistical methods and using artificial neural networks.

4. Testing of methods for determining the contribution to improving the accuracy of prediction by climatic and meteorological factors, technological, social and other features of the analyzed systems with the subsequent creation of a balanced set of regulatory, environmental and economic measures to stimulate the implementation of the identified reserves for improving energy efficiency.

The scientific novelty of the research resides in the development of a methodology consisting of methods and approaches, the use of which through the digitization of energy balances provides a solution to a major applied task of improving the energy efficiency of industrial and municipal heat and power systems in conjunction with industrial and regional economic development.

IOP Conf. Series: Materials Science and Engineering **552** (2019) 012013 doi:10.1088/1757-899X/552/1/012013

The practical value of conducting industrial experiments and energy audits of a number of enterprises of energy-intensive industries is to accomplish the state goal of a 40% reduction in the energy intensity of GDP by 2020 compared with 2007 and to determine priorities for using approaches to digitization of the energy sector and reducing the cost of energy balances by automating the procedure for forecasting the volume of consumption of fuel and energy resources of industrial and municipal consumers in the conditions of insufficient data.

3. General formulation of the problem and solution methodology

In the period from 2008 to 2016 12 regulatory documents were issued on the items of calculating and forecasting energy-saving effect, incl. taking into account comparable conditions (figure 1). Let us consider in more detail some provisions relating to the issues of forecasting the fuel and energy resources consumption.

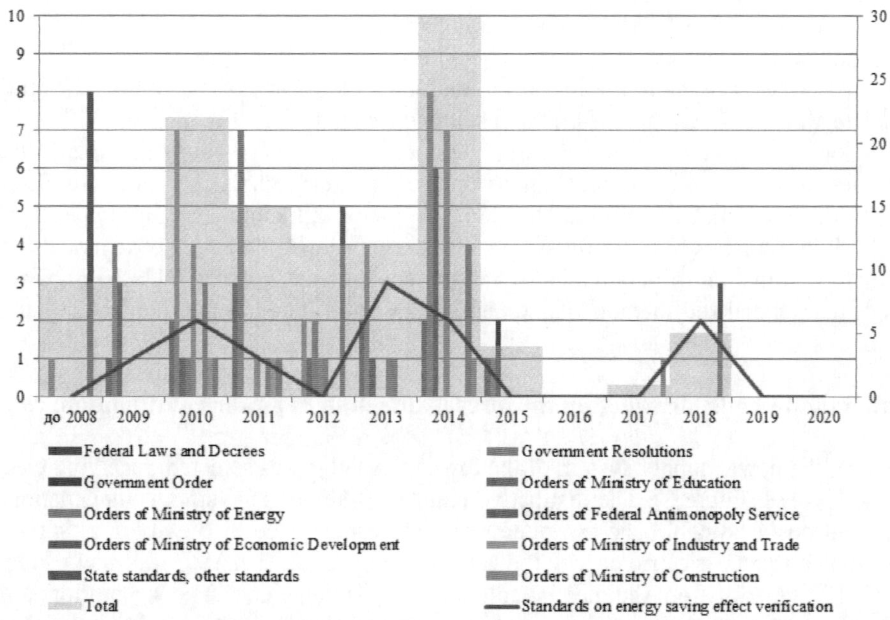

Figure 1. The intensity of the publication of the regulatory framework for energy saving and energy efficiency [4]

In order to improve the performance of the Federal Law # 261 [5], in 2010 a method was introduced for calculating the values of target indicators in the field of energy saving and energy efficiency, including in comparable conditions [6]. This document defines the procedure for calculating the target indicators of regional and municipal programs in the field of energy saving and energy efficiency. Target indicators are determined by calculating the shares and ratios of indicators of the estimated year and the year preceding the year when the implementation of regional and municipal programs began. The mathematical apparatus proposed in the document is primarily focused on analytics [7] of the activities included into regional and municipal programs of the subjects of the Russian Federation, and is not intended for use in the preparation of energy consumption forecasts of the analyzed regions. At the same time, the inaccuracy of the data, their partial unreliability leads to the need to use mathematical tools suitable for prediction in conditions of insufficient data [8].

The method of direct counting [9] used in a number of documents in combination with the method of direct extrapolation, which consists in predicting through the transfer of events and states of the past to the future, makes it possible to determine the need for energy resources. To use the method, the stability of the process of energy consumption of the object and the absence of a sharp influence of qualitative and quantitative changes in power/weight ratio are necessary. In determining the maximum load [10], this method uses a technique based on the calculation of specific values as a function of the number of power receivers: apartment, workshop, factory, city, region. In general, the technique is determined by the expression:

$$P_{flat\ sp.} = a + \frac{b}{\sqrt{n}}, \quad (1)$$

where n is the number of apartments in the object;
a, b - coefficients corresponding to housing with one of V following types of electrification:
I. with natural gas stoves;
II. with LPG / solid fuel stoves;
III. with electric stoves up to 8.5 kW;
IV. with electric stoves up to 10.5 kW; with luxury apartments;
V. with household air conditioners.

This approach, when moving to the level of a microdistrict and above, is not very useful due to the almost complete absence of reliable information suitable for analysis.

Forecasting methods based on statistical processing of time series [11] carry out forecasting of loads and modes of energy system objects interconnected in a hierarchical tree structure. StatPlus [12], Statgraphics [13], Exel [14], StatSoft Statistica [15], Deductor academic [16], Gretl [17], JMulTi [18], EViews [19], SPSS Statistics [20] - software packages regression and econometric models of time series, realizing the estimation of parameters by the method of least squares [21], the maximum likelihood method, the generalized method of moments [22], the autoregressive model [23], the moving average, etc.

4. Materials and Methods: drawing up the forecast function of power consumption, city of Anadyr

For the analysis, the power supply system of the town of Anadyr was selected, receiving electrical energy from two sources (figure 2). The distinctive feature of the power system is its isolation from the country's unified power system. The not quite exact correspondence of the volumes supplied to the network and the consumed electric energy indicates the absence of the ASTUE system. According to the data for 2017, the error from the manual collection of data on electrical consumption is 4.1%. We would apply the mathematic apparatus for analyzing statistical data in the conditions of their insufficiency and partial unreliability for further analysis.

An analysis of the total electricity consumption of the town of Anadyr including utilities, street lighting, road infrastructure, enterprises [24], etc. was carried out on the basis of hourly data for 2017. Regression analysis was carried out on the basis of determining the correlation for 7 factors:

1. T - air temperature at a height of 2 meters above the ground, ºC;
2. P0 - atmospheric pressure at the station level, mm Hg;
3. U - relative humidity at a height of 2 meters above the surface of the earth, %;
4. DD - the angle of the wind direction at a height of 10-12 meters above the earth's surface, averaged over a 10-minute period immediately preceding the observation period, m / s;
5. FF - wind speed at a height of 10-12 meters above the earth's surface, averaged over a 10-minute period immediately preceding the time of observation, m / s;
6. N - total cloudiness, %;
7. P - duration of daylight hours, min.

The use of regression analysis with $R^2 = 0.8448$ allowed to describe the following dependence of the power consumption of the city:

$$W = 2923.2\text{-}53.1 \cdot T + 3.4 \cdot P0 + 2.6 \cdot U\text{-}0.4 \cdot DD + 21.8 \cdot FF + 267.9 \cdot N + 1.4 \cdot P \quad (2)$$

Using a polynomial definition with $R^2 = 0.8669$ allowed to reveal a similar relationship (figure 3):

$$W = -5E\text{-}06x4 + 0.0041x3 - 0.9798x2 + 62.434x + 5321.7 \quad (3)$$

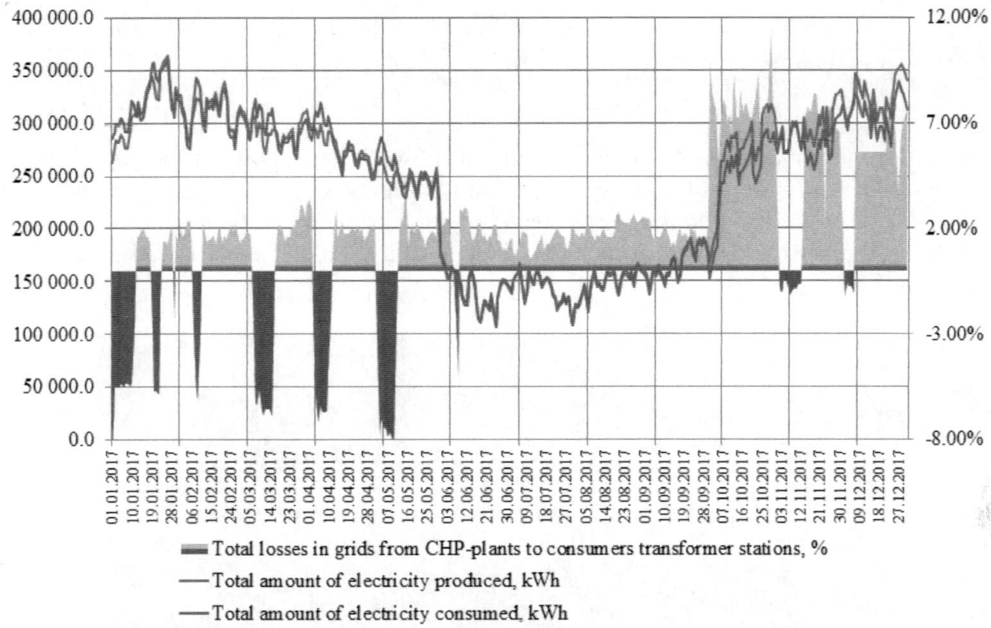

Figure 2. The production and consumption of electricity in town of Anadyr

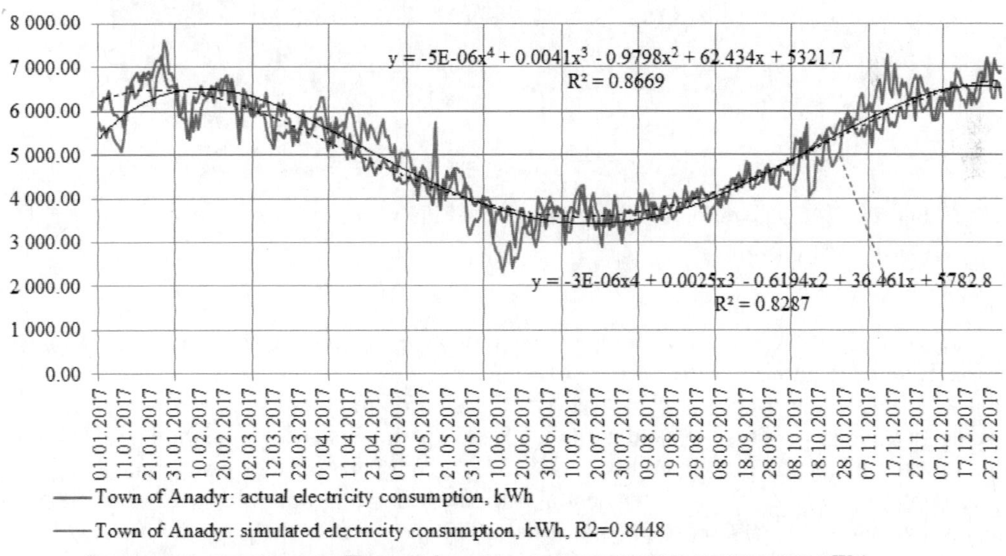

Figure 3. Analysis of electricity consumption in Anadyr

Despite the rather high correlation coefficient of the correspondence of the polynomial to the actual values, the use of such an approach to predict power consumption is impossible, due to the lack of consideration of a number of influencing factors [25].

IOP Conf. Series: Materials Science and Engineering **552** (2019) 012013 doi:10.1088/1757-899X/552/1/012013

The significance of the impact of climatic factors has been analyzed based on the data received (table 1).

Table 1. The significance of the impact of climatic factors on the power consumption of the town of Anadyr

T, °C	P_0, mm Hg	U, %	DD	FF, m/c	N, %	P, min
15.1%	1.0%	0.8%	0.1%	6.2%	76.3%	0.4%

Obviously, the largest dependence in 76.3% with the power consumption of the city is shown by the factor N - total cloudiness. At the same time, an additional analysis of the correlation between the power consumption of the city and the wind direction with a breakdown of the wind rose with a frequency of 22.5° (figure 4), taking into account the directional pattern of the city's roads, did not reveal such a connection.

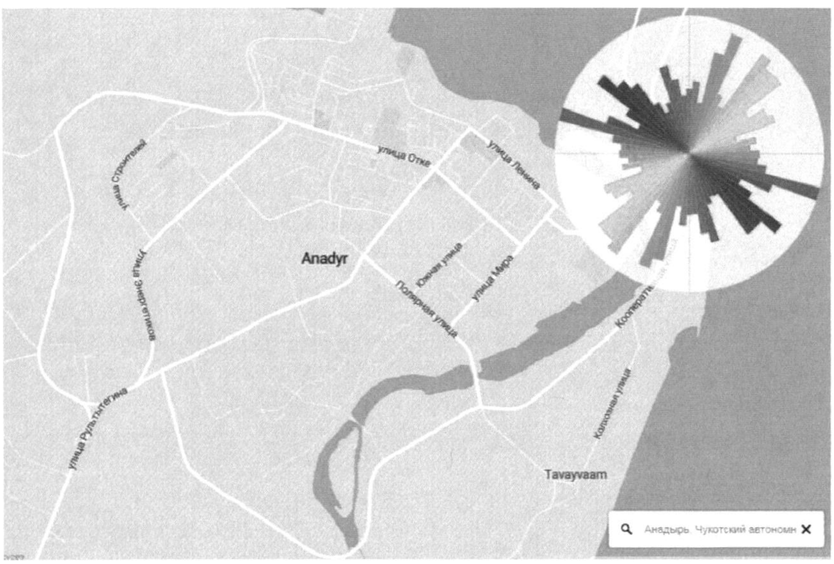

Figure 4. Visualization of the orientation of the streets of Anadyr on an interactive map [26]

5. Results: Analysis of trends in the change of specific loads of electric consumers in Moscow microdistrict and the administrative center of Anadyr

The total area of housing in the town of Anadyr [27] is 335.8 thousand square meters. The housing stock of the city has 157 houses [28] and a total of 5,405 apartments. The population is 15604 people. The total power consumption of the city is 1 849 332.41 kWh per year. For comparison, we took the electricity consumption data from 12 residential buildings in the Lyublino district of Moscow. The total area of housing is 111 394.00 sq.m., and it has 1935 apartments. The population is 5347 people. The total electricity consumption is 387 339.59 kWh per year, taking into account the fact that this figure does not take into account other electric consumers in the area other than electricity consumption by residential buildings. The results of the calculation of various specific values of power consumption are presented in table 2.

Table 2. The calculated specific values of the consumed electrical energy of the town of Anadyr and the microdistrict of Lyublino of Moscow city

Specific power consumption	Wh / 1 м^2	Wh / 1 pers.	Wh / 1 flat	Wh /(1pers. 1 м^2)
Anadyr	0,628	345,86	342,15	0,35293
Moscow	0,396	291,96	806,77	0,06589

Such a discrepancy of values can be explained by the working hypothesis that Moscow has now achieved a higher level of energy intensity of residential buildings due to a higher standard of living [29]. At the same time, a higher standard of living leads to the purchase of appliances that have lower power consumption due to a higher energy efficiency class, which somewhat reduces the specific power consumption.

From the point of view of the current regulatory documentation [30] on the design of newly constructed and reconstructed electrical networks of cities, districts and microdistricts, there are approved specific values of energy consumption (table 3), similar to those calculated.

Table 3. Normative specific values of the consumed electric energy of the town of Anadyr and the microdistrict of Lyublino, city of Moscow

		Anadyr	microdistrict of Moscow
кW / 1 м^2	Actual specific value, including all loads, кW / 1 м^2	0,628	0,3969
	Specific value according to RD 34.20.185-94, кW / 1 м^2	0,093	0,109
кW / 1 pers.	Actual specific value, кW/pers.	0,34586	0,29196
	Specific value according to RD 34.20.185-94, кW/pers.	0,528 (allowable range 0,396 – 0,528)	~0,27 (allowable range 0,24-0,32)
кW / 1 flat	Actual specific value, кW/ flat	0,34215	0,80677
	Specific value according to RD 34.20.185-94, кW/flat	0,620125	0,620125

These figures take into account the following electric loads: dwellings, public buildings (administrative, educational, scientific, medical, commercial, entertainment, sports), communal utilities, outdoor lighting, electric transport (non-metro), water supply and sewerage systems, heat supply systems. The central areas of the city include established areas with a significant concentration of various administrative institutions, educational, scientific, design organizations, trade, public catering, entertainment enterprises, etc. Due to incomparable conditions, the specific electricity consumption per 1 m^2 is not analyzed, because in the data of the town of Anadyr, all the loads of the town have been taken into account, while the normative specific indicators show values that take into account only the load of residential buildings.

The forecast of trends in the change in specific loads of electricity consumers in microdistricts and towns under conditions of insufficient data shows that, according to the parameter "specific electricity consumption per person" (figure 5), the housing stock of the city of Anadyr has a growth potential of 52.6%. At the same time, the Moscow microdistrict demonstrates the excess of specific power consumption by 8.1%.

According to the parameter "specific electricity consumption per apartment" (figure 6), the housing stock of the city of Anadyr has a growth potential of 81.5%. The difference between the levels of electricity consumption for an apartment for the administrative center of Anadyr and the microdistrict of Moscow, which are similar to housing standards, is 136%. At the same time, the Moscow microdistrict demonstrates the excess of specific power consumption by 29.9%.

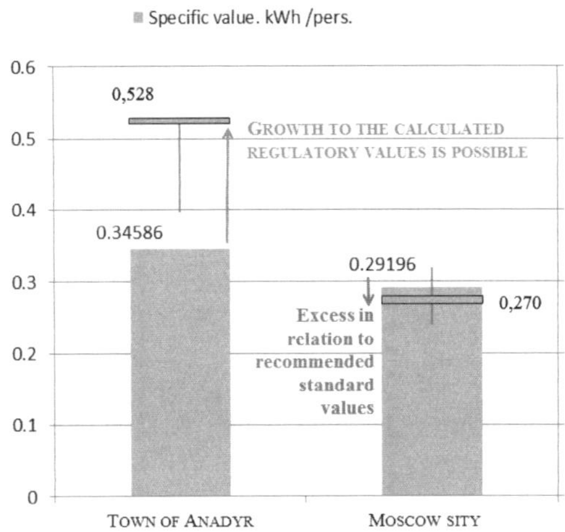

Figure 5. The results of the analysis of specific energy consumption per person.

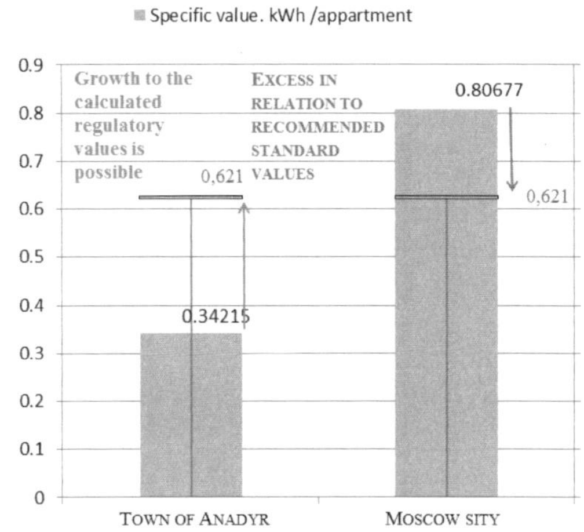

Figure 6. The results of the analysis of specific energy consumption per apartment

6. Conclusions:

1. The approach of using regression models is suitable to solve the problems of forecasting the volume of electricity consumption of mixed loads of housing estates and cities in the conditions of insufficient data accuracy.

2. The assessment of the impact of the quality of life of the population on the level of electricity consumption by housing is given. The difference in the actual specific power consumption for Anadyr and Moscow is 136%.

3. The discrepancy between the values of specific amounts of electrical consumption in RD 34.20.185-94 "Instructions for the design of urban electrical networks" and the actual levels of specific power consumption is shown. The above method can be used to update obsolete specific values.

4. The proposed approach can be used to predict energy demand, taking into account the aging factors of buildings and networks, increasing comfort and loads, renovation, urbanization of suburbs, etc.

Acknowledgments
The study was supported by the grant of the Russian Science Foundation No. 16-19-10568.

References

[1] Bagbleyter O I 2006 *Razrabotka matematicheskikh modeley i metodov dlya otsenki vliyaniya uchastnikov sistemy elektrosnabzheniya na kachestvo elektro-energii* [Development of mathematical models and methods for assessing the impact of participants in the power supply system on the quality of electricity] (Irkutsk: Irkutsk State Technical University) [In Russian]

[2] Khomenko T V 2016 *Metody i modeli prinyatiya resheniy na etape poiskovogo proyektirovaniya tekhnicheskikh sistem* [Methods and models of decision-making at the stage of search design of technical systems] (Astrakhan: Astrakhan. gos. tekhn. un-t) [In Russian]

[3] Gusynina Y S 2010 *Modelirovaniye upravleniya proizvodstvennymi sis-temami s uchetom sluchaynosti sostoyaniya formiruyushchikh parametrov* [Modeling the management of production systems, taking into account the randomness of the state of the forming parameters] (Penza: Penz. gos. un-t) [In Russian]

[4] 2018 Normativno-pravovaya baza 2015 Federal'nyye zakony. Prikazy [Legal and regulatory framework 2015 Federal laws. Orders] URL: http://energoeducation.ru/normativno-

pravovaya-baza.php [In Russian]

[5] 2018 Federal'nyy Zakon ot 23 noyabrya 2009 g. N 261-FZ Ob energosberezhenii i o povyshenii energeticheskoy effektivnosti i o vnesenii izmeneniy v otdel'nyye zakonodatel'nyye akty Rossiyskoy Federatsii [Federal Law of November 23, 2009 N 261-ФЗ On Energy Saving and Improving Energy Efficiency and External Changes in Certain Legislative Acts of the Russian Federation] URL: http://www.consultant.ru/document/cons_doc_LAW_93978/ [In Russian]

[6] 2018 Metodika № 273 2010 Raschot znacheniy tselevykh pokazateley v oblasti energosberezheniya i povysheniya energeticheskoy effektivnosti, v tom chisle v sopostavimykh usloviyakh [Method № 273 2010 Calculated the values of indicators in the field of energy saving and energy efficiency, including in appropriate conditions] URL: http://www.garant.ru/products/ipo/prime/doc/12077409/ [In Russian]

[7] 2018 Postanovleniye Pravitel'stva RF ot 01 iyunya 2010 g. № 391 2010 O poryadke sozdaniya gosudarstvennoy informatsionnoy sistemy v oblasti energosbe-rezheniya i povysheniya energeticheskoy effektivnosti i usloviy dlya yeye funktsionirovaniya [Decree of the Government of the Russian Federation of june 01, 2010 N 391 2010 on the procedure for creating a state information system in the field of energy saving and energy efficiency Improvement and conditions for Its operation] URL: http://base.garant.ru/12176205/ [In Russian]

[8] Gmurman V Y 2013 *Teoriya veroyatnostey i matematicheskaya statistika* [Theory of Probability and Mathematical Statistics] (Moscow: Izdatel'stvo Yurayt) [In Russian]

[9] Shagidullin A V 2016 *Planirovaniye elektropotrebleniya na promyshlen-nykh predpriyatiyakh s uchetom poter' elektroenergii, opredelyayemykh po ekviva-lentnomu soprotivleniyu tsekhovykh setey* [Planning of electricity consumption at industrial enterprises, taking into account the losses of electricity, determined by the equivalent resistance of shop networks] (Cheboksary: Chuvash. gos. un-t im. I.N. Ul'yanova) [In Russian]

[10] Solomakho K S 2015 *Primeneniye metoda glavnykh komponent dlya progno-zirovaniya ob"yemov elektropotrebleniya energosbytovogo predpriyatiya* [The use of the principal component method for predicting the volume of electricity consumption of a power supply company] (Chelyabinsk: Yuzhno-Ural'skiy gosudarstvennyy universitet) [In Russian]

[11] 2018 GOST R 56743 2015 Izmereniye i verifikatsiya energeticheskoy effektivnosti. Obshchiye polozheniya po opredeleniyu ekonomii energeticheskikh resursov [Measurement and verification of energy efficiency. General provisions on energy resources of energy resources]. URL: http://www.energosovet.ru/news.php?zag=1449568516 [In Russian]

[12] 2018 StatPlus dlya Windows StatPlus-6 URL: http://www.analystsoft.com/ru/products/statplus/

[13] 2018 Statgraphics URL: http://www.statgraphics.com/

[14] 2018 Ofitsial'nyy sayt Microsoft Office URL: https://products.office.com/ru-RU/

[15] 2018 StatSoft [Elektronnyy resurs]. URL: http://statsoft.ru/

[16] 2018 BaseGroup Labs URL: https://basegroup.ru/deductor/description

[17] 2018 Gretl URL: https://ru.wikipedia.org/wiki/Gretl

[18] 2018 JMulTi URL: https://en.wikipedia.org/wiki/JMulTi

[19] 2018 Tsentr statisticheskogo analiza [Center for statistical analysis] URL: http://www.statmethods.ru/trainings/eviews.html [In Russian]

[20] 2018 SPSS Statistics dlya Windows 2018 [SPSS Statistics for Windows 2018] URL: http://spss-soft.ru/ [In Russian]

[21] Ded A V and Andreyeva Y G 2017 Resheniye zadachi dlya opredeleniya zavisimo-stey poter' moshchnosti v nesimmetrichnykh rezhimakh metodom naimen'shikh kvadratov [Solution of the problem for determining the dependences of power losses in asymmetric modes by the least squares method] *Omskiy nauchnyy vestnik* **2** pp 66-9 [In Russian]

[22] Kryanev A V, Lukin G V and Udumyan D K 2012 *Metricheskiy analiz i ob-rabotka dannykh* [Metric analysis and data processing] (Moscow: Fizmatlit)

IOP Conf. Series: Materials Science and Engineering **552** (2019) 012013 doi:10.1088/1757-899X/552/1/012013

[23] Kalitkin N N 2008 *Approksimatsiya: chislennyye metody* [Approximation: Numerical Methods] (Moscow: Informatsionnyye sistemy v ekonomike) [In Russian]

[24] Lisiyenko V G and Shchelokov Y M 2011 *Energeticheskiy analiz khozyaystvennoy deyatel'nosti* [Energy analysis of economic activity] (Yekaterinburg) [In Russian]

[25] Valeyev S G 2003 *Sistema poiska optimal'nykh regressiy* [Search system for optimal regressions] (Kazan) [In Russian]

[26] 2018 Vizualizatsiya oriyentatsii ulits na interaktivnoy karte [An Interactive Map Of Road Orientations] URL: https://blog.mapbox.com/visualizing-street-orientations-on-an-interactive-map-1eefa6002afc [In Russian]

[27] 2018 Postanovleniye Administratsii Gorodskogo Okruga Anadyr' ot 28 dekabrya 2015 goda N 716 Ob utverzhdenii munitsipal'noy programmy gorodskogo okruga Anadyr' [Resolution of the Administration of the City District of Anadyr dated December 28, 2015 N 716 On approval of the municipal program of the city district of Anadyr] URL: http://docs.cntd.ru/document/440587098 [In Russian]

[28] 2018 Statistika po godam postroyki domov v Anadyre [Statistics on the years of construction of houses in Anadyr] URL: http://dom.mingkh.ru/chukotskiy-ao/anadyr/year-stats [In Russian]

[29] Rodygina S V 2017 *Modeli prognozirovaniya elektricheskoy nagruzki energoyemkikh predpriyatiy, electrical load forecasting models for industrial facilities with high power consumption* [Models for predicting the electrical load of energy-intensive enterprises] (Moscow: Energobezopasnost' i energosberezheniye) **2** pp 22-6 [In Russian]

[30] Lordkilanidze V D, Akimkin A F, Antipov K M and Fajbisovich D L 1995 *Instruktsiya po proyektirovaniyu gorodskikh elektricheskikh setey RD 34.20.185* [Instructions for the design of urban electrical networks RD 34.20.185] (Moscow: E`nergoatomizdat) [In Russian]

Three-phase current sensor for motor protection against turn-to-turn and phase-to-phase short circuits

I K Ishmukhamedov

Department of Electrical Engineering and Electrical Facilities of Enterprises, Ufa State Petroleum Technological University, Kosmonavtov St., 1, Ufa, 450062, Russia

E-mail: islamius2008@yandex.ru

Abstract. The main causes of electric motor failures are transients at power supply interruption and short circuit in stator windings. Most of the failures are due to short circuits. Phase-to-phase short circuits refer to the most severe damage of electrical installations, which are also dangerous for other intact electric consumers. Therefore, electrical plants require a high-speed protection against phase-to-phase short circuits in the stator windings and connections with the commutator. The most difficult to detect are turn-to-turn (or inter-turn) faults in stator windings, especially when a small number of winding turns is shorted. Currently, there is no standard way to detect turn-to-turn short circuits in stator windings. It is difficult to detect turn-to-turn fault in magnitude of phase currents, so it is complicated to provide sufficient protection sensitivity. Thus, the development of new protection devices for detect turn-to-turn short circuit is considered relevant. The article presents a three-phase current sensor based on measuring the magnetic field from the three phases simultaneously. This method will identify turn-to-turn and phase-to-phase short circuits by magnetic-field pattern. The presented method increases the protection sensitivity and speed of protection response. The effectiveness of the proposed method is demonstrated by mathematical modeling.

Keywords: electric motor, turn-to-turn short circuit, phase-to-phase short circuit, short-circuit protection, three-phase current sensor, magnetic field, magnetosensitive element, magnetic-field pattern, finite element method, simulation.

1. Introduction

Electric motors are the main consumers of electrical energy in industrial enterprises. Therefore, failure of the electric motor can cause heavy damage. Early detection of abnormalities during motor operation would eliminate subsequent damage and reduce repair cost and motor outage time [1].

Stator faults are about 30-40% of failures of motors [2]. It is necessary to highlight turn-to-turn and phase-to-phase short circuits among other failure causes.

A turn-to-turn fault is a short circuit between two turns of same phase [3]. Turn-to-turn faults are usually the result of the development of a frame fault, local insulation overheating and defects of the active stator steel or the result of the breakdown of the stator phase in the case. They are a dangerous type of damage and are characterized by the flow of high currents in shorted turns with a slight change of the current in the undamaged part of the winding.

In contrast to phase-to-phase short circuits, turn-to-turn short circuits occur only in the motor frame. A phase-to-phase fault is a short circuit between any two phases [4]. They are accompanied by high currents, several times higher than the rated stator current, and voltage reduction in the supply main. In order to prevent failures, it is necessary to use motor protection.

Differential protection is used to protect a motor against phase-to-phase short circuits [5]. Differential protection provides high sensitivity to short circuits on the terminals and in the windings of an electrical installation, but cannot detect turn-to-turn short circuits [6].

Standard motor protection is generally not designed to detect a turn-to-turn short-circuit. The operating values of such devices are usually limited by necessity of offset from the phase current unbalance, caused by other factors of line voltage unbalance, stator winding asymmetry, etc.

Many different methods have been developed for detecting turn-to-turn short-circuits. Motor Current Signature Analysis (MCSA) is the most preferred method [7] for diagnosing failures in both induction [8] and synchronous motors [9].

In [10], described the use of the extended Park's vector approach (EPVA) for diagnosing the occurrence of stator winding faults in operating three-phase motors. A comparative study a new method, called Park–Hilbert (P–H), with the MCSA and EPVA method is performed in [11].

In [12], a method is described, based vibration signature analysis.

Modeling, study and determining the parameters of motors with turn-to-turn insulation fault is first step in the development of fault diagnosis and fault tolerant machine design [13].

Finite element method (FEM) is used for parameter identification of the motor under fault condition in [14-16].

However, there is still no single method for detection of turn-to-turn short circuits.

Accordingly, the development of new principles and protection devices that detect turn-to-turn faults is considered relevant.

The key problem in the development of turn-to-turn fault protection is the difficulty to detect these short circuits in magnitude of phase currents in the traditional phase-by-phase measurement.

As a result of the research, the three-phase current sensor was developed for control and protection system of an electric motor, based not on phase-by-phase current measurement, but on measuring the magnetic field from three phases simultaneously.

The purpose of this article is to present the developed three-phase current sensor for motor protection against turn-to-turn and phase-to-phase short circuits and its effectiveness. This is achieved by modeling of the magnetic field in ELCUT software using calculation of phase currents in MATLAB/Simulink software.

2. Materials and Methods

The patent has been obtained for the developed three-phase current sensor [17]. The description of this device is given below.

The sensor case is made of ferromagnetic material to reduce the influence of external magnetic fields on the sensor readings. In the center of the case at the same distance from the grooves for the power cables of the phases there is a sensor device for measuring the magnetic induction vector. Magnetoresistors KMZ 10A and KMZ 10C are used as a sensor device.

According to electromagnetic field theory, a rotating magnetic field with a rotation frequency equal to a frequency of the supply voltage will be created in the center of the device, where the magnetoresistors are located orthogonal.

Magnetoresistors of the current sensor perceive the magnetic component of the three-phase system that rotates in space. The "rotating in space" vector of the generalized current from three phases is proportional to the magnetic component.

The microprocessor presents output in the form of a fixed vector in the coordinate system that rotates with the frequency of three-phase currents [18]. The generalized current will be equal to the value of the phase current increased by one and a half times for a symmetric system of three-phase currents [19].

The change of magnetic-field pattern, which the current sensor registers during one complete period, is a consequence of the emergency mode of the monitored electric network.

Figure 1 shows the three-phase current sensor for motor protection against turn-to-turn and phase-to-phase short circuits.

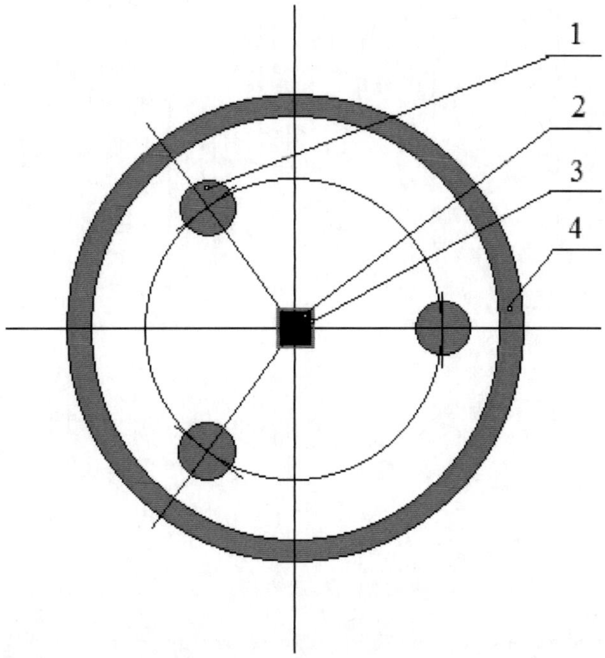

Figure 1. Three-phase current sensor for motor protection against turn-to-turn and phase-to-phase short circuits: 1 – power cable of the phase; 2 – sensor device; 3 – microprocessor; 4 – ferromagnetic case.

Due to the use of this device, there is no need to install a current transformer for each phase, but only one as a magnetic field sensor, which measures the total magnetic field of the three phases.

The speed of protection response will be provided by reducing current unbalance that usually caused by non-identical characteristics of current transformers.

The device operation principle is described below.

The sensor device measures the magnetic field that produced by conductors. The magnetic-field pattern will identify the type of emergency. The magnetic field at the center of the device will be rotating field in balanced conditions, unlike emergency conditions.

3. Results

The mathematical model of calculation of phase currents for a three-phase device (an induction motor) in MATLAB/Simulink software, taking into account the presence of turn-to-turn short circuits in phase A (figure 2) is developed. The basis is the method of determining the parameters of the motor and phase currents described in [20-21].

Due to turn-to-turn short circuits, an unbalance of the phase resistances and stator current unbalance occur. In shorted turns, a high current appears. In this, the main magnetic field in the motor remains close to circular.

The analysis of currents for phase-to-phase short circuit is performed for a solid short circuit. The currents of faulty phases will be equal, increased in amplitude and opposite in phase in a two-phase circuit, and the current of fault-free phase will be zero. Phase currents will be equal and increased in amplitude in a three-phase circuit [22].

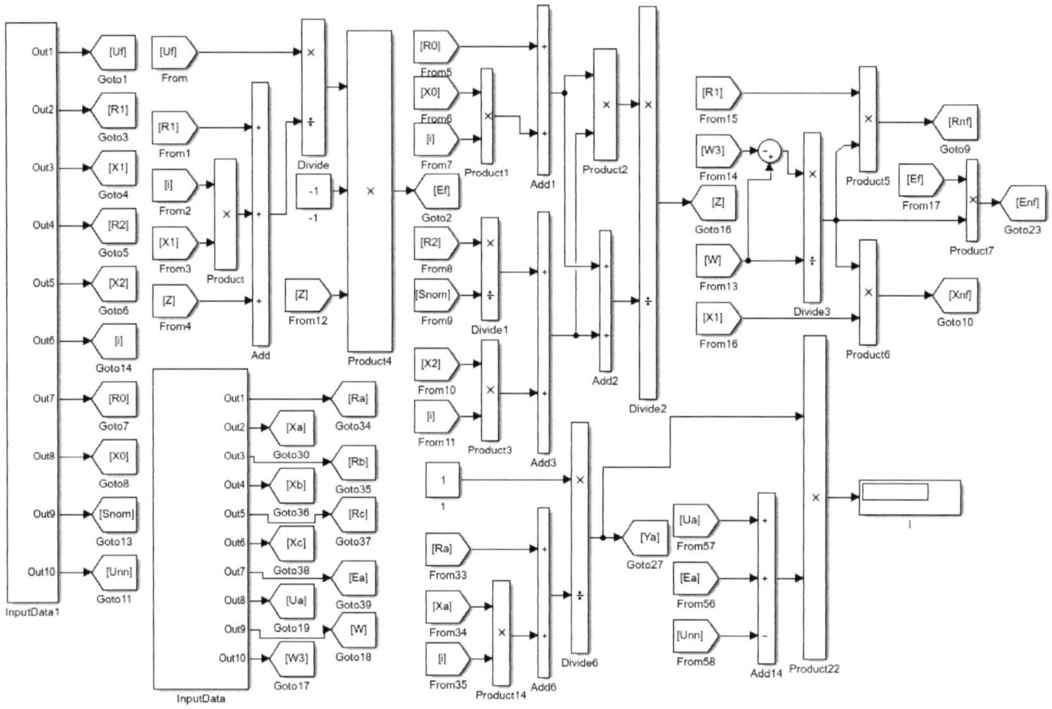

Figure 2. Mathematical model for calculating phase currents in MATLAB/Simulink software.

The model of the magnetic field in the center of the current sensor in ELCUT software in balanced conditions and in emergency conditions using the results of the model in MATLAB/Simulink software and vector diagrams [23] is developed. At low frequencies, the magnetic field in a first approximation can be considered quasi-stationary. Therefore, the vector magnetostatic model can be used.

Figure 3 shows the finite element model of the proposed device. Figure 4 shows the pattern of magnetic-field vectors (H) in balanced conditions for one complete period of system simulation.

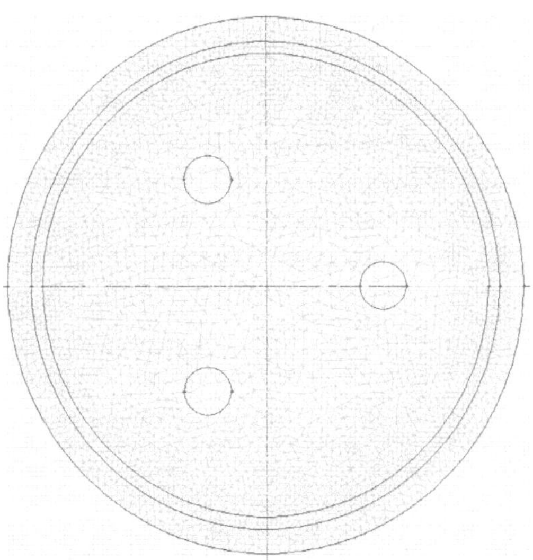

Figure 3. Finite element model.

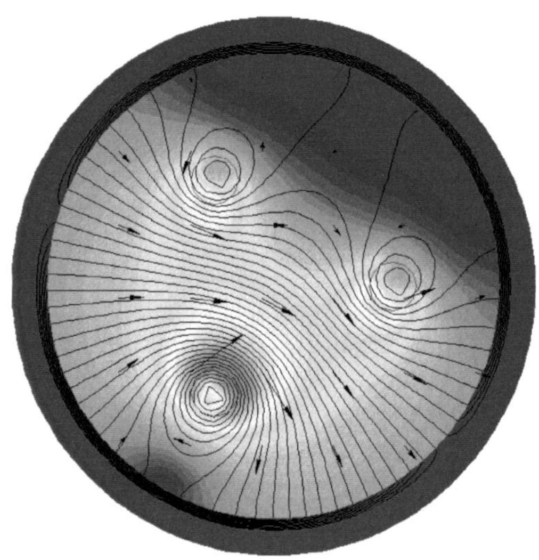

Figure 4. Magnetic-field pattern.

Table 1 shows the results of measuring the local values of the module of the magnetic field strength (H) and its projections (H_X and H_Y) for balanced conditions. Input data is current densities in phases (δ_A, δ_B and δ_C). The phases of the magnetic-field vector (Ψ_H) were calculated. The measurements were conducted at 13 stages to obtain the hodograph of the magnetic-field vector [24] for one complete period in all the studied modes.

Table 1. Measurement results of local values (balanced conditions).

Stage	Phase (°)	δ_A (A/ m^2)	δ_B (A/m^2)	δ_C (A/m^2)	H (A/m)	H_x (A/m)	H_y (A/m)	ψ_H (°)	B (mT)
1	0	0	-2206542	2206542	4620,9	4620,9	-17,1	359,79	5,81
2	30	1273948	-2547895	1273948	4626,1	4002,0	-2320,6	329,89	5,81
3	60	2206542	-2206542	0	4621,4	2310,7	-4002,2	300,00	5,81
4	90	2547895	-1273948	-1273948	4611,5	0,4	-4611,5	270,00	5,80
5	120	2206542	0	-2206542	4606,0	-2310,1	-3985,2	239,90	5,79
6	150	1273948	1273948	-2547895	4611,0	-4001,6	-2291,0	209,79	5,79
7	180	0	2206542	-2206542	4620,9	-4620,9	17,1	179,79	5,81
8	210	-1273948	2547895	-1273948	4626,1	-4002,0	2320,6	149,89	5,81
9	240	-2206542	2206542	0	4621,4	-2310,7	4002,2	120,00	5,81
10	270	-2547895	1273948	1273948	4611,5	-0,4	4611,5	90,00	5,80
11	300	-2206542	0	2206542	4606,3	2310,1	3985,2	59,90	5,79
12	330	-1273948	-1273948	2547895	4611,0	4001,6	2291,0	29,79	5,79
13	360	0	-2206542	2206542	4620,9	4620,9	-17,1	-0,21	5,81

Simulation results for turn-to-turn short circuit (10% shorted turns in phase A) are presented in Table 2. The current density is calculated using the currents of mathematical modeling in MATLAB/Simulink software.

Table 2. Measurement results of local values (turn-to-turn short circuit).

Stage	Phase (°)	δ_A (A/ m^2)	δ_B (A/m^2)	δ_C (A/m^2)	H (A/m)	H_x (A/m)	H_y (A/m)	ψ_H (°)	B (mT)
1	0	833114	-2626609	2179253	5195,9	5032,4	-1293,1	345,59	6,529
2	30	2083993	-2651294	1258192	5302,2	4094,1	-3369,1	320,55	6,663
3	60	2776469	-1965567	0	4987,1	2059,9	-4542,3	294,39	6,267
4	90	2724991	-753168	-1258192	4529,3	-528,1	-4498,4	263,30	5,692
5	120	1943355	661042	-2179253	4404,5	-2973,6	-3249,2	227,54	5,535
6	150	640998	1898126	-2516384	4758,2	-4622,3	-1129,3	193,73	5,979
7	180	-833114	2626609	-2179253	5195,9	-5032,4	1293,1	165,59	6,529
8	210	-2083993	2651294	-1258192	5302,2	-4094,1	3369,1	140,55	6,663
9	240	-2776469	1965567	0	4987,1	-2058,9	4542,3	114,38	6,267
10	270	-2724991	753168	1258192	4529,3	529,1	4498,4	83,29	5,692
11	300	-1943355	-661042	2179253	4404,5	2973,6	3249,2	47,54	5,535
12	330	-640998	-1898126	2516384	4758,2	4622,3	1129,3	13,73	5,979
13	360	833114	-2626609	2179253	5195,9	5032,4	-1293,1	-14,41	6,529

Similarly, the modeling was performed in other modes. The results are presented in the form of hodographs of the magnetic-field vector in figure 5.

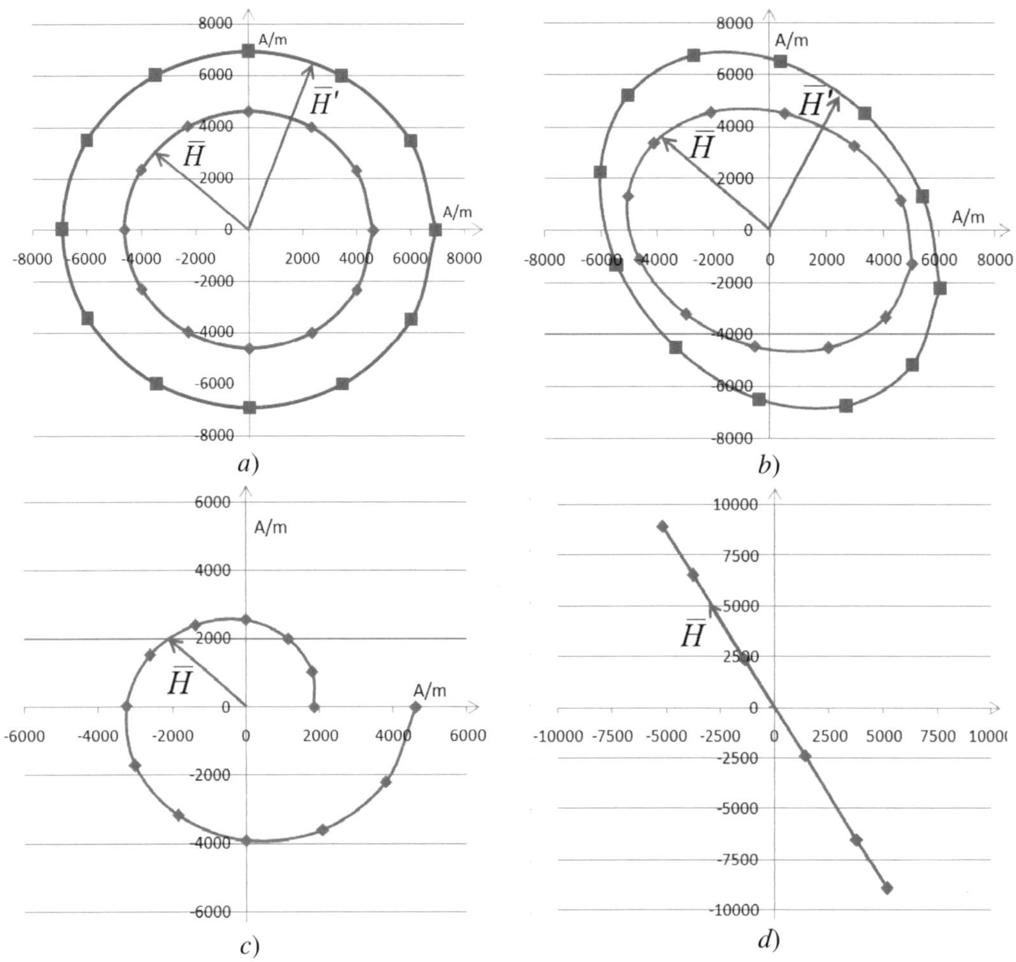

Figure 5. Hodographs of the magnetic-field vector in balanced conditions (*a*), turn-to-turn short circuit (*b*), supply fail (*c*) and external two-phase short circuit (*d*).

4. Discussion

The simulation results show that the three-phase current sensor able to detect turn-to-turn and phase-to-phase short circuits.

Changing magnetic-field pattern to elliptical, spiral or linear means that there is an accident in the measured three-phase system.

The hodograph of the magnetic-field vector will be circular with an amplitude $|H|$ in balanced conditions, and in a three-phase short circuit - with an increased amplitude $|H'|$ (figure 5 (a)).

The hodograph of the magnetic-field vector will be elliptic with an amplitude $|H|$ for 10% shorted turns, and for 30% shorted turns - with an increased amplitude $|H'|$ (figure 5 (b)).

The hodograph of the magnetic-field vector will be spiral with an amplitude $|H|$ for supply fail or overload (figure 5 (c)).

The hodograph of the magnetic-field vector will be linear with variable amplitude $|H|$ for two-phase short circuit (figure 5 (d)).

In the mathematical model, the phase currents were determined without taking into account influence of current unbalance on the main flux of the motor.

This assumption, firstly, allows studying the effect of reducing the number of turns in faulty phase on unbalance of the phase currents without taking into account the influence of other factors. Secondly, it simplifies the determination of phase currents without using method of symmetrical components. Thirdly, it introduces a reliability coefficient into the developed relay protection when calculating currents.

The proposed three-phase current sensor can be used in the construction of relay protection and automation devices.

A balanced protection device based on the three-phase current sensor has been developed. But instead of the power cables of the phases, coaxial conductors are used. The input cables of the electrical installation are connected to the current carrying wire, and the return cables of the electrical installation are connected to the outer conductor [25].

Currently, the experimental setup is under development to confirm theoretical conclusions and simulation results.

5. Conclusion

It is necessary to protect a motor against phase-to-phase and turn-to-turn short circuits. Differential protection detects phase-to-phase faults, but does not detect turn-to-turn short circuits. Currently, there is no single method for detecting turn-to-turn faults.

This article presents the three-phase current sensor, using which it is possible to detect turn-to-turn and phase-to-phase short circuits. The proposed device consists of a ferromagnetic case, a microprocessor, power cables of phases and magnetoresistors in the center of the sensor.

The three-phase current sensor measures the total magnetic field from three phases simultaneously. Speed of protection response will increase due to the design of the sensor. The operating principle of the device is to determine the conditions (balanced or emergency) by a magnetic-field pattern.

The model of the magnetic field for balanced conditions and the models of the magnetic field for emergency conditions were developed in ELCUT software. The elliptical, spiral or linear magnetic-field pattern means a violation of the balanced conditions.

The three-phase current sensor can also be used in the construction of relay protection and control system devices of electric motors.

Acknowledgments
The article is written in the framework of the research with the financial support of the grant from the Ufa State Petroleum Technological University.

References

[1] Joksimovic G M and Penman J 2000 The detection of inter-turn short circuits in the stator windings of operating motors *IEEE Trans. Ind. Electron.* **47** pp 1078-84
[2] Siddique A, Yadava G S and Singh B 2005 A review of stator fault monitoring techniques of induction motors *IEEE Trans. Energy Convers.* **20** pp 106-14
[3] Karmakar S, Chattopadhyay S, Mitra M and Sengupta S 2016 *Induction motor fault diagnosis: approach through current signature analysis* (Singapore: Springer) p 161
[4] Badri Ram and Vishwakarma D N 2001 *Power system protection and switchgear* (New Delhi: TataMcGraw-Hill Education) p 456
[5] IEEE Std 3004.8-2016: IEEE Recommended practice for motor protection in industrial and commercial power systems
[6] Ziegler G 2012 *Numerical differential protection: principles and applications* (Erlangen: Publicis Publishing) p 287
[7] Nandi S, Toliyat H and Li X 2006 A review of stator fault monitoring techniques of induction motors *IEEE Trans. Energy Convers.* **20** pp 719-29
[8] Thomson W T and Fenger M 2001 Current signature analysis to detect induction motor faults *IEEE Ind. Appl. Mag.* **7** pp 26-34
[9] Otava L 2016 Implementation of PMSM inter-turn short fault detection using frequency analysis of stator currents *IFAC-PapersOnLine* **49** pp 86-91

[10] Cruz S M A and Cardoso A J M 2001 Stator winding fault diagnosis in three-phase synchronous and asynchronous motors, by the extended Park's vector approach *Trans. Ind. Appl.* **37** pp 1227-33
[11] Sahraoui M, Guedidi S, Ghoggal A and Zouzou S 2014 Detection of inter-turn short-circuit in induction motors using Park–Hilbert method *Int. J. Syst. Assur. Eng. Manag.* **5** pp 337-51
[12] Vishwanath H and Maruthi G S 2017 Detection of stator winding inter-turn short circuit fault in induction motor using vibration signals by MEMS accelerometer *Electric Power Components and Systems* **45** pp 1463-73
[13] Vaseghi B, Takorabet N and Meibody-Tabar F 2009 Transient finite element analysis of induction machines with stator winding turn fault *Prog. Electromagn. Res.* **95** pp 1-18
[14] Fitouri M, Bensalem Y and Abdelkrim M N 2016 Modeling and detection of the short-circuit fault in PMSM using finite element analysis *IFAC-PapersOnLine* **49** pp 1418-23
[15] Pietrowski W and Konrad G 2017 Detection of inter-turn short-circuit at start-up of induction machine based on torque analysis *Open Phys.* **15** pp 851-6
[16] Fireteanu F 2013 Detection of the short-circuit faults in the stator winding of induction motors based on harmonics of the neighboring magnetic field *J. Phys.: Conf. Ser.* **450** 012021
[17] Chigvintsev S V and Ishmukhamedov I K 2017 Ustrojstvo dlya balansnoj zashhity`, sovmeshhennoj s poperechnoj differencial`noj dlya trexfazny`x vy`sokovol`tny`x e`lektroustanovok Patent RU 176605 [Supply current for overcurrent protection of three phase electrical installations Patent RU 176605] [In Russian]
[18] Chigvintsev S V and Ishmukhamedov I K 2016 Trekhfaznyy datchik toka i napryazheniya [Three-phase current and voltage sensor] (*Neftegazovoe delo)* **14** pp 200-5
[19] Edward M Purcell and David J Morin 2013 *Electricity and magnetism* (Cambridge: Cambridge University Press) p 839
[20] Gamazin S I, Stavtsev V A and Tsyruk S A 1997 *Perekhodnye protsessy v sistemakh promyshlennogo elektrosnabzheniya, obuslovlennye elektrodvigatel'noy nagruzkoy* [Transients in industrial power supply systems due to motor load] (Moscow: Izdatel'stvo MEI) [In Russian]
[21] Korogodskiy V I, Kuzhekov S L and Paperno L B 1987 *Releynaya zashchita elektrodvigateley napryazheniem vyshe 1 kV* [Relay protection of motors with voltage above 1 kV] (Moscow: Energoatomizdat) [In Russian]
[22] Chernobrovov N V and Semenov V A 1998 *Releynaya zashchita energeticheskikh sistem* [Relay protection of energy systems] (Moscow: Energoatomizdat) [In Russian]
[23] Atabekov G I 1960 *The relay protection of high voltage networks* Pergamon (Oxford: Pergamon Press) p 564
[24] Apostolatos T A 2003 Hodograph: a useful geometrical tool for solving some difficult problems in dynamics *Am. J. Phys.* **71** pp 261-6
[25] Chigvintsev S V and Ishmukhamedov I K 2018 Ustroystvo dlya zashchity balansa v sochetanii s poperechnoy differentsial'noy zashchitoy dlya trekhfaznykh vysokovol'tnykh elektricheskikh ustanovok Patent RU 2652381 [Device for balance protection combined with transverse differential protection for three-phase high-voltage electrical installations Patent RU 2652381] [In Russian]

Mathematical and visual models of asynchronous electric machines energy fields

S P Kurillin[1], V N Denisov[2] and M I Dli[3]

[1] Department of Electromechanical Systems, Branch of the National Research University «Moscow Power Engineering Institute» in Smolensk, Energetichesky proyezd, 1, Smolensk, 2014013, Russia
[2]Department of Higher Mathematics, Branch of the National Research University «Moscow Power Engineering Institute» in Smolensk, Energetichesky proyezd, 1, Smolensk, 2014013, Russia
[3]Scientific Department, Branch of the National Research University «Moscow Power Engineering Institute» in Smolensk, Energetichesky proyezd, 1, Smolensk, 2014013, Russia

E-mail: kurilinsp@inbox.ru

Abstract. The research considers the development of topological theory of asynchronous electric machines (AEM). The results of the development of energy aspects of the topological theory are presented in the article. The facts proving the actuality of the research are also given. The central notion giving the topological theory the characteristic of being systematic is the vector space notion. The definition of AEM vector space, its dimension, canonical basis, power and energy field of AEM are presented. Mathematical and visual models of AEM energy fields are analyzed. Energy fields of electric losses and magnetic field dissipation having the spherical symmetry are analyzed in particular. Energy and visual models of the main magnetic field are analyzed using AEM with a short circuited rotor and linear AEM as examples. It is established that this field obtains cylindrical or elliptical symmetry. In general the main magnetic field and the main matrix of winding make AEM vector space fully or partially heterogeneous. Energy efficiency is analyzed and its differences for vector space currents are revealed. The areas of vector spaces with high and low energy currents efficiency are defined. According to this indicator longitudinal and transvers subsets of current having, correspondingly, high and zero energy efficiency are distinguished in the vector space. The description of practical application spheres of AEM topological theory is given.

Keywords: asynchronous electric machine, topological theory, vector space, basis, the canonical basis, energy field, mathematical models, visual models, energy efficiency, a subset of currents

1. Introduction

One of the actual aim of the modern electromechanics is to create a theory reflecting systematic properties and systematic reactions of asynchronous electric machines (AEM) on external actions and changes of internal properties. There is a number of factors concerning AEM internal heterogeneous demonstration which are ignored or interpreted indefinitely by the existing theoretical models.

These factors are:

- the difference in energy efficiency of some types of systems of AEM phase current;
- the efficiency of mathematical methods for AEM with homogeneous internal properties and their inefficiency for AEM with heterogeneous properties;
- the appearance and energy consequence of compensating currents and interchange power in the heterogeneous electromechanical structures;
- the generation of not basic frequencies or additional mechanical power by AEM heterogeneous rotor.

Besides the number of problematic questions are connected with the correctness of mathematical models of heterogeneous AEM.

Among the earliest works affecting aspects of the systemic nature of electrical objects, the works of N. Hancock and G. Kron should be mentioned [1-3], referring to the second half of the last century. G. Kron introduced the term "topological theory". Among the late publications is the work of A.D. Artym [4]. Today the significant progress has been achieved in this study. The terminological apparatus of the theory [5 - 7] is formed, definitions of the vector space of its dimension, phase basis, canonical basis, longitudinal and transverse current subsets are given to determine the rules and properties of basis transformations [8 - 10]. Based on them, a method for analyzing vector space has been created that focuses on a multidimensional vector-matrix representation of electromagnetic states and energy fields of AEM [11,12], and it describes the phenomenon of excessive energy dissipation in AEM [13 - 15]. The study of the vector spaces of parametric winding matrices of the AEM [16 - 18] has been carried out and a number of mathematical models of the AEM have been developed taking into account their topological properties [19 - 22]. It should be noted that, when applying aspects of topological theory, work in the field of monitoring of energy indicators and technical condition of AEM [23 - 25] takes into account the specifics of their energy processes and characteristics [26, 27].

In this article the authors present the results of their developments of energy aspects of topological theory.

2. Main definitions

The central notion giving the topological theory the property of being systematic is the notion of AEM vector space. Let us introduce the definition of this term and some accompanying definitions.

The vector space of AEM polyphase winding is understood as a set of vectors of electromagnetic values (currents), voltages, electromotive force (EMF) and flux linkage corresponding all its possible electromagnetic conditions.

The dimension of the winging vector space fully corresponds to the number of its phases. The vector space having the greatest dimension is considered to be AEM vector space. However the electromechanical transformation of energy in AEM is possible only in case the electromagnetic values of rotor and stator windings have a common vector subspace.

The basis of the vector subspace of three–phase AEM considers any of orthonormal three vectors belonging to this subspace. The vectors of AEM windings currents are physically given in the basis of phase axes. It means that their coordinates in these basis are the physically phase currents of windings. Every vector can be written in any of other vector subspace basis by the operation of basis transformation. This operation performing corresponds to the change of the real winding on the equivalent one situating along the axes of the new basis and also the change of the winding parametric matrix on the similar one. In this case the power of the real and equivalent windings are equal.

Among the basis of vector space canonic basis are distinguished. The canonic basis of the winding vector space considers the orthonormal three eigenvectors of its parametric matrix.

Not only the vector of winding current corresponds to every point of the vector space but the scalar value – the power of the winding vector space as well

$$p=\left(\overline{\imath},Z\,\overline{\imath}\right)$$

where - winding parametric matrix connecting the vectors of its current and power (EMF).

In canonic basis matrix obtains a diagonal form

$$Z=\begin{pmatrix} \lambda_x & 0 & 0 \\ 0 & \lambda_y & 0 \\ 0 & 0 & \lambda_z \end{pmatrix}$$

where λ_x , λ_y , λ_z - the eigenvalues of the matrix along the axes of basis and quadratic form obtains the canonical form

$$p=\lambda_x i_x^2+\lambda_y i_y^2+\lambda_z i_z^2$$

where i_x, i_y, $\bar{i}_z$ - the coordinates of current vector in canonical basis.

Parametrical AEM matrixes, excepting the matrixes of mutual induction, are symmetric [5, 6]. The transformation of quadratic form to the canonical form is always defined for them. So the basis transformation is invariant with the respect to the power

$$p=(\bar{i},Z\bar{i})=(\bar{i}_v,Z_v\bar{i}_v)$$

where index v shows that the object belongs to the canonical basis.

The scalar function of the vector argument $p(\bar{i})$ or $W(\bar{i})$ expressing the power or energy of some winding physical field is called the winding energy field. The energy field is usually represented by surfaces of the level in the form of $p(\bar{i})=C=\text{const}$ or $W(\bar{i})=C=\text{const}$.

3. Energy fields of asynchronous electric machines

Let us write down in accordance with [6] the balance of power of AEM stator three–phase winding.

$$\begin{aligned}(\bar{i}_S,\bar{u}_S)+\left(\bar{i}_S,M\frac{d\bar{i}_m}{dt}\right)&=(\bar{i}_S,r_S E\bar{i}_S)+\\+\left(\bar{i}_S,l_{\sigma S}E\frac{d\bar{i}_S}{dt}\right)&+\left(\bar{i}_m,M\frac{d\bar{i}_m}{dt}\right)\end{aligned} \quad (1)$$

where $\bar{u}_S,\bar{i}_S,\bar{i}_{rS},\bar{i}_m$ - vectors of supply voltage, stator winding current, rotor winding current and magnetizing current in the canonical basis of stator winding $\{\bar{x},\bar{y},\bar{z}\}$, $r_S E$, $l_{\sigma S}E$, M - diagonal matrixes of active phase resistance, inductances of phase dissipation , the main matrix of stator winding; E - unity matrix; $r_S,l_{\sigma S}$ - active resistance and inductance of stator winding dissipation phase.

In the left part (1) there are powers of external, for stator, source (drain) of energy: the power of the source of supply and electromagnetic power. In the right part (1) there are powers of stator winding energy fields: energy losses, magnetic field of dissipation and the main magnetic field.

The power field of electrical losses is a scalar field of the vector argument $\bar{i}_S$. In accordance with (1),

$$p_{el}=p_{el}(\bar{i}_S)=(\bar{i}_S,r_S E\bar{i}_S)=r_S|\bar{i}_S|^2$$

In vector space the surfaces of the field level are the surfaces of the fixed power C

$$p_{el}(\bar{\imath}_S)=r_S\left(\imath_{Sx}^2+\imath_{Sy}^2+\imath_{Sz}^2\right)=C,$$

representing the spheres by the radius $|\bar{\imath}_S|$.

Two of them, for the values of power and electrical losses C and 4C, in canonical coordinates are shown in figure1. The vector space multiple currents $\bar{\imath}_S$ are shown here by the arrows

The field of energy of the magnetic field of a stator winding dissipation is a scalar field of vector argument $\bar{\imath}_S$. In accordance with (1), the power of the field is equal to

$$\left(\bar{\imath}_S,l_{\sigma S}E\frac{d\bar{\imath}_S}{dt}\right)=\frac{d}{dt}(W_{\sigma S})=\frac{d}{dt}\left(\frac{l_{\sigma S}\left|\bar{\imath}_S\right|^2}{2}\right)$$

In vector spaces the surfaces of the field level $W_{\sigma S}(\bar{\imath}_S)$ are the surfaces of the fixed energy C

$$W_{\sigma S}(\bar{\imath}_S)=\frac{l_{\sigma S}}{2}\left(\imath_{Sx}^2+\imath_{Sy}^2+\imath_{Sz}^2\right)=C,$$

representing the spheres of radius $|\bar{\imath}_S|$. Two of them, which are for the values of energy of the magnetic field dissipation C and 4C, are shown in figure 1.

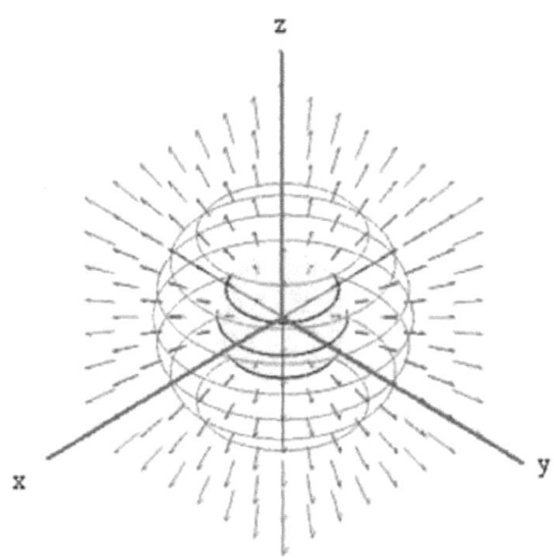

Figure 1. Energy field with spherical symmetry.

The field of energy of the main magnetic field is a scalar field of vector argument $\bar{\imath}_m$. In accordance with (1), the power of the field is equal to

$$\left(\bar{\imath}_m,M\frac{d\bar{\imath}_m}{dt}\right)=\frac{d}{dt}W_m.$$

For the AEM with a short – circuit rotor the main matrix along the axis x is degenerated, and the plane yz is parametrically homogeneous [6]

$$M = \begin{pmatrix} 0 & 0 & 0 \\ 0 & \lambda_m & 0 \\ 0 & 0 & \lambda_m \end{pmatrix}.$$

With this connection its energy field has the cylindrical symmetry and the surfaces of the level

$$W_m(\overline{\iota}_m) = \frac{\lambda_m}{2}\left(\iota_{my}^2 + \iota_{mz}^2\right) = C \qquad (2)$$

represent a cylinder of the infinite length along the axis x having the radius of rolling circle $\sqrt{2\times C/\lambda_m}$. For the values of energy of the main field C and 4C they are shown in figure 2. The arrows show the multiple currents $\overline{\iota}_m$.

For linear AEM the main matrix is not degenerated and parametrically heterogeneous along all axes of basis [6]

$$M = \begin{pmatrix} \lambda_{mx} & 0 & 0 \\ 0 & \lambda_{my} & 0 \\ 0 & 0 & \lambda_{mz} \end{pmatrix}.$$

In accordance with this its energy field has elliptic symmetry and the surfaces of level

$$W_m(\overline{\iota}_m) = \frac{1}{2}\left(\lambda_{mx}\iota_{mx}^2 + \lambda_{my}\iota_{my}^2 + \lambda_{mz}\iota_{mz}^2\right) = C \qquad (3)$$

represent ellipsoids with semi-axes $\sqrt{2C/\lambda_{mx}}, \sqrt{2C/\lambda_{my}}, \sqrt{2C/\lambda_{mz}}$. For the values of energy of the main field C and 4C they are shown in figure 3. The arrows show the multiple currents $\overline{\iota}_m$.

Thus, the main matrix of the stator winding makes AEM vector space fully or partially heterogeneous.

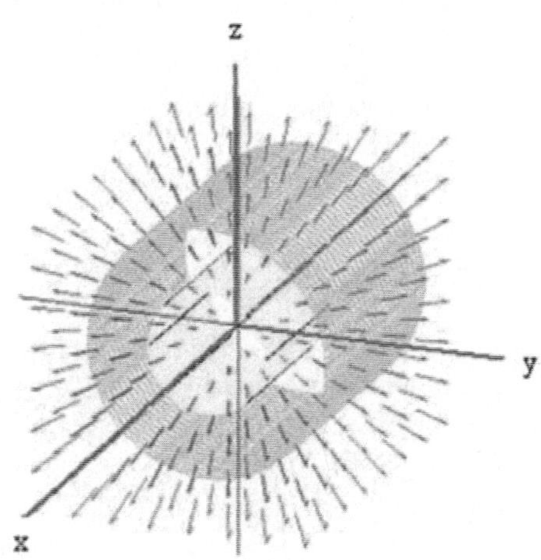

Figure 2. Energy field with cylindrical symmetry.

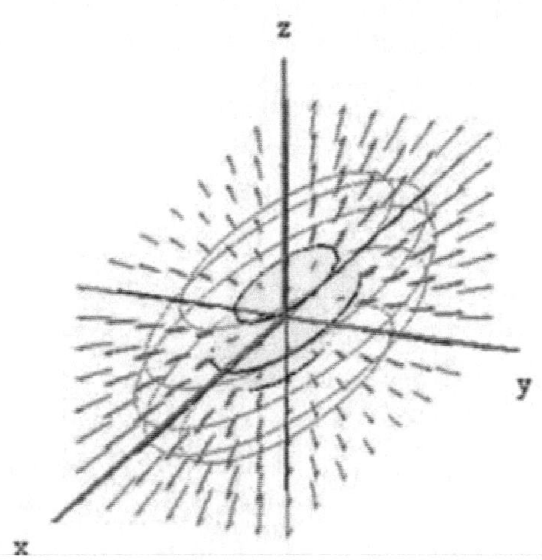

Figure 3. Energy field with elliptic symmetry.

The energy efficiency of vector space. In accordance with (2) and (3) eigenvalues of the matrix have the sense of specific energy of the main magnetic field for the squared current unit. They can be regarded as the characteristics of vectors current efficiency orienting along the axes of canonical basis. In accordance with such an interpretation and figure 2, figure 3 the area of the most energy efficiency of the vector space is the plane yz and the currents oriented along the axis x are less efficient from the point of their contribution to the energy of the main magnetic field.

In vector space two sets of current vectors having correspondingly maximum and zero efficiency are distinguished in general. The authors called them longitudinal and transvers subsets of currents.

In figure 4 the surface trace of level $W_m(\bar{i}_m) = C$ along (3) in plane xz and three of its characteristic points a, b, c are shown. The figure corresponds to the general case of parametric heterogeneous of AEM vector space at which in plane xz the inequalities are valid

$$\lambda_{mx} \neq 0, \lambda_{mx} < \lambda_{mz}$$

The vectors of the main flux linkage $\bar{\psi}_m$ and currents $\bar{i}_m$ for each point are shown in the figure. Vectors of flux linkage being the gradient vectors of the energy field are orthogonal to the level surfaces.

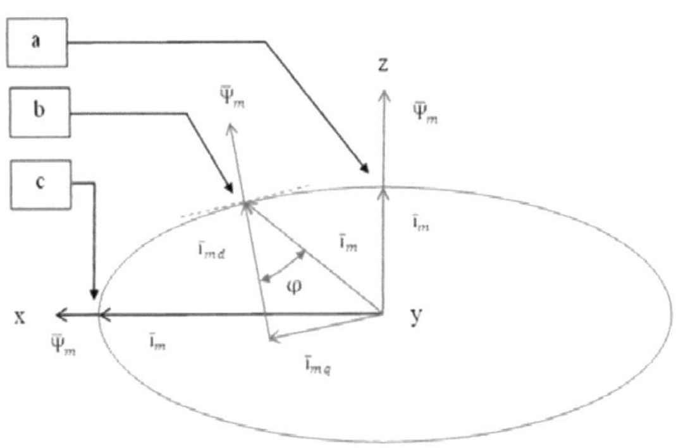

Figure 4. Vectors of electromagnetic values in heterogeneous vector space.

Points a and c are on canonic axes z and x and point b is between them. The position provides the space coincidence of flux linkage vectors for points a and c:

$$\bar{\psi}_m = \lambda_{mz}\bar{i}_m$$

– for point a and

$$\bar{\psi}_m = \lambda_{mx}\bar{i}_m$$

– for point c. These equations express the fact that the space is homogeneous along the canonic axes but has different inductances.

In the direction of current $\bar{i}_m$ of point b the space is parametrically heterogeneous, that is why

$$\bar{\psi}_m = M\bar{i}_m = \begin{pmatrix} \lambda_{mx} & 0 & 0 \\ 0 & \lambda_{my} & 0 \\ 0 & 0 & \lambda_{mz} \end{pmatrix} \begin{pmatrix} i_{mx} \\ 0 \\ i_{mz} \end{pmatrix} = \begin{pmatrix} \lambda_{mx} i_{mx} \\ 0 \\ \lambda_{mz} i_{mz} \end{pmatrix},$$

and between vectors $\bar{\psi}_m$ and $\bar{i}_m$ there is a space angle ϕ.

In figure 4 collinear ($\bar{\imath}_{md}$) and orthogonal ($\bar{\imath}_{mq}$) to vector $\bar{\Psi}_m$ current components are given. The energy of the main magnetic field in point b is made only by current component $\bar{\imath}_{md}$

$$W_m = \frac{(\bar{\imath}_m, \bar{\Psi}_m)}{2} = \frac{(\bar{\imath}_{md} + \bar{\imath}_{mq}, \bar{\Psi}_m)}{2} = \frac{(\bar{\imath}_{md}, \bar{\Psi}_m)}{2}.$$

The energy of current component $\bar{\imath}_{mq}$ is equal to zero

$$\frac{(\bar{\imath}_{mq}, \bar{\Psi}_m)}{2} = 0.$$

The last equation testifies that equivalent phases of AEM along axes x and z interchange the energy

$$\frac{\lambda_{mx} i_{mqx} i_{mx}}{2} = -\frac{\lambda_{mz} i_{mqz} i_{mz}}{2}.$$

Current vector $\bar{\imath}_{mq}$ does not take part in making the energy of the main magnetic field of AEM. At the same time quite certain powers of dissipation correspond to it, they are equal to

$$r_S \left|\bar{\imath}_{mq}\right|^2$$

and

$$\frac{d}{dt}\left(\frac{l_{\sigma S} \left|\bar{\imath}_{mq}\right|^2}{2}\right)$$

These powers should be brought to the category of energy overdissipation. It is defined by the fact that they are not connected with doing effective work and caused only by the heterogeneity of AEM vector space.

The set of vectors $\bar{\imath}_{md}$ of vector space forms the longitudinal subset of currents. The set of vectors $\bar{\imath}_{mq}$ of vector space forms the transvers subsets of currents. The subsets of currents $\bar{\imath}_{md}$ and $\bar{\imath}_{mq}$ of heterogeneous vector space are shown in figure 5 and figure 6 by the arrows.

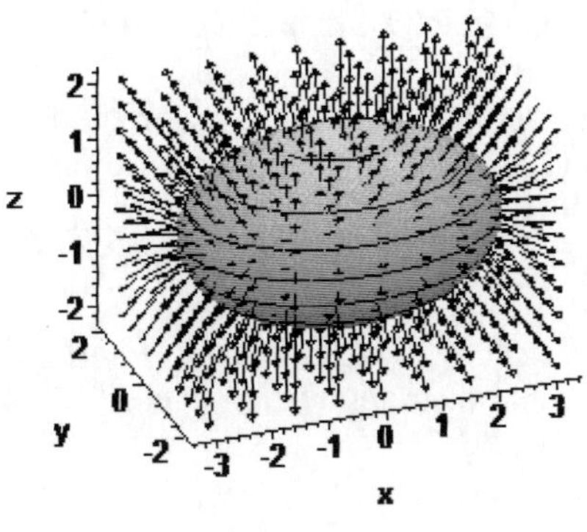

Figure 5. Vectors of longitudinal subsets of currents.

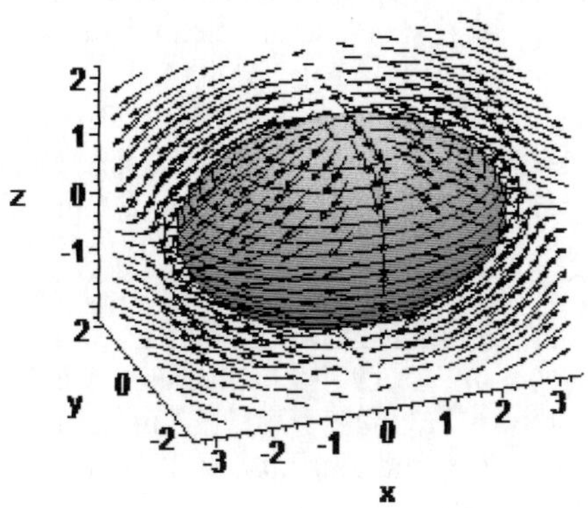

Figure 6. Vectors of transvers subsets of currents.

4. Some practical applications

Particular application of topological theory are the energy processes in asynchronous motors with the alternating form of a rotor bar. Rotor cages of such motors represent a parametrically heterogeneous structures and the question about their rational design is open at present.

Wider area of application is given by the problem of diagnostic of AEM and their operating. For such problems it is necessary to have information factors pointing authentically to the object's saving or loosing parametrical homogeneity because of operational imperfection, machine aging, violation of technological and operating standards etc.

Minimization of overdissipation for linear AEM is one of the criteria of rational design together with such criterias as the highest efficiency and maximum starting force.

5. Conclusion

AEM vector space is partially or fully parametrically heterogeneous.

Parametrical heterogeneity is appeared in AEM energy fields configuration, in effects of internal energy interchange and energy overdissipation.

Different fields of AEM vector spaces have different energy efficiency. Thus, it is reasonable to supply power and operate AEM the way to concentrate its currents vectors in the area of the highest energy efficiency.

The apparatus of the energy fields gives visual images of energy conditions and energy processes of AEM.

Topological theory gives methods and tools for solving a number of problems of development and operation of heterogeneous AEM.

References

[1] Hancock N 1967 *Matrichny`j analiz e`lektricheskix mashin* [Matrix Analysis of Electric Machines] (Moscow: Energiya) [In Russian]

[2] Kron G 1972 *Issledovaniye slozhnyh sistem po chastyam (diakoptika)* [The development of complex systems by components (Dacoptics)] (Moscow:Nauka) [In Russian]

[3] Kron G 1978 *Tenzorny analiz setey* [Tensor analysis of circuits] (Moscow: Sovetskoye radio) [In Russian]

[4] Artym A D 1987 *Matrichno-topologicheskiye metody analiza elektricheskih tsepei s primeneniyem EVM* [Topological methods of electrical circuits with computer application] (St. Petersburg: Leningr.elektrotehn.inst-t svyazi im.M.A. Bonch-Bruyevicha) [In Russian]

[5] Kurilin S P, Denisov V N and Kruglov V V 2008 *Matrichnaya teoriya elekricheskih mashyn* [Matrix theory of electric machines] (Moscow: Rossiiskiy un-t kooperatsii) [In Russian]

[6] Denisov V N and Kurilin S P 2011 *Matrichnoye modelirovaniye elekromagnitnyh i energeticheskig protsessov v elektricheskih mashinah* [Matrix modelling of electromagnetic and power processes in electric machines] (Smolensk: RIO filiala GOUVPO "MEI(TU) v Smolenske") [In Russian]

[7] Denisov V N and Kurilin S P 2007 Preobrazovaniye koordinat i analiz parametricheskih svoistv elektricheskih mashin [Coordinate conversion and analysis of the parametric properties of electric machines] *Elekrichestvo* **6** pp 45 – 50 [In Russian]

[8] Zheng J, Huang S, Rong F and Lye M 2018 Six-phase space vector PWM under stator one-phase open-circuit fault condition *Energies* **7** 1796

[9] Kurilin S P and Denisov V N 2013 *Matimaticheskiye osnovy modelirovaniya i ih prilozheniya v elektromehanike* [Mathematical basis in modelling and their application in electromechanics] (Smolensk: Universum) [In Russian]

[10] Denisov V N and Kurilin S P 2014 *Metody i prilozheniya matimaticheskogo modelirovaniya v elektrotehnike. Monografiya* [Methods and applications of mathimatic modeling in electromechanics] (Smolensk: Smolensky filial Rossiiskogo universiteta kooperatsii) [In Russian]

[11] Kurilin S P and Denisov V N 2010 Visualnye modeli vektornyh prostranstv i problemy energosberezheniya [Visual models of vector spaces and energy saving problems] *Sistemy komp'juternoj matematiki i ih prilozhenija: materialy XI mezhdunarodnoj nauchnoj konferencii, posvjashhennoj 70-letiju professora V.P. D'jakonova.* vol 11 (Smolensk: Izdatel`stvovo SmolGU) pp 45-6 [In Russian]

[12] Denisov V N and Kurilin S P 2004 Visualnye modeli moshchnosti mnogofaznyh obmotok [Visual power models of multi-phase windings] *Perspektivy i tendencii razvitija jelektrotehnicheskogo oborudovanija Trudy pjatogo mezhdunarodnogo simpoziuma JeLMASh* vol 1 (Moscow) pp 105 – 11 [In Russian]

[13] Kurilin S P and Denisov V 2013 Izbytochnoye rasseyaniye v obmotkah nesemetrichnyh elektricheskih mashin [Overdissipation in windings of asynchronous electric machines] *Sbornik trudov 3 Mezhdunarodnoj nauchno-tehnicheskoj konferencii "Jenergetika, informatika, innovacii - 2013"* vol 1 (Smolensk: filial MEI v g. Smolenske) pp 215-20 [In Russian]

[14] Kurilin S P and Denisov V N 2014 *Matematicheskaya model neyavnopolyusnoi elektricheskoy mashiny v matrichnoy forme* [Mathematical model of a non-salient pole machine in the matrix form] *Elektrichestvo* **4** pp 43 – 9 [In Russian]

[15] Kurilin S P and Denisov V N 2015 *Energeticheskiye polya i izbytochnoye rasseyaniye energii v neyavnopolusnoy elektricheskoy mashine* [Energy fields and excess energy dissipation in a Non-Salient Pole Machine] *Electrichestvo* **3** pp 35-41 [In Russian]

[16] Zoric I, Jones M and Levi E 2017 Arbitrary power sharing among three-phase winding sets of multiphase machines *IEEE Transactions on Industrial Electronics* vol 65 **2** pp 1128-1139

[17] Denisov V N and Kurilin S P 2006 Vidy i svoistva parametricheskih matrits matematicheskih modelei elektricheskih mashin [Types and properties of parametric matrices of mathematical models of electric machines] *Vserossiiskaya nauchno-tekhnicheskaya konferentsiya (*Tula: Izdatel'stvo TulGu*)* pp 115-7 [In Russian]

[18] Mamyedov F A, Denisov V N and Kurilin S P 2002 Parametricheskiye svoistva i osobennosti energetiki assimetrichnyh obmotok elektricheskih mashin [Parametric properties and energy features of asymmetric windings of electrical machines] *Trudy chetvertogo mezhdunarodnogo simpoziuma "ELMASh – 2002"* vol 2 (Moscow) pp 85-8 [In Russian]

[19] Kosmatov V I, Sarvarov A S and Danilov E I 2018 Mathematical model of energy-saving asynchronous motor in the rotating coordinates system. *Proc.of the 17th International Ural Conference on AC Electric Drives* (ACED-April,8341702) pp 1-5 [In Russian]

[20] Nos O V 2008 Matimaticheskaya model asynhronnogo dvigatelya v lineinyh prostranstvah,svazannyh so statorom i rotorom [Mathematical Model of Asynchronous Motor in Linear Spaces Connected with Stator and Rotor] *Izv. vuzov. Electromehanika* **2** pp 14-20 [In Russian]

[21] Denisov V N and Kurilin S P 2011 Inzhenernaya model lineinogo asinhronnogo dvigatelya [Engineering model of a linear asynchronous motor] *Elektrichestvo* **3** pp 52-4 [In Russian]

[22] Mamedov F, Kurilin S and Tchutorov D 2001 Ideal no – load operation and asymmetry of phases of low speed linear induction machine *Proc. of the 5th Int. Conf. on unconventional electromechanical and electrical systems* vol 2 (Poland:Szczecin) pp 323-8

[23] Smolin V I, Topolskaya I G and Volovich G I 2016 The Energy Method for Monitoring the Instantaneous State and the Formation of a Synchronous Motor Control Variables *The 2nd Int. Conf. on Industrial Engineering, Applications and Manufacturing (ICIEAM)* (Russia)

[24] Bitoleanu A, Popescu M, Linca M and Mitran A 2011 Energetical Performances Analysis of PWM Asynchronous Motor and Voltage Inverter Driving System *The 7th International Symposium on Advanced Topics in Electrical Engineering (ATEE)* (Romania)

[25] Farhani F, Zaafouri A and Chaari A 2017 Real time induction motor efficiency optimization *Journal of the Franklin Institute* vol 354 **8** pp 3289-304

[26] Mamyedov F A , Denisov V N and Kurilin S P 2002 Osobennosty energeticheskih protsessov v

elektricheskih mashinah s assimetrichnymi obmotkami [The Features of Power Processes in Electric Machines with Asymmetric Windings] *Electrichestvo* **9** pp 36 – 43 [In Russian]
[27] Mamyedov F A and Kurilin S P 2001 Ob opredelenii aktivnoi i reaktivnoi moshchnosti dlya assimetrichnyh mashin [About the detection of active and reactive power for asymmetric electric machines] *Elektrotehnika* **4** pp 17 – 21 [In Russian]

Mathematical model of managing of the generating company on the criterion of the profit maximization

B N Moshkin[1], T V Myatezh[1], Y A Sekretarev[1] and M A Averbukh[2]

[1]Industrial Power Supply Systems Department, Novosibirsk State Technical University, Karl Marks Ave.,20, Novosibirsk, 630073, Russia
[2]Department of Electric Power Engineering and Automation, Belgorod State Technological University of V.G. Shukhov, Kostyukov St., 46, Belgorod, 308012, Russia

E-mail: tatianamyateg@mail.ru

Abstract. The urgency of the work is conditioned by the electric power industry reforming, which implies a gradual transition to a competitive market model. So each producer will be interested in maximizing his own profit and independently determine the volumes of electric and heat energy production. On the basis of the developed criterion of maximizing profits, the following problems are considered in the article: the most advantageous distribution of electric energy of thermal power plants at given values of the rate for electricity; distribution of thermal energy between the plant's units, taking into account the forced heating mode of operation of the stations; finding the optimal modes of operation of the stations for the combined method of producing electric and thermal energy; appointment of the most favorable operating modes of the generating company; substantiation of tariff rates for the products sold, depending on the optimal production volumes at thermal power plants that are part of the generating company, determining the optimal operating conditions of SIBEKO JSC (Novosibirsk, Russia). The results: methodology of optimization of CHPP for electric and heat energy has been developed in the work. And also a new criterion has been proposed for maximizing profits for managing the functioning of a generating company.

Keywords: mathematical model, optimization of operating modes of CHP, characteristics of relative increments of fuel consumption, criterion of profit maximization

1. Introduction

The urgency of the problem of energy saving, environmental friendliness, efficiency and ensuring rational work of the heat and power systems of cities, as well as their reliability throughout the life of the operation is dictated by the priority directions of development of Russia and the world community as a whole. Recently developed documents in the energy sector - "Target vision of the strategy for the development of the power industry of Russia for the period up to 2030", the adjusted General Layout of the Power Industry Objects until 2030, the Power Strategy for the period until 2030 - as one of the main areas The development of the country's energy sector considered a wider use of coal-fired power plants, the share of power generation of which was to increase from 17.5 (2010) to 23% in the most likely variant and to 25% in the maximum. This growth predetermines new technologies of coal-fired power plants, which are increasingly used in many countries around the world. As noted by the Institute of World Resources, in 59 countries 1199 coal-fired power plants with a total capacity of more than 1,400 GW have been proposed for construction, among which 77% of the capacity falls on China and India.

Since cities and the majority of industrial enterprises receive electric and heat energy from CHP, the solution to the task of energy saving set in this project is to develop and implement criteria and methods that significantly reduce energy costs and optimize operating modes of generating companies (GC) and thermal power plants (CHP, as previously existing, and the new generation). The perspective of scientific directions in the energy sector of the economy of the GC today depends on the actual application of the results of scientific research and approbation in practice. It should be noted that one of the current areas is the use of high-tech, low-energy-intensive, energy-saving techniques to optimize the operating modes of the GC and CHP with a high economic effect. In conditions when the power industry was the state branch of management, the operation consisted in fulfilling the following requirements. There are given volume and schedule of supplying electric and thermal energy to consumers and observing the reliability conditions of the energy system, and observing the system constraints, determine the optimal mode that satisfies the given economic criterion understood the minimization of operating costs [1].

Currently, a strategy of reforming the electric power industry has been developed and is being implemented. It involves a gradual transition to a competitive market, where each business entity will independently determine the volumes of production of electrical and thermal energy and modes of operation.

Among the characteristic limitations of the existing principle of control of functioning, the following can be distinguished: the mismatch between the goal of management and modern conditions. Every business entity is interested in increasing its own profit, the ineffectiveness of the former control criterion, which is focused on setting higher levels of control of electricity generation and thermal energy [2,3]. Therefore, the old criteria and methods of management have become inadequate for management purposes.

The presence of these limitations necessitates the development of a different principle and methods for managing the operation of an energy facility.

There is a need to determine the principles and methods of management of the functioning of energy facilities that are acceptable in the new economic conditions. Through the work of several generations of scientists, unique methods of managing energy production were created and successfully put into practice, which ensured its high reliability and economic efficiency. Of course, they should form the basis of the approaches proposed in the work, become its intellectual environment and, in combination with economic levers, create prerequisites for the emergence of competitive relations that increase the economic efficiency of energy production.

The solution to this problem is the development of a fundamentally new criterion for managing the operating modes of GC and CHP, as well as creating an integrated methodology for separating fuel costs between the types of energy produced (electrical and heat), which will allow GC and CHP to take a competitive position in the electricity and heat market.

Urgency of the problem of energy saving, environmental friendliness, efficiency and ensuring rational work of the heat and power systems of cities, as well as their reliability throughout the life of the operation is dictated by the priority directions of development of Russia and the world community as a whole [1,2]. Recently developed documents in the energy sector - "Target vision of the strategy for the development of the power industry of Russia for the period up to 2030", the adjusted General Layout of the Power Industry Objects until 2030, the Power Strategy for the period until 2030 - as one of the main areas The development of the country's energy sector considered a wider use of coal-fired power plants, the share of power generation of which was to increase from 17.5 (2010) to 23% in the most likely variant and to 25% in the maximum [3,4]. Since cities and the majority of industrial enterprises receive electric and heat energy from CHP, the solution to the task of energy saving set in this project is to develop and implement criteria and methods that significantly reduce energy costs and optimize operating modes of generating companies GC) and thermal power plants (CHP, as previously existing, and the new generation) [5, 6].

IOP Conf. Series: Materials Science and Engineering **552** (2019) 012016 doi:10.1088/1757-899X/552/1/012016

The purpose of the study is to develop a mathematical model of the optimal loading of a generating company in terms of electric power by the criterion of maximizing profit [7, 8].

In conditions when the power industry was the state branch of management, the operation consisted in fulfilling the following requirements. There are given volume and schedule of supplying electric and thermal energy to consumers and observing the reliability conditions of the energy system, and observing the system constraints, determine the optimal mode that satisfies the given economic criterion understood the minimization of operating costs [9, 10]. Currently, a strategy of reforming the electric power industry has been developed and is being implemented. It involves a gradual transition to a competitive market, where each business entity will independently determine the volumes of production of electrical and thermal energy and modes of operation. Among the characteristic limitations of the existing principle of control of functioning, the following can be distinguished: the mismatch between the goal of management and modern conditions. Every business entity is interested in increasing its own profit [11, 12]. Therefore, the old criteria and methods of management have become inadequate for management purposes [13, 14]. The presence of these limitations necessitates the development of a different principle and methods for managing the operation of an energy facility. In this paper, it is proposed to use the economic criterion of profit maximization for optimal control of the load of thermal power plants in modern conditions, and also developed a strategy for managing a generating company [15, 16].

2. A mathematical model for controlling the operation of a generating company

As is known, it is necessary to solve the problem of constructing a mathematical model to solve optimization problems. Modeling can be defined as a method of mediated cognition, in which the object under study i.e. the original is in some relationship with another object i.e. a model. The model object is able in some way to replace the original at some stages of the cognitive process [4].

We represent the control object in the form of a black box (figure 1). Mode parameters are divided into several categories. We will distinguish between input and output parameters. In addition, we divide the parameters (input and output) into controlled (index "c") and non-controlled ones (index "nc"). The latter include those parameters that we either cannot control e.g. weather, power system load, or do not want to control, translating them into uncontrollable, for example, insensitive parameters that have little effect on the objective function. In addition, process parameters can be measurable (index "m") or non-measurable (index "nm").

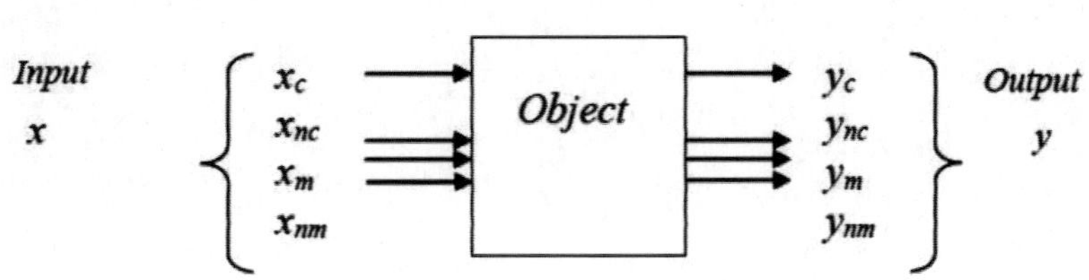

Figure 1. System parameters.

We construct a mathematical model of the most favorable distribution of electrical power between stations or aggregates. We assume that the system has i = 1,2, ..., n thermal power plants, for which the flow characteristics $B_i(P_{Ti})$ and total load P_n are known. For this case:

1. The goal equation

$$B = B_1(P_{T1}) + B_2(P_{T2}) + ... + B_n(P_{Tn}) \Rightarrow \min \tag{1}$$

2. The equation of connection $B_i(P_{Ti})$.

IOP Conf. Series: Materials Science and Engineering **552** (2019) 012016 doi:10.1088/1757-899X/552/1/012016

3. Limitations are balance power equations

$$\sum_i P_{Ti} - P_n - \pi = 0 \tag{2}$$

where π is the total loss of active power.

4. To derive the optimization equation, we use the Lagrange method. The use of the method of indeterminate Lagrange multipliers makes it possible to reduce the task to solving a system of algebraic equations.

In some cases, the analytical solution of the problem uses linearization techniques or other idealization of nonlinear dependencies.

$$\Phi = (B_1 + B_2 + ... + B_n) + \lambda * \sum_i P_{Ti} - P_n - \pi = 0 \tag{3}$$

where λ is a constant Lagrange multiplier.

It should be noted that such a system of equations is compiled for each hour of the day, i.e. there are 24 systems in total.

An analysis of modern methods of mathematical programming (linear and nonlinear, dynamic and the theory of optimal processes) and a general form of the constructed model shows that the only acceptable and possible way to solve this problem is to use non-linear programming methods, namely the Lagrange method.

Let us consider the main provisions of this method as applied to the set task.

Since the expression in the second brackets in expression (3) is zero, the minima of the Lagrange function and the objective function (1) are the same.

Let is differentiate the Lagrange function with respect to variables $P_{T1,...,}P_{Tn}$ and equate the derived zero. It should be borne in mind that when we take the partial derivatives of the Lagrange function on the load of the *i* station, the partial derivative of the consumption of the fuel of the *i-th* station is $b_i = \frac{dB_i}{dP_{Ti}}$ only on its load P_{Ti}, and on the loads of all other stations is zero. Then we get:

$$\left.\begin{aligned} \frac{\partial \Phi}{\partial P_{T1}} &= \frac{\partial B_1}{\partial P_{T1}} + \lambda * \left(1 - \frac{\partial \pi}{\partial P_{T1}}\right) = 0; \\ \frac{\partial \Phi}{\partial P_{Tn}} &= \frac{\partial B_n}{\partial P_{Tn}} + \lambda * \left(1 - \frac{\partial \pi}{\partial P_{Tn}}\right) = 0. \end{aligned}\right\} \tag{4}$$

The number of equations of the system corresponds to the number of thermal power plants. It is easy to see from (4) that:

$$\frac{\dfrac{\partial B_1}{\partial P_{T1}}}{1 - \dfrac{\partial \pi}{\partial P_{T1}}} = ... = \frac{\dfrac{\partial B_n}{\partial P_{Tn}}}{1 - \dfrac{\partial \pi}{\partial P_{Tn}}} \tag{5}$$

Let us introduce the following denotation: $b_i = \frac{dB_i}{dP_{Ti}}$ is a relative increase in fuel consumption of electric stations, which shows how the fuel consumption of the *i-th* station will change if its load changes by the value $\sigma_i = \frac{\partial \pi}{\partial P_{Ti}}$, which is a relative increase in active power losses in networks, i.e. the value indicating how much network losses will change if the power of only the *i-th* station changes by ∂P_{Ti}.

Applying these notation, we obtain the conditions for the most favorable load distribution:

$$\frac{b_i}{1-\sigma_i} = idem. \tag{6}$$

If we assume that the active power loss is zero, that is to neglect them, then we get the following expression:

$$b_1 = b_2 = ... = b_n \tag{7}$$

In this paper, it is proposed to use the economic criterion of profit maximization for optimal control of the load of thermal power plants in modern conditions, and also developed a strategy for managing a generating company [5, 6].

According to the criterion of maximizing profits, the manufacturer will maximize profits by producing products at the point where marginal revenues equal marginal costs. This guiding principle of profit maximization is called the marginal cost rule of marginal revenue. A graphic illustration of this condition is presented in figure 2.

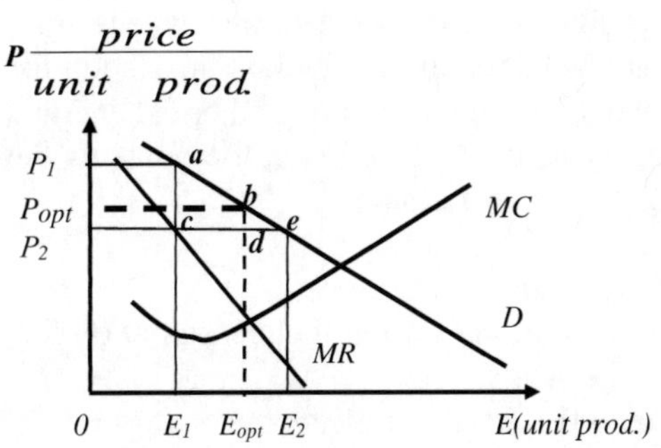

Figure 2. Determination of the optimal production output: D is an energy demand for a specified time interval; E_{opt} is an optimal output for a specified time interval; P_{opt} is an optimal price at the optimal production output.

Marginal revenue (*MR*) from the sale is determined by the differential demand curve for energy products, and marginal cost (*MC*) is a differential component of the cost of energy production, which can be represented as a characteristic of the relative increase in fuel costs for thermal power plants. All these values have the same dimension (price / unit prod.), therefore, they can be compared in the calculations [6].

The optimal amount of energy production (E_{opt}) allows the energy company to maximize its profits. Suppose that a smaller number of E_1 products is produced compared to the optimum number, but at a higher price P_1. In this case, the marginal income of the manufacturer exceeds marginal costs, and by increasing the production volume to E_{opt}, at which an additional profit derived from the release of another unit of production is zero, it would increase the total profit by an amount equal to the area *abc*.

Larger with respect to the optimal production volume of E_1 also does not maximize profit, since marginal costs exceed marginal revenue. The increase in profit by reducing the volume of production to the value of E_{opt} instead of E_2 corresponds to a value equal to the area *bde*.

In modern conditions, each economic entity is interested in increasing its own profit, therefore, the proposed method allows to solve the problem of determining the volume of production at the station adequately to the modern conditions of the domestic energy industry.

On the basis of the developed criterion, the following tasks are solved in the paper:

1. The most advantageous distribution of electric energy of thermal power plants for the given values of the tariff for electricity.
2. Distribution of thermal energy between the station's units, taking into account the forced heat and power plants operation mode.
3. Finding the optimal operating modes of the stations for the combined method of producing electric and thermal energy.
4. Appointment of the most favorable operating modes of the generating company.
5. Justification of tariff rates for the products sold, depending on the optimal production volumes at thermal power plants that make up the generating company.

The formulated tasks and stages of their implementation can be represented in figure 3.

Since the main mode of operation of thermal power plants is heat-generating, their operation should be managed taking into account the forced schedule for the release of thermal energy [7]. Therefore, the work examines the question of the possibility of applying the approach to the most advantageous distribution of electricity between the station's units to determine the optimal modes for the production of both types of energy, taking into account the "hard constraints" imposed by the forced operation of the CHP.

3. An optimal distribution of electrical and thermal energy at the stations based on the criterion of profit maximization

The most profitable distribution of electrical and heat energy at the stations in accordance with the conducted analytical analysis. At the same time, various types of restrictions are taken into account, in particular, the forced operation mode of the station according to the heat-generating cycle [8,9].

This problem is solved by several stages. At the first level, the optimal load on the station for electric power, taking into account the forced mode of its work on the generation of thermal energy. Then assignment of optimal operation modes of the stations by thermal energy is carried out. At the final stage, the corresponding work and their mutual coordination is performed.

Let us dwell in more detail on the solution of these issues.

To determine the optimal power plant, the following tasks must be solve:

Construction of the characteristics of the relative increase of the station fuel consumption for given compositions of operating equipment for the seasons of the year.

Getting the dependencies of the marginal costs of stations for each of the seasons.

Finding the optimal electrical power and the corresponding values of the stated price.

To obtain the dependencies of the marginal costs of the station, we used the characteristics of relative increments in fuel consumption, taking into account the average seasonal prices for purchased fuel.

The characteristics of marginal revenue are built on the basis of the demand curves for electrical energy as follows:

$$\frac{\Delta TR}{\Delta E_{ou}} = MR = p + E_{ou}\frac{\Delta p}{\Delta E_{ou}} \tag{8}$$

where $\frac{\Delta p}{\Delta E_{ou}}$ is a slope of the demand curve, i.e. marginal revenue represents the derived demand for electricity;

ΔTR is an increase in total revenue from power generation;

Δp is a change in electricity prices;

ΔE_g is a change in power generation.

For the station, as a characteristic of demand, it is possible to take the characteristic of the cost of production of electric energy, which can be adjusted by the value of the rate of profit. The real demand curve is piecewise due to the variable nature of energy consumption. However, in practice, an

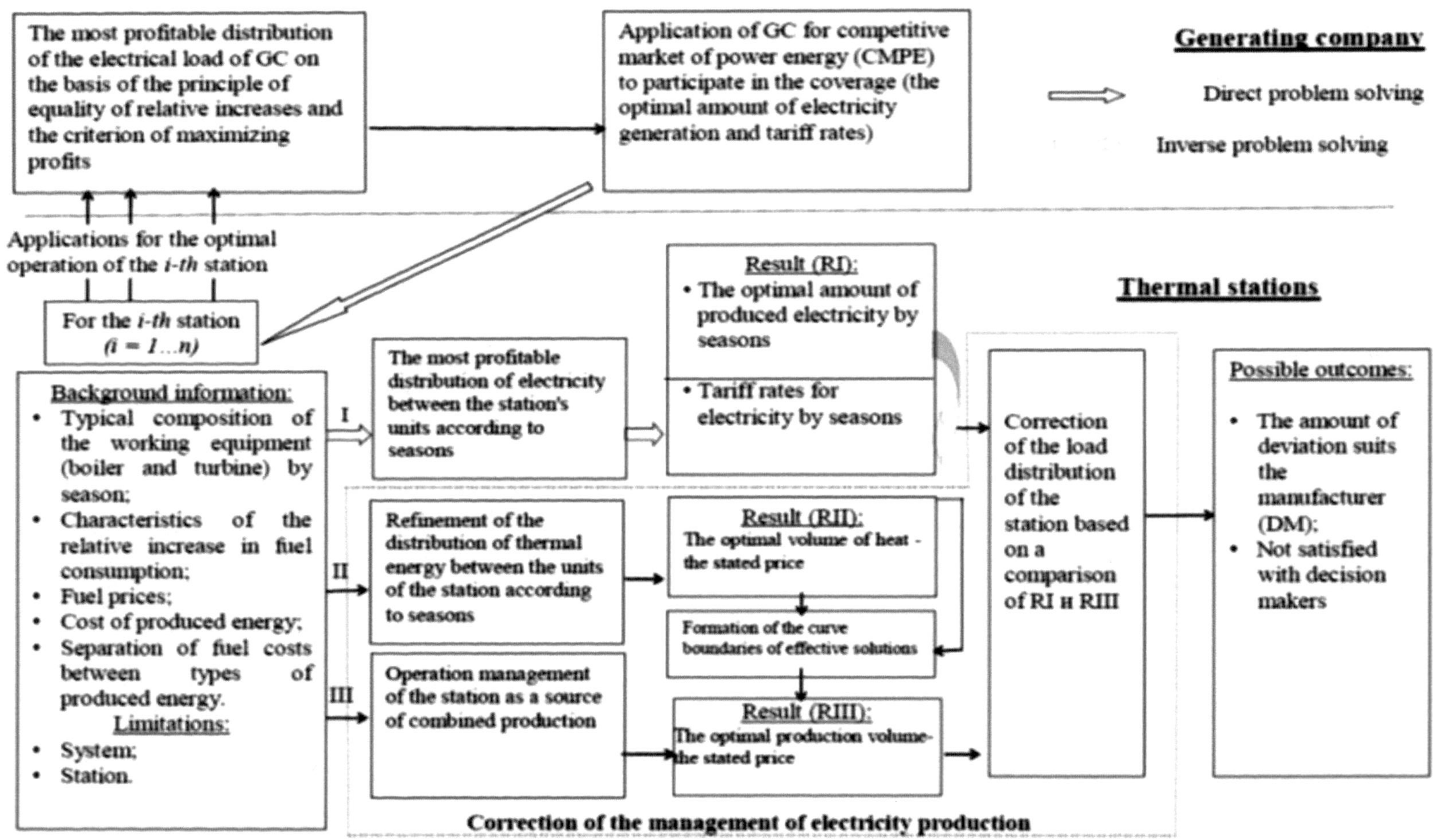

Figure 3. Logic diagram of market generation company functioning.

approximation of these dependencies is used, selecting for this the corresponding polynomials. Approximate dependencies of cost and demand for electricity are shown in figure 4.

Since energy consumption has a pronounced seasonal nature, it is advisable to consider three demand curves: Zone I (figure 4) corresponds to summer consumption; Zone II (figure 4) is for the transition period (spring-autumn) and, finally, Zone III (figure 4) is for the consumption of the winter period. Each curve can be approximated by the corresponding polynomials [10, 11].

Jointly solving the system of equations describing the curves of marginal costs and marginal revenue, you can determine the optimal values of the average monthly production (E_{opt}) for each season and the average daily capacity:

$$N_{opt} = E_{opt} / t_{month}, \tag{9}$$

where t_{month} is an average number of hours per month (720 h).

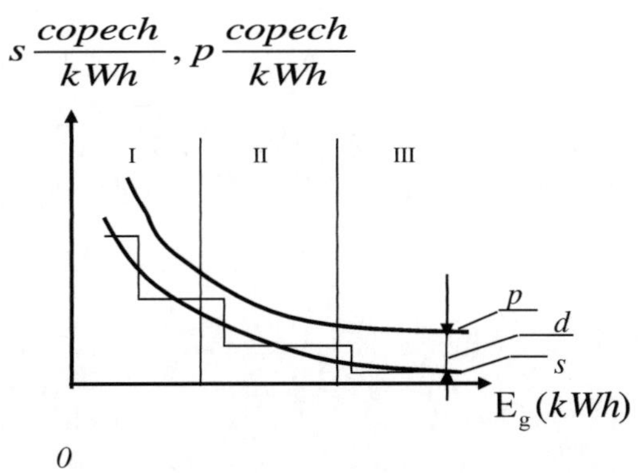

Figure 4. Curves of electricity production cost and electricity demand: s is an electricity production cost, p is an electricity price, E is an electric power released from the station buses, s= f(E) is the electricity production cost function, D = p = f(E) is the electricity demand function derived as the sum of electricity production cost and profit rate, i.e. p = s + d.

On the basis of the proposed approach, the work shows the possibility of solving two important and interrelated tasks of controlling station operation modes:

To determine the optimal range of its generation at the power plant, with the tariff for electricity, which is formed in the modern conditions of functioning of the CHP.

It is possible to substantiate the size of the stated price in the power system, based on the optimal power generation at the stations.

As already noted, the calculation of the optimal load of a thermal power plant by electric power is carried out taking into account the forced mode of operation of a thermal power plant by thermal energy [12].

However, the use of this approach allows you to check the loading of the station by thermal energy.

The problem of optimal distribution of thermal energy at the station is solved step by step according to a model developed for the most advantageous distribution of electrical energy [13].

Let us consider the possibility of using the developed model of the optimal distribution of electricity between the station's units to determine the modes for the production of both types of energy, taking into account the "hard constraints" imposed by the forced operation mode of the CHP [14].

To solve this problem, it is necessary to determine the optimal range within which further optimization of station modes is allowed due to a change in turbine selections and variation of electric power generation volumes. A graphic illustration of the solution to this problem is shown in figure 5. Suppose, for example, point 1 corresponds to the optimal mode for the production of electricity, and point 2 - by thermal energy. Then the modes, characterized by a change in the selection of turbines and the variation of electric power, correspond to intermediate points 3, 4 and 5.

Consider the following options for changing the selections at the station.

The first boundary variant is characterized by the optimal loading of the station by electric power at given thermal energy selections. In this case, the optimization criterion is the maximum profit from the production of electric energy.

The second boundary variant is characterized by optimal loading of the station by thermal energy. At the same time, the proposed principle of station operation control is used.

It should be noted that the value of electrical power in the optimization of thermal energy (N_{el2}) is based on the following expression:

$$N_{el2} = N_{el20} + N_{const,} \tag{10}$$

where N_{el20} is the power value corresponding to the optimal load of the station for thermal energy;

N_{const} is the constant component of electric power corresponding to the difference (according to the consumption characteristic) difference of the average for the season of the actual value of the supply of thermal energy and heat load, which is determined by the optimal amount of electricity generation.

Intermediate calculations of the station operation modes vary within the first and second boundary variants [15].

The obtained points form a curve of effective solutions, which is proposed to be used to find the optimal operating modes of the station with the combined generation of electrical and thermal energy [16].

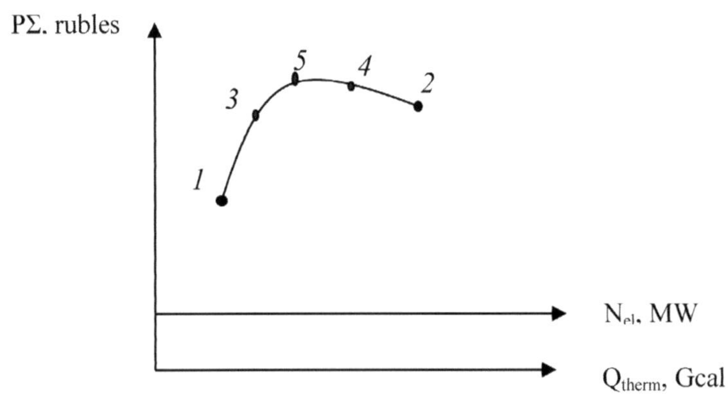

Figure 5. Dependence of total station profit on electric power (heat energy).

The optimal criterion of the combined method of producing electric and thermal energy will be the maximum of the total profit PΣ:

$$P_{\Sigma} = P_E + P_Q \Rightarrow \max \tag{11}$$

where P_E is a profit from the production of electricity;
P_Q is a profit from the production of heat energy.

In this case, it may be that the solution obtained will give serious deviations in the production of electrical and thermal energy at the station.

Therefore, the paper proposed a method for adjusting the production of electric power by the optimal value of heat [17].

To do this, it is necessary to calculate the deviation (Δ) of the optimal generation of electric energy (obtained with a strictly specified selection) with the optimal mode, which is calculated based on criterion (11):

$$\Delta = \frac{\Delta_w * Э_w + \Delta_s * Э_s + \Delta_t * Э_t}{Э_w + Э_s + Э_t} \tag{12}$$

where $\Delta_w, \Delta_s.\Delta_t$ is the deviation for the winter, summer and transition periods, respectively;

$Э_w, Э_{s,} Э_t$ is an optimal power values for the characteristic season of the year.

The proposed model allows distributing fuel costs between the types of energy produced by the criterion of profit maximization, taking into account mode constraints.

4. Management of operating modes of thermal power plants as a combined production sources

The next iteration is the practical implementation of the models proposed in the article for determining the optimal operating modes of the CHP-2, CHP-3, CHP-4, CHP-5 in Novosibirsk.

In accordance with the proposed approach, it is necessary to calculate the boundary and intermediate energy production options at the stations (figure 5). Then, according to criterion (4), determine the optimal mode of operation of each of the above stations by electrical and thermal energy.

As a general note, it should be noted that all calculations are carried out for given compositions of operating equipment [18-20].

Table 1 shows the composition of the operating equipment of NCHP-4 by seasons of the year:

Table 1. Typical operating equipment structure at the Novosibirsk CHP-4.

Season	Boilers			Turbines		
	ТП-170	ТП-81	ЦКТИ-75-39	ПТ-22-90	Т-100-130	Т-24,5-90
Winter (months XI,XII,I,II,III)	4	3	–	2	3	1
Summer (VI,VII,VIII)	2	1	1	2	2	–
Transition period (IX,X,IV,V)	2	3	–	2	3	–

In figures 6, 7 and 8 showed characteristics of the relative increase in fuel consumption, marginal costs and marginal revenue for the winter period of NCHP-4, respectively.

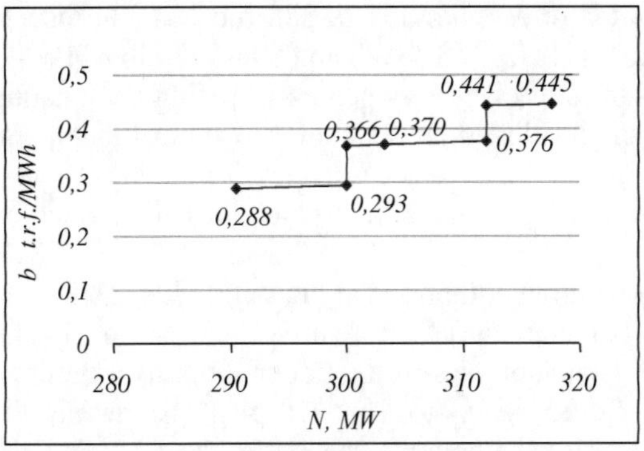

Figure 6. Characteristics of the relative increase in fuel consumption of NCHP-4 for the winter period.

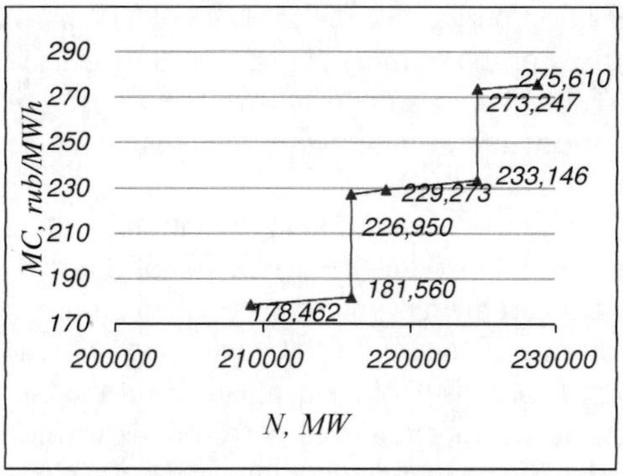

Figure 7. Characteristics of the marginal costs of NCHP-4 for the winter period.

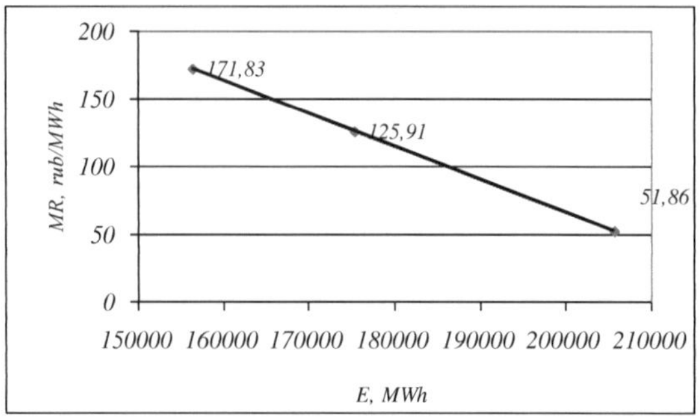

Figure 8. Marginal revenue curveof NCHP-4 for the winter season.

The results of calculations according to the developed principle of controlling the operation of a power generation station are presented in table 2.

Table 2. Optimum values of power for the seasons of the year for NCHP-4.

Season	Winter		Transition period		Summer	
Profit rate	0%	12%	0%	12%	0%	12%
N, MW	220	234	151	166	39	41
E, MWh	158702	168394	109058	119635	28412	29716
Stated price, rub/MWh	560	650	600	800	1500	1700
Revenue, rubles	285807009	331995964	152955789	185715182	76410172	88707757
Profit, rubles		46188955		32759394		12297585

The range of changes in the generation of electricity of NCHP-4 for the winter period is 158702-168394 MWh. In this case, the stated price of electricity will be from 360 to 394 rub/ MWh; for the transition period of 109058-119635 MWh with the stated price from 351 to 388 rub / MWh, respectively. For the summer period, these values are equal to 28411-29715 MWh and 896-995 rub / MWh.

Table 3 shows the results of verification of the volumes of electricity generation using the existing (the principle of minimizing fuel costs) and the criterion developed by the author to control the operation of stations.

Under the second boundary option is understood this mode of operation, which is characterized by the optimal load for the production of thermal energy.

This problem is solved for a given composition of operating equipment at the station [21-23].

In accordance with the developed approach, the characteristic of the relative increase in the station's fuel consumption is obtained on the basis of the characteristics of the boiler shop, by extracting from it a component that corresponds to the supply of heat energy to the CHP plant. Below are the results of the implementation of the developed approach for calculating the station heat load on the example of the winter period of the NCHP-2 (table 4).

A comparison of actual and optimal volumes of electricity generation shows that the use of the principle of profit maximization allows us to obtain solutions belonging to the range that took place with the existing criteria for managing the operation of thermal power plants.

According to the results of the implementation of the developed approach at NCHP-4, the specific fuel consumption for power generation was saved on the order of 5, and in some cases 10 g / kWh.

For NCHP-2, the range of change in heat supply for the winter period is 215450-229629 Gcal. In this case, the stated price of thermal energy will be from 142 to 163 rub / Gcal. Similarly, for the transition period is 165126-191910 Gcal with the stated price ranging from 156 to 175 rub / Gcal, respectively. For the summer period, these values are 52232-56255 Gcal and 195-216 rub / Gcal.

Table 5 summarizes the results of verification of heat generation volumes using the existing and proposed by the author optimization criterion. A comparison of actual and optimal volumes of electricity generation shows that the use of the principle of profit maximization allows us to obtain solutions belonging to the range that occurred with the existing criteria for controlling the operation of thermal power plants.

As a result of the calculation of intermediate optimal variants of station operation modes as a combined production, an efficient decision curve is obtained, which allows making decisions on the most advantageous distribution of electrical and thermal load between units at CHP. For an example in figure 9 shows the curve of effective solutions for the winter period of NCHP-4.

Table 3. Verification of electricity generation volumes.

Season	**Winter**		**Transition period**		**Summer**	
CHP	Rate of production E, MWh	E_{opt}, MWh	Rate of production E, MWh	E_{opt}, MWh	Rate of production E, MWh	E_{opt}, MWh
NCHP-2	147300-238000	155685	85600-150000	88213	23900-87400	65500
NCHP-3	176000-257040	191890	89280-221760	150057	23600-76320	62554
NCHP-4	156486-234964	168393	38518-121769	119635	23577-32766	29715
NCHP-5	295200-666000	408000	207360-547200	280000	115200-288000	158000

Table 4. Optimal value of power output at the Novosibirsk CHP-2 for seasons of the year.

Season	**Winter**		**Transition period**		**Summer**	
Profit rate	0%	12%	0%	12%	0%	12%
Q_{opt}, Gkal	215450	229630	165126	191910	52232	56255
Stated price, rub./ Gkal	145	163	156	175	195	216
Revenue, rubles	153101233	186633440	103032019	134347042	30607245	36408433
Profit, rubles		33532207		31315023		5801188

Analysis of the obtained results shows that for the NCHP-2 in the winter period it is advisable to produce 216 MW (348942 Gcal), in the summer is 91 MW (79149 Gcal) and for the transition period is 146 MW (191910 Gcal); for NTEC-4 in winter, the optimal load is 253 MW (356589 Gcal), in the summer is 59 MW (81282 Gcal) and for the transition period is 210 MW (264725 Gcal). It is these modes of operation that will allow stations to get the maximum profit for each season of the year.

The possibility of applying the approach to managing the operation of CHP for the production of electricity to find the optimal operating conditions of the plant as a combined production in modern conditions allows us to judge the amount of deviation found by expression (11). For NCHP-4, it is 13%, for NCHP is 2 - 4%.

Table 5. Thermal power production volumes comparison.

Season	**Winter**		**Transition period**		**Summer**	
CHP	Rate of production Q, Gcal	Q_{opt}, Gcal	Rate of production Q, Gcal	Q_{opt}, Gcal	Rate of production Q, Gcal	Q_{opt}, Gcal
NCHP-2	223400-352600	229630	62700-196900	191910	45000-70000	56255
NCHP-4	336081-499962	356589	93126-269552	264725	533331-109444	106640

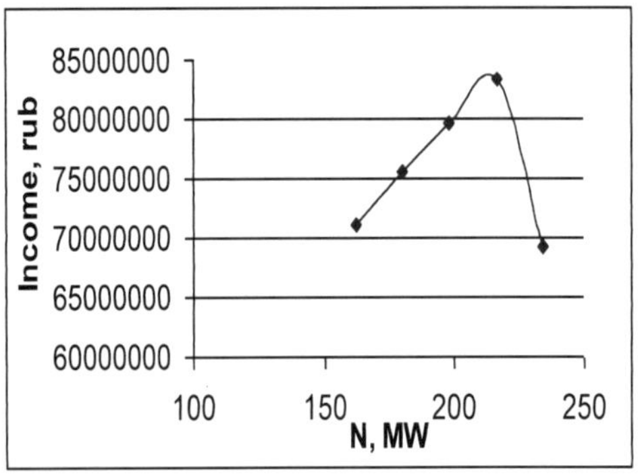

a)

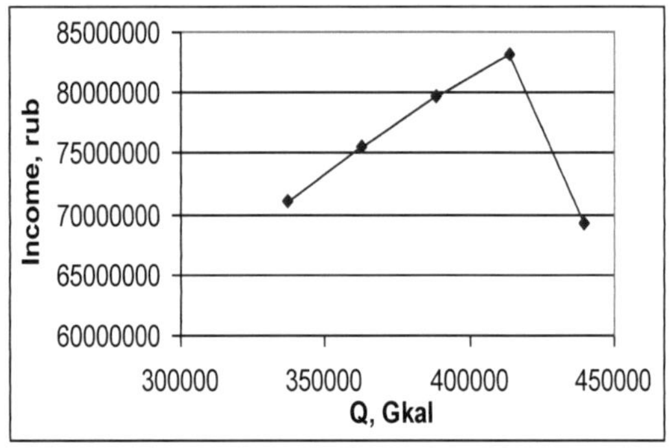

b)

Figure 9. Determination of the optimal operation mode of the NCHP-4 (winter) by: a) electric power, b) thermal energy.

5. Management of operating modes modes of the generating company

The task of managing the operation of the generating company (GC) is solved on the basis of the obtained optimal operating conditions of thermal power plants that are part of it.

The principle of interaction within the generating company is as follows:

each station included in the group of companies calculates and transmits its optimal mode of operation (for each of the seasons of the year) taking into account the station and system limitations, indicating the volume of possible supplies and the stated price for electricity;

GC, having collected applications from CHP, loads the stations in an optimal way, guided as an optimization criterion by the rule of profit maximization;

on the basis of regimes, energy supply contracts are concluded with manufacturers with detailed deliveries by month, working day, holidays and weekends.

For the described principle of interaction within the GC, an approach has been developed that allows one to calculate the range of optimal volumes of electricity generation, based on the criterion of profit maximization.

The implementation of the approach is carried out sequentially according to the model proposed for controlling the operation of CHP.

As a result of the optimization of the operating mode, the Communications Department of the GC submits its application for participation in the coverage of the total load schedule of the consumer.

It should be noted that this application has the character of "solid supplies", which the management of the GC undertakes to deliver to the CMPE [24-26].

All calculations were made for the generating company on the basis of the Novosibirsk CHPs based on data on the operating modes of the stations, as well as information on the average monthly values of fuel prices and the cost price of the products sold. The results of the calculation according to the developed methodology are given in Table 6.

Table 6. Optimal power generation for the generating company.

Season	**Winter**		**Transition period**		**Summer**	
Profit rate	0%	12%	0%	12%	0%	12%
Electricity output, MWh	489300	554900	129250	143690	343370	388100
Stated price, rub/MWh	1150	1275	2500	2750	1670	1800

Analysis of the obtained results shows that the range of changes in the power generation of the station for the winter period is 489300-554900 MWh. In this case, the stated price of electricity for the group of companies will be from 52 to 57 copecks/ kWh. Similarly, for the transition period 343370-388100 MWh with the stated price from 89 to 99 copecks/kWh, respectively. For the summer period, these values are equal to 129250-143690 MWh and 127-142 copecks/kWh. It is with this proposal that the GC can enter the CMPE. Thus, the developed mathematical model allows to provide significant competitive advantages of the Russian energy industry [27].

6. Conclusion

The most significant results obtained are formulated, which should include the following:

A mathematical model has been developed for controlling the functioning of the GC in modern conditions;

A critical analysis of the existing criteria for the management of electric power system, which are not suitable in market conditions, has been carried out;

A criterion for controlling the operating modes of an energy facility that combines the technological features of the energy industry with new economic controls is proposed;

Principles and methods for managing the operation of a thermal power plant for the production of heat and electricity based on the principle of profit maximization have been developed;

A comprehensive methodology has been created for distributing fuel costs between the types of energy produced at CHP by the criterion of profit maximization;

A method has been developed for obtaining optimal operating conditions for thermal power plants that form a generating company, based on the principle of maximizing profits;

The calculated and experimental verification of the developed approaches and methods was carried out, as well as the implementation of the main provisions of the research at specific sites was carried out.

Acknowledgments

The article was prepared within development program of the Flagship Regional University on the basis of Belgorod State Technological University named after V.G. Shukhov.

Reference

[1] W G Chandler, P L Dandeno, A F Glimn and L K Kirchmayer 1953 Schort-Range economic operation a combined thermal and hydroelectric power system *AIEE Transactions* vol 72 part III p 1057

[2] C W Watchorn 1955 Coordination of hydro-and steam generation *AIEE Transactions* vol 74 part III p 142

[3] T V Chekalina 2003 Ensuring of the generating company competitiveness at the energy market due to the assignment of the optimum states of thermal stations *Proceedings of the 7th Russian-Korean International Symposium on Science and Technology KORUS-2003* (Ulsan: Korea) pp 62-4

[4] U A Sekretarev, T V Chekalina and B V Malosemov 2011 Administration Functioning Power Generation Companies by Criterion of Maximization Profit *Symposium of papers the 6th International Forum on Strategic Technology IFOST-2011* (Harbin University of Science

and Technology: Harbin China) vol 1 pp 491-4
[5] A Noskes 1963 Arismunender Bibliography on optimum operation of systems 1919 1959 *IEEE Trans. Power Apparatus and Systems* **3**
[6] E Y Handschin 2001 Deregulation and transmission management in the German electricity market *IEEE Liberalization and modernization of power systems operation and control problems* (Irkutsk: Energy Systems Institute)
[7] W F Reinke, B Meyer and C Ray 2000 *Power systems of the future and consequences of Re regulation* (CIGRE)
[8] S Sun and J K Jensen 2000 *Power system planning and development* (CIGRE)
[9] N G Ram 2000 The single-shaft combined cycle *Proceeding of ASME TURBOEXPO*
[10] A Pasha and R Allen 2003 Designing and modifying HRSG's for cycling operation *Power*
[11] Yu A Sekretarev, T V Myateg and B N Moshkin 2016 Optimizaciya rezhimov raboty` generiruyushhej kompanii na baze TE`Cz po vy`rabotke e`letkroe`nergii na osnove kriteriya maksimizacii priby`li [Optimization of operating modes of a generating company based on a heat and power plant for generating electricity on the basis of a criterion of profit maximization] *Novosti vysshikh uchebnykh zavedeniy Elektromekhanika* **4(546)** pp 82-8 [In Russian]
[12] 2015 A multiple objective decision making model for energy generation portfolio under fuzzy uncertainty Case study of large scale investor-owned utilities in Florida *Renewable Energy* vol 75 pp 224 -42
[13] A Fallahia, R Ebrahimib and S F Ghaderic 2011 Measuring efficiency and productivity change in power electric generation management companies by using data envelopment analysis A case study *Energy* vol 36 pp 6398 - 405
[14] S F Ghaderi, A Azadeh, B Pourvalikhan and N E Fathi 2012 Behavioral simulation and optimization of generation companies in electricity markets by fuzzy cognitive map *Expert Systems with Applications* vol 39 pp 4635- 46
[15] Seif Azghandi Kenneth Mark Hopkinson and Kennard Robert Laviers 2016 Benchmarking approach for empirical comparison of pricing models in DRMS *The Journal of Engineering* p 8
[16] A Christos 2017 Cogeneration Technologies *Optimization and Implementation* p 360
[17] A M Rosen and Dr S Koohi-Fayegh 2016 Cogeneration and District Energy Systems Modelling *Analysis and Optimization* p 344
[18] T V Chekalina, Yu A Sekretarev, B N Moshkin, V S Karmanov and K N Yakovchenko 2013 Povy`shenie e`nergeticheskoj e`ffektivnosti generiruyushhej kompanii za schet vy`bora optimal`ny`x rezhimov funkcionirovaniya po kriteriyu maksimizacii priby`li [Improving the energy efficiency of the generating company by choosing the optimal operating modes according to the criterion of profit maximization] *Nadezhnost' i bezopasnost' elektroenergetiki* **120** pp 35-40 [In Russian]
[19] B N Moshkin, V S Karmanov, Yu A Secretarev, T V Chekalina and K N Yakovchenko 2013 Optimizaciya rezhimov funkcionirovaniya TE`Cz kak sposob povy`sheniya e`nergeticheskoj e`ffektivnosti [Optimization of the functioning modes of heat and power plants by the way to increase energy efficiency] *Energiya Tatarstana* **3** pp 61 -7 [In Russian]
[20] T V Chekalina, Yu A Sekretarev, B N Moshkin and V S Karmanov 2013 Upravlenie rabotoj generiruyushhej kompanii na osnove kriteriya maksimizacii priby`li [Management of the generating company operation on the base of the maximization profit criterion] *Glavnyy energetik* **3** pp 39-45 [In Russian]
[21] V Y Olhovskiy, T V Myateg and M A Klepche 2017 The influence of high harmonics generation of low power consumers on up to 1000 V networks operation *IEEE Dynamics of Systems Mechanisms and Machines Dynamics 2016 proc of a meeting* (Omsk: Dynamics 2016) vol 1 pp 456-62
[22] Yu A Sekretarev, T V Myateg and B N Moshkin 2018 Matematicheskaya model` upravleniya funkcionirovaniem generiruyushhej kompanii v sovremenny`x usloviyax [Mathematical

model of the generating company operation in modern circumstances] *Novosti Tomskogo politekhnicheskogo universiteta. Inzhenerno-geologicheskiye resursy Vestnik Tomskogo politekhnicheskogo universiteta Geo As Inzhenernyye nauki* vol 329 pp 146-58 [In Russian]

[23] Yu A Sekretarev et el. 2017 *Optimal`noe upravlenie nagruzkoj e`nergosistem v sovremenny`x usloviyax* [Optimum load management of power systems in modern circumstances monograph] (Karaganda: KSTU) p 162 [In Russian]

[24] D N Haiman 1992 E`konomika e`nergetiki: osnovy` teorii [Modern microeconomics: the analysis and application] (Moscow: Finance and statistica) p 384 [In Russian]

[25] L B Melamed and N I Suslov 2000 *Sovremennaya mikroe`konomika: analiz i primenenie* [Economy of energetic bases of theory] (Novosibirsk: Rossijskaya akademiya nauk) p 180 [In Russian]

[26] V M Sincov and A V Bogoslovsky 1973 Optimizaciya rezhimov e`nergeticheskix sistem [Optimization of power systems modes] (Kiev: Vyshskaya shkola) p 274 [In Russian]

Adaptive correction of current longitudinal differential protection of power equipment

M G Popov

Department of Power Plants and Automation of Energy Systems, Peter the Great St.Petersburg Polytechnic University , Polytechnicheskaya St., 29, St.Petersburg, 195251, Russia

E-mail: PopovMG@eef.spbstu.ru

Abstract. The report reflects the long-term solutions for the modernization of main means of emergency control and protection of integrated power systems in the implementation of adaptive control algorithms. The substantiation of new structures and algorithms for the formation of corrective signals using non-stationary filters of the displacement current of overhead power lines is performed. Calculated, using the bias currents of the phases are used for dynamic correction of the working and brake signals of the longitudinal differential protection. The basis of adaptive means of differential protection of power equipment of power systems is developed. They are based on the use of adaptive correction of operating and brake protection signals. A mathematical description of the optimization problem of determining the response parameters of sensitive longitudinal current differential protection with adaptive correction of operating and braking signals is performed.

Key Words: power equipment, power pools, adaptive means, current differential protection, measuring elements, sensitivity enhancement, dynamic correction, Hilbert transform.

1. Introduction.

The existing financial and economic, and as well operational and technical interrelations between the electric power engineering entities are characterized by a reduction in costs for depreciation, ageing of the resource and poorly justified prolongation of the operating lifetime of the main equipment of power plants and substations. In these complicated conditions, ensuring the reliability and stability of the power supply in the integrated power system (IPS) can be partially solved by introducing into the operation practice a new electrical power equipment for regulation and control of the interchange power (FACTS), as well as the by creating of ring circuits of electric networks [1-8]. However, this approach significantly complicates the problem of fast and high-quality supervisory control of transient operation modes of power systems [9-11]. The appearance of new measuring and information systems, monitoring means (WAMS) of transient modes and systems of centralized (system) control allowed to some extent to improve the efficiency of dispatch and emergency control of the integrated power systems modes [12, 13]. An important role in this has the information and technical perfection of communication devices.

2. Methods and relevance of research area.

Many researches and scientific publications have been devoted to the issues of stability and reliability growth of electrical power equipment operation of power systems. By means of these studies, advanced control means of the excitation of synchronous machines have been created [2, 4, 11, 14, 15], the technical possibilities and conditions for the transmission of electric power by alternating current in extended power pools have been scientifically justified. Despite significant advances in research on sustainability and the principles of emergency control of the integrated power systems, there is much concern about the issue of ensuring the speed performance, selectivity and operating reliability of the protection and automation means. It must be emphasized that practically all (with rare exception) scientific research and publications of authors are based on the use of an idealized mathematical description and linear models of power equipment [1, 2, 4, 14, 16, 17], and of a robust description of the filter synthesis problem [15]. And by extension, such an approach is characterized not only by the presence of a method error, but also by errors in practical and quantitative assessments of the limit conditions and, as a consequence, in the choice of parameters of technical protection means and other means of emergency control that do not satisfy basic regulatory requirements.

The elimination of this substantial defect is possible only with the development of the theory of nonlinear filtration [3, 18-21], the application of numerical methods and rigorous mathematical models in the study of quasi-stationary asynchronous and oscillatory synchronous modes of the integrated power systems. Such approach becomes possible now due to growth of computing capacities and functional capabilities of industrial microcontrollers, and as well due to the increase in the automation level of modern power facilities. Obviously, the validity of the new obtained research results should be confirmed based on carrying out and improving physical, full-scale experiments and experimental-industrial trials.

All these current problems, the issues of qualitative and reliable control of modern integrated power systems, predetermined the subject of complex research, reflected in this report on the author's research.

3. Results of researches of adaptive protection and emergency control of electrical power equipment of extended power pools.

According to modern estimates, the degree of reactive power compensation in backbone networks is on average no more than 50% for 500-1500 kV lines [1, 3, 22]. The acuteness of this problem has increased even more recently due to a certain decrease in power consumption and, as a result, a decrease in active power flows with a significant undercompensation of the reactive power of extended electrical networks. Due to this situation, most of the large integrated power systems (IPS) have a very difficult situation with reliability and sensitivity of the main (differential) protections of electrical power equipment of the backbone network of 500-1500 kV - the minimum operating current is about 0.7-1.0 p.u. In addition, the problem of reliability of modern microprocessor differential protection systems is exacerbated by the large nomenclature and variety of used devices (half-sets), as well as the insufficient experience (lifetime) in operation of digital protections and the absence of a scientifically grounded approach to the selection (calculation) of their response parameters.

In connection with the abovementioned negative factors, cases of non-selective protection operation, accompanied by overloading of the lower voltage class lines entering the controlled sections, have become more frequent, which, eventually can cause a breakdown in the stability of the parallel operation of power pools.

To ensure the selectivity, increase the speed performance and sensitivity of the improved current lengthway differential protection of power equipment inter-system (interstate) power transmission lines, the adaptive correction of the response characteristic must be performed (figure 1). Approbation and evaluation of the efficiency of correction of the response characteristic of the current lengthway differential protection is performed using a database of digital oscillograms of full-scale experiments on the study of non-stationary modes of electrical power equipment of extended power pools.

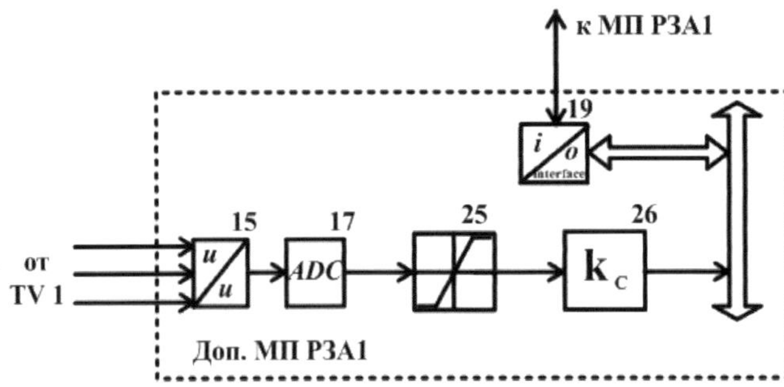

Figure 1. Generalized diagram of correction signals formation

To ensure the selectivity, increase the speed performance and sensitivity of the improved current lengthway differential protection of power equipment inter-system (interstate) power transmission lines, the adaptive correction of the response characteristic must be performed (figure 1). Approbation and evaluation of the efficiency of correction of the response characteristic of the current lengthway differential protection is performed using a database of digital oscillograms of full-scale experiments on the study of non-stationary modes of electrical power equipment of extended power pools.

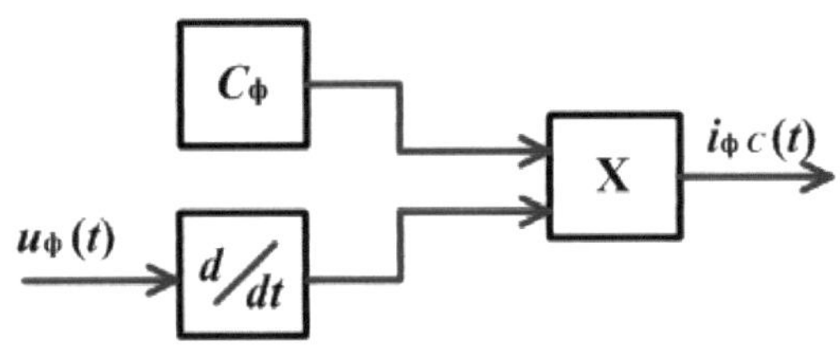

Figure 2. The filter circuit of the phase charging |current line with the control of the derivative of the phase voltage

In the block diagram of figure 2, the phase voltage is numerically differentiated. This method is most effective for increasing the sensitivity and selectivity of the line protection in the modes of its inclusion at idle.

An alternative is the use of nonlinear capacitive current filters of the line, implemented according to the scheme of figure 3.

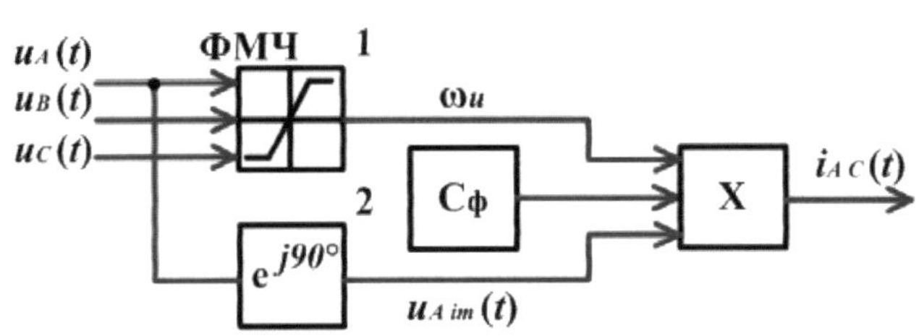

Figure 3. The filter circuit of the bias current of transmission line orthogonal control additions by Hilbert for the phase (shown only for phase A) voltage

In this scheme, to obtain the imaginary part of the $u_{A\ im}(t)$ analytical phase signal (the figure shows the example of phase A) voltage filter is used Hilbert [23, 24], which is not physically feasible. However, its application in the practice of digital signal processing is widely known [25]. At the output of the phase-reversal link (block 2, figure 3), an orthogonal complement $u_{A\ im}(t)$ of the phase A voltage is formed with a shift of the instantaneous spectral density $S(\omega)$ by ninety degrees. Instantaneous frequency measurement (block 1, figure 3) is also performed using the Hilbert transform [23-25].

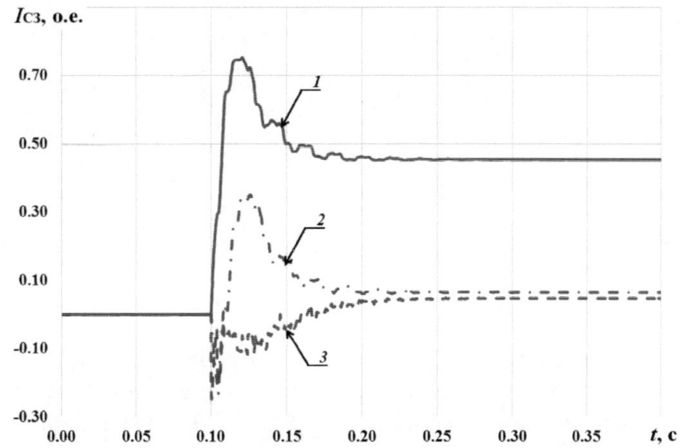

Figure 4. Change of the phase C protection tripping current in the power line idle mode from the opposite side of the substation in the absence of (1, solid line) and the presence (2, dash- dotted line and 3, dashed line) of correction signals

The calculated characteristics of the change in the protection response current (dashed and dashed lines characteristics 2, 3), as well as the differential signal (solid line, characteristic 1) are shown in figures 4, 5. Due to the limitations on the volume of illustrative material, these figures reflect the change in only one, the most characteristic phase current of the protection operation. Since the C phase current has the largest of the three phases of the value for the mode of switching on the idle transmission line, its calculated dependences are shown in figures 2, 3. Accordingly, to assess the selectivity and sensitivity in the mode of interphase (AB) short circuit, the most characteristic are the calculated oscillograms of the differential current of phase B (figure 5) smaller in comparison with phase A values current.

The change in the current of protection operation during the control of the derived voltage corresponds to the characteristics 3 (dashed line, figures 4, 5). The practical implementation of this scheme (figure 2) introduces significant difficulties associated with the need to increase the sampling rate over time (ADC survey), as well as to ensure the selectivity of the protection operation in case of internal short circuits. As follows from the characteristics 3 presented in figure 5, the application of the considered method of correction can lead to a deep (more than one and a half), non-selective overstrain in case of short circuits.

The only measures that allow to justify the admissibility of this method of forming correction signals are the introduction of blocking, after the successful inclusion of the line at idle or the introduction of a limit of its operating time in the test cycle. Taking into account the uncertainty of the criteria for the successful inclusion of the line, as well as the difficulties of implementing the logic of the blocking body, this issue is not considered in the publication and can be the subject of future, promising research.

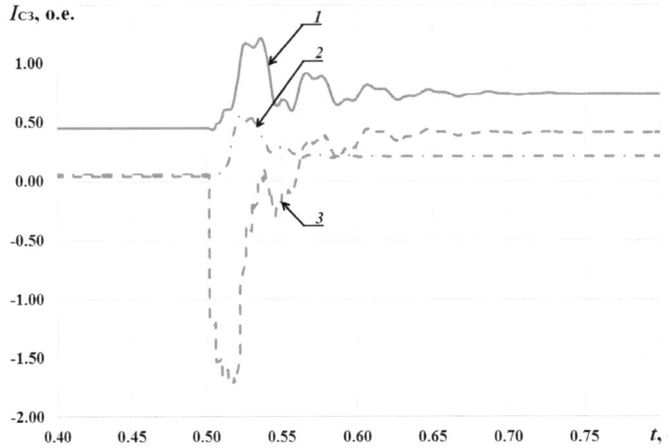

Figure 5. Change of the phase *B* protection tripping current in the internal two-phase (*AB*) short circuit mode near opposite side substation buses in the one-way power supply circuit in the absence of (1, solid line) and the presence of (2, dash-dotted line and 3, dashed line) corrections

Analysis of the change in the current of protection operation in the correction of its dynamic properties with the use of bias current filters (figures 3) showed that the proposed method can be effective in a variety of modes of operation of intersystem power lines with proper and reasonable selection of the parameters of operation of differential protection. The choice of the parameters of operation of the protections meeting the requirements of standard sensitivity is made according to the below methodical instructions.

Figure 6 shows the conditions of normative sensitivity and selectivity, which corresponds to the space of the protection operation parameters for which the white surface (the vector function of the protection operation parameters) prevails over the shaded for the damaged phases (phase A, figure 6,a). The selective operation of the measuring elements of undamaged phases (phase C, figure 6,b) is characterized by opposite conditions for the arrangement of surfaces.

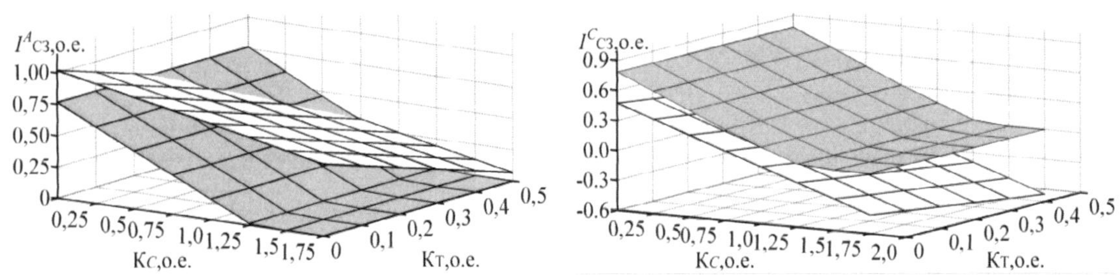

a) protection tripping current of phase A b) protection tripping current of phase C

Figure 6. Protection tripping currents of phases A (a) and C (b) in the space of parameters K_T and K_C when the line is switched to idle (shaded surface) and in short-circuit mode of phases A and B (white surface)

Taking into account the complexity of the perception of new principles for calculating the triggering parameters of current differential protections, the author, in the concluding part, presents a mathematical description of the method for selecting the operating values ($I_{prot\ min}$, K_T, K_C) of advanced current lengthway differential protections with dynamic correction of their operating and brake signals.

Then, the vector functions of the selectivity of protection in idle mode $W_{I_{\text{prot select idle}}}$ and short circuit mode $W_{I_{\text{prot select. fault}}}$ (figure 7) are described by expressions (with explanations only for phase A):

$$W_{I_{\text{prot select idle}}} = I_{\text{prot }min} - \max\left\|\begin{matrix} I^{A}_{\text{prot idle }max}\left(K_C\right) \\ I^{B}_{\text{prot idle }max}\left(K_C\right) \\ I^{C}_{\text{prot idle }max}\left(K_C\right) \end{matrix}\right\| \geq 0, \qquad (1)$$

$$W_{I_{\text{prot select fault}}} = \min\left\|\begin{matrix} I^{A}_{\text{prot fault }min}\left(K_T, K_C\right) - I^{A}_{\text{prot idle }max}\left(K_C\right) \\ I^{B}_{\text{prot fault }min}\left(K_T, K_C\right) - I^{B}_{\text{prot idle }max}\left(K_C\right) \\ I^{C}_{\text{prot fault }min}\left(K_T, K_C\right) - I^{C}_{\text{prot idle }max}\left(K_C\right) \end{matrix}\right\| \geq 0, \qquad (2)$$

where $I^{A}_{\text{prot idle }max}$, $I^{A}_{\text{prot fault }min}$ - the maximum and minimum values of the operating current of phase A protection in the idle and short circuit modes, p.u., respectively.

The requirements for protection according to the conditions of a given sensitivity ($K_{sense\ set}$) relative to the target set point ($I_{prot\ min}$) are described by a vector function:

$$W_{I_{\text{prot sense fault}}} = \frac{I^{min}_{\text{prot fault }min}\left(K_T, K_C\right)}{I_{\text{prot }min}} - K_{\text{sense set}} \leq 0, \qquad (3)$$

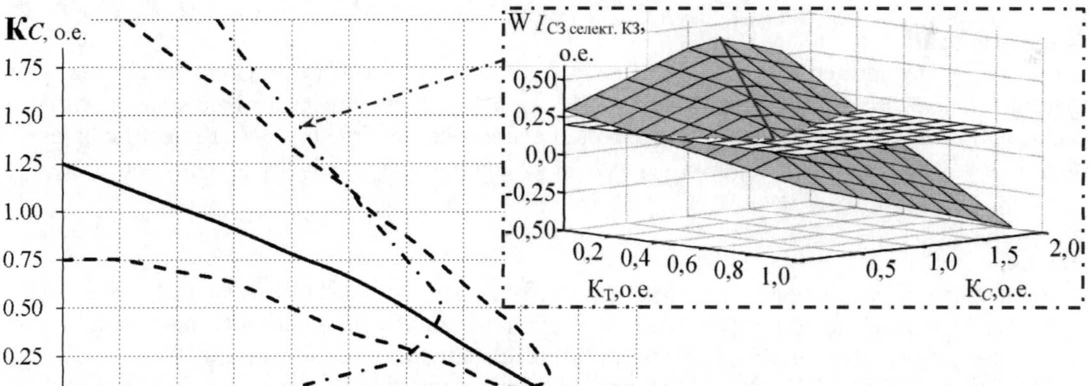

Figure 7. The selectivity area (dot-dash line), the sensitivity area (dashed line) and the characteristic (solid line) of the maximum selectivity and sensitivity of protection in the space of the braking parameters (K_T) and correction (K_C)

Absolute selectivity of the protection, characterized by zero value of the minimum operating current (Iprot min = 0), is achieved with 100-120% correction of operating signals (figure 7, small area K_T = 0-0.2 p.u.). From the analysis of figure 7 it follows that the normative sensitivity is achieved with the braking coefficient $K_T \leq 0{,}2$ p.u. and the correction coefficient K_C, equal to 0,70-1,25 p.u. The increase of the braking coefficient more than K_T = 0,35-0,40 p.u. leads to a narrowing of the selectivity area (figure 7).

4. Conclusion

The research and development carried are aimed at improving the basic technical high-response means of emergency control of power systems that ensure the continuity of their operation, the quality of electrical energy, economy, technical, information security and automation of its production.

1. The author has developed and justified new structures and algorithms for the formation of correction signals using non-stationary filters of the bias current of overhead power lines. Phase bias currents calculated with their help are used for dynamic correction of operating and braking signals of longitudinal differential protection.

As a result of the comparative analysis of digital filters of bias current it is established that the best indicators in terms of ease of implementation, sensitivity and speed have filters, which are based on the Hilbert transform. All further evaluation of the effectiveness of the correction of the dynamic properties of protection obtained in its application.

2. The basis of adaptive means of differential protection of power equipment of power systems is developed. They are based on the use of adaptive correction of operating and brake protection signals. Self-tuning of protection means is achieved as a result of applying gradient parametric numerical methods for identifying the parameters of power equipment that affect the value of the minimum operating protection current. Absolute selectivity of protection, characterized by zero value of the minimum operating current ($I_{prot\ min}$ = 0), is achieved only by correction of operating signals in the volume of 100-120%.

3. A mathematical description of the optimization problem of determining the response parameters (*I*prot min, K_T and K_C) of sensitive longitudinal current differential protection with adaptive correction of operating and braking signals is performed.

As a result of testing of this technique, it is shown that the absolute selectivity of the protection, characterized by a zero value of the minimum operating current ($I_{prot\ min}$ = 0), is achieved at the level of correction signals of 100-120% or more ($K_C \geq$ 1,2 p.u.).

Acknowledgments

The research was conducted with the financial support under contract Peter the Great Saint-Petersburg Polytechnic University (SPbPU) No. 13.8886.2017/warhead "Improvement of the theory of working processes of production, conversion and transport of energy and the development of scientific bases of design of highly efficient thermal power plants, energy and transport machines, systems and complexes".

References

[1] Zakharova E V 2016 Kriteriy bystroy otsenki staticheskoy ustoychivosti integrirovannykh energosistem, dis.kand.tekhn.nauk [Criterion of rapid assessment of static stability of integrated power systems, PhD thesis] (St. Petersburg: SPbPU) [In Russian]
[2] Lyubarskiy D 2006 Povysheniye ustoychivosti funktsionirovaniya ustroystv releynoy zashchity i protivoavariynoy avtomatiki v usloviyakh elektromagnitnykh i elektromekhanicheskikh perekhodnykh protsessov, dis.d-ra.tekhn.nauk [Improving the stability of relay protection and emergency control devices in electromagnetic and Electromechanical transients, ScD thesis] (Moscow) [In Russian]
[3] Popov MG 2018 Sovremennye sredsva protivoavariynogo upravleniya obyedineniykh energosistem, dis.d-ra.tekhn.nauk [Modern means of emergency control of integrated power systems, ScD thesis] (St. Petersburg) [In Russian]
[4] Ragozin AA 1998 Obobshchennyy analiz dinamicheskikh svoystv energoobyedineniy na osnove strukturnogo podkhoda: dis. d-ra tekhn.nauk [The generalized analysis of dynamic properties of power connections on the basis of the structural approach, ScD thesis] (St.

Petersburg) [In Russian]
[5] Akke M Some. 1997 Control Applications in Electric Power Systems. Doctoral Dissertation. Department of Industrial Electrical Engineering and Automation (Sweden: Lund University).
[6] E.Martinez M 2002 *Proc. Int. Conf. on Power Systems and Communication Systems Infrastructures for the Future* (Beijing) p 517.
[7] Ragozin AA and PopovMG 2002 An analysis of effectiveness of using controllable shunting reactors in system-forming networks of power pools *Power of Russia* **2**
[8] Ragozin AA and PopovMG 2002 On the effect of the moments of nonconservative forces on the static stability of power systems *Power of Russia* **10**
[9] Ellithy K 2014 Design of power system stabilizers based on µ-controller for power system stability enhancement *International journal of electrical power & energy systems* **63**
[10] Saravana Ilango G 2009 Control algorithms for control of real and reactive power flows and power oscillation damping using UPFC *Electric power systems research* **79**
[11] Mahmud M 2014 Transient stability enhancement of multimachine power systems using nonlinear observer-based excitation controller *International journal of electrical power and energy systems* **58**
[12] Perez S 2006 Modeling relay for power system protection studies Doctoral Dissertation Department of Electrical Engineering (Canada: University of Saskatchewan Saskatoon).
[13] Phadke A 2009 Synchronized Phasor and Frequency Measurement Under Transient Conditions *IEEE Transactions on Power Delivery*
[14] Yusufov I X 2003 Metodicheskiye ukazaniya po ustoychivosti energosistem Utverzhdeny prikazom Minenergo Rossii № 277 [Guidelines for the stability of power systems Approved by the order of the Ministry of Energy of Russia No. 277] [In Russian]
[15] Mokeyev AV 2011 Metody i algoritmy obrabotki signalov v intellektualnykh elektronnykh ustroystvakh energosistem razlichnogo funktsionalnogo naznacheniya: dis. d-ra tekhn. nauk [Methods and algorithms of signal processing in intelligent electronic devices of power systems for various functional purposes, ScD thesis] (St. Petersburg) [In Russian]
[16] Bessolitsyn AV 2010 Polevyye metody opredeleniya elektricheskikh parametrov vozdushnykh liniy elektroperedachi, dis. kand. tekhn. Nauk [Field methods for determining electrical parameters of overhead power lines, PhD thesis] (St. Petersburg) [In Russian]
[17] Popov MG, Zakharova EV and Sinyanskiy IV and Goryachevskiy KS 2015 Issledovaniye staticheskoy ustoychivosti slozhnykh obyedinennykh energosistem proizvolnoy struktury [Study of static stability of complex interconnected power systems of arbitrary structure] *Elektricheskiye stantsii* **11** pp 61-9 [In Russian]
[18] Vanin VK, Popov MG and Popov SO and Ambrosovskaya T 2015 Povysheniye dostovernosti raboty izmeritelnykh tsepey releynoy zashchity [The reliability of the operation of measuring circuits of the relay protection] *Elektricheskiye stantsii* **11** pp 30-5 [In Russian]
[19] Mokeev AV et al. 2011 *Digital Filters InTech*
[20] Popov MG, Vanin VK and Popov SO. 2016 Increasing the reliablity of the measuring circuits of protection relays *Power Technology and Engineering* **50**
[21] Qing-quan Jia 2010 A novel error compensation algorithm for three-current transformer connection *Electric Power and Systems* **32**
[22] Popov MG 2018 Differentsialnaya zashchita mezhsistemnykh liniy elektroperedachi s kompensatsiyey zaryadnoy moshchnosti [Differential protection of intersystem power lines with charging power compensation] *Nauchno-tekhnicheskiye vedomosti SPbPU Estestvennyye i inzhenernyye nauki* [In Russian]
[23] Harkevich A 1962 *Spektry i analiz* (Moscow: GIFML) [In Russian]
[24] Hemming R 1980 *Cifrovye fil'try* (Moscow: Sov. Radio) [In Russian]
[25] Gonorovskij I 2006 *Radiotekhnicheskie cepi i signaly* (Moskva: Drofa) [In Russian]

Methodology of GHG emissions assessment caused at the construction of energy facilities. Case study: Hydropower

I P Avilova, A E Naumov and M O Krutilova

Department of Assessment and Management of Real Estate, Belgorod State Technological University of V.G. Shukhov, Kostyukov St., 46, Belgorod, 308012, Russia

E-mail: kafeun@mail.ru

Abstract. As energy is the main 'fuel' for social and economic development, and since energy-related activities have significant environmental impacts, it is important to lower emissions and stabilize atmospheric CO_2 levels to avoid the worst predicted effects of climate change. Reducing energy consumption and costs is becoming central to planning, construction, and use of energy construction facilities from an environmental and economic point of view. Each energy source has advantages and disadvantages referred to: operating costs, environmental impact, and other factors. Each generation method produces some greenhouse gases (GHG) in varying quantities through construction, operation, and decommissioning. Some generation methods like coal fired plants release the majority of GHGs during operation. Others, such as wind power and hydro power release the majority of emissions during construction and decommissioning. Normalizing the lifecycle emissions with electrical generation allows a fair comparison of different generation methods on a per gigawatt-hour basis.

Keywords: GHG emissions, hydropower plant construction, eco-oriented assessment of the estimated cost, sustainability, green building.

1. Introduction

The changes in the energy industry over the past 20 years have been significant. The growth in energy consumption has been higher than anticipated even in the high-growth scenarios. The energy industry has been able to meet this growth globally assisted by continuous increases in reserves' assessments and improving energy production and consumption technologies. The results of the 2013 WEC World Energy Resources survey show that there are more energy resources in the world today than 20 years ago, or ever before [1].

Providing the benefits of electricity to hundreds of millions of people around the World is a key challenge of this century. In the International Energy Agency's World Energy Outlook 2010, global energy demand was expected to rise 1.4% per year on average to 2035, assuming no change in current business-as-usual energy policy [2]. In 2010, actual global energy use jumped by 5.6%, the largest single year increase since 1973. The current global energy mix remains heavily weighted towards conventional fossil fuels. Coal's share of global energy consumption was 29.6%, the highest since 1970. By 2030, it is expected that World energy consumption will rise from just under 12 btoe (billions of tonnes of oil equivalent) to over 16 btoe, with much of this growth occurring in non-OECD countries, particularly China and India [3]. As energy is the main 'fuel' for social and economic development, and since energy-related activities have significant environmental impacts, it is important to lower emissions and stabilize atmospheric CO_2 levels to avoid the worst predicted effects of climate change. From an environmental and economic point of view, reducing energy consumption and costs is therefore becoming central to planning, construction, and use of energy construction facilities.

Energy facilities' emissions should be analyzed according to a life cycle analysis (LCA) approach, including emissions at the construction stage [4]. Life cycle GHG emissions for an energy generation facility include emissions associated with the construction and eventual decommissioning of the facility which called indirect emissions, as well as any emissions resulting from the facility's operation, which called direct emissions. Power plants consist of:

- Energy construction property (ECP),
- Energy construction equipment (ECE).

Power plant's emissions should be analyzed according to a life cycle analysis (LCA) approach, including emissions at the construction phase of the facility (indirect emissions).

Table 1. Categories of LCA emissions.

Energy construction property (ECP)	**Energy construction equipment (ECE)**
Indirect emissions	**Direct emissions**
Infrastructure	Combustion of fuels
Construction and installation work	Operational fuel use
Building Materials	Other emissions from operation (e.g., flooded land)
Transport	Goods and services consumed during operation
Decommissioning and waste disposal	

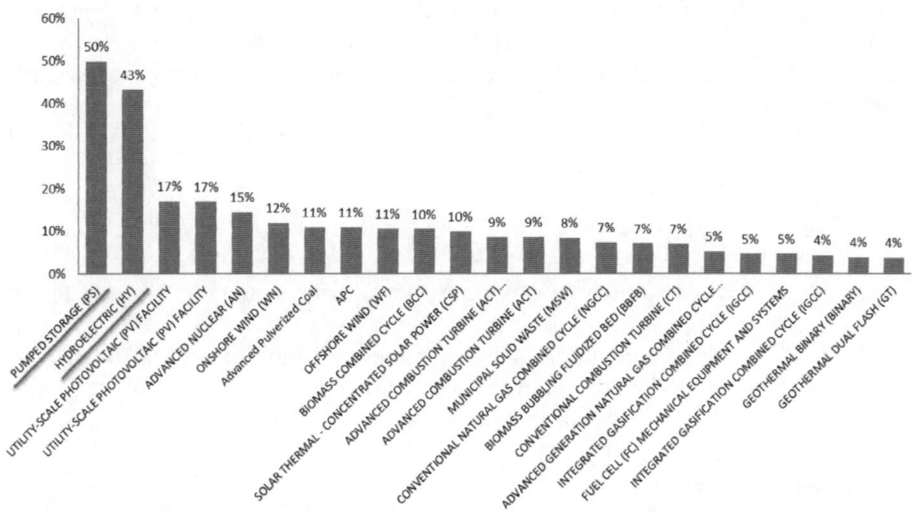

Figure 1. Rank of power plant types by Construction Material and Installation Cost (CMIC) in Total Project Cost (TPC).

Sources of indirect emissions for power plants include emissions from: development of infrastructure (e.g., roads and transmission lines), construction and installation work on the facility itself, manufacturing of building materials and equipment, transportation of materials and workers, decommissioning and waste disposal. According to Dones et al. [5], the major sources of GHG emissions for hydropower within these categories include cement and steel production, and the use of diesel and electricity. Raadal et al. [6] states that "the major contributing factors to the infrastructure GHG emissions are

IOP Conf. Series: Materials Science and Engineering **552** (2019) 012018 doi:10.1088/1757-899X/552/1/012018

concrete production and the transportation of rocks in the construction of dams and tunnels”. In order to substantiate the relevance of the topic, we should to compare the civil structural material and installation cost (CSMIC) in relation to total project cost (TPC) for a range of different power plants types. It’s important to note that there are many energy facilities that require high construction costs. Using the data U.S. Energy Information Administration [7] a ranking was conducted figure 1.

2. Materials and Methods

We propose using the following methodologies:

- Methodology for mathematical relation of the scope of main energy-consuming works and building materials, and the capacity of the hydropower plants;
- Methodology assessment of GHG emissions per unit for basic building materials applied to construction of hydropower plants (concrete, steel framework, coarse aggregates for backfilling, etc.);
- Methodology for determining of GHG emissions per unit for the main works carried out in-situ to construct the structures of hydropower stations (soil development, stone development, soil and cofferdam filling, pouring the concrete, resources delivering, etc.).

2.1. Methodology of determining the scope of main energy-consuming works and building materials, depending on the capacity of the hydropower plants.

Approximation method. For determining of possible correlation among volumes of basic construction materials and construction works consumed at modern HPS’ construction and generating power of HPS in MW non-linear regression analysis was carried out. In Nonlinear regression analysis observational data are modeled by a function which is a nonlinear combination of the model parameters and depends on one analyzed variable. For approximation in techno-economical processes what are always consist of non-linear (irregular, stochastic) and linear (regular, systematic) parts, K-polynomials proposed by author are useful, relevant and confident [9].

K-polynomial of nth degree means the symmetrical mathematical expression of normal and inverse powered variables as follows:

$$Y = a_{(-n)}x^{n} + a_{(-(n-1))}x^{n-1} + \ldots + a_0x^0 + \ldots + a_{n-1}x^{-(n-1)} + a_nx^{-n}, \quad (1)$$

where a_i – constants, x – single variable, x^0 – dummy term (always equal to 1), used for structure’s clearness.

Left part of K-polynomials (before dummy) used for approximation of non-linear parts in approximated processes, right one (after dummy) used for approximation of linear parts in approximated processes. Proposed K-polynomial could be easily converted to:

1. linear function ($i = 1$; $a_{(-1)}\ldots a_0 = \text{const}$, $a_1 = 0$);
2. polynomial of nth degree ($a_{(-n)}\ldots a_0 = \text{const}$; $a_1\ldots a_n = 0$);
3. exponential of nth degree ($a_n\ldots a_0 = \text{const}$; $a_{(-n)}\ldots a_{(-1)} = 0$);

or remain complex to unite advantages of all above types. These confirms utility of the proposed expression in techno-economical analysis to handle multifactorial processes.

Correlation among volumes of basic construction materials/works and generating power of HPS offered to be expressed as K-polynomial of 1st degree as follows:

$$Y = ax + bx^{-1}, \quad (2)$$

where a, b are constant, x – analyzed variable.

IOP Conf. Series: Materials Science and Engineering **552** (2019) 012018 doi:10.1088/1757-899X/552/1/012018

All variables analyzed separately (number of freedom degrees is 1, variables row consist of 8 numbers), according initial suggestion for significant correlation of designed power of HPS and following:
a) basic construction works:

- Earth Excavation,
- Rock Excavation,
- Earthfill Dam,
- Cofferdam construction,
- RCC Dam construction.

b) basic construction materials:

- Structural steel,
- Concrete.

The data are fitted by a least squares method — standard approach in regression analysis to the approximate solution. Significance of models checked by R^2 adjusted determination. Models characterized by $R^2_{adj} > 0.75$ are significant. Utility of models was verified by Fisher's testing at fixed confidence level 0.95. Models characterized by F-test $<$ F (0.05;1;6) are utility and could be taken into account for further analysis.

2.2. Methodology assessment of GHG emissions per unit for basic building materials applied to construction of hydropower plants (concrete, steel framework, coarse aggregates for backfilling, etc.).
The second methodology considers GHG emissions per unit for basic building materials such as concrete and steel. Using the emission data, we can determine the certain amount of emissions per unit volume. To do this better use a special calculator, allowing to get the current value of GHG emissions by entering the value of building material and selecting parameters [10].

2.3. Methodology for determining of GHG emissions per unit for the main works carried out in-situ to construct the structures of hydropower.
The third methodology considers determining GHG emissions per unit for the main works carried out in-situ to construct the hydropower station. Using the data of U.S. Environmental Protection Agency [11] GHG emissions for excavators per 1 m^3 were determined as following in table 1.

3. Results

3.1. Approximations for Russian power plants.

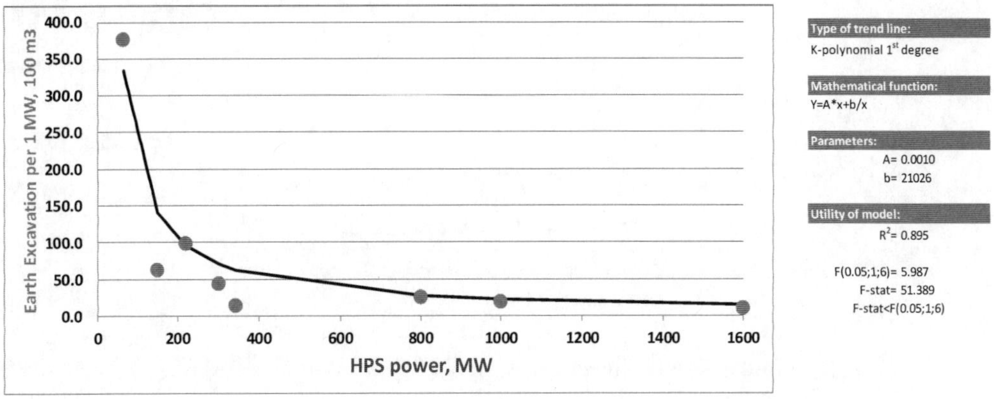

Figure 2. Non-linear regression for Earth Excavation per 1 MW.

IOP Conf. Series: Materials Science and Engineering **552** (2019) 012018 doi:10.1088/1757-899X/552/1/012018

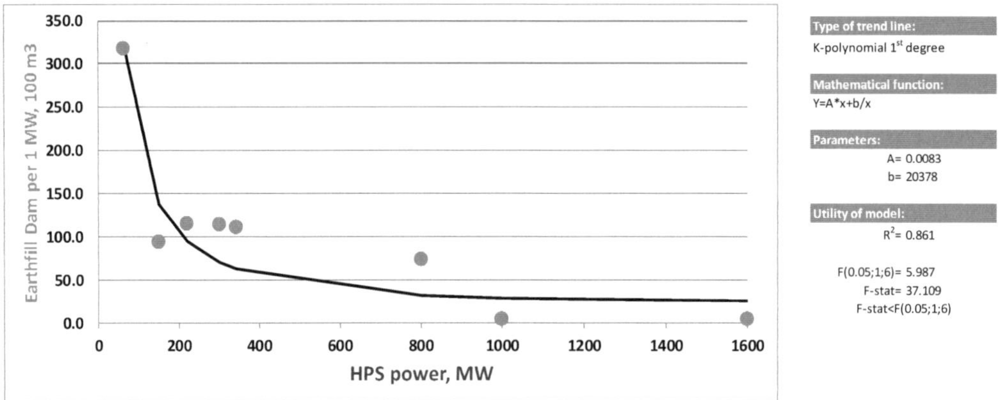

Figure 3. Non-linear regression for Earthfill Dam per 1 MW.

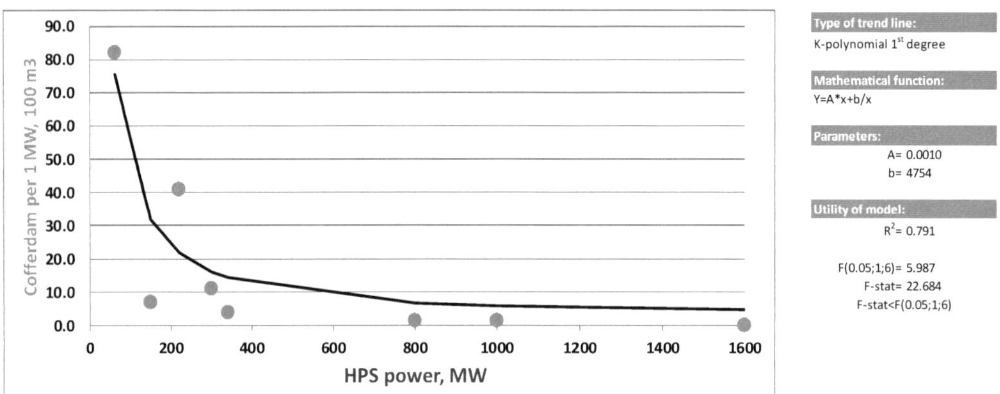

Figure 4. Non-linear regression for Cofferdam per 1 MW.

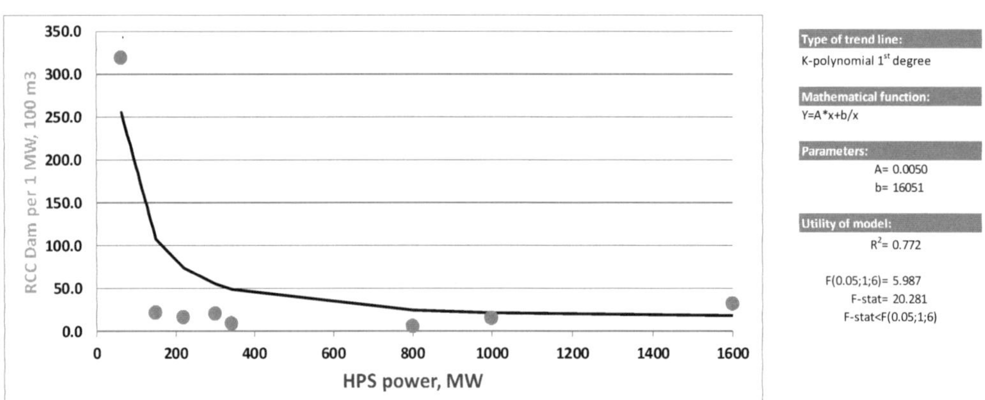

Figure 5. Non-linear regression for RCC Dam per 1 MW.

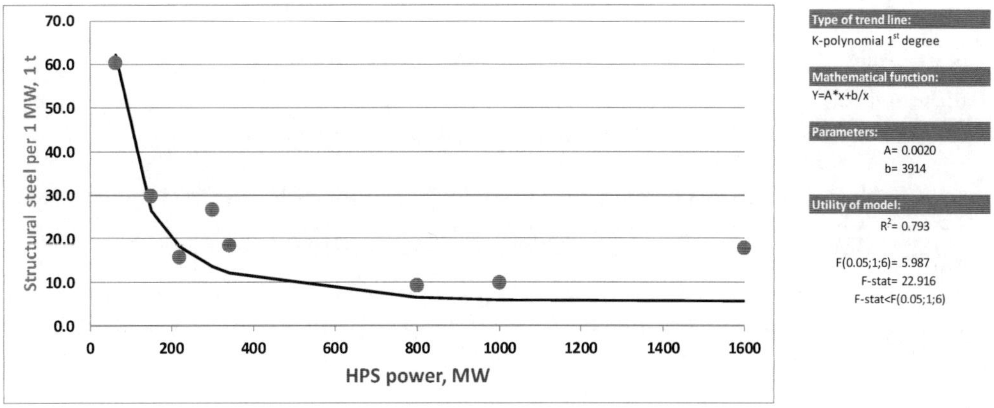

Figure 6. Non-linear regression for Structural steel per 1 MW.

3.2. Approximations for Russian and foreign power plants.

Table 2. Earth excavation consumed at modern HPS' construction, general [13].

HPS power, MW	63	150	220	300	342	400	476	600	707	800	824	1000	1600	1224	1800
Earth Excavation m^3	1000	2370	922	2148	1309	480	200	200	688	2500	2022	1000	1850	1550	4000	4000

* blue marked — foreign plants (USA, Canada), not marked – Russian plants.

Table 3. Earth excavation consumed at modern HPS' construction, per 1 MW.

HPS power, MW	150	220	300	342	400	476	600	707	800	824	1000	1600	1224	1800
Earth Excavation per 1 MW, 100 m^3	61.5	97.6	43.6	14.0	5.0	4.2	11.5	35.4	25.3	12.1	18.5	9.7	32.7	22.2

* blue marked — foreign plants (USA, Canada), not marked – Russian plants.

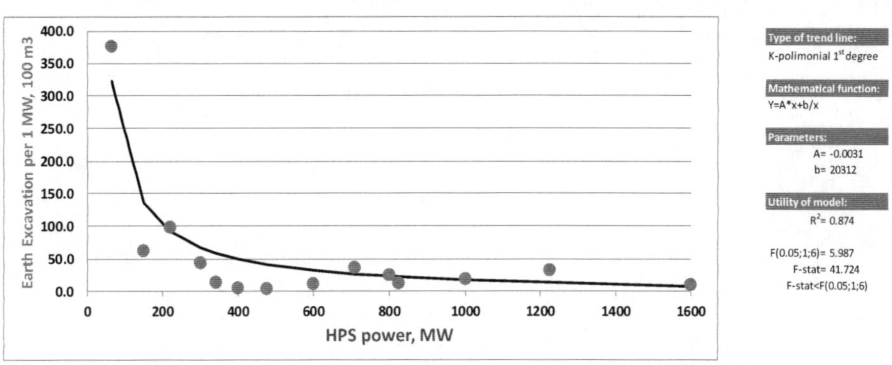

Figure 7. Non-linear regression for Earth Excavation per 1 MW.

Table 4. Cofferdam construction consumed at modern HPS' construction, general [14].

HPS power, MW	63	150	220	300	342	400	476	800	824	1000	1224	1600
Cofferdam, 1000 m^3	517	105	901	335	129	180	180	107	300	133	600	500

* blue marked — foreign plants (USA, Canada), white marked – Russian plants

Table 5. Cofferdam construction consumed at modern HPS' construction, per 1 MW.

HPS power, MW	63	150	220	300	342	400	476	800	824	1000	1224	1600
Cofferdam per 1 MW, 100 m^3	82.1	7.0	41.0	11.2	3.8	4.5	3.8	1.3	3.6	1.3	4.9	3.1

* blue marked — foreign plants (USA, Canada), white marked – Russian plants

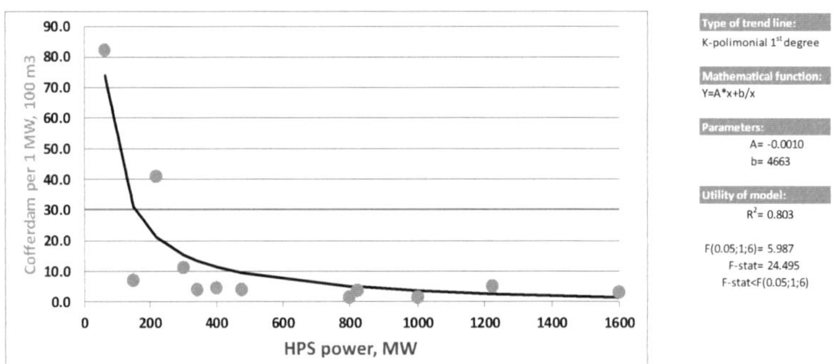

Figure 8. Non-linear regression for Cofferdam per 1 MW.

Table 6. RCC Dam construction consumed at modern HPS' construction, general [15].

HPS power, MW	63	65	150	220	300	342	600	707	800	824	1000	1224	1600	2028
RCC Dam, 1000 m^3	2010	1790	320	354	602	283	4893	700	447	300	1491	700	5119	700

* blue marked — foreign plants (USA, Canada), white marked – Russian plants

Table 7. RCC Dam construction consumed at modern HPS' construction, per 1 MW.

HPS power, MW	63	65	150	220	300	342	600	707	800	824	1000	1224	1600	2028
RCC Dam per 1 MW, 100 m^3	319	276	21.3	16.1	20.1	8.3	81.6	9.9	5.6	3.6	14.9	5.7	32.0	3.5

* blue marked — foreign plants (USA, Canada), white marked – Russian plants

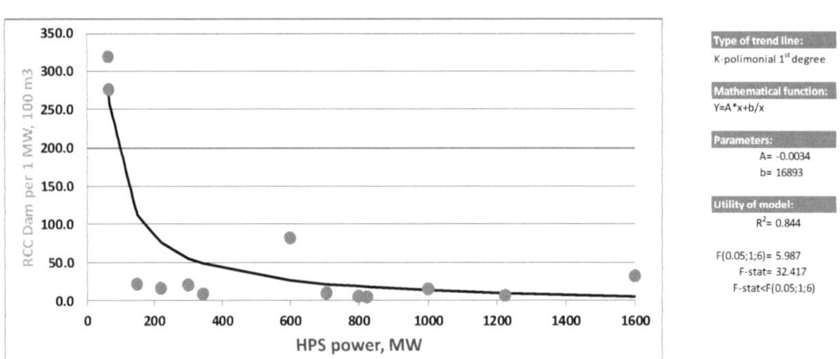

Figure 9. Non-linear regression for RCC Dam per 1 MW.

Table 8. Concrete consumed at modern HPS' construction, general [16].

HPS power, MW	**63**	**65**	**400**	**476**	**600**	**707**	**1224**	**1800**	**2028**
Concrete, 1000 m3	30	40	200	200	268	290	400	550	600

* blue marked — foreign plants (USA, Canada), white marked – Russian plants

Table 9. Concrete consumed at modern HPS' construction, per 1 MW.

HPS power, MW	**63**	**65**	**400**	**476**	**600**	**707**	**1224**	**1800**	**2028**
Concrete per 1 MW, m3	476	615	500	420	447	410	327	306	296

* blue marked — foreign plants (USA, Canada), white marked – Russian plants

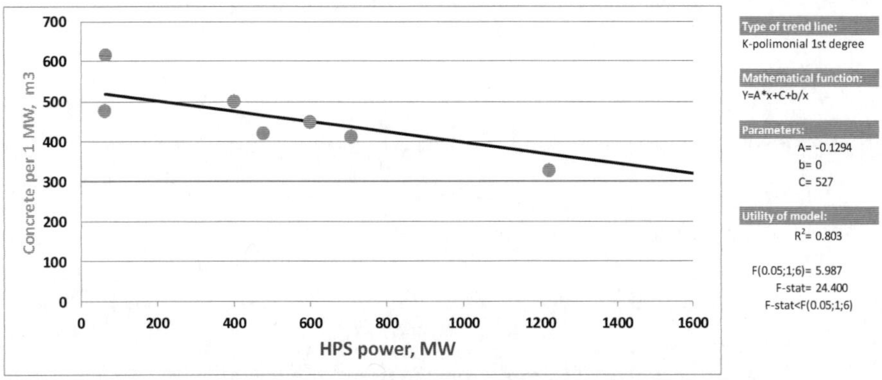

Figure 10. Close to linear regression for RCC Dam per 1 MW.

Volumes of basic construction materials and construction works consumed at modern HPS' construction have significant correlation with volume of generating power of designated HPS in MW. Correlations are approximated by proposed K-polynomial of 1st degree, suitable for further analysis and management [17]. Volumes of basic construction materials and construction works per 1 MW of generated power inversely depend on total designated power of HPS, decrease by power increasing drastically till designated power 400 MW and slightly after 400 MW till 1000-1500. Each obtained MW of HPS energy after 600-800 MW of designated power costs equally, so the more power HPS designed for the more profitable and less environmentally harmful energy is.

3.3. Methodology assessment of GHG emissions per unit for basic building materials applied to construction of hydropower plants (concrete, steel framework, coarse aggregates for backfilling, etc.)

Table 10. GHG Emission calculator: Basic materials [18, 19].

Initial data		**GHG Calculation**	
Steel volume, t	150		
Production process	ISM	Estimated GHG emission per t, t	2,0
		Total GHG emission for the whole scope of material, t	**300**
Concrete volume, cub.m	58000	Estimated GHG emission per 1 cub.m, kg	1,51
		Total GHG emission for the whole scope of material, t	**87,6**
		Total GHG emission, t	**387,6**

3.4. Methodology for determining of GHG emissions per unit for the main works carried out in-situ to construct the structures of hydropower.

Table 11. Power-productivity of excavators [20].

Excavator	Type 1	Type 2	Type 3	Type 4	Type 5
Power, kW	150	250	350	500	700
Bucket capacity, m^3	1,5	3	5	10	20
Full cycle time, s	25	35	45	55	60
Productivity, m^3 / hour	216	309	400	655	1200

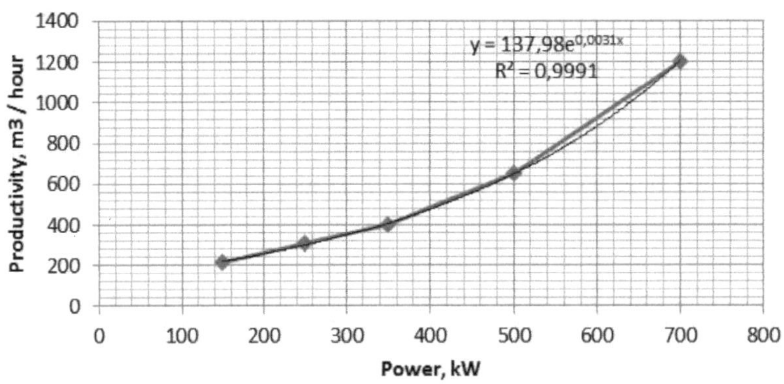

Figure 11. Power-productivity of excavators.

Table 12. GHG Emission for excavators [21].

Excavator	**Komatsu GD655**	**Cat D4CXL**	**Cat D4CXL**	**Comatsu C340C**	**John Deere 310G**
Productivity, cub.m/hr	360	432	409	525	514
Power. kW	128	232	255	330	349
GHG Emission, g/kW-hr	10,372	11,84	8,715	16,145	11,83
GHG Emission, g/hr	1327,62	2746,88	2222,33	5327,85	4128,67
GHG Emission per 1 m^3, g	3,688	6,359	5,434	10,148	8,032

4. Conclusion

Thereby using all three engineering techniques, we can independent and simultaneously get an estimation of GHG emissions at the construction stage of HY power plant of different capacities, constructed with different building equipment and distances for material's delivery. As a main result of the research we have GHG Emission calculator for different kinds of main construction works and materials. Using the technique engineer in general can calculate the total GHG emissions caused by construction of different types, using a design and estimate documentation of a power plant and GHG Emission calculator.

For large hydropower projects, the capital costs are dominated by the civil works. The cost of civil works is influenced by numerous factors pertaining to the site, the scale of development and the technological solution that is most economic [22]. Nearly 500 hydropower projects totaling more than 50,000 MW have been served by Black & Veatch worldwide [23]. The Black & Veatch historical database incorporates a good understanding of hydroelectric costs. Black & Veatch used this historical background to develop the cost estimates vetted in the WREZ stakeholder process and to subsequently update that pricing and adjust owner's costs as necessary [24].

Table 13. GHG Emission calculator: Hydropower plant construction.

INITIAL DATA			
Plant's parameters		**Volume**	
Designated power, MW		250	
Materials' transportation average distance, km		25	
Construction works	**Volume**	**Machines type**	**Machines' model**
Excavated Earth, cub.m	2380000	H.Excavator	Case CX210B
Earfilled Dam, cub.m	3220000	Bulldozer	Hitachi FD255
Cofferdam, cub.m	750000	H.Grab	Terex MHL364
Construction Materials	**Volume**	**Producer**	**Technology**
RC, cub.m	250000	EuroCon	Dry
Structural Steel, t	3 650	Mittal	ISM
GHG CALCULATION			
Machines		**Materials**	
Excavators	**Volume**	**RC**	**Volume**
Excavator's engine power, kW	110	Estimated GHG for cement production, g/t	15,00
Excavator's bucket, cub.m	0,75	Estimated GHG emission for coarse aggregate, g/cub.m	8,50
Estimated productivity, cub.m/hr	205	Estimated GHG emission for fine aggregate, g/cub.m	2,50
Estimated GHG emission, g/cub.m	5,30	Estimated GHG emission for mix production, g/cub.m	0,80
Estimated GHG emission total, t	12,61	Estimated GHG emission for mix transportation, g/cub.m/km	3,50
		Estimated GHG emission total, t	25,39
Bulldozers		**Construction Steel**	
Bulldozer's engine power, kW	175	Estimated GHG emission for steel production, g/t	2 500
Buldozer's blade, cub.m	2,50	Estimated GHG emission for steel transportation, g/t/km	2,50
Estimated productivity, cub.m/hr	150	Estimated GHG emission total, t	9,35
Estimated GHG emission per 1cub.m	4,15	**Total GHG emissions caused by materials, t**	**34,74**
Estimated GHG emission total, t	13,36		
Grabs			
Grab's engine power, kW	190		
Grab's bucket, cub.m	2,00		
Estimated productivity, cub.m/hr	130		
Estimated GHG emission per 1cub.m	8,40		
Estimated GHG emission total, t	6,30	**Total GHG emissions caused by construction, t**	**67,02**
Total GHG emissions caused by machines, t	**32,28**	**Total GHG emissions caused by construction, t/MW**	**0,268**

IOP Conf. Series: Materials Science and Engineering **552** (2019) 012018 doi:10.1088/1757-899X/552/1/012018

Similar to geothermal technologies, the cost of hydropower technologies can be site-specific. Numerous options are available for hydroelectric generation; repower in an existing dam or generator, or installing a new dam or generator, are options. As such, the cost estimates shown in this report are single-value estimates and may not be representative of any individual site. 2010 capital cost for a 500 MW hydropower facility was estimated at 3,500 \$/kW +35%. Table 22 presents cost and performance data for hydroelectric power technology [25]. For the most types of ECP necessary volumes of basic construction materials and machine-hours could be estimated approximately using some confident margins under given level of confidence.

Table 14. Eco-penalties

Group of eco-penalties	**Reasons for eco-penalties**	**Eco-penalties (additional cost), \$ for ECP type**				
		wind	**NG**	**Solar PV**	**Hydro**	**Fossil-L**
Construction material (for 1 measurement unit of consumed material)	Concrete, m3	12	20	16	12	12
	Steel for rebars, t	114	139	145	118	192
	Steel for framework, t	156	156	140	165	112
	Ceramics, etc, t	32	34	40	23	27
Transportation (for 1000 t/km of average distance to construction plants)	up to 20 km	9	8	7	9	10
	21-49 km	9	9	6	6	8
	more than 50 km	7	8	8	6	5
Installation (for 100 machine-hours of machine/vehicle in installation processes)	up to 20 machines/vehicles	6	6	6	8	7
	21-49 machines/vehicles	5	6	6	7	6
	more than 50 machines/vehicles	5	4	5	4	4
In-situ energy consumptions for facilities and workers accommodation (for 10 worker-day at construction site)	up to 50 workers	15	15	10	12	11
	51-199 workers	8	9	11	11	11
	more than 200 workers	9	6	8	7	9
In-situ energy consumptions for tools and engines (for 1 day of 10 tool/engine at construction site)	up to 20 tools/engines	20	20	21	21	26
	21-49 tools/engines	18	16	16	16	17
	more than 50 tools/engines	13	14	12	12	13

For example, constructing of typical hydro power station ECP requires for each MW of energy 30-50 m3 of ready-mix concrete. Each exact number in the margins is acceptable according to construction details: region, season of year, engineering and transportation level of neighborhood's development, etc. The same reasons makes transportation distances, in-situ energy consumers' number and other significant details of calculation equally uncertain. Due to this, best mathematical procedure to handle uncertain source data and calculate confident figures of energy consumption per 1 unit of ECP is Monte-Carlo's imitation modelling and statistical operation with generated variables in data-panels [26].

All kinds of penalties for usability purposes are to be separated into 4 main groups: 1) construction materials; 2) transportation; 3) installation; 4) in-situ energy consumptions.

IOP Conf. Series: Materials Science and Engineering **552** (2019) 012018 doi:10.1088/1757-899X/552/1/012018

To avoid negative influence of money power deprivation and discounting all the penalties could be expressed in basic money equivalent fixed for 2018 year. All further estimation could be easily converted to current prices according to single deflator-index, estimated annually by means of dynamic construction market analysis (subject of additional researches). After calculating all penalties per 1 measurement unit for each used material, construction item and machine-hour one can recalculate average construction cost of energy (ACCE), taking caused (embodied) environmental harm into account. According to this calculations ACCE could be ranked like this:

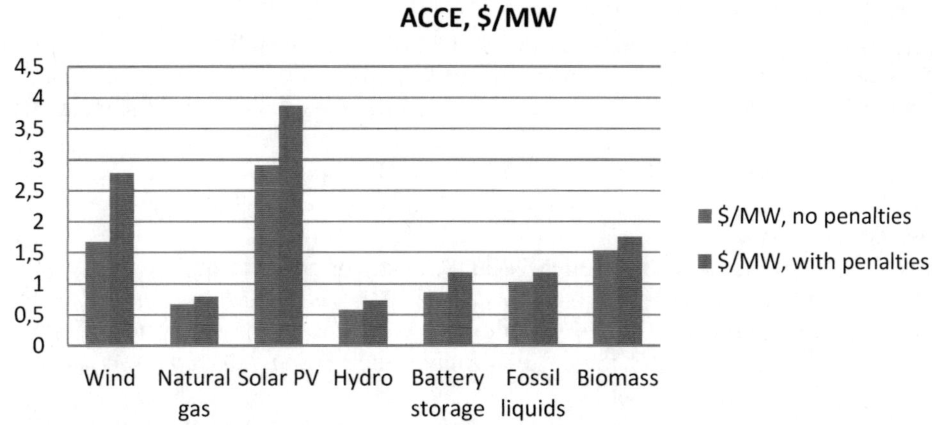

Figure 12. Average construction cost of energy

Table 15. Rank of energy station types by ACCE

Before penalties		After penalties	
Power plant type	**ACCE, $/MW**	**Power plant type**	**ACCE, $/MW**
1. Hydro	0.58	1. Battery storage	0.92 ↑58%
2. Natural gas	0.67	2. Hydro	1.12 ↑67%
3. Battery storage	0.86	3. Fossil liquids	1.19 ↑38%
4. Fossil liquids	1.02	4. Natural gas	1.30 ↑27%
5. Biomass	1.53	5. Biomass	1.90 ↑24%
6. Wind	1.67	6. Wind	2.79 ↑67%
7. Solar PV	2.91	7. Solar PV	3.46 ↑19%

Thus, some types of ACCE will increase its self-cost, other – decrease, and one can more confident and relevant estimate the ACCE or desirable or planned energy type according to embodied environmental harm caused by ECP construction process.

Acknowledgments
The work was carried out in the framework of the Program of flagship university development on the base of Belgorod State Technological University named after V.G. Shukhov.

References
[1] World Energy Council 2013 *World Energy Resources: A Summary*
[2] IEA 2010 Energy Technology Perspectives *International Energy Agency - OECD Style Guide: Rules and Conventions*

[3] Avilova I, Naumov A and Krutilova M 2017 Methodology of cost-effective eco-directed structural design *International Multidisciplinary Scientific GeoConf. SGEM* vol 53 pp 255-61
[4] Avilova I, Krutilova M and Peresypkina E 2017 Economic incentives of green standards in civil and municipal engineering *International Multidisciplinary Scientific GeoConf. SGEM* vol 53 pp 551-7
[5] Dones R, Bauer C, Bolliger R, Burger B, Faist Emmenegger M, Frischknecht R, Heck T, Jungbluth N and Röder A 2007 Life Cycle Inventories of Energy Systems: Results for Current Systems in Switzerland and other UCTE Countries *Final report ecoinvent data* vol 2.0 (5)
[6] Raadal H L, Gagnon, L and Hanssen O J 2011 Life cycle greenhouse gas (GHG) emissions from the generation of wind and hydro power *Renewable and Sustainable Energy Reviews*
[7] U.S. Energy Information Administration (EIA) 2016 Updated Capital Cost Estimates for Utility Scale Electricity Generating Plants
[8] National Hydropower Association (NHA) and the Hydropower Research Foundation (HRF) 2012 Renewable energy technologies: Cost analysis series, Hydropower, vol 1: Power Sector, Issue 3/5
[9] Zharikov I S, Laketich A and Laketich N 2018 Impact of concrete quality works on concrete strength of monolithic constructions *Construction Materials and Products* vol 1(1) pp 51-8
[10] Avilova I P and Krutilova M O 2018 Methodology of ecooriented assessment of constructive schemes of cast in-situ RC framework in civil engineering *IOP Conf. Series: Earth and Environmental Science* **3** 012127
[11] Abakumov R G and Naumov A E 2018 Building information model: advantages, tools and adoption efficiency *IOP Conf. Series: Materials Science and Engineering* vol 11 p 022001.
[12] U.S. EPA. Report on the 2011 U.S. Environmental Protection Agency (EPA) Decontamination Research and Development Conference. U.S. Environmental Protection Agency, Washington, DC, EPA/600/R/12/557, 2012
[13] U.S. Energy Information Administration (EIA) 2013 Annual Energy Outlook, 2001
[14] International Energy Agency (IEA) 2000 The IEA Greenhouse Gas R&D programme Greenhouse Gas Emissions From Major Industrial Sources - III Iron and steel production
[15] International Energy Agency (IEA) 2013 CO2 emissions from fuel combustion (2013 Edition)
[16] Rubino L, Bonnel P, Hummel R, Krasenbrink A and Manfredi U 2007 Mobile Measurement of Pollutant Emissions and Fuel Consumption of Road Vehicles in Real-world Driving Situations Using Portable Emission Measurement Systems (PEMS)
[17] Kara K A, Dolzhenko A V and Zharikov I S 2018 Influence of processing factors over concrete strength *IOP Conf. Series: Materials Science and Engineering* vol 327 032027
[18] Lewis M 2009 *Estimating fuel use and emission rates of nonroad diesel construction equipment performing representative duty cycles* (North Carolina State University) 3357743
[19] Dimoudi A and Tompa C 2008 Energy and environmental indicators related to construction of office buildings *Resources Conservation and Recycling* vol 53 pp 86–95
[20] Gubareva V V 2018 Reduction of Energy Consumption During Building Materials Production *IOP Conf. Series: Journal of Physics* vol 1066 012022
[21] U.S. National Energy Technology Laboratory (NETL) 2010 Cost Estimation Methodology for NETL Assessments of Power Plant Performance
[22] National Hydropower Association (NHA) and the Hydropower Research Foundation (HRF) 2012 *Renewable energy technologies: cost analysis series* vol 1: Power Sector Issue 3/5
[23] Black & Veatch 2017 *Strategic Directions: Electric Industry Report*
[24] R Pletka and J Finn 2009 National Renewable Energy Laboratory (NREL) *Western Renewable Energy Zones, Phase 1: QRA Identification Technical Report*
[25] National renewable energy laboratory (NREL) 2012 *Cost and performance data for power generation technologies*
[26] Climate Change 2013 *The Physical Science Basis* (Cambridge University Press) p 697

Evaluation of energy-saving effect during the installing process of automated heating stations in educational institutions

S A Bychikhin, M V Svirin and P A Trubaev

Department of Energy Engineering of Heat Technologie, Belgorod State Technological University of V.G. Shukhov, Kostyukov St., 46, Belgorod, 308012, Russia

E-mail: trubaev@gmail.com

Abstract. The aim of the work is to assess the energy-saving effect after the installation of automated individual heat points that carry out weather regulation of the heating load. The analysis was carried out for 18 educational institutions of the Belgorod region, in which energy service contracts were implemented in 2016. The data of the actual monthly consumption of heat energy during the basic heating season before the implementation of energy-saving measures and data of daily accounting of heat consumption during the two heating seasons after the installation of automated individual heat points were used. The average unit savings for all institutions was 14 per cent. It is established that the greatest value of economy, 20-30%, provides decrease in temperature in buildings to optimum level. Elimination of" overflows " in the autumn-spring period allows to save 5-10% of thermal energy. The decrease in temperature on non-working days is characterized by a small amount of savings, not more than 3%, or its absence.

Keywords: heating, energy saving, individual heating station, educational institutions

1. Introduction

In practice, there is a noticeable reduction in energy consumption for heating buildings after the installation of automated individual heat points (IHP), which carry out weather regulation of the heating load. Similar behavior of consumption characteristics trends is observed in a number of publications [1-3], which proves the effectiveness of the method.

Energy is the lifeblood of modern societies. In recent decades, world energy consumption and related CO2 emissions grew rapidly due to the increase in population and meet the needs of people [4]. Due to the high energy consumption, energy efficiency in buildings has become a hot topic since 1970. As the effects of human-induced climate change became more pronounced, local and Federal governments began to pursue more aggressive policies to reduce energy use in existing buildings [5]. Energy saving in buildings is a globally important issue also due to the growing attention to the quality of indoor heat, which is directly related to the quality of life. In areas with a cold climate, there is an even greater problem in increasing the energy efficiency of buildings due to the high demand for heat [6].It is believed that good heat supply from heating companies and the mode of operation of institutions plays an important role in reducing energy consumption in buildings [7].

In modern society, people spend more than 90% of their time indoors. Students spend more time at school, so it is important to ensure comfortable temperature conditions in the room. Thermal comfort is associated with the characteristics of the building and energy saving, which provides the necessary conditions and budget savings, so the analysis of consumption and improvement of the characteristics of buildings have become important in recent years [8]. Low-energy buildings are spreading rapidly throughout Europe. The basic concept of energy saving is based on the improvement of building envelope and the use of high-performance equipment. However, thermal comfort in such buildings will not be fully achieved [9].

The authors of many publications believe that the traditional behavioral model, which does not provide for the need to save energy, is one of the reasons for inefficient heating of buildings [10]. In this regard, modeling of energy consumption in buildings is necessary to mitigate current impacts and to develop policies and technical solutions for the future [11]. At the same time, measures taken in developed countries are insufficient to guarantee a significant reduction in energy consumption in buildings in the coming years. Under the current scenario, by 2050 there will be at least a doubling of global energy demand in buildings compared to today. To avoid this prognosis, it is necessary to disseminate and adopt cost-effective methods and technologies, as well as to change the behavior and lifestyle of the population [12]. The creation of a scientific and rational method of comparative analysis of energy consumption in buildings is the key to an objective, fair and accurate assessment of the energy characteristics of the building [13]. In order to successfully save energy, modernization activities should take into account structural transformations of buildings (e.g. improvement of building envelope, automation of energy systems). The allocation of heating costs according to actual consumption can be one possible solution to save energy. The introduction of automated meters allows consumers to pay only for the delivered heat and allows to analyze the actual heat consumption and regulate the supply depending on the weather conditions in real time and thereby reduce the heat consumption in institutions [14].

Forecasting energy consumption in buildings is essential for planning, managing, and conserving energy. Data-based models provide a practical approach to predicting energy consumption [4]. Forecasting is of great importance for heating, ventilation and air conditioning control tasks, such as optimal operation strategies, etc. compared to traditional methods, data-based methods have received more attention because of their flexibility and efficiency [15].

Conventional approaches to energy management have proven to be inadequate for public buildings. Systematic observations of numerous energy-saving projects have shown that four pillars can be identified that are critical to the success of these projects. They should be used as often as possible and can be specified as follows:

1. Technical expertise: knowledge of the specifics of the object and possible savings on the basis of experience.
2. Quality operational data: on-site data collection.
3. Simulation, simulation and optimization: evaluate different options and choose the most profitable one.
4. Methodology: a systematic approach involving all three remaining pillars.

The application of these recommendations provides an effective way of conducting energy-saving projects [16].

Energy consumption in buildings accounts for approximately 35% of the world's final energy. Educational institutions provide 20-40% of heat energy consumption, which makes it necessary to study their energy characteristics [17]. In some cases, they contribute significantly to local government expenditure. The high level of energy consumption is usually due to the low characteristics of buildings, mode of operation, etc. [18]. Reducing energy consumption in educational institutions is a priority specified in the legislative documents [19].

One of the alternatives to reduce energy consumption in buildings is the conclusion of energy service contracts to improve energy efficiency [20, 22]. A number of publications are devoted to the study of energy efficiency for school and pre-school buildings after the introduction of energy service contracts [21]. An energy service contract can provide a cost-effective way to overcome barriers to energy efficiency. Energy service contracts reduce operating costs. Production costs will be determined by a combination of physical characteristics of the power system and the technical efficiency of the relevant organizational measures [22]. Thus, the automation of heating load regulation, implemented on the basis of the energy service contract, allows for energy saving with minimal cost to the consumer [23]. But for development of model of the power service contract it is necessary to have data on size of economy after installation of the automated individual calorific point (AICP) [24].

The estimation of thermal energy saving depends on the efficiency of the reconstructed heating system and requires detailed information on the value of heat energy consumption depending on climatic parameters. Therefore, in practice, the amount of savings is estimated by analogy using the actual results of the installation of AICP. In the literature and regulatory sources the value of savings is estimated in a very wide range-from 3 to 30% (table 1)

Table 1. Value of saving on heating costs with the installation of IHP

Origin	Value of saving
Methodical documents in construction 13-7.2000. Recommendations about the priority low-cost actions providing energy resources saving in housing and public utilities of the city. Are approved by the order of the State Committee for Construction of the Russian Federation of 17.01.2000 No. 5. [In Russian]	3-5%
Eremkin A I, Koroleva T I, Danilin G V, By`zev V V and Averkin A G 2008 *E`konomika e`nergosberezheniya v sistemax otopleniya, ventilyacii i kondicionirovaniya vozduxa* [Energy saving economy in heating systems, ventilation and air conditionings] (Moscow: Izd-vo Associacii stroitel`ny`x vuzov) [In Russian]	8%
Shilkin N V 2007 E`konomicheskie aspekty` vnedreniya individual`ny`x teplovy`x punktov [Economic aspects of introduction of individual thermal points] *E`nergosberezhenie* **3** 12–5 [In Russian]	15%
Gasho E G (general edition) 2016 *Metodicheskie rekomendacii po raschetu e`ffektov ot realizacii meropriyatij po e`nergosberezheniyu i povy`sheniyu e`nergeticheskoj e`ffektivnosti: Spravochno-analiticheskij document* [Methodical recommendations about calculation of effects of realization of actions for energy saving and increase in power efficiency: Help and analytical document] (Moscow: Russian Government Analytical Centre) [In Russian]	20…30%
Vasil`ev G P et al 2015 *Prakticheskoe posobie po povy`sheniyu e`nergeticheskoj e`ffektivnosti mnogokvartirny`x domov (MKD) pri kapital`nom remonte* [Practical grant on increase in power efficiency of apartment houses at major repair.] vol 1 (Moscow: .: GK Fond Sodejstviya Reformirovaniyu ZhKX; OAO «Insolar-invest») [In Russian]	15%
2004 *E`nergosberezhenie v zhilishhnom fonde: problemy`, praktika i perspektivy* [Energy saving in housing stock: problems, practice and prospects] (Moscow: Dena, Fond «Institut e`konomiki goroda») [In Russian]	5-10%
Prakticheskie rezul`taty` ustanovki ITP v tipovom 9-e`tazhnom 4-pod``ezdnom panel`nom dome [Practical results of the ITP installation in the type 9-storey 4-access apartment house of bearing-wall construction]. URL http://dom-pogoda.ru/automation/primer-ekonomicheskogo-effekta-raboty [In Russian]	25%

A number of publications indicate that there is a large difference in the specific energy consumption between different types of schools and pre-school institutions, classified by mode of operation, nature and level [25].

The aim of the work was to assess the actual energy-saving effect after the installation of AICP in educational institutions of the Belgorod region.

IOP Conf. Series: Materials Science and Engineering **552** (2019) 012019 doi:10.1088/1757-899X/552/1/012019

The analysis of thermal energy consumption in the spring and autumn seasons [26]. Special attention is paid to the comparison of heat consumption in identical buildings. Awareness of the difference in heat consumption in similar houses can strengthen confidence in energy saving tools and improve the thermal reconstruction of residential and public buildings [27, 28].

2. Materials and Methods

The reconstruction of heating systems was carried out in the summer and autumn of 2016 within the framework of energy service contracts. 18 buildings (12 schools and 6 kindergartens) were selected for the analysis, where before the events the heat consumption was taken into account by the instrument method. The buildings had the characteristics shown in the table 2.

Table 2. Characteristics of buildings

Building number	Name of institution	Floors	Area, m^2	Volume, m^3	Wall material*	Start date ESK
1	Nursery school # 20 (Razumnoye area)	2	999	3702	P	01.12.2016
2	Nursery school # 7 (Belovskoye area)	2	1334	3787	P	20.12.2016
3	Nursery school # 14 (Golovchino area)	2	1658	5706	SB	14.10.2016
4	Nursery school # 13 (Politotdel'skii area)	2	1984	3806	P	01.10.2016
5	Nursery school # 29 (Krasnyy oktyabr)	3	2700	12231	SB	20.12.2016
6	School of Streletskoye	2	1755	5266	SB	01.10.2016
7	School Pushkarnoe area	2	2150	6542	SB	16.10.2016
8	School Belovskoye area	2	2615	8783	P	11.11.2016
9	School Zhuravlevka area	2	2676	6560	SB	01.10.2016
10	School of Khokhlovo	2	3036	7275	SB	01.12.2016
11	School of Myasoedovo	2	3121	9524	SB	26.11.2016
12	School of Red October	3	4163	15654	P	01.10.2016
13	School # 1 of Razumnoye area	3	5420	16285	P	15.10.2016
14	School of Bessonovka area	3	4780	15175	SB	20.12.2016
15	Otradnenskaya school	2	4854	14564	SB	20.12.2016
16	School #2 of Razumnoe area	3	4389	18489	P	20.12.2016
17	School of Schetinovka area	3	5614	22370	SB	20.12.2016

* P – reinforced concrete panels; SB – silicate brick

The heating season 2015/2016 is adopted as the base period, the estimated periods are the heating seasons 2016/2017 and 2017/2018.

The actual climatic data of the weather station of Belgorod received from the site were used for the calculation of the DDHP https://rp5.ru. The considered periods were characterized by a significant difference in climatic conditions (figure 1).

Therefore, the calculation of savings was given with the reduction of data to comparable conditions using the value of the parameter degree-day heating period Dd, °C·day:

$$Dd = D(t_{\text{in}} - t_{out}), \tag{1}$$

where D – the number of days(days) in the period; t_{in} – the average temperature inside the building, °C; t_{out} – the average ambient temperature in the period, °C.

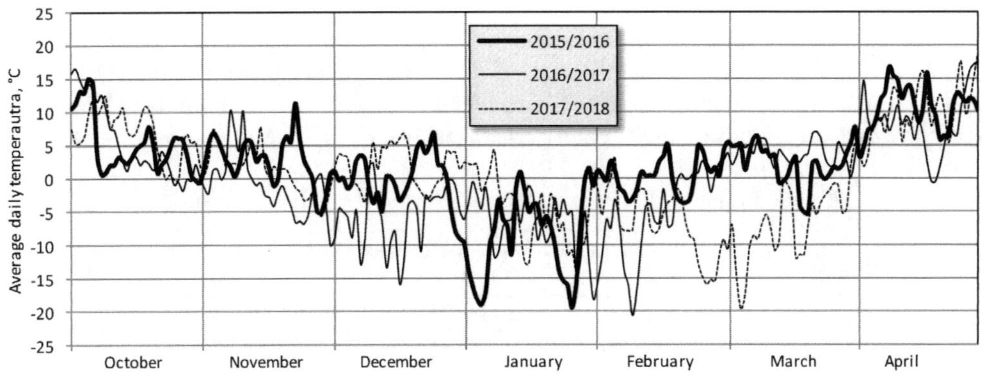

Figure 1. Temperature changes during heating periods t_{out}

If the period consists of several parts (for example, several heating seasons, or with interruptions in heating or data storage), the total value of the degree-day can be found as the sum of the values for each part of the period:

$$Dd = \Sigma\, Dd_j.$$

If to recalculate the regulatory conditions, the relative thermal energy saving Δq, %, it will have a look

$$\Delta q = \frac{\frac{Q_b}{Dd_b} Dd_n - \frac{Q_i}{Dd_i} Dd_n}{\frac{Q_b}{Dd_b} Dd_n} 100\% , \qquad (2)$$

Where Q_b and Q_i – heat consumption in the base and estimated period; Dd_b, Dd_i and Dd_n – degree-day of the base period, the estimated period and the normative value for the city of Belgorod 4807 °C•day for preschool and 4183 °C •day for school.
After conversion expression (2) will look like:

$$\Delta q = \left(1 - \frac{Q_i}{Q_б} \frac{Dd_b}{Dd_i}\right) 100\%. \qquad (3)$$

The expression (3) will not change if instead of normative conditions recalculation is made on conditions of the base or estimated period, or other conditions.

Information and consumption of thermal energy in buildings for each month of the base period and the duration of the heating period, common to all institutions, were obtained in the district heat supply organization (beginning of the heating season 07.10.2015, ending 11.04.2016, actual Dd_b = 3676 °C•day).

Daily consumption of thermal energy for the estimated periods was obtained using the energy management System of the Belgorod region, which carries out the accumulation and processing of remotely obtained data from heat metering devices. According to the daily consumption, the beginning and the end of heating periods were determined individually for each building. Since the launch of the AICP in different institutions was carried out at different times in the period from October 1, 2016 to December 20, 2016, and for individual metering devices there were interruptions in operation, lasting up to a month, the estimated periods included only those days in which the metering devices worked, so the consumption of thermal energy and DDHP used to calculate the savings could be less than these indicators in the heating season. In the calculations, the air temperature in tin buildings was assumed to be constant and equal to 20°C.

IOP Conf. Series: Materials Science and Engineering **552** (2019) 012019 doi:10.1088/1757-899X/552/1/012019

3. Results

The estimated (taken into account in the calculation of savings) heat consumption for the three heating seasons and the resulting specific actual savings per year after the installation of the AICP compared to the base heating season is shown in table 3. The average value of the economy was 14%, the maximum-31%. Savings were not available for individual agencies.

Table 3. Thermal energy consumption and the value of saving

Building number	Volume, m^3	The consumption for the billing period, Gcal (Dd, °C·day)			Specific savings Δq
		2015/2016	2016/2017	2017/2018	
1	3702,2	164,8 (3676)	143,5 (3168)	163,0 (4145)	5,6%
2	3786,6	242,0 (3676)	139,3 (2656)	200,0 (3820)	20,4%
3	5705,6	277,7 (3676)	288,2 (4129)	255,2 (3995)	11,4%
4	3806,3	214,8 (3676)	227,0 (4171)	195,5 (3864)	10,1%
5	12231,0	350,7 (3676)	243,9 (2656)	404,0 (4102)	0,3%
6	5266,2	299,9 (3676)	215,0 (4094)	259,3 (3728)	25,2%
7	6542,4	253,1 (3676)	267,5 (3987)	268,6 (4133)	4,1%
8	8783,0	352,7 (3676)	276,5 (3628)	307,7 (3832)	18,4%
9	6559,8	280,8 (3676)	300,9 (4242)	275,1 (4087)	9,4%
10	7275,0	586,8 (3676)	372,1 (3175)	320,7 (3188)	31,8%
11	9523,9	348,6 (3676)	296,3 (3290)	354,3 (4066)	6,6%
12	15653,6	675,4 (3676)	483,3 (3208)	573,7 (4151)	21,4%
13	18489,3	616,3 (3676)	381,9 (2656)	599,9 (4119)	13,7%
14	15175,4	522,9 (3676)	360,5 (2666)	437,7 (4104)	15,0%
15	14564,4	357,6 (3676)	185,8 (2656)	321,0 (4133)	24,1%
16	16284,8	617,3 (3676)	695,9 (3987)	690,1 (4073)	-2,4%
17	22370,4	707,9 (3676)	405,6 (2656)	617,4 (3987)	20,1%

There is no dependence of the saving value Δq on the area, volume or heat consumption (correlation coefficients, respectively, 0,04; 0,08; 0,25), but the saving depends on the specific heating characteristics q_{sp}, Вт/(m^3·K), figure 2, which for the building should be a constant value:

$$q_{sp} = \frac{4{,}186 \cdot 10^9}{24 \cdot 60 \cdot 60} \frac{Q}{V\ Dd} = 48449 \frac{Q}{V\ Dd}, \tag{4}$$

where Q – heat consumption for heating, Gkal; V – volume, m^3.

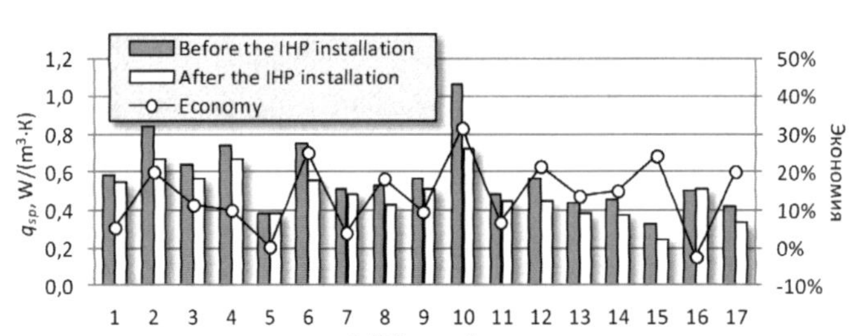

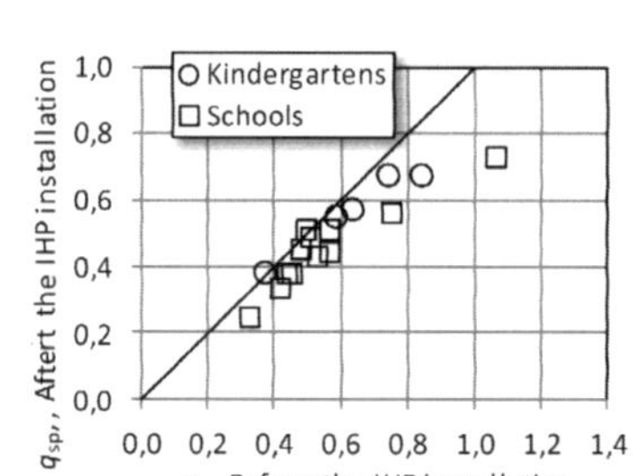

Figure 2. The change of the specific heat characteristics of buildings (taken at a constant temperature of the internal air 20°C)

The correlation coefficient between economy and specific heat characteristic is 0, 93 for kindergartens and 0,45 for schools (figure 3).

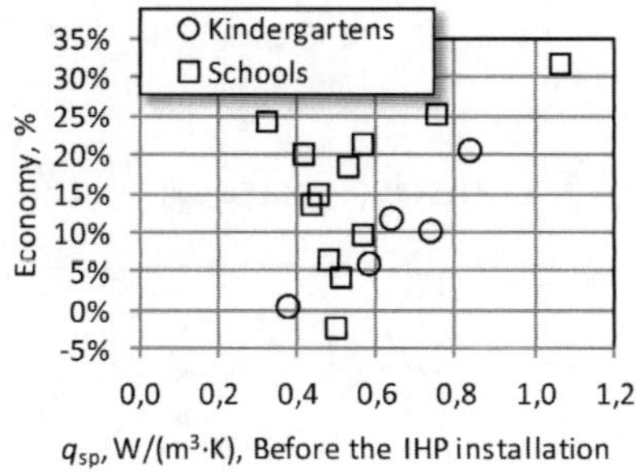

Figure 3. The relationship between savings and specific heat characteristics

4. Discussion

Analyzing the changes in characteristics before and after the installation of AICP, three cases can be distinguished.

1. The heating load of the building (the dependence of the heating consumption on the outside temperature) is regulated both in the base period and in the estimated period, but the heat energy consumption for heating is reduced (figures 4, 5), which is typical for buildings # 2 (economy is 20 %); 6 (25%); 8 (18%); 10 (32%); 15 (24%).

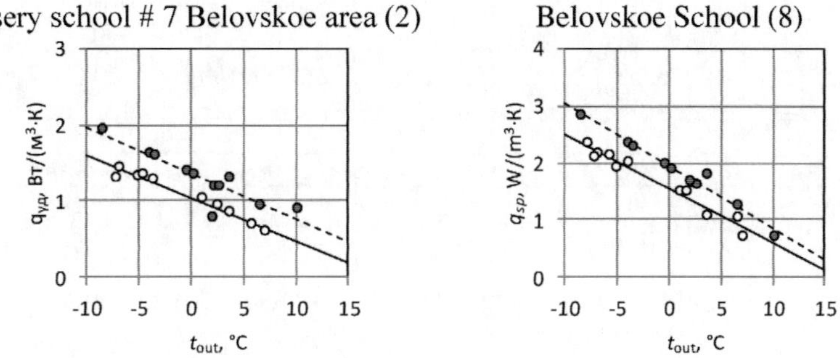

Figure 4. The heating feature: •, --- – before installing AICP; ◦, — – after installing AICP

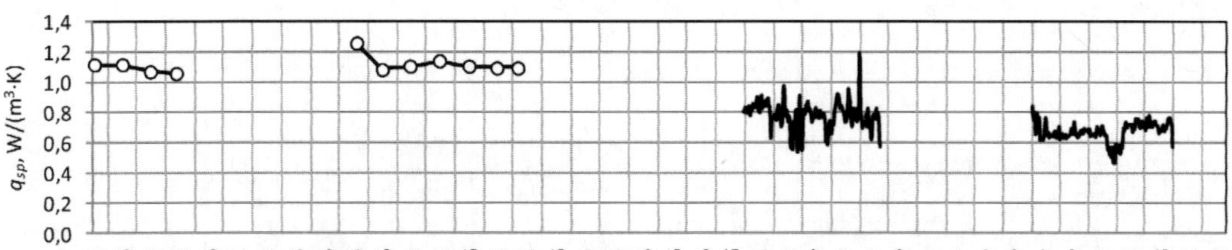

Figure 5. Change in specific heating characteristics for Khokhlovskaya secondary schools (# 10)

In this case, the savings achieved by the elimination of the constant "overheating" and reduce the temperature in the building increased to the optimum. For the conditions of the city of Belgorod the average outdoor temperature during the heating season close to 0°C, so the change of temperature inside the building 1 °C with leads to a change in consumption for heating by approximately 5%. Therefore, we can assume that prior to the installation AICP the temperature in the buildings was about 24 °C.

2. The heating load of the building was not adjusted before the events (Fig. 6, 7), which is typical for buildings # 3 (economy 11%); 4 (10%); 5 (0%); 7 (4%); 9 (9%); 11 (7%); 12 (21%); 13 (14%); 17 (20%)

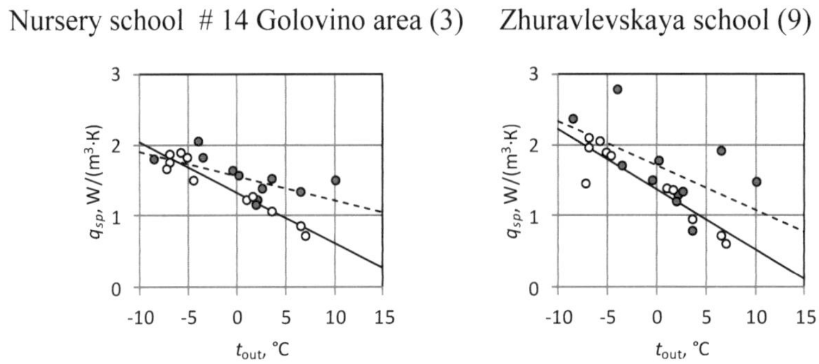

Figure 6. Heating characteristic:
•, --- – before installing AICP; ◦, — – after installing AICP

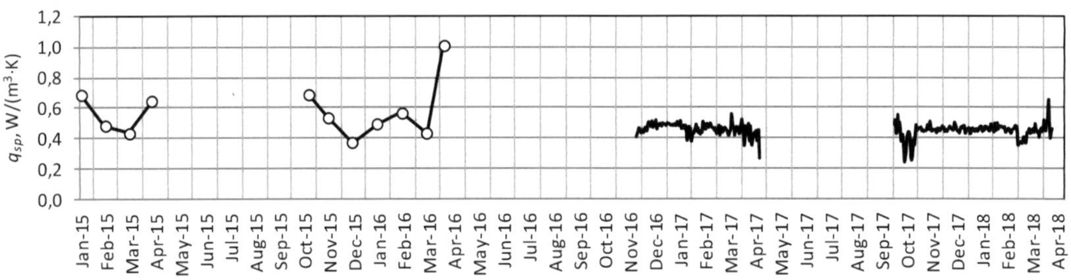

Figure 7. Change of specific heating characteristic for Myasoyedov Secondary School (# 11)

In this case, the savings achieved by the elimination of periodic "overheating" at the beginning and end of the heating season. On parts of buildings additional savings is also determined by the reduction of the constantly high temperature in the building to the optimum.

3. There were no savings in buildings 5 and 16 (Figure 6).

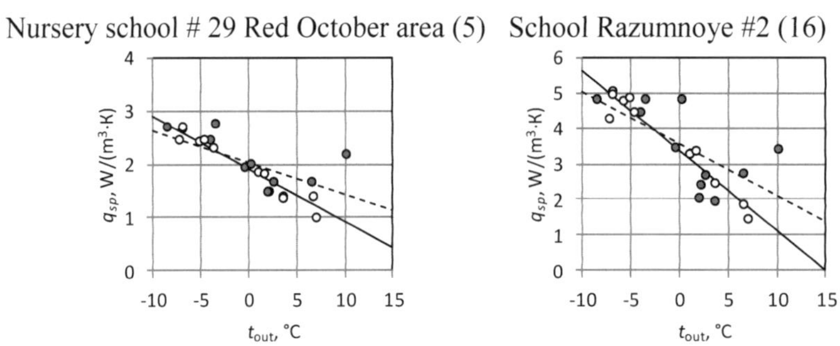

Figure 8. The heating feature: •, --- – before installing AICP;
◦, — – after installing AICP

IOP Conf. Series: Materials Science and Engineering **552** (2019) 012019 doi:10.1088/1757-899X/552/1/012019

In this case, it can be assumed that before the events in the buildings was below the standard temperature.

Setting the AICP involves reducing the temperature on holidays and weekends. For each building for the period of two years, the average value of the specific heating characteristics was calculated separately for non-working and working days, as well as the specific heating characteristics on the first non-working day after the working day and on the first working day following the non-working day (table 4).

Table 4. Average specific heating characteristics (calculation periods 2016/2017 and 2017/2018)

Building number	The specific heating characteristics q_{sp}, W/(m^3·K)					Relation to the average heating characteristic			
	average	working days	on non-working days	on the first non-working day	on the first working day	working days	on non-working days	on the first non-working day	on the first working day
1	0,542	0,556	0,551	0,560	0,547	103%	102%	103%	101%
2	0,668	0,662	0,679	0,667	0,670	99%	102%	100%	100%
3	0,568	0,575	0,553	0,589	0,554	101%	97%	104%	98%
4	0,672	0,669	0,679	0,701	0,664	100%	101%	104%	99%
5	0,384	0,388	0,376	0,386	0,383	101%	98%	101%	100%
6	0,557	0,558	0,555	0,564	0,548	100%	100%	101%	98%
7	0,484	0,486	0,479	0,499	0,472	100%	99%	103%	98%
8	0,423	0,430	0,411	0,428	0,407	101%	97%	101%	96%
9	0,499	0,508	0,498	0,531	0,490	102%	100%	106%	98%
10	0,719	0,726	0,706	0,731	0,711	101%	98%	102%	99%
11	0,449	0,450	0,448	0,460	0,442	100%	100%	102%	98%
12	0,448	0,446	0,453	0,451	0,446	99%	101%	101%	100%
13	0,374	0,382	0,360	0,387	0,355	102%	96%	103%	95%
14	0,375	0,390	0,362	0,400	0,358	104%	97%	107%	95%
15	0,248	0,251	0,243	0,250	0,243	101%	98%	101%	98%
16	0,497	0,501	0,489	0,507	0,478	101%	98%	102%	96%
17	0,332	0,344	0,319	0,333	0,308	103%	96%	100%	93%
Average	0,485	0,489	0,480	0,497	0,475	101%	99%	102%	98%

Economy is characterized by the relation of heating characteristic in the non-working days (weekends) $q_{sp\ nw}$ and in the working days $q_{sp\ w}$. This ratio makes from 0.93 to 1. The value 0.93 corresponds to saving 7%. The value 1 shows lack of economy. For objects number 1, 7, 10 a ratio more than 1 (respectively 1.03; 1.02; 1.01).

Economy was in standard conditions as a difference the difference of expenses without decrease in a heating load to the actual costs of heating:

$$\Delta q = \frac{q_{sp\ w}(n_w + n_{nw}) - (q_{sp\ w}n_w + q_{sp\ nw}n_{nw})}{q_{sp\ w}(n_w + n_{nw})}100\% = \left(1 - \frac{q_{sp\ w}n_w + q_{sp\ nw}n_{nw}}{q_{sp\ w}(n_w + n_{nw})}\right)100\%,$$

where n_w – number of the working days; n_{nw} – number of the non-working days (weekends).

Results of definition of economy are presented on Figure 9.

IOP Conf. Series: Materials Science and Engineering **552** (2019) 012019 doi:10.1088/1757-899X/552/1/012019

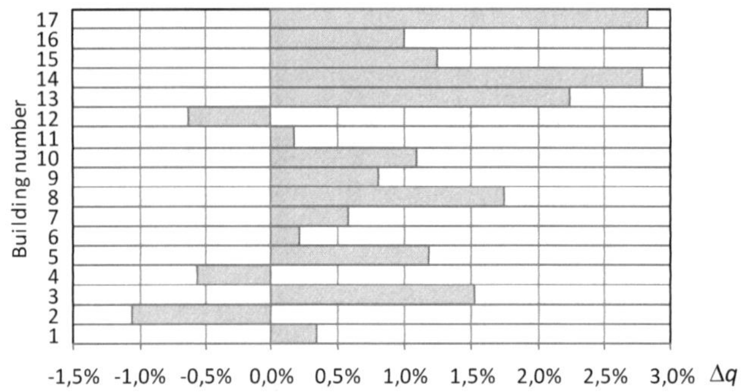

Figure 9. Economy at decrease in a heating load on the non-working days (weekends)

Average economy at decrease in temperature in buildings in down time has made 0.9%.

The expense of heat raised in the first working day in comparison with average (Figure 10). The over expenditure has made 6-7% for the most part of buildings. The over expenditure did not depend on economy size.

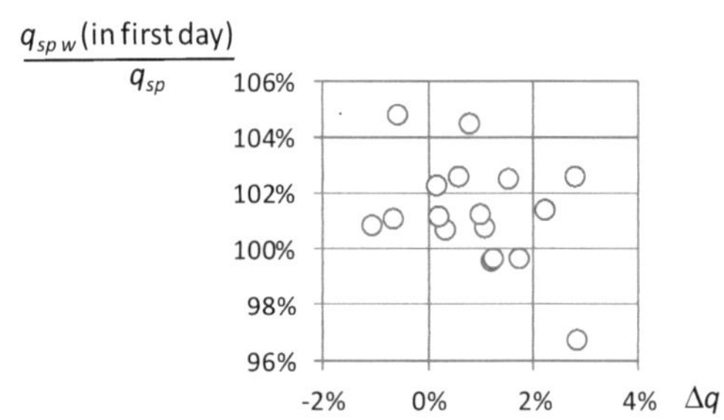

Figure 10. The relation of economy of thermal energy of Δq to an over expenditure of thermal energy in the first working day after weekends

The average decrease in heating heat consumption on the first non-working day was accompanied by an increase in heating consumption on the first working day after non-working days. That is, the resulting savings were subsequently offset by increased heating costs to warm up the building after non-working days.

5. Conclusion

In work for 18 educational institutions of the Belgorod region on the basis of the actual data of metering devices the analysis of economy of thermal energy after installation in organizations of the automated individual calorific points performing weather regulation of heating loading is made. The average unit savings for all institutions was 14 per cent. It is established that the greatest value of economy, 20-30%, provides temperature reduction in buildings to an optimum level. Elimination of" overflows " in the autumn-spring period allows to save 5-10% of thermal energy. The decrease in temperature on non-working days is characterized by a small amount of savings, not more than 3%, or its absence.

Acknowledgments

The article was prepared within development program of the Flagship Regional University on the basis of Belgorod State Technological University named after V.G. Shukhov.

References

[1] Belova I N, Ginzburg A S and Krivenok L A 2018 Heating seasons length and degree days trends in Russian cities during last half century *Energy Procedia* **149** 373-9
[2] Ruusala A, Laukkarinen A and Vinha J 2018 Energy consumption of Finnish schools and daycare centers and the correlation to regulatory building permit values *Energy Policy* **119** 183-95
[3] Tumanova K and Borodinecs A 2016 The Extensive Analysis of Circumstances Between Heat Consumption of Multi-apartment Buildings and Information Campaigns *Energy Procedia* **96** 945-52
[4] Amasyali K and El-Gohary N M 2018 A review of data-driven building energy consumption prediction studies *Renewable and Sustainable Energy Reviews* **81** 1192-205
[5] Papadopoulos S and Kontokosta C E 2019 Grading buildings on energy performance using city benchmarking data *Applied Energy* **233** 244-53
[6] Nord N 2017 Building Energy Efficiency in Cold Climates *Encyclopedia of Sustainable Technologies* 149-57
[7] Chen C, Neng Z, Kun Y and Fan Y 2018 Data and analytics for heating energy consumption of residential buildings: The case of a severe cold climate region of China *Energy and Buildings* **172** 104-15
[8] Zomorodian Z S, Tahsildoost M and Hafezi M 2016 Thermal comfort in educational buildings: A review article *Renewable and Sustainable Energy Reviews* **59** 895-906
[9] Hawila A, Merabtine A, Chemkhi M, Bennacer R and Troussier N 2018 An analysis of the impact of PMV-based thermal comfort control during heating period: A case study of highly glazed room *Journal of Building Engineering* **20** 353-66
[10] Ivanova Z, Pichugin I and Naimaviciene J 2015 Behavioral Aspects of Energy and Heat Consumption (Basing on Opinion Poll Data) *Procedia Engineering* **117** 154-161
[11] Leroy Y and Yannou B 2018 An activity-based modelling framework for quantifying occupants' energy consumption in residential buildings *Computers in Industry* **103** 1-13
[12] Berardi U 2017 A cross-country comparison of the building energy consumptions and their trends *Resources, Conservation and Recycling* **123** 230-41
[13] Dongmei S 2018 Research and Application of Energy Consumption Benchmarking Method for Public Buildings Based on Actual Energy Consumption *Energy Procedia* **152** 475-83
[14] Ziemele J, Pakere I, Blumberga D and Zogla G 2015 Economy of Heat Cost Allocation in Apartment Buildings *Energy Procedia* **72** 87-94
[15] Jiangyu W, Guannan L, Huanxin C, Jiangyan L, Yabin G, Shaobo S and Yunpeng H 2018 Energy consumption prediction for water-source heat pump system using pattern recognition-based algorithms *Applied Thermal Engineering* **136** 755-66
[16] Máša V, Stehlík P, Touš M and Vondra M 2018 Key pillars of successful energy saving projects in small and medium industrial enterprises *Energy* **158** 293-304
[17] Ouf M M and Issa M H 2017 Energy consumption analysis of school buildings in Manitoba, Canada *International Journal of Sustainable Built Environment* **6** 359-71
[18] Bourdeau M, Guo X and Nefzaoui E 2018 Buildings energy consumption generation gap: A post-occupancy assessment in a case study of three higher education buildings *Energy and Buildings* **159** 600-11
[19] Magrini A, Gobbi L, d'Ambrosio F R 2016 Energy Audit of Public Buildings: The Energy Consumption of a University with Modern and Historical Buildings. Some Results *Energy Procedia* **101** 169-75
[20] Fossati M, Scalco V A, Linczuk V C and Lamberts R 2016 Building energy efficiency: An overview of the Brazilian residential labeling scheme *Renewable and Sustainable Energy Reviews* **65** 1216-31

[21] Allab Y, Pellegrino M, Guo X, Nefzaoui E and Kindinis A 2017 Energy and comfort assessment in educational building: Case study in a French university campus *Energy and Buildings* **143** 202-19

[22] Sorrell S 2007 The economics of energy service contracts *Energy Policy* **35** 507-21

[23] Belousov A, Koshlich Y, Grebenik A and Gvozdevskiy I 2017 Lower-level communication line analysis of automated dispatch control systems of distributed facilities *Challenges for Technology Innovation: An Agenda for the Future - Proceedings of the International Conference on Sustainable Smart Manufacturing S2M 2016* 31-6

[24] Belousov A, Koshlich Y, Grebenik A and Gvozdevskiy I 2017 An approach of lower-level communication line implementation of automated dispatch control systems of distributed facilities *International Multidisciplinary Scientific GeoConference Surveying Geology and Mining Ecology Management SGEM* **Vol 17(21)** 183-90

[25] Zhou X, Yan J, Zhu J and Cai P 2013 Survey of energy consumption and energy conservation measures for colleges and universities in Guangdong province *Energy and Buildings* **66** 112-8

[26] Tsynaeva A A, Nikitin M N and Tsynaeva E A 2016 The Technology of Automatic Control of Heat Consumption in Buildings *Procedia Engineering* **153** 785-90

[27] Juodis E, Jaraminiene E and Dudkiewicz E 2009 Inherent variability of heat consumption in residential buildings *Energy and Buildings* **41** 1188-94

[28] Seminenko A S, Sheremet E O, Gushchin S V, Elistratova J V and Kireev V M 2018 Hydraulic stability of heat networks for connection of new consumers *IOP Conference Series: Materials Science and Engineering* **327(4)** 042105

IOP Conf. Series: Materials Science and Engineering **552** (2019) 012020 doi:10.1088/1757-899X/552/1/012020

Using DEA models to measure the efficiency of energy saving projects

M Yu Derevyanov[1], Yu E Pleshivtseva[1], L S Kordyukova[2], A V Rapoport[3]

[1] Department of Heat and Power Engineering, Samara State Technical University, Molodogvardeyskaya Str., 244, Samara, 443100, Russia
[2] VakuummashElectro Ltd., Proezd Deryabina, 2/52, Izhevsk, 426057, Russia
[3] EvroGaz Ltd., Solnechnogorskaya, 4a, Moscow, 125413, Russia

E-mail: mder2007@mail.ru

Abstract. The paper discusses the approach for multi-criteria estimation of the efficiency of energy saving projects by the Data Envelopment Analysis (DEA) method. The measurement of efficiency of considered projects and the identification of sources of their inefficiency is a very important condition to put the projects into practice in industrial competitive environment. The term "energy saving project" refers to a project resulting in certain output improvements with respect to energy consumption of the technological process by spending certain inputs. Thus, the energy saving project can be treated as a production unit or Decision Making Unit (DMU). The estimation of the efficiency of energy-saving projects by DEA is illustrated here via example of an industrial enterprise for the production of cellular autoclaved aerated concrete. The nine typical inputs/outputs are used and the twelve DEA models are developed that characterize the technological, environmental, economic and comprehensive estimators of the efficiency of energy saving projects. The developed approach for multi-criteria estimation of the efficiency of energy saving projects by the DEA method has the potential of widespread use in practice for making effective management decisions on the formation of investment programs for industrial enterprises.

1. Introduction

The current state of any industrial production is directly related to the efficiency of the use of fuel and energy resources (FER). Implementation of energy saving projects at the enterprises leads to reducing the consumption of fuel and energy resources, thereby decreasing the cost of production, which allows enterprises to remain competitive in the market [1-4]. However, the use of the enterprise asserts or, moreover, borrowed investment funds requires a well-reasoned decision in the choice of the most efficient energy saving projects. For the decision making not only the economic component, but also the efficiency of projects with respect to technological, environmental, logistical and other criteria (indicators) should be taken into account.

In recent years, the investment in the energy saving projects has grown continuously, and there is increasing recognition that measuring the efficiency of these projects becomes a vital important problem [5-8]. A large number of publications devoted to the problem of multi-criteria estimation of efficiency of industrial energy saving projects have appeared in Russia and abroad [9-12]. The use of certain estimation methods described in these publications is determined mainly by the features of the specificity of implementation objects and the projects themselves, as well as by the characteristics of investment sources [13, 14].

In Russia, «Guidelines for evaluating the effectiveness of investment projects» [15] and the «Guideline for the estimation of the investments efficiency of the United Nations Industrial Development Organization UNIDO» are most often used for measuring the economic efficiency of energy saving projects [16]. The Guideline, published by UNIDO in 1978, is accepted worldwide as a standard. It is relevant to this day both for new investments and for projects devoted to extension of an enterprise, modernization and business process re-engineering. The estimation methods proposed in [17-19] are based on this approach, which determines the benefit by the cost of the projected reduction in energy consumption. However, the implementation of energy saving projects also releases a capacity, the effect of which sometimes turns out to be more significant than the saving of fuel and energy resources (FER); this fact is not always taken into account in the technoeconomic study. Thus, while estimating the efficiency of energy-saving projects, it is necessary to consider not only the savings in fuel and energy resources, but also, for example, the technological effect of the release of capacity, the environmental effect of reducing CO_2 emissions, the social effect of easy operation of equipment, reliability, durability and other criteria.

Another method for measuring the efficiency of an energy saving project is the rapid assessment of equipment and technology [17]. The contribution of energy saving equipment is determined by including the coefficient of its' efficiency into the equation of heat balance of an industrial object (building); this coefficient characterizes the relative value of the reduction in energy consumption. This method is used to determine a class of buildings' energy efficiency and is universal for most projects, however, it gives fairly generalized results and does not take into account economic and operational factors, affecting only the energy indicators of the object of implementation (DMU).

Another approach to estimating the efficiency of implementation of energy saving technologies and energy saving measures is based on analysis of the environmental component of the Kyoto Protocol methodology [20]. For estimation of the contribution of energy saving technologies into the environmental component of efficiency, the author proposes to be guided by the following criteria: greenhouse gas emissions into the atmosphere; total energy consumption; consumption of material resources; emissions of ozone-depleting gases into the atmosphere; water consumption. In addition to the proposed generally applicable criteria, specialized criteria for toxic effects on humans and animals were also considered. The proposed indicators of environmental performance allow one to analyze new technological solutions in terms of their global and local environmental impacts.

In this paper, it is proposed to use the Data Envelopment Analysis (DEA) method for multi-criteria analysis of the efficiency of energy saving projects that allows evaluating projects using heterogeneous criteria having different units of measure [21, 22]. The energy saving project is proposed to be treated as a production unit or Decision Making Unit (DMU). Usage of DEA models provides the values of generalized efficiency indexes, which determine the relative efficiency or degree of inefficiency of each project. Projects with a maximum efficiency index of 1 form the efficiency boundary or efficiency frontier, and, therefore, the objects belonging to this frontier can be considered as the most efficient ones [23]. DMU located below the efficiency maximum in the interval [1], depending on the size of the estimation, are considered as less effective or not effective projects.

Analysis of the criteria for measuring efficiency of energy saving projects by the methods listed above demonstrates that the DEA method provides the most complete estimation (table 1).

The aim of the presented study is the development of the procedure for multi-criteria estimation of the efficiency of energy saving projects by the DEA method on the example of an industrial enterprise for the production of cellular autoclaved aerated concrete.

2. Object of investigation

In the presented researches an industrial enterprise for the production of cellular autoclaved aerated concrete is considered as an object of investigations. It is located in the Krasnoyarsk district of Samara region, Russia. The total industrial square of the enterprise is 96600 m^2 (figure 1).

Table 1. Criteria for estimation of efficiency of energy saving projects.

Method of measuring efficiency	Estimation criterion
Rapid assessment	Reducing of energy consumption
Estimation based on the Kyoto Protocol methodology	General energy consumption; Greenhouse gas emissions; Consumption of material resources.
Technoeconomic study	Capital commitments; Return on investment; Pay-off period; Return on average assets; Generalized index of energetic efficiency of enterprise.
DEA	Reducing of energy consumption; Lifetime of equipment; Capital commitments; Exploitation costs; Operating convenience; Service personnel; Reliability growth; Increasing quality of energy resources, etc.

Figure 1. Location of main buildings at the enterprise

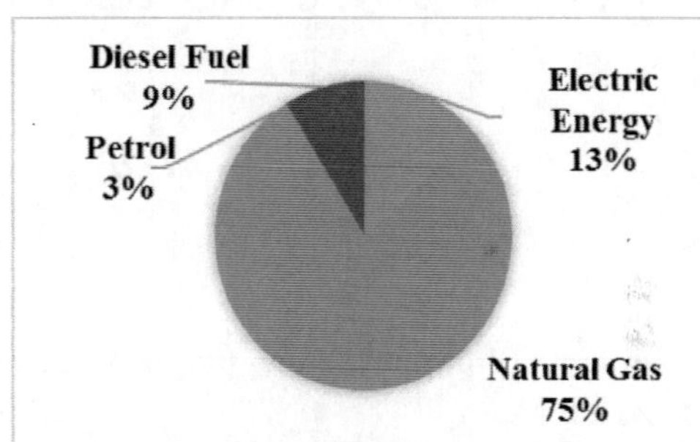

Figure 2. Balance of fuel and energy resources consumption at the enterprise in 2016 year

The enterprise contains 15 buildings and units for various purposes (Figure 1): the main workshop (1); finished products storage area (2); emulsol warehouse building (3), boiler house building and compressor station (4), transformer substation building (5), autoscales (6), sand warehouse (7), security posts (8, 15), air drying unit building (9), a lime warehouse with a grinding department (10), a forklift parking (12), special equipment garages (11, 13), an administrative building (14). The production technology of cellular aerated concrete by the autoclaved hardening method causes a large amount of natural gas consumption (figure 2). Due to the high energy intensity of production, the need to reduce the consumption of energy resources by an enterprise through the implementation of energy saving projects is obvious.

To identify the potential for energy saving and energy efficiency increasing at the enterprise, an energy consumption survey of the enterprise has been conducted. The analysis of the results shows necessity of implementation of several energy-saving projects (table 2):

1. Implementation of a frequency-controlled drive of an electric motor of a pump for supplying water to the technological chain;

2. Replacing the insulation of the process steam line;
3. Implementation of the energy management system according to ISO 50001 standard;
4. Replacing the burner devices of water boilers;
5. Installation of a heat exchanger of a plate type at the site of concrete mixing;
6. Reconstruction of outdoor lighting of the enterprise;
7. Reconstruction of the internal lighting of the main workshop;
8. Installation of thermal insulation of the heating system pipeline;
9. Warming of the facade of the main workshop;
10. Insulation of the roof of the main workshop;
11. Development of automated system of electric power technical metering (ASTME).

The listed projects allow to save electric and thermal energy, and also natural gas. However, in real-life conditions of limited investments, not all projects are reasonable to be implemented and it becomes necessary to determine the most effective and low-cost ones. Table 2 shows the main technical and economic parameters of energy-saving projects, as well as the specific production indicators and the predicted values of CO_2 emission reduction, which may be considered as criteria for estimation of the projects' efficiency.

The following notations in table 2 are used.

I is investments (in thousands of rubles); this economic index displays the amount of investment required to implement an energy saving project.

C_{OP} is operating expenses (in thousands of rubles); this economic index takes into account production costs associated with maintaining the operating systems, machines and equipment in working conditions.

DPP is discounted payback period (in months); this parameter takes into account the period of refund depending on the time value of money (discount rate) that can be calculated according to expression:

$$DPP = \frac{I}{\sum_{t=1}^{T} \frac{D_t}{(1+\alpha)^t}}, \tag{1}$$

D_t is the flow of funds, including net income and depreciation for the certain period $t = \overline{1,T}$; $\alpha = 20\%$ is a discount rate.

B is an annual savings of energy resources in terms of their costs (in thousands of rubles); this economic criterion describes the value of saved fuel and energy resources, which can be achieved as a result of the energy-saving project implementation.

NPV is a net present value calculated as follows:

$$NPV = \sum_{t=1}^{T} \frac{D_t}{\left(1+\alpha\right)^t} - I. \tag{2}$$

ΔW is a reduction of power capacity of equipment (in tons of fuel oil equivalent per m^3); this technological parameter can be defined as the following ratio of the saved energy resource E (in tons of oil equivalent (TOE)) to the volume of production V_{prod} (m^3):

$$\Delta W = \frac{E}{V_{prod}}. \tag{3}$$

Economic parameter ΔP describes increasing of economic efficiency (in thous. Rub·m^{-3}) as the ratio of money saved to the volume V_{prod} of end product:

$$\Delta P = \frac{B}{V_{prod}}. \tag{4}$$

Parameter *MTBF* means a middle time between equipment failures (in hours), characterizing the average time of equipment failure-free operation and determines its reliability.

Em_{CO2} is a reduction of CO_2 emission (in tons). In addition to reducing the consumption of fuel and energy resources, the energy saving problem solution also leads to the reduction of CO_2 emissions into the atmosphere.

Table 2. Inputs and outputs of energy saving projects.

Pr. No.	I, (th.RUB)	C_{OP}, (th.RUB)	DPP, (months)	B, (th.RUB)	NPV, (th.RUB)	ΔW, (TOE·m^{-3})	ΔP, (th.RUB·m^{-3})	$MTBF$, (hours)	Em_{CO2}, (tons)
	X_1	X_2	X_3	Y_1	Y_2	Y_3	Y_4	Y_5	Y_6
1	313.08	24.08	37	201.02	458.253	60.66	2,801.4	9,635	15.93
2	79.64	6.13	46	39.7	92.162	63.70	553.3	32,604	7.33
3	1650	137.5	48	749.55	1,854.48	224.55	10,445.8	47,424	25.83
4	226.22	17.4	55	94.61	155.16	151.46	1,318.5	10,640	17.42
5	176.3	13.56	63	65.35	59.27	104.87	910.8	11,742	12.06
6	258.46	17.23	41	147.75	360.99	44.60	2,059.1	100,000	11.71
7	1,581.35	105.42	71	533.85	635.82	161.12	7,439.9	100,000	42.3
8	114.64	8.81	48	54.34	120.51	87.20	757.3	96,360	10.03
9	1,713.6	122.36	104	454.63	412.03	653.61	6,335.9	131,400	52.26
10	1,531.87	109.42	78	483.14	727.03	775.23	6,733.1	131,400	66.25
11	2,278	189.833	45	1,148.16	2,534.96	346.94	16,001	105,120	90.9

3. Statement of the problem of multi-criteria estimation using DEA method

Integral efficiency in the original formulation of the DEA method proposed by A. Charnes, W.W. Cooper, E. Rhodes [24] (CCR-model), for the estimated simple one-dimensional DMU [25] with one input X and one output Y, can be characterized as the ratio of the efficient result Y of the DMU to the resources expended X (figure 3):

$$f = \frac{Y}{X}. \tag{5}$$

Based on the relation (5) the numerical values of efficiency f can be defined for the estimated DMU.

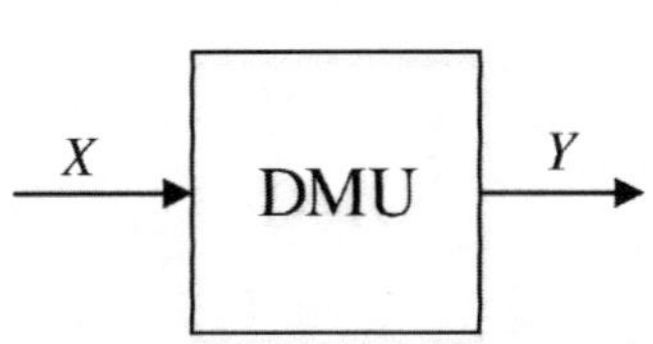

Figure 3. Model of one-dimensional DMU

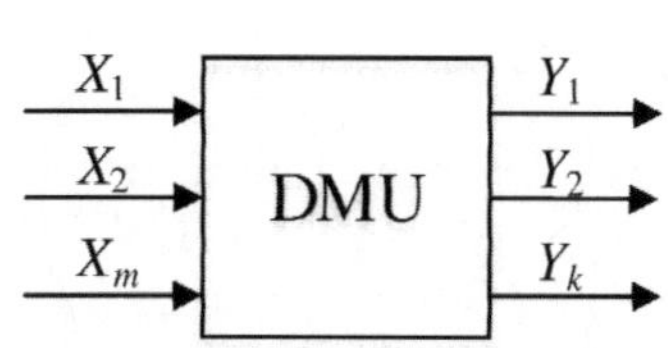

Figure 4. Model of multi-dimensional DMU

Using multi-dimensional DEA model, it is possible to evaluate n productive units, DMUs, where each DMU takes m different inputs to produce k different outputs (figure 4).

Values of inputs X_j, j=1,2,…,m represent the costs of resources involved in the implementation of the production process of DMU, the reduction of which leads to an increase in the final index of efficiency. Thus, capital, financial, labor, material, information, energy and other resources can be considered as inputs.

An increase in the input value leads to a decrease in the efficiency of DMU, i.e.:

$$\frac{\partial f(X_1, X_2, ..., X_m)}{\partial X_j} < 0, \ \mathrm{j} = 1, 2, ..., m. \tag{6}$$

The choice of outputs Y_i, i=1,2,…,k is carried out in such a way that they characterize the factors that positively affect to the overall efficiency of the DMU. This, the income, profitability, volume and quality of the final product, reliability, durability, environmental friendliness, sustainability and other similar characteristics can be considered as outputs. The values of outputs may be not numerically comparable and not related to each other, however, they must be quantified, and their increase should lead to an increase in the overall efficiency:

$$\frac{\partial f(Y_1, Y_2, ..., Y_k)}{\partial Y_i} < 0, \ \mathrm{i} = 1, 2, ..., k. \tag{7}$$

According to (5), based on the choice of m inputs and k outputs, the efficiency rate of such a multidimensional DMU can be generally expressed as:

$$f = \frac{u_1 \cdot Y_1 + u_2 \cdot Y_2 + ... + u_k \cdot Y_k}{v_1 \cdot X_1 + v_2 \cdot X_2 + ... + v_m \cdot X_m}, \tag{8}$$

where u_i, i=1,2,…,k are weights assigned to i^{th} output Y_i, i=1,2,…,k; v_j, j=1,2,…,m are weights assigned to j^{th} input X_j, j=1,2,…,m.

For the definition of numerical indexes of efficiency of each estimated DMU$_z$, z=1,2,…,N it is necessary to demand that the efficiency value belongs to an interval [0;1]. The essence of DEA models in measuring the efficiency of productive unit DMU lies in maximizing its' efficiency rate, subject to the condition that the efficiency rate of any other units in the population must not be greater than 1:

$$\frac{\sum_{i=1}^{k} u_{iz} Y_{iz}}{\sum_{j=1}^{m} v_{jz} X_{jz}} \le 1, \ \mathrm{z} = 1, 2, ..., N; \ i = 1, 2, ..., k; \ \mathrm{j} = 1, 2, ..., m; \ u_{iz} > 0; v_{jz} > 0 \ / \tag{9}$$

Then the condition of achievement of maximum efficiency f_z for z^{th} DMU can be written as follows:

$$f_z = \frac{\sum_{i=1}^{k} u_{iz} Y_{iz}}{\sum_{j=1}^{m} v_{jz} X_{jz}} \to \max_{u_{iz}, v_{jz} \in G} \tag{10}$$

Solving the zth, $z = \overline{1, N}$ problem (10) under the constraints (9) gives the zth value of the complex efficiency index f_z and the corresponding set of weights $u_{iz} \in G, \mathrm{i} = \overline{1, k}$ and $v_{jz} \in G, j = \overline{1, m}$ providing the maximum of the functional (10).

The formulated problem relates to the problems of mathematical programming that can be solved by standard optimization methods (the method of least squares, the method of conjugate gradients, etc.).

The application of the DEA method allows to:
- identify efficient and inefficient energy saving projects using a quantitative measures of their efficiency (inefficiency);
- indicate effective goals for each project;
- find the best ways to achieve these goals;
- determine the most promising further investment directions for energy saves and energy efficiency improvements at the enterprise while analysing the efficiency of energy saving projects.

4. Algorithm for multi-criteria estimation of energy saving projects based on DEA method

For multi-criteria evaluation based on DEA method, the energy saving project is proposed to be treated as a production unit or DMU.

Then a following algorithm including seven main steps can be used:

1) Definition of a group including N projects, DMUs, for multi-criteria estimation;

2) Selection of DEA models for the zth energy saving project, DMU; choice of input parameters $X_{jz}, j=\overline{1,m},$ and output parameters $Y_{iz}, \mathrm{i}=\overline{1,k}, z=\overline{1,N}$;

3) The formulation of mathematical programming problem in the form (9) - (10) for each DEA model;

4) Solving mathematical programming problem (9) - (10);

5) Analysis of the results;

6) DMUs ranking;

7) Decision making.

At the first stage, a group of N energy-saving projects (DMUs) is formed; the technical and economic justification of the projects is carried out. All projects in the group should meet the same efficiency criteria.

Further, at the second stage, the DEA models are compiled, which are characterized by sets and combinations of input and output parameters (criteria). In this case, the input and output criteria can have different dimensions and estimates (Table 2).

At the third stage, the problems of mathematical programming are formulated: a functional is written in the form (10) under conditions of constraints (9). The solution of mathematical programming problem includes definition of the weights $u_{iz} \in G, \mathrm{i}=\overline{1,k}, z=\overline{1,N}$ and $v_{jz} \in G, j=\overline{1,m}, z=\overline{1,N}$, which correspond to the searched values of the relative efficiency indexes $f_z, z=\overline{1,N}$ for the every considered DEA model.

Solving N problems of mathematical programming using the well-known MATLAB software package allows one to present the result as a set of values of the relative efficiency indexes of the investigated projects.

Based on these values, the analysis of the computational results can carried out depending on the direction of the energy saving program at the enterprise (based on economic, technological, environmental effects separately or on a combination of all of them). Then, in the final stage, projects are ranked according to their indexes of efficiency $f_z, z=\overline{1,N}$.

The developed approach allows identifying the advantages and drawbacks of each energy saving project. However, such estimation is relative and must be compared with the corresponding absolute indexes calculated during the feasibility study.

5. DEA models for measuring efficiency of energy saving projects

The nine typical input and output parameters have been selected and the twelve DEA models are developed that characterize the technological, environmental, economic and comprehensive estimators of the efficiency of N=11 energy saving projects at industrial enterprise for the production of cellular autoclaved aerated concrete (Figure 5-16).

The input parameters, $X_j, \mathrm{j}=\overline{1,3}$ which increasing has a negative impact on the final value of the relative efficiency of the project are the follows: X_1 is investment (I), X_2 is operating expenses (C_{OP}), X_3 is a discounted payback period (DPP). The output parameters, Y_i, $i=\overline{1,6}$ which increasing has a positive impact on the final value of the relative efficiency of the project are follows: Y_1 is the annual savings of energy resources in terms of their costs (B), Y_2 is a net present value (NPV), Y_3 is reduction of power capacity of the equipment (ΔW), Y_4 is increasing of economic efficiency (ΔP); Y_5 is time between equipment failures ($MTBF$); Y_6 is a reduction of CO_2 emission (Em_{CO2}). Numerical values of inputs $X_{jz}, \mathrm{j}=\overline{1,3}$ and output parameters Y_{iz}, $i=\overline{1,6}$ for $z=\overline{1,11}$ projects are reported in table 2.

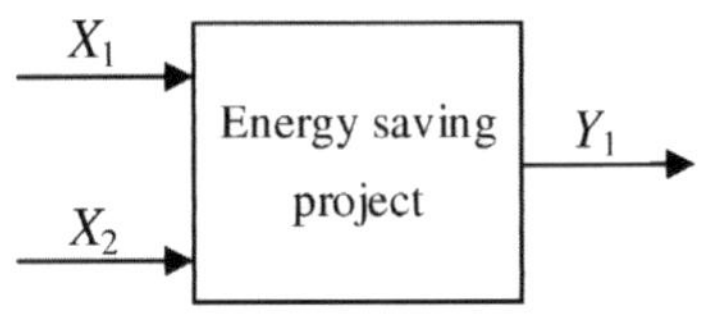

Figure 5. DEA model No. 1 (estimation of annual savings of energy resources).

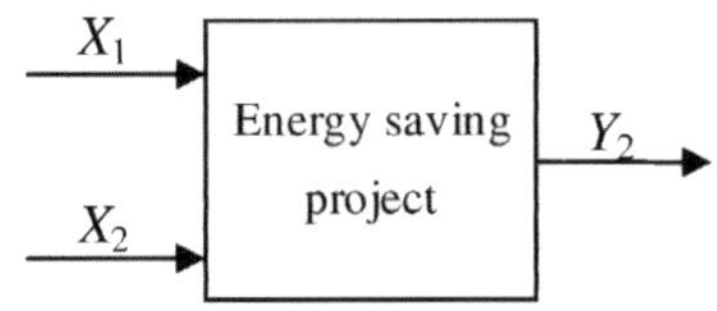

Figure 6. DEA model No. 2 (estimation of NPV).

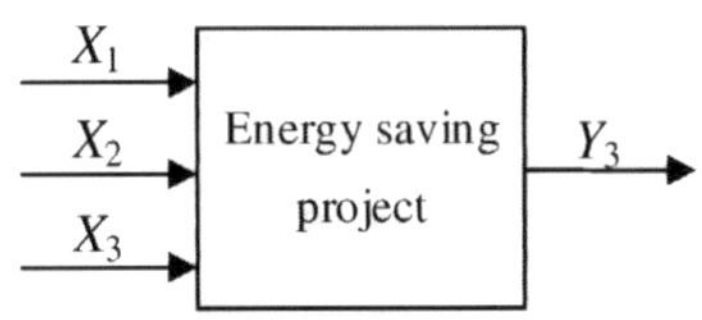

Figure 7. DEA model No. 3 (estimation of decreasing energy consumption by equipment)

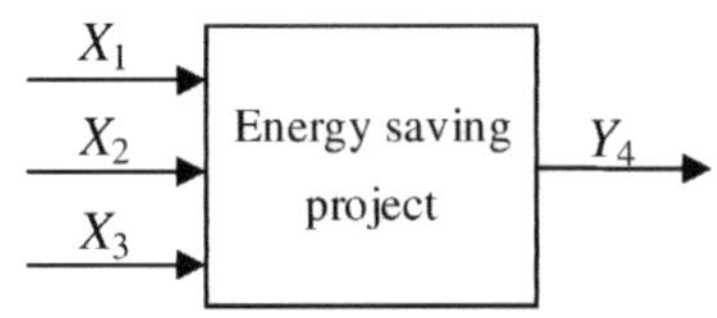

Figure 8. DEA model No. 4 (estimation of increasing economical efficiency)

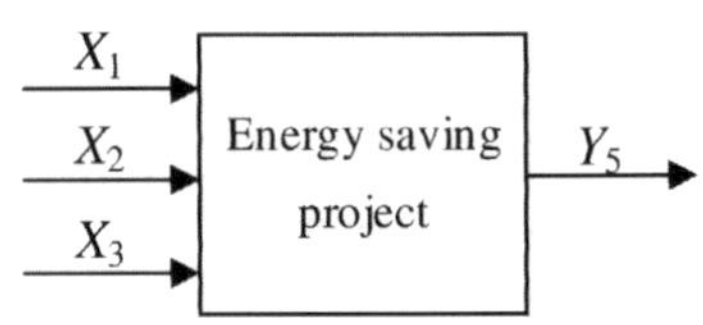

Figure 9. DEA model No. 5 (estimation of increasing reliability)

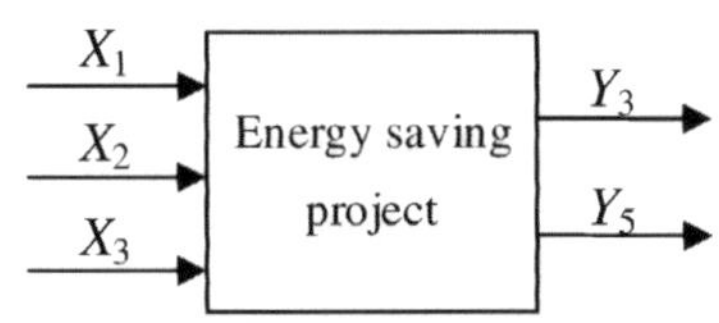

Figure 10. DEA model No. 6 (estimation of decreasing energy consumption by equipment and increasing reliability)

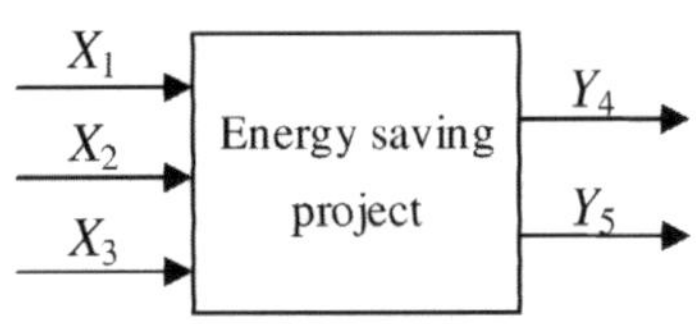

Figure 11. DEA model No. 7 (estimation of increasing economical efficiency and reliability)

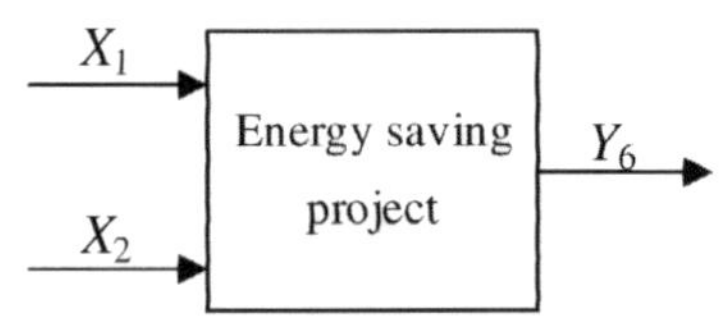

Figure 12. DEA model No. 8 (estimation of decreasing CO_2 emission)

IOP Conf. Series: Materials Science and Engineering **552** (2019) 012020 doi:10.1088/1757-899X/552/1/012020

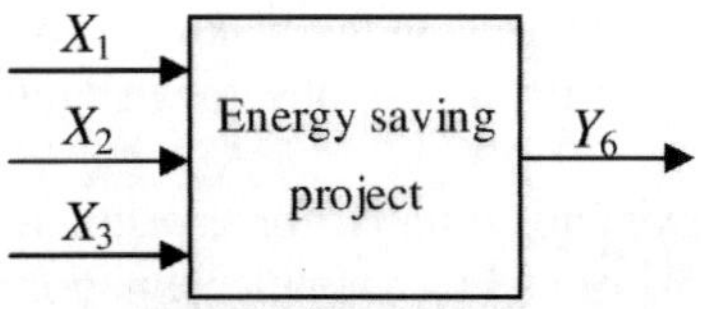

Figure 13. DEA model No. 9 (estimation of decreasing CO_2 emission taking into account DPP)

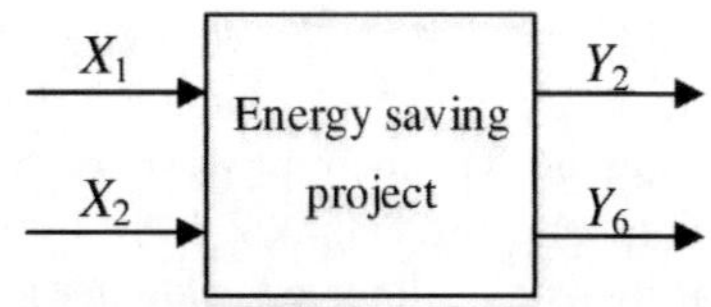

Figure 14. DEA model No. 10 (measuring of economical and ecological efficiency)

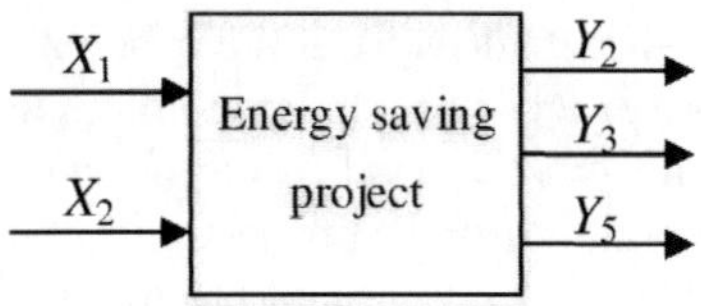

Figure 15. DEA model No. 11 (measuring economical and technological efficiency).

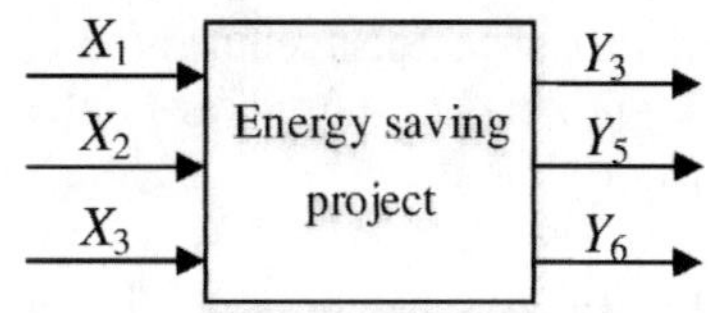

Figure 16. DEA model No. 12 (measuring the ecological and technological efficiency)

For all presented DEA models (figures 5-16) the problem of mathematical programming can be formulated in the form (9)-(10).

6. Results analysis

Mathematical programming problems in the form of (9), (10) have been solved for each energy saving project with respect to all 12 models and the results are shown in table 3. According to the data in table 3, the priority for implementation should be given to energy saving projects having the maximum value of the DEA-index Σf. The most effective projects No. 2, 8, 11 and 6 (from 1 to 4 places in the ranking) demonstrate the highest value of the relative efficiency in the most models. However, each of these projects has its own drawbacks. Thus, projects No. 6 for the reconstruction of outdoor lighting of the enterprise's territory and No. 8 for installing insulation of the heating system's pipeline are not expedient from the point of view of reducing CO_2 emissions. Project No. 2 on the replacement of the steam line insulation should be improved in terms of reliability improving.

Table 3. Computational results for the efficiency of energy saving projects with respect to selected models based on the DEA method.

Pr. No.	Model DEA												**DEA-index** Σf	Rank.
	1 (Fig.5)	2 (Fig.6)	3 (Fig.7)	4 (Fig.8)	5 (Fig.9)	6 (Fig.10)	7 (Fig.11)	8 (Fig.12)	9 (Fig.13)	10 (Fig.14)	11 (Fig.15)	12 (Fig.16)		
1	1	1	0.344	1	0.107	0.345	1	0.553	0.914	1	1	0.914	**9.177**	5
2	0.776	0.791	1	0.776	0.487	1	0.896	0.952	1	1	1	1	**10.678**	1
3	0.708	0.768	0.471	0.878	0.405	0.513	0.878	1	0.378	0.768	0.769	0.513	**8.049**	8
4	0.651	0.469	1	0.651	0.088	1	0.660	0.336	1	0.838	0.838	1	**8.531**	7
5	0.577	0.230	0.816	0.577	0.091	0.816	0.592	0.568	0.816	0.744	0.744	0.816	**7.387**	9
6	1	1	0.316	1	1	1	1	0.170	0.777	1	1	1	**10.263**	4
7	0.591	0.288	0.228	0.742	0.577	0.605	0.762	0.357	0.674	0.370	0.314	0.710	**6.218**	11
8	0.739	0.718	1	0.739	1	1	1	0.507	1	0.952	1	1	**10.655**	2
9	0.439	0.168	0.744	0.521	0.518	0.745	0.547	0.434	0.681	0.357	0.514	0.745	**6.413**	10
10	0.522	0.332	1	0.641	0.691	1	0.698	0.838	1	0.507	0.682	1	**8.911**	6
11	0.785	0.760	0.776	1	0.958	1	1	0.744	1	0.765	0.765	1	**10.553**	3

Projects No. 5, 7, 9 turned out to be ineffective (from 9th to 11th places in the ranking). For none of the models, these projects provide the maximum value of relative efficiency. Thus, they should be excluded from the investment program of the enterprise.

The remaining projects No. 1, 3, 4, 10 (from 5th to 8th places) are suitable for investment only if additional funds are allocated, they need to be improved in terms of technical implementation and the use of better materials.

Conclusions

The paper proposes a successfully tested procedure of multi-criteria evaluation of the efficiency of energy-saving projects using the DEA method on the example of an industrial enterprise for the production of cellular autoclaved aerated concrete. This approach allows measuring the efficiency of energy saving projects proposed for implementation at the enterprise, according to economic, technological and environmental criteria. The developed procedure for multi-criteria evaluation of the efficiency of energy saving projects based on the DEA method has the potential of widespread use in practice for developing effective management decisions regarding the formation of investment programs in industrial enterprises.

Acknowledgments

This research was partially supported by the Ministry of Education and Science of the Russian Federation (project part of government contracts, Project №10.3260.2017/4.6).

References

[1] Máša V, Stehlík P, Touš M and Vondra M 2018 *Energy* **158** 293-304
[2] Borisova N, Borisov A and Dushko O 2017 *MATEC Web of Conferences* **129** 01040
[3] Trubaev P and Tarasyuk P 2017 *Int. J. of Energy Economics and Policy* **7** 201-8
[4] Máša V, Stehlík P and Havlásek M 2017 *Chemical Engineering Transactions* **61** 1081-6
[5] Marchuk T, Ryzhakov D, Ryzhakova G and Stetsenko S 2017 *Investment Management and Financial Innovations* **14** 12-20
[6] Abeelen C, Harmsen R and Worrell E 2016 *Energy Efficiency* **9** 153-69
[7] Kvon G, Khamidullin F, Samysheva E, Vaks V and Mararov N 2016 *Int. J. of Environmental and Science Education* **11** 7441-52
[8] McCabe J 2008 *Chemical Engineer* **807** 37-9
[9] Vidunova A and Evstafjeva O 2012 *Sistema klyuchevy`x pokazatelej e`ffektivnosti predpriyatiya v oblasti e`nergosberezheniya* [The system of key performance indicators of the enterprise efficiency in the field of energy saving] (E`kologiya i promy`shlennost` Rossii) **6** pp 46-9 [In Russian]
[10] Vodyannikovoв V 2010 *E`konomicheskie osnovy` e`nergosberezheniya* [Economic fundamentals of energy saving] (Vestnik FGOU VPO MGAU) **5** pp 69-73
[11] Fahri N 2010 *E`nergosberezhenie. Konflikt interesov* [Energy saving. Conflict of interests] (Stroitel`ny`e nauki) **3** pp 416-8 [In Russian]
[12] Plotnikov A and Plotnikov D 2010 *Obzor innovacionny`x texnologij ucheta rasxoda i snizheniya poter` e`nergii na predpriyatii, napravlenny`x na povy`shenie ego e`nergoe`ffektivnosti* [Review of innovative technologies for metering and reducing energy losses in the enterprise, aimed at improving its energy efficiency] (Izdatel`skij otdel fakul`tetea VMiK imeni M.V. Lomonosova) p 208 [In Russian]
[13] Semenov V 2009 *Zachem e`nergosberezhenie Rossii ili pochemu imenno 40 %?* [Why energy saving in Russia or why exactly 40%?] (E`nergosovet) **3** pp 9-10 [In Russian]
[14] Gumenyuk N 2012 *Sistema chastny`x i integral`ny`x pokazatelej, kriterii i metody` rascheta e`ffektivnosti e`nergosberegayushhix texnologij* [The system of private and integral indicators, criteria and methods for calculating the efficiency of energy saving technologies] (Aktual`ny`e voprosy` e`konomicheskix nauk) **24** pp 132-8 [In Russian]

IOP Conf. Series: Materials Science and Engineering **552** (2019) 012020 doi:10.1088/1757-899X/552/1/012020

[15] Guide to Practical Project Appraisal 1986 Social Benefit-Cost Analysis in Developing Countries (Vienna: UNIDO)
[16] Mokrousov V 2012 Rezul`taty` mnogokriterial`nogo ocenivaniya meropriyatij po e`nergosberezheniyu pri provedenii e`nergeticheskogo obsledovaniya prompredpriyatiya [The results of a multi-criteria evaluation of energy saving measures during an energy survey of an industrial enterprise] *Vestnik voronezhskogo gosudarstvennogo universiteta. Seriya: sistemny`j analiz i informacionny`e texnologii* **2** pp 117-23 [In Russian]
[17] Tabunschikov Yu, Shilkin N and Miller Yu 2013 E`kspress-ocenka e`ffektivnosti e`nergosberegayushhego oborudovaniya, texnologij i meropriyatij [Express assessment of the efficiency of energy-saving equipment, technologies and events] *Izvestiya vy`sshix uchebny`x zavedenij. Stroitel`stvo* **6** pp 57-63 [In Russian]
[18] *STO 17532043-001-2005 «Normy` teplotexnicheskogo proektirovaniya ograzhdayushhix konstrukcij i ocenki e`nergoe`ffektivnosti zdanij»* [STO 17532043-001-2005 Norms of heating designing building envelopes and evaluation of energy efficiency in buildings] (Moscow:ORNTO) p 69 [In Russian]
[19] *RD 153-34.1-09.321-2002 Metodika e`kspress-ocenki e`konomicheskoj e`ffektivnosti e`nergosberegayushhix meropriyatij na TE`S* [RD 153-34.1-09.321-2002 Methodology for rapid assessment of the economic efficiency of energy saving measures at TPPs] (Moscow: ORGRES) p 77 [In Russian]
[20] Lugin V, Zigantovich A, Klyukvin T, Kornilov A, Novikov S, Khotchenkov E and Sverdlov I 2009 Variant ocenki e`ffektivnosti texnologij e`nergosberezheniya na osnove metodologii Kiotskogo protokola [Option evaluation of the effectiveness of energy saving technologies based on the methodology of the Kyoto Protocol] *Nauka v neftyanoj i gazovoj promy`shlennosti* **2** pp 12-4 [In Russian]
[21] Lissitsa A and Babicheva T 2003 *Analiz obolochki danny`x (DEA) – sovremennaya metodika opredeleniya e`ffektivnosti proizvodstva* [Data Envelopment Analysis (DEA) – a modern method for determining production efficiency] (IAMO, Obsuzhdeniye) p 56 [In Russian]
[22] Pleshivtseva Yu, Derevyanov M, Kashirskikh D, Pimenov A, Kerov A and Tyan V 2018 *Oil Industry* **8** 104-9
[23] Pleshivtseva Yu, Derevyanov M and Setin S 2014 *Oil Industry* **8** 124-8
[24] Charnes A, Cooper W and Rhodes E 1978 *European Journal of Operation Research* **2** pp 429-44
[25] Banker R, Charnes A and Cooper W 1984 *Management Science* **30** pp 1078-90

Creation of a new energy-efficient design of the dustexhaust system

V M Kireev, A B Goltsov, A S Seminenko and Y G Ovsyannikov

Department of Heat and Gas Supply and Ventilation, Belgorod State Technological University of V.G. Shukhov, Kostyukov St., 46, Belgorod, 308012, Russia

E-mail: vit31rus@mail.ru

Abstract. Aspiration systems of processing enterprises (local exhaust ventilation) consume a lot of electricity. The article discusses the new energy-efficient design of the aspiration system when using inclined overload gutters. Energy efficiency is the use of the principle of partial recirculation of aspirated air and the Coanda effect when it is fed directly into the transshipment chute. The article discusses in detail the new effective unit for supplying recirculated air to the reloading chute. The use of the proposed design will reduce more than twice the energy consumption of aspiration systems and the energy intensity of general ventilation systems that supply air to the room. The article provides theoretical information on the operation of systems, the results of the experiment.

Keywords: dust-extracting ventilation, energy efficiency, aspiration system, recirculation, Coanda effect, bulk solids reloading

1. Introduction.

Different technological processes of production and processing of bulk solids in many industrial sectors including construction, mining, metallurgy, coal industry etc. are accompanied with intensive dust emission to the operating area [1-3]. Despite its local character, the indicated (concentrated) sources of dust emission have a dominant influence on the amount of dust in the working areas. The major reason for spreading of the dust is ejection, i.e. air entrainment by the flow of the reloaded bulk solid [4-8].

The only effective way to remove dust when producing bulk materials is to use integrated dedusting systems, which include dust collection systems, means to combat secondary dust emission (Central Vacuum Systems) and general ventilation. Dust collection systems remove dusty air in the places of its occurrence, the Central Vacuum System removes settled dust on surfaces in the room, general ventilation is necessary both to compensate for the volumes of air removed by dust collection systems and to maintain the required parameters of the air in the working area.

The energy intensity of the aspiration systems is determined by the power consumed by the fan, which depends on the volume of aspirated air moved through the air ducts of the system and their hydraulic resistance, a significant part of which (more than 50%) consists of pressure losses in cyclones - the most common dust collectors of the system [9]. Also, the energy intensity of the aspiration systems depends on the cost of cleaning the aspiration air [10].

2. Actuality.

According to the researches, a number of enterprises spend up to 20% of their circulating assets [11–13] for the maintenance of dust-extraction systems and that influences their competitiveness. The reason for that is, first of all, the use of non-effective engineering solutions and the faults in design.

The most promising direction for the improvement of traditional aspiration systems, which makes it possible to significantly reduce the dust emission of the system, is recirculation (return to the reloading chute) of the part of aspirated air (the volume of ejected air). This technical solution allows, due to low power consumption, to reduce both the required performance of the dust collection system and the gas treatment station and the system of general ventilation.

Analysis of the aerodynamics of existing aspiration systems indicates a lack of attention paid to the issues of optimal organization of the supply of recirculated air into the reloading chute. In this connection, the relevance of the development of the method of supplying recirculation air and the technical means of its implementation, which allow to influence the process of formation of the volume of ejection air, is obvious

Thus, the main idea of the presented scientific article is to reduce the energy intensity of both the aspiration systems and the entire dedusting complex due to the supply of recirculated air into the reloading inclined chutes.

3. Task setting. Computation methods

In the works of O.D. Neykov and I.N. Logachev [14, 15], V.A. Minko [16, 17] the movement of air through the chute under the action of ejection of bulk material is considered as the movement of air in the channels with a certain hydraulic characteristic under the action of pressure drop. In general cases, when transferring unheated bulk materials, the volumes of ejection air flowing through the chute Q_e into the lower cover will be:

$$Q_e = F\sqrt{\frac{P_e \pm P_{ob} + \Delta P_u}{0{,}5\rho\sum\xi}}, \qquad (1)$$

where F is the cross-sectional area of the chute; $\Sigma\xi$ is the sum of local resistance coefficients (LRC) of the chute and cover; P_{ob} is the pressure in the executive devices of the aspirated equipment; ΔP_u -is the pressure difference caused by the effect of local exhaust (negative pressure in the cover ΔP_u); P_e is the ejection pressure.

The ejection pressure is a function of the drag coefficient of the material particles, the volume concentration of the material, the average particle diameter, the material flow and its density, the relative velocity of the air and the material in the chute, depending on the height and angle of material incidence.

Most of the above parameters are either uncontrollable (density of air and material, particle size distribution, particle shape, etc.), or poorly-controlled under production conditions (vacuum in the cover, height of material incidence, tilt angle of the overload chute, cross-sectional area of chute etc.).

The only parameter that allows reducing the volume of aspiration under production conditions is the increase in the hydraulic resistance of the "upper cover - chute - lower cover" channel along which the ejected air flow moves.

The hydraulic resistance of the specified channel is determined by the total coefficient of local resistance, calculated by the formula:

$$\sum\xi = \xi_v + \xi_g + \xi_n, \qquad (2)$$

where ξ_v, ξ_g, ξ_n is LRC respectively of the upper cover (of the feed conveyor), chute and lower cover (receiving conveyor), related to the air velocity in the chute.

Sealing the upper cover, contributing to an increase in ξv, is currently the most difficult task under production conditions due to the design of the boot part of the gutter and the reasons for its operational nature. The most promising solution is to reduce the volume of driving air by increasing the hydraulic resistance of the channel "chute - lower cover".

To increase the LRC of the lower cover, various methods are used [18-24], which consist of installing various mechanical obstacles on the path of the driving air flow (partitions, chains, zigzag plates, etc.), some of which can also serve as dust precipitators.

To increase the LRC of the reclaim chute, along with the reduction of its cross section, the installation of various mechanical valves and gates in the path of bulk material is used [25].

However, it should be noted that the above methods of increasing the hydraulic resistance of the "chute - lower cover" channel are not widely used due to some complication of the structures of the cover and chutes, reducing their operational reliability (wear and possibility of jamming of control elements, hanging of the overloaded material, etc.).

The second direction to achieve a significant reduction in the amount of dust-loaded air emitted by the dust-exhaust system into the atmosphere is the use of air recirculation.

More optimal is the partial recirculation of aspirated air, which can be natural or forced.

Forced recirculation is carried out by an external activator of draft, natural - due to the pressure gradient that occurs in the chute as a result of the ejection of air by the bulk material [13]. The areas of overpressure and vacuum of the chute are connected by an aerodynamic channel. This channel (bypass) can be placed inside the chute (coaxially with it or at its side walls), as well as outside the chute.

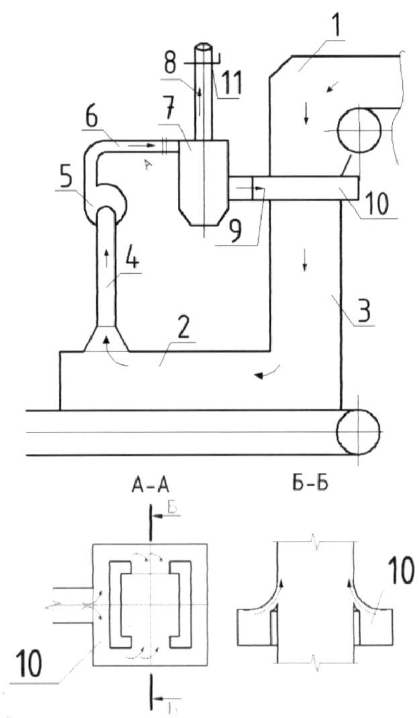

Figure 1. Schematic diagram of the recirculation dust-exhaust system using the Coanda effect for with vertical reloading chute [26]: 1. upper aspiration hood; 2. lower cover; 3. reloading chute. 4. aspiration tube; 5 fan; 6. pressure tube; 7 dust cleaning device (cyclonic dust collector); 8 purified air exhaust; 9. bypass pipe; 10. distributor of air-dust concentrate; 11. gate valve.

The most promising solution is the use of forced recirculation, carried out by an external activator of draft, which is used as the AC fan, the principle of operation of which is given below. To compensate for the amount of air ejected by the bulk material, the fan 5 'Figure 1' boosts the air, ejected from the cover 2 of the receiving conveyor, through the recirculation duct 6 into the upper cover of the loading chute 3. The amount of recirculated air injected into the upper cover should be such that it does not create positive pressure. The regulation of the volume of recirculated air is carried out by a flap installed on the outlet duct of the fan 5. Moreover, to prevent air emissions through the upper cover, the adjustment should be carried out during the period of the absence of material supply. The aspirated air is cleaned in the cyclone 7, and then divided with the help of the flap 11 into the recirculation air and air removed from the system. The analysis of the aerodynamic parameters of dust-exhaust systems with forced recirculation indicates a lack of attention paid to the issues of optimal organization of the recirculation air supply to the reclaim chutes. Existing methods, as a rule, are reduced to a kind of replacement of the driving air flow by the recirculation air. In this connection, the relevance of the development of the method of supplying recirculation air and the technical means of its implementation, which allow to influence the process of formation of the volume of ejection air, is obvious.

There are a number of studies related to the issue [13]. However, in our opinion, the most promising direction today is the use of the Coanda effect when supplying th air to the reclaim chutes. There is a development proposed by the chair of "Heat and gas supply and ventilation" of the Belgorod State University of Technology n.a. V.G. Shukhov 'Figure 1' [26], however, this solution is suitable for material transferring along a vertical chute ($\alpha = 90^{\circ}$), with certain concerns for the possibility of material entering the distribution channel. At the same time, in practice, in most cases, there is the use of inclined chutes; in this regard, the use of the design [26] is unacceptable due to the blockage of the Coanda channels.

4. The results of the research and its discussion

We have carried out a search experiment as a result of which we proposed a method for supplying recirculation air using the Coanda effect [27] when material is transferred through inclined chutes figure 2 '. The operation of the system is organized in such a way that air is sucked out from the lower cover 2 by the fan 5 through the aspiration tube 4, this air flows through the pressure tube 6 to the inlet of the dust-cleaning device 7 to separate the air-dust concentrate. The purified air through the tube 8 enters the atmosphere or the next stage of cleaning, and the air-dust concentrate due to the bypass pipe 9 is discharged into the distributors 10 of air-dust concentrate located on the upper edge of the reloading chute in series. Distributors 10 of air-dust concentrate provide the supply of air-dust concentrate in the chute 3 through the Coanda air ducts 11. At the same time, due to the special form of the Coanda air ducts 11, ensuring the occurrence of the Coanda effect, the flow of air-dust concentrate "sticks" to the chute, which allows to increase the angle between the directions of movement of the air-dust concentrate and the air dusty stream moving in the flow. This is explained by the fact that on the one side of the jet of air-dust concentrate the surface of the chute 3 prevents free flow of the air dusty stream, and on the other side the part of the air dusty stream is carried along. As a result, the aerodynamic resistance of the system increases, which reduces air consumption. It should be noted that the number of consecutive distributors 10 'figure 2' may vary depending on the flow rate of recirculation air and the speed of its entry into the reloading chute.

The principal difference between the design and 'figure 1' [26] is that the recirculation air is supplied through the Coanda channels, which are located successively in the upper part of the reloading chute. This eliminates the possibility of channel blockage with overloaded material 12 'figure 2' moving along the bottom of the chute.

We have conducted studies of the effectiveness of the proposed design using the COSMOSFloWorks software package [28], which is based on the continuity equation; Navier-Stokes and stationary three-dimensional flow in accordance with the k-ε turbulence model. Studies were carried out for the dust-exhaust system containing two distributors, at the air velocity in the Coanda channel of 10 m / s. Some graphical experimental results of the proposed design are presented in figure 3, a '.

Analyzing the obtained graphical results, we can distinguish two extensive zones of vortex formation in the chute, before and after the supply of recirculation air, which significantly increase the aerodynamic resistance ξ_g equation (2), thereby reducing the volume of the ejected air Q_e. Also, the used Coanda effect can be seen at the upper edge of the reloading chute.

To evaluate the performance of the proposed design, a series of comparative experiments were carried out with the design for a vertical chute [26] under the same initial conditions. The results of the experiment 'figure 3' showed that the arrangement of the Coanda channels in a row in the proposed

IOP Conf. Series: Materials Science and Engineering **552** (2019) 012021 doi:10.1088/1757-899X/552/1/012021

structure is more efficient than from opposite sides. For example, in 'figure 3, a, c, e', there is a more extensive impact of the Coanda effect [28] (beyond the calculated area) compared to the similar 'figure 3, b, d, f'. Also one of the tasks was to determine the flow rate of ejected air passing through the chute Q_e, which turned out to be more than two times lower when using the proposed one-way air supply.

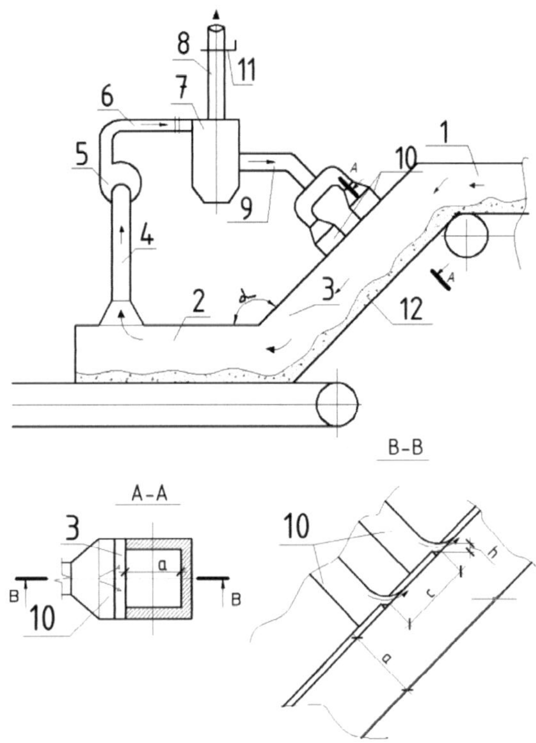

Figure 2. Schematic diagram of the dust-exhaust system with forced recirculation: 1. upper aspiration hood; 2. lower cover; 3. reloading chute. 4. aspiration tube; 5 fan; 6. pressure tube; 7 dust cleaning device (cyclonic dust collector); 8 purified air exhaust; 9. bypass pipe; 10. distributor of air-dust concentrate with the Coanda channel; 11. gate valve. 12 overloaded material.

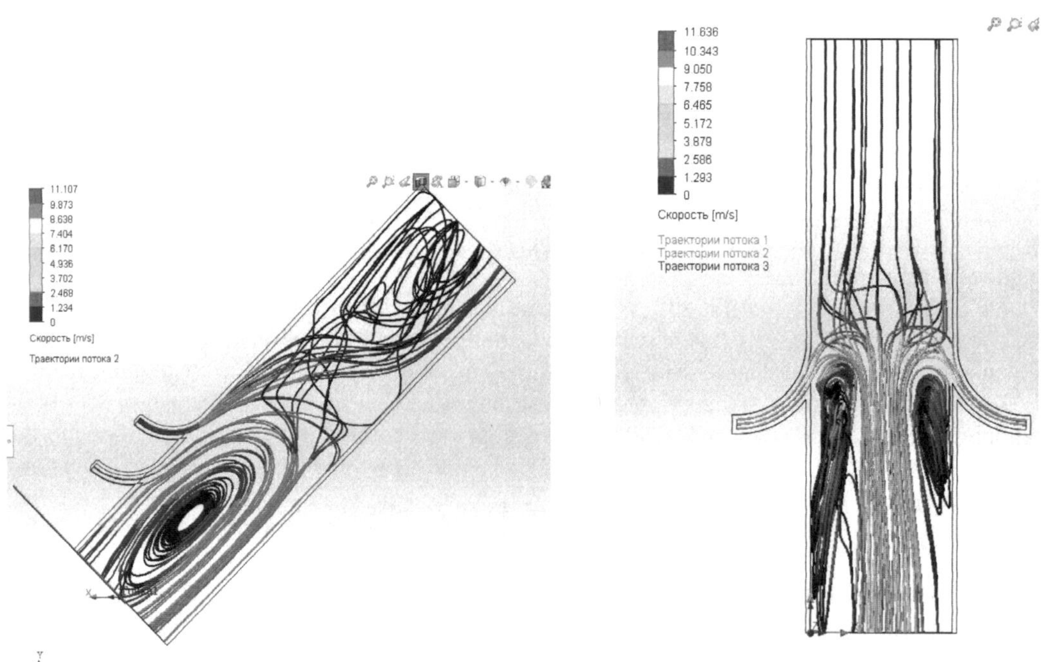

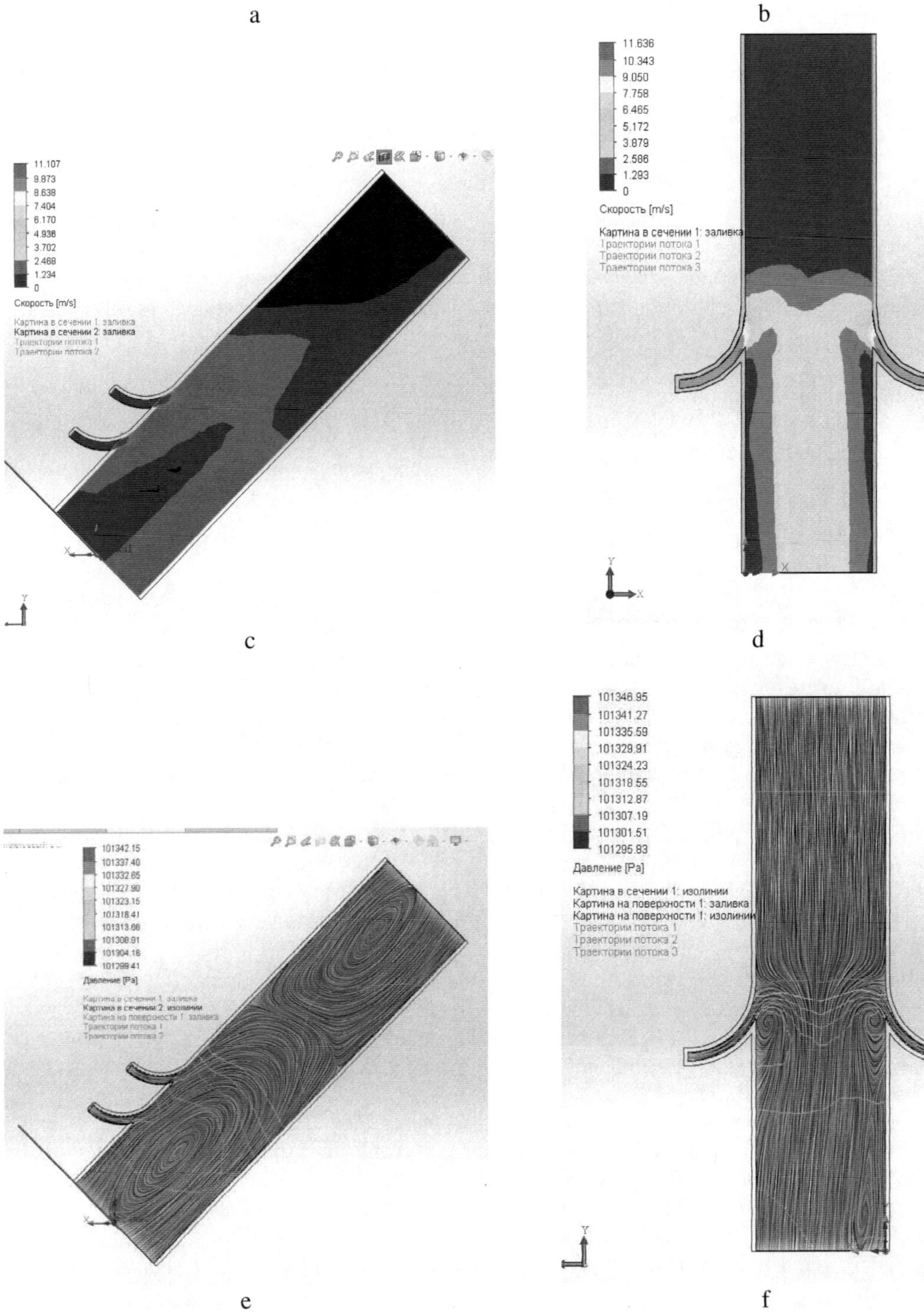

Figure 3. Graphical results of the experiment (air inlet velocity v = 10 v m / s, pressure drop in the lower and upper cross sections of the reloading chute 20 Pa, identical cross sections of the chute and Coanda channels, chute height 1.2 m: a, b) air flow pattern in the developed and similar aspiration cover construction, respectively; c, d) distribution of the velocity field in the cross section of the chute in the developed and similar design of the aspiration hood, respectively; e, f) air flow lines in the de-

veloped and similar aspiration hood, respectively.

It should be noted that the effect described by us was obtained at an air speed in the Coanda channel of 10 m / s, a distance between Coanda channels *c* 150 mm and a channel height of 20 mm, figure 2, and, therefore, for the construction to work under different conditions (air velocity, number of channels, geometric parameters) more research is required.

5. Conclusion

The article presents a new design of the dust-exhaust system, which not only meets the structural requirements to prevent blockage of the channel, but also more than twice reduces the volume of ejection air in relation to the analogue, which in turn will reduce the energy consumption of both dust-exhaust systems and general ventilation systems by 30-70% depending on the parameters of the transfer.

Acknowledgments.
The article was prepared within a development program of the Base University on the basis of Belgorod State Technological University named after V.G. Shukhov.

References

[1] Howard L Hartman, Jan M Mutmansky, Raja V Ramani and Y J Wang Mine 1997 *Ventilation and Air Conditioning* (Wiley-Interscience)
[2] Logachev I N and Logachev K I 2014 *Industrial air quality and ventilation: controlling dust emissions* (Boca Raton: CRC Press)
[3] Huque S T, Donecker P, Rozentals J J and Benjamin C 2010 *The Transfer Chute Design Manual: For Conveyor Belt Systems* (Conveyor Transfer Design Pty. Limited)
[4] Mallick S and Wypych P W 2010 *Evaluation of scaleup procedures using "system" approach for pneumatic conveying of powders*. (Particulate Sci. Technol) vol 28 **1** pp 41–50
[5] Hastie D and Wypych P 2009 *Researching conveyor belt trajectories* (Engineers Australia*)* vol 81 **8** p 50
[6] Li Xiaochuan, Wang Qili, Liu Qi and Hu Yafei 2016 *Developments in studies of air entrained by falling bulk materials* (Powder Technology) vol 291 pp 159-69
[7] Chenchen Fang, Jianliang Xu, Hui Zhao, Weifeng Li and Haifeng Liu 2015 *Experimental investigation on particle entrainment behaviors near a nozzle in gas–particle coaxial jets* (Powder Technology) vol 286 pp 55-63
[8] Koichiro Ogata, Katsuya Funatsu and Yuji Tomita 2001 *Experimental investigation of a free falling powder jet and the air entrainment* (Powder Technology) vol 115 pp 90-5
[9] Ovsyannikov Y G, Gol'Tsov A B, Seminenko A S, Logachev K I and Uvarov V A 2017 Reducing the power consumption of ventilation systems through forced recirculation *Refractories and Industrial Ceramics* **57(5)** 557-61
[10] Sheremet E O, Seminenko A S, Gol'Tsov A B and Podporinov B F 2018 Efficiency of dust separating devices in ventilation emission cleaning systems *IOP Conference Series: Materials Science and Engineering* **327(4)** 042081
[11] Howard L Hartman, Jan M Mutmansky, Raja V Ramani and Y J Wang Mine 1997 *Ventilation and Air Conditioning* (Wiley-Interscience)
[12] Logachev I N and K I Logachev 2014. *Industrial air quality and ventilation: controlling dust emissions* (Boca Raton: CRC Press)
[13] Ovsyannikov Yu G 2000 Recirkulyacionny`e sistemy` aspiracii oborudovaniya mexanicheskoj pererabotki sy`puchix materialov: dissertaciya kandidata texnicheskix nauk: 05.02.13. [Recycling systems of aspiration of the equipment of mechanical processing of bulk materials: the dissertation of a candidate of technical sciences: 05.02.13.] (Belgorod) [In Russian]
[14] Neykov O D and Logachev I N 1981 *Aspiraciya i obespy`livanie vozduxa pri proizvodstve*

poroshkov [Aspiration and dedusting of air in the production of powders] (Moscow: Metallurgiya) [In Russian]

[15] Logchev I N Aspiraciya peregruzok sy`puchix materialov aglomeracionny`x fabrik [Aspiration of overloads of bulk materials of agglomeration factories] *Sb. nauchny`x trudov "Mestnaya vy`tyazhnaya ventilyaciya"* (Moscow: MDNTP) pp 93-100 [In Russian]

[16] Minko V A 1981 *Obespy`livanie texnologicheskix processov proizvodstva stroitel`ny`x materialov* [Dust removal of technological processes of production of building materials] (Voronezh: izd-vo VGU) [In Russian]

[17] Minko V A, Kuleshov M I, Plotnikova L V and others 1987 *Obespy`livanie v litejny`x cexax mashinostroitel`ny`x predpriyatij* [Dust removal in foundry shops of machine-building enterprises] (Moscow: Mashinostroenie) [In Russian]

[18] Kireev V M 2013 *Sovershenstvovanie konstrukcij aspiracionny`x ukry`tij s cel`yu snizheniya zapy`lyonnosti pri peregruzke formovochny`x mass v litejny`x cexax: dissertaciya kandidata texnicheskix nauk.* [Perfection of designs of aspiration shelters for the purpose of decrease in dust content at an overload of forming masses in foundry shops: the dissertation of a candidate of technical science] (Belgorod: BSTU named after V.G. Shukhov) [In Russian]

[19] Kireev V M, Goltsov A B and Minko V A 2010 Raschet i razrabotka aspiracionny`x ukry`tij mest peregruzok [Calculation and development of aspiration shelters for overload sites] *Vestnik BGTU im. V.G. Shuxova* **3** pp 114-17 [In Russian]

[20] Kireev V M and Minko V A 2013 *Razrabotka aspiracionny`x ukry`tij i inzhenernoj metodiki ix raschyota* [Development of aspiration shelters and engineering methods of their calculation] (Bezopasnost` truda v promy`shlennosti) **2** pp 42-46 [In Russian]

[21] Kireev V M 2010 Patent RU no. RU11 9743813 U1 Aspiracionnoe ukry`tie mest peregruzki sy`puchego materiala [Aspiration shelter beds overload of bulk material] [In Russian]

[22] Zhaberov S V 1987 O raschete gidravlicheskogo soprotivleniya cepnoj zavesy` v ukry`tii [On the calculation of the hydraulic resistance of the chain veil in the shelter] *Tez. dokl. Vsesoyuznoj konf. Uskorenie nauchno-texnicheskogo progressa v promy`shlenno-sti stroitel`ny`x materialov i stroitel`noj industrii* (Belgorod: BTISM) **3** p 103 [In Russian]

[23] Minko V A Shaptala V G and Selivanov G G 1989 Avtorskoye svidetel'stvo №1488521 Aspiracionnoe ukry`tie mesta peregruzki sy`puchix materialov [Copyright certificate №1488521 Aspiration Shelter of the Place of Reloading Bulk Materials] (USSR) [In Russian]

[24] Grashchenkov N F, Kharkov V S, Tsai B and others Avtorskoye svidetel'stvo №947O15 1985 No.13 *Ukry`tie mesta zagruzki lentochnogo konvejera* [Copyright certificate №947O15 1985 No.13 Shelter loading places belt conveyor] (USSR) [In Russian]

[25] Abramkin N G, Naumov V P and Lapin O F 1990 Zagruzochny`j zhelob peremennogo secheniya [Variable cross section loading chute] *Sb. nauchny`x trudov "Bor`ba s py`l`yu na predpriyatiyax po pererabotke sy`puchix materialov"* (Belgorod: BTISM) pp 36-41 [In Russian]

[26] Ovsyannikov Yu G , Goltsov A B, Semenenko A S, Kireev V M, Burianov I and A Zhilina 2018 Patent 177236 No.: 2017112163 Ukry`tie s recirkulyaciej aspiriruemogo vozduxa [Shelter with recirculation of aspirated air] [In Russian]

[27] Semenenko A S 2018 №3 E`ksperimental`ny`e ustanovki dlya opredeleniya uslovij realizacii koandovskix techeni. [Experimental facilities for determining the conditions for the implementation of the Coanda current] *Belgorod: Vestnik BGTU im. V.G. Shuxova* pp 52-9 [In Russian]

[28] Kireev V M, Minko V A and Logachev I N 2012 Ispol`zovanie programmny`x kompleksov pri issledovanii raboty` aspiracionny`x ukry`tij [Using software systems in the study of the work of aspiration shelters] *E`nergosberezhenie i e`kologiya v zhilishhno-kommunal`nom xozyajstve i stroitel`stve gorodov* (Belgorod: BSTU) pp 45-9 [In Russian]

The increasing of the energy efficiency of nuclear power plants with fast neutron reactors by utilizing waste heat using heat pumps

V S Kostarev, O L Tashlykov and V A Klimova

Department of Nuclear Power Plants and Renewable Energy Sources, Ural Federal University named after the first President of Russia B. N. Yeltsin, Mira St., 19, Yekaterinburg 620002, Russia

E-mail: slavakostarev@yandex.ru

Abstract. Today, an important task for the economy is to increase the efficiency of using primary fuel. At present, for nuclear power plants with VVER reactors, thermal efficiency is about 33%, with fast neutron reactors - more than 40%. This means that most of the heat produced is released into the atmosphere. In this paper, we consider methods to solve the problem of increasing energy efficiency by the example of NPPs with fast neutron reactors due to the utilization of low-grade waste heat. The results of practical implementation of utilization of low-potential thermal energy of thermal and nuclear power plants using heat pumps were generalized. The analysis of the utilization efficiency of waste heat from various NPP systems was performed.

1. Introduction

The creation of a competitive electric power market in Russia inevitably leads to the fact that energy companies are forced to look for new ways to improve the efficiency and cost-effectiveness of their activities. Therefore, today, the issue of developing new energy-saving technologies for the purpose of more efficient use of energy resources remains a pressing problem in the power industry of Russia.

Today, nuclear and thermal power plants constitute a large part of the energy system of Russia (about 67%). Therefore, in the operation of thermal power plants and nuclear power plants, it is necessary to take into account the factor of thermal pollution of the environment, which leads to significant environmental consequences, both on a local and a global scale. Reducing thermal discharges is an urgent task facing thermal and nuclear energy.

One of the possible ways to reduce thermal discharges from nuclear power plants is the use of heat pumps in the design of a steam turbine plant, which helps reduce environmental burden by cutting down thermal emissions from disposing of waste heat.

A heat pump is a thermodynamic installation in which heat from a low-potential source is transferred to the consumer at a higher temperature. This is done by mechanical work. The principle of operation of a heat pump is that the refrigerant evaporates in the chamber with low pressure and temperature and condenses in the chamber with high pressure and temperature, thereby transferring heat from a cold body to a heated one, that is, in a direction in which spontaneous heat exchange is impossible.

Heat of technological processes and industrial discharges can be used as a source of low-potential thermal energy. Low-grade heat can be transformed into sufficient for beneficial use heat by increasing the temperature of the medium with heat pumps. Such heat can be directed to internal and external consumers.

Several ways of heat pump inclusion in the scheme of steam turbine are described in the literature.

In [1] it is proposed to direct a part of the steam that has been exhausted in the turbine to the condenser, and to supply its' another part to the heat pump evaporator. Steam condensate after the turbine condenser enters the heat pump, where it is heated. Then it goes to the regenerative heating system. The network water returning from consumers is also heated with a heat pump. The basic flow diagram of the heat pump connection is shown in figure 1.

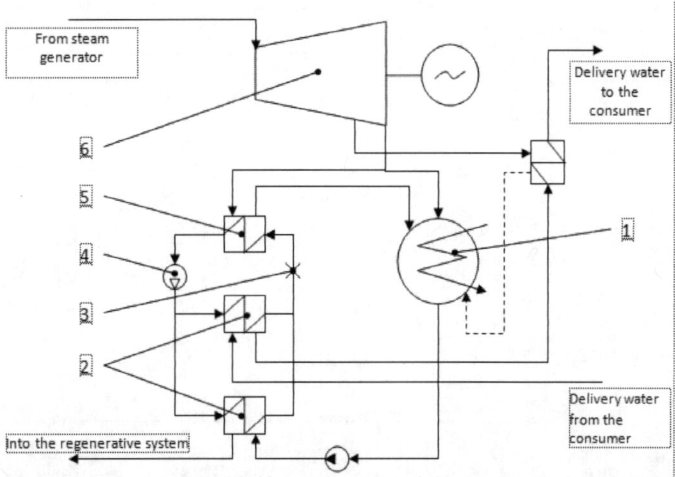

Figure 1. The basic flow diagram of connecting of the heat pump: 1 - turbine condenser, 2 - heat pump condensers, 3 - choke, 4 - compressor, 5 - heat pump evaporator, 6 – turbine

In [2], it is proposed to connect the heat pump evaporator directly to the set of turbine condenser tubes. Here, the operation of the heat pump is used to heat the network water. The basic flow diagram of the heat pump connection is shown in figure 2.

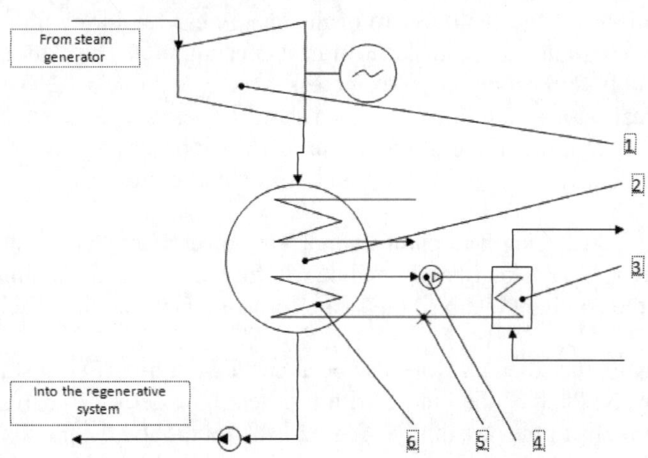

Figure 2. The basic flow diagram of connecting of the heat pump: 1 - turbine, 2 - turbine condenser, 3 - heat pump condenser, 4 - compressor, 5 - throttle, 6 - bundles of heat pump evaporator tubes

In the scheme proposed by the development team of South-Russian State Technical University, it is proposed to use the cooling water of a steam turbine condenser as a source of low-grade heat. Heat can be taken from both water entering the condenser inlet and water leaving the condenser. The heat pump is connected to the regenerative heating system and replaces the first low pressure heater. The basic flow diagram of the heat pump connection is shown in figure 3.

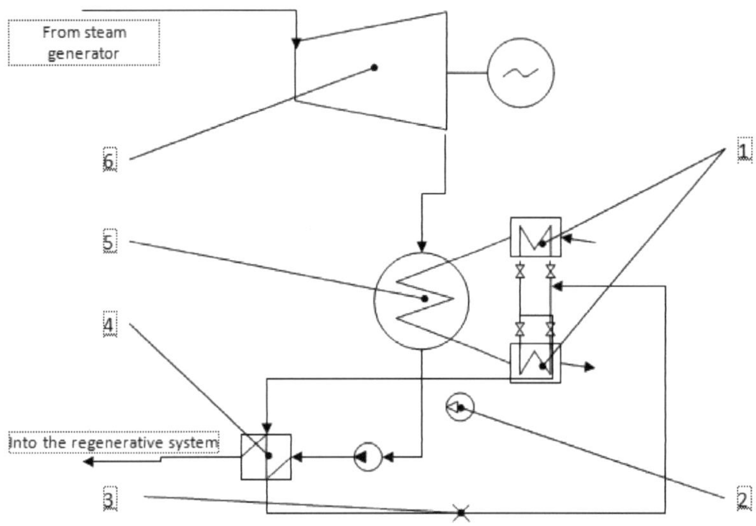

Figure 3. The basic flow diagram of connecting of the heat pump: 1 - heat pump evaporator, 2 - heat pump compressor, 3 - choke device, 4 - low pressure heater, 5 - turbine condenser, 6 - turbine

2. Materials and Methods

One of the currently considered modes of heat supply from nuclear power plants of settlements located at a considerable distance can be cited as an example of the use of heat pumps at nuclear power plants. The project of St. Petersburg heat supply from Leningrad NPP using adsorption heat pumps can significantly reduce the temperature and the pressure of the energy carrier during transportation, and bring them to the required by consumer near the places of heat consumption. The authors of this project state that using heat pumps only for heating purposes increases the revenue of a 1.2 GW NPP unit from 6 to 12 billion rubles per year with 16.2 million megawatt-hour of heat supplied per year depending on its cost in the region [3]. Also in the future of this project, it is supposed to grow microalgae in bioreactors with large-scale production of various bioproducts, including biofuels, by utilizing low-potential heat [4].

Consider the possibility of using heat pumps at nuclear power plants with fast neutron reactors. To date, thermal efficiency is 39.4% for BN-800 and 42.6% for BN-600, the remaining heat (about 60%) is irretrievably lost in the cooling system of the steam turbine condenser, i.e. discharged into the environment.

For a number of years, the research work has been carried out at UrFU to study ways to improve the energy efficiency of NPPs, including those with fast neutron reactors. In particular, in [5], [6], [7], [8], the results of an analysis of the possibility of using heat pumps and the possibility of utilizing low-grade heat for a BN-600 power unit with are presented. According to the results of these works, it was concluded that the use of heat pumps in the main condenser cooling circuit with the aim of improving the economic performance of the plant today is not economically justified, but its use reduces the thermal discharges of nuclear power plants. For a power unit with a BN-600 reactor, the reduction of thermal pollution was from 5 to 6%, depending on the season and the operation mode of the heat pump (figure 4) [5].

A variant was also proposed for heat pumps installing in a system of make-up water preparing for heat grids in Zarechny [5]. It was calculated that in case of installation of 3 heat pumps of 300 kW each (1 per loop), the net profit would be 2 845 888 rubles per year, with a 4.7 years payback period.

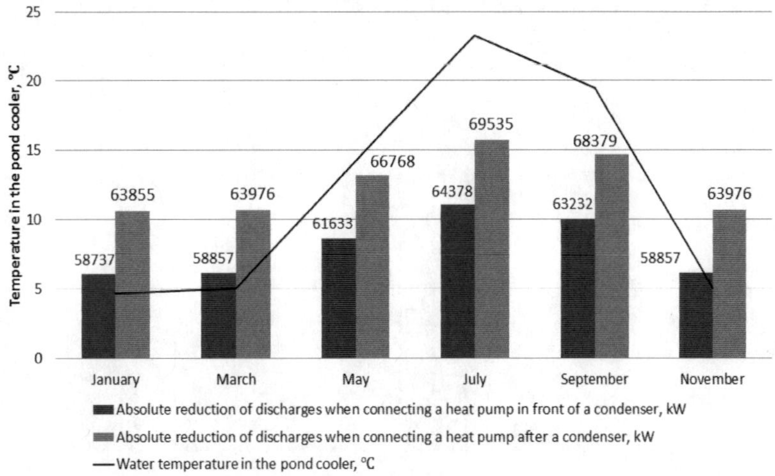

Figure 4. Absolute reduction of thermal discharges of a power unit with a BN-600 reactor.

The purpose of this work is to investigate the possibility of energy efficiency increasing of NPPs with fast neutron reactors due to the utilization of low-grade heat discharged from auxiliary systems.

The fourth power unit of Beloyarsk NPP with BN-800 reactor was considered as a possible source of low-grade heat.

Additional heat generated by heat pumps can be used on:

- heating of industrial and administrative-household NPP buildings;
- heating of water used for the heating and hot water supply systems of Zarechny.

The heat capacity consumed by industrial and administrative buildings of a nuclear power plant can reach several mega-watts. Heat consumers of a power unit with fast neutron reactor are shown in figure 5

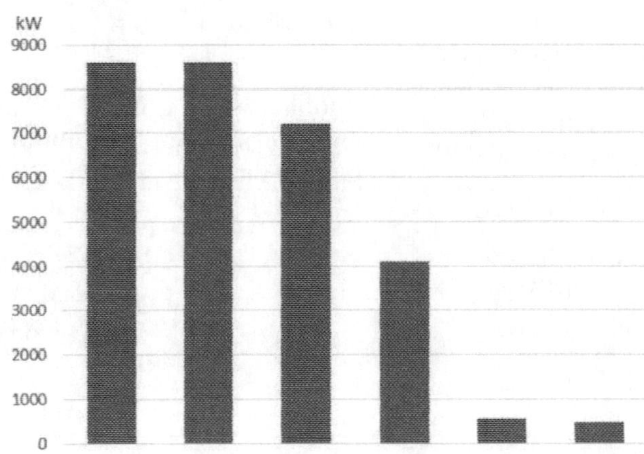

Figure 5. Heat consumers at Beloyarsk NPP: 1 - forced ventilation, air heating systems, etc. (computer room), 2 - inlet ventilation (room A), 3 - inlet ventilation (room B), 4 - inlet ventilation (room B), 5 - air curtains (railway corridor), 6 - other consumers. Designation of premises is conditional.

Currently, heat for these consumers is taken from the turbine operating steam, which is economically unprofitable, while the utilization of waste heat using heat pumps can improve the plant's energy efficiency and reduce thermal pollution of the environment.

The calculation results are shown in figure 6. As a result of research, it was found that the most efficient possible source of heat is the cooling system of the stator winding of the generator (thermal power 14 905.1 kW with a temperature difference at the inlet and outlet 21.4 °C).

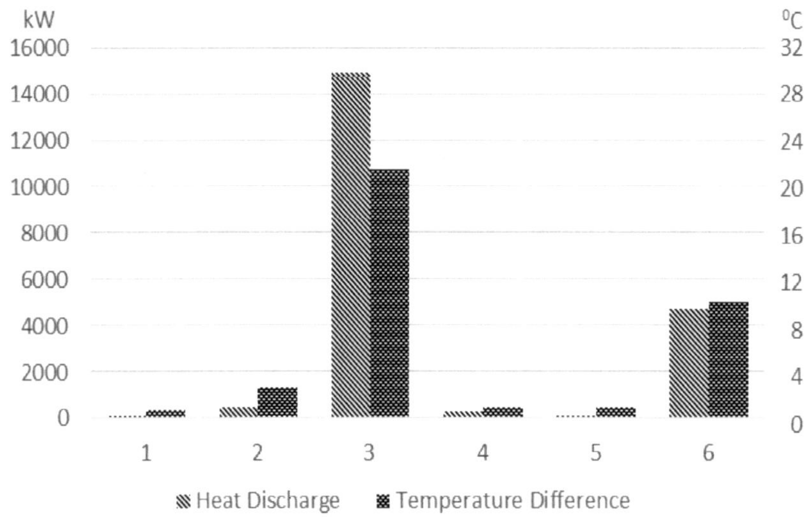

Figure 6. Heat discharges and temperature drops of the auxiliary systems of the BN-800 power unit: 1 - evaporator condenser of drainage expander, 2 - exciter air coolers, 3 - generator stator coil coolers, 4 - cooling of the main feed pumps PEN, 5 - cooling of the main upstream pumps PEN, 6 - cooler dirty condensate.

3. Results

Sources of low-grade heat at the BN-800 power unit are auxiliary systems cooled by an intermediate circuit. Data on possible sources of low potential heat are shown in table 1. Measurements of the parameters were carried out while the power unit was operating at rated power, using instruments at the location of the equipment, IVS, TX-5.11 thermal probe. Also data of design materials and [9] were used.

Table 1. Expenses of cooling water and temperature characteristics of some consumers of the industrial circuit system

Consumer name	Input / output temperature, °C	Number of working / standby units	Cooling water consumption, m^3/h
Drain expansion trap cooler	16,5/17,2	1	140
Causative air coolers	16,5/19,1	4	150 (4x37,5)
Generator stator coil coolers	16,5/37,9	1/1	600
Cooling of the main feed pumps PEN	17,1/18,1	3/1	262,4 (4x65,6)
Cooling upstream PEN feed pumps	17,4/18,3	3/1	105,6 (4x26,4)
Dirty condensate cooler	16,5/26,5	1	400

Calculations to determine thermal discharges were made according to the formula:

$$Q = G \cdot c_p \cdot (t_{inp} - t_{outp}), \qquad (1)$$

where G is the flow rate of process water, c_p is the specific heat capacity of water, t_{inp} and t_{outp} are the inlet and outlet temperatures of the source, respectively.

4. Discussion

To increase the potential of industrial water, which removes heat from the coolers of the generator stator windings, it is proposed to use a heat pump that raises the temperature to 50-55 °C, which is sufficient for heating or hot water.

As the German experience shows, it is most efficient to use the rear heating with the help of heat pumps for underfloor heating. This experience can be used in the design of buildings under construction and design of NPP power units, including BN-1200.

Obviously, from the point of view of the production efficiency of electric and thermal energy, the use of heat pumps in auxiliary systems will be justified. According to preliminary calculations, when installing a single 1.5 MW heat pump with a BN-800 reactor into a thermal circuit of a power unit, the payback period will be about 4.8 years.

5. Conclusion

Thus, the use of heat pumps for low-grade heat utilizing at the BN-800 power unit has the following positive points:

- reduction of steam extraction from the thermal circuit of the unit, which leads to additional energy production at constant production costs;
- reduction of thermal emissions into the environment.

At present, work is underway on the design substantiation of the economic efficiency of utilization of the waste heat of the auxiliary systems of the NPP power unit with the BN-800 fast neutron reactor and the designed unit with the BN-1200 reactor.

References

[1] Efimov N N, Lapin I A, Malyshev P A, Popov R V, Radaev P N, Cherni A V, Lemeshev A A, Karatayev G B, Skubienko S V and Oschepkov A S 2007 Teplovaya elektrostantsiya: Patent RF № 81259 23052007 [Thermal Power Plant: RF Patent Number 81259 23052007]

[2] Stenin V A 2005 Sposob ekspluatatsii teploelektrostantsii: Patent RF № 2247840 10032005 [The Method of Operation of a Thermal Power Plant: RF Patent Number 2247840 10032005] [In Russian]

[3] Saviczkij A I, Pustovalov S B, Egorov S V, Leguenko S K and Vasil`ev A P 2013 Ispol`zovanie sbrosnogo tepla ae`s dlya regional`nogo teplosnabzheniya i proizvodstva bioprodukcii [The use of waste heat from a nuclear power plant for regional heating and bioproduction] URL http://www.osatom.ru/mediafiles/u/files/zasedania_2013/Doklad_Vasilev_2013.pdf [In Russian]

[4] Egorov 2012 Sbrosnoe teplo AE`S – vy`sokoe`ffektivny`j e`kologicheski bezuprechny`j e`nergoresurs dlya regional`nogo teplosnabzheniya [The Waste Heat of Nuclear Power Plants is a Highly Efficient, Environmentally Friendly Energy Resource for Regional Heating] URL http://www.osatom.ru/mediafiles/u/files/VII_forum_2012/Egorov.pdf [In Russian]

[5] Semenov M Yu, Kovin I V and Tashlykov O L 2010 Analysis of the Possibility of Utilization of Low-Grade Heat Using a Heat Pump in the Steam-Turbine Circuit of a Nuclear Power Plant with a Fast-Neutron Reactor E*nergy and Resource Saving. Power Supply. Non-Traditional and Renewable Energy Sources: Proc. of the All-Russian Scientific Conf. of Students, Graduate Students and Young Scientists* (Yekaterinburg: USTU-UPI) pp 399-402

[6] Semenov M Yu and Tashlykov O L 2010 Evaluation of the Use of Heat Pumps in the NPP Scheme to Reduce Thermal Discharges into the Environment *Energy of the Present and the Future: Proc. of the I Euro-Asian Exhibition and Conf.* (Yekaterinburg: UGTU-UPI) pp 28-30
[7] Tashlykov O L, Tolmachev E M, Semenov M Yu and Sapozhnikov B G 2012 *Alternative Energy and Ecology* **3** pp 16-21
[8] Tashlykov O L Kovin I V and Kokorin V V 2012 *Alternative energy and ecology* **3** pp 22-5
[9] Beltyukov A I, Karpenko A I, Poluyaktov S A, Tashlykov O L, Titov G P, Tuchkov A M and Shcheklein S E 2013 *Nuclear Power Plants with Fast Neutron Reactors with Sodium Coolant: a Tutorial* (Ekaterinburg: UrFU)
[10] Tashlykov O Shcheklein S Sesekin A Chentsov A Nosov Y and Smyshlaeva O 2014 *WIT Transactions on Ecology and the Environment* **190** pp15-20
[11] Tashlykov O L and Scheklein S E 2015 International Scientific Journal Alternational Energy and Ecology **8-9(172-173)** pp 50-8
[12] Scheklein S E, Tashlykov O L and Dubinin A M 2015 *Proc. of Higher Educational Institutions. Nuclear energy* **4** pp 15-25.
[13] Tashireva I A, Lazarenko O V and Tashlykov O L 2015 Energy Efficiency of Nuclear Power Plants with Different Types of Reactors *Energy and Resource Saving. Power supply. Non-Traditional and Renewable Energy Sources: Proc. of the All-Russian Scientific and Practical Conf. of Students, Postgraduates and Young Scientists* (Yekaterinburg: UrFU) pp 229-31
[14] Bessonova K N, Chernykh D O and Tashlykov O L 2015 On the Possibility of Utilizing the Low-Potential Waste Heat of Nuclear Power Plants and Thermal Power Plants *Energy and Resource Saving. Power supply. Non-Traditional and Renewable Energy Sources: Proc. of the All-Russian Scientific and Practical Conf. of Students, Postgraduates and Young Scientists* (Yekaterinburg: UrFU) pp 349-51
[15] Tashlykov O et al 2016 *Int. J. of Energy Prod. & Mgmt.* **2** pp133-40
[16] Ivankov G E and Tashlykov O L 2016 Analysis of the Impact of the Technical Water Supply Scheme on the Energy Efficiency of NPPs with Fast and Thermal Neutron Reactors *Proc. of the First Scientific and Technical Conf. of Young Scientists of the Ural Energy Institute* (Yekaterinburg: UrFU) pp 270-3
[17] Tashlykov O L 2016 *Fundamentals of Nuclear Energy* (Yekaterinburg: UrFU)
[18] Tsiglevkina K N, Chernykh D O and Tashlykov O L 2016 The Possibility of Using the Waste Heat of Nuclear Power Plants in Greenhouses *Prospective Energy Technologies. Ecology, Economics, Safety and Personnel Training. Proc. of the Scientific and Practical conference* (Yekaterinburg: UrFU) pp 176-7
[19] Balakin D Yu and Tashlykov O L Reduction of Thermal Discharges at IRM 2016 Energy and Resource Saving. Power Supply. Non-traditional and Renewable Energy Sources: Proc. of the All-Russian Scientific and Practical Conf. of Students, Postgraduates and Young Scientists (Yekatherineburg: UrFU) pp 74-7
[20] Novikov G A, Tashlykov O L and Scheklein S E 2017 *Ensuring Safety in the Field of Atomic Energy Use* (Yekaterinburg: UfFU)
[21] Popov S A and Tashlykov O L 2017 About the Possibility of Utilization of Low-Potential Heat of NPPs with Fast Neutron Reactors *Proc. of the Sec. Scientific Conf. of Young Scientists of the Ural Energy Institute* (Yekaterinburg: UrFU) pp 361-4.
[22] Popov S A and Tashlykov O L 2017 Combined Production of Heat and Electricity Using a Nuclear Power Plant Energy and Resource Saving. Power Supply. Unconventional and Renewable Energy Sources: Proc. of the Int. Scientific. Conf. Danilov Readings (Yekaterinburg: UrFU) pp 327-31
[23] Tashlykov O L, Klimova V A, Shcheklein S E and Popov S A 2018 Ecological and Economic Advantages of Remote Heat Supply from Reactor Plant of Advanced Safety BN-1200 *Sino-*

Russian ASRTU Conf. Alternative Energy: Materials, Technologies, and Devices (Dubai: Knowledge E) pp 243–9

[24] Tashlykov O L and Scheklein S E 2017 The experience of Nuclear Heating in the City of Zarechny and the Prospects for the Nuclear Heating of Yekaterinburg *Russian Regions in the Focus of Change: A Collection of Procs. of the XII International Conf.* (Yekaterinburg: Publishing House of the UI Medical Center) vol 2 pp 525-35

[25] Puzakov V S 2011 *Heat News* **02(126)** pp 2-5

[26] Yakovlev B V 2008 *Improving the Efficiency of Heat Supply Systems* (Moscow: News of Heat Supply)

[27] Shlapakov V I 2011 *News of Heat Supply* **2(126)** pp 1-3

The thermal and physical aspects of designing facade heat insulation at capital repairs and reconstruction of civic buildings

M M Kosukhin and A M Kosukhin

Department of Construction and Municipal Economy, Belgorod State Technological University of V.G. Shukhov, Kostyukov St., 46, Belgorod, 308012, Russia

E-mail: mkosuhin@mail.ru

Abstract. The brief analysis of civic buildings condition has been presented in this paper. It has been demonstrated that at present the housing stock doesn't meet the up-to-date heat engineering, energy saving and energy performance requirements. To create the necessary indoor climate of living accommodations and to provide the comfortable living environment, most of civic buildings, built at different times, are in need of large-scale reconstruction. The leading role in solving these problems is assigned to capital repairs. For its satisfactory implementation an integrated approach to repairs and construction works with the use of up-to-date energy-efficient technologies and materials is needed. The efficient heat insulation of en-closing structures should be preceded with professional engineering survey of structures and with providing the scientifically-grounded recommendations, based on present-day scientific achievements in this sphere, for the subsequent making of design specification. It has been pointed out that the economic and processing efficiency of the selection and installation of heat insulation can be achieved only by means of scientific approach and the knowledge of physical and chemical processes and phenomena, taking place in enclosing structures under the influence of internal and external operational environment.

Key words: housing and utility sector, housing stock, civic buildings, enclosing structures, capital re-pairs, energy efficiency, energy saving, heat insulation, heat transfer, heat conductivity, convection, radiation, heat flux, heat-transfer resistance of an enclosing structure.

1. Introduction

Nowadays, in order to provide the comfortable living environment and comply with the requirements of improving the energy efficiency, energy saving, operational reliability and architectural expressiveness of civic buildings, the special norms are specified for these buildings. Meeting these requirements is the issue of national importance in terms of construction sector and housing and utility sector of Russia [1].

At present the capital housing facilities stock make up over 26% of the total amount of the Russian economy's funds, most of it presented by civic buildings. About 20 mln of housing stock units, with total area about 3 bln m^2, consume over 20% of energy resources of the country [2]. The yearly turnover in the sphere of housing and utilities sector is over 4,1 tln rubles and makes up more than 5,7% gross domestic product of Russia. Nevertheless, it is the most problematic branch of economy. The depreciation of its fixed assets is over 60%, and according to some indices even higher [3].

Due to the fact that the state of housing and utility complex directly determines the level of the country's welfare, the sector faces a very acute problem of finding a way out of the critical situation. Over the past few decades, during which the sector has been constantly reformed, numerous legislative and regulatory legal acts in this sphere have been developed, including the «Strategy of Housing and Utilities Sector Development in the Russian Federation for the Period up to 2020», adopted on January, 26, 2016 [4]. The development of this strategy intended improving the comfort of the living environment, modernization and increasing the energy efficiency of buildings and constructions, switching to the principles of using advanced and efficient materials and technologies in construction and modernization of communal infrastructure facilities and housing stock, availability of apartments in multi-apartment residential buildings for incapacitated persons and other people with limited mobility [5].

But due to the fact that the approaches specified in the strategy, are mostly of subjective nature, the efficiency of its implementation is far from what was expected. For the satisfactory performance of tasks, faced by this sector of industry, it is necessary to carry out a number of measures, taking into account its objective state. The examples of the domestic experience of the past clearly demonstrate the possibilities of using the little-known and unconventional mechanisms and technologies. The work [6] elaborates on the topics, which are rather urgent for implementing reforms in urban housing and utilities sector.

2. The main part

The civic buildings, which are under construction at the moment, mostly comply with the requirements of energy performance. As for the other buildings, built at different times, most of them are in need of large-scale modernization [7]. A special role in this process is played by capital repairs. But the program of capital repairs suggested by the Ministry of Housing and Building, turned out to be a politically charged material for various uneconomical ventures [8]. The program of capital repairs, as the author notes, should be considered not only in terms of social initiatives, but also of the design terminology. It means that not only the sources and order of financing should be discussed, but, first of all, such basic elements as the state of housing stock, the further use of land resources and faulted facilities. An important role in this is played by preventive measures – the engineering survey of the housing stock, certification of buildings and forming on their basis the plans of capital repairs conduction. The key element in solving this problem is the availability of highly-qualified specialists. Taking these factors into account would allow achieving the desired efficiency in carrying-out capital repairs and, first of all, by means of improving the level of repair and construction industry itself, by basing it on the results of professional engineering survey of buildings before reconstruction [9]. To solve this problem an integrated approach to building and repair works technology is needed, with the use of up-to-date energy-saving technologies and differing from the traditional, conventional ways of organizing construction and repair operations [10].

The experience of carrying-out capital repairs has shown that dealing with fragmentary, random or local problems didn't help solving the global problem of reducing the excessive energy consumption of constructions. This resulted only in the extra costs for these works; and the ageing of housing stock and its insufficient repairs only increased the amounts of hazardous dwellings and dilapidated housing and reduced the comfort level [11].

At carrying out capital repairs, apart from providing the required parameters of the indoor climate and service reliability, a special role is assigned to improving the energy performance and energy saving of the repaired buildings. The main tool of implementing this task is the outer enclosing structures [12]. Since the enactment of the Federal Law № 261-ФЗ «About energy saving and energy efficiency improvement … » [13], virtually all regions of Russia have started activities in carrying out capital repairs of outer enclosing structures of all types of buildings as part of implementation of the requirements under this law. But the main efficiency indicators of construction and repair works were the number of repaired houses and the amount of cash disbursement. The issues of quality and durability of facades heat insulation were relegated to the background, and the scientific justification for performing these works was not done at all. The performance of repair works must be preceded with the

IOP Conf. Series: Materials Science and Engineering 552 (2019) 012023 doi:10.1088/1757-899X/552/1/012023

professional and detailed engineering survey of structures and with providing the scientifically-grounded recommendations, based on up-to-date scientific achievements in the sphere of enclosing structures heat insulation, for the subsequent making of design specification. But the financing of such survey works have always been a great problem [14].

Nevertheless, the conduction of works in making the efficient facade heat insulation requires serious scientific approach with the purpose of providing the economic and processing efficiency of its selection and installation. It should be borne in mind that the direct source of innovative and groundbreaking technologies, including those in construction sector, is the applied and fundamental scientific research [15].

Enclosing structures, exposed to the action of external and internal operational environment, are a complex thermodynamic system, subject to physical and chemical laws, processes and phenomena. So, its thermal-physical properties can be efficiently managed only on the basis of the scientific approach. This becomes especially important when it comes to using advanced building technologies, materials and heat-insulation systems [16].

The durability of enclosing structures, especially their thermal-physical properties, are greatly influenced by the climatic conditions: atmospheric precipitations, daily and seasonal temperature fluctuations, wind loads, solar radiation, biocorrosion processes, as well as the processes and phenomena within the constructions, caused by these conditions. The influence of these factors results in the reduction of service reliability of both the enclosing structures themselves, and of the building in general. In the course of time, their thermophysical properties considerably deteriorate as well [17].

An exterior enclosing structure is a phase contact area of the interaction between the cold external and the warm internal operational environments. Both on the surfaces of this contact area and within it the new processes and phenomena take place, which considerably influence the quality of heat insulation, the indoor climate and the service reliability and durability of the building [18]. So, the exterior enclosing structures of civic buildings must meet a number of requirements. They should be low-cost in construction and operation, sturdy and durable, should fulfill their functional purpose, providing the energy efficiency of the building with the required indoor climate and architectural expressiveness. This paper considers the heat transfer processes, taking place in the enclosing structures of buildings.

The sense of using thermal insulation consists in minimizing the heat exchange with the environment, which results in reducing the heat energy losses and cutting costs for heating. Thermal insulation has low heat conductivity and reduces heat transfer by means of heat conduction and radiation. Applying heat-insulating coatings on the surface of walls increases the heat transfer resistance of the structure in general; this, in its turn, reduces the heat flux from the structure to the environment (heat losses). At this, the heat-transfer resistance of the structure is directly proportional to the insulating layer thickness. But there is also such concept as critical insulation thickness – the thickness of insulation, over which the increase of insulation thickness doesn't cause any substantial reduction of heat losses and is economically unviable.

Following from the conditions of outer wall's operation, the heat within wall structure is moved from the warm inner surface to the cold external surface under the action of temperature gradient. In classical thermodynamics this phenomenon is called heat transfer and is performed by means of three ways of heat exchange: heat conductivity, convection and emission (radiation) [19].

Heat conductivity is the main thermal characteristic of any material. In the enclosing structures of buildings heat exchange is carried out mostly by heat conductivity. The process of one-dimensional steady-state heat conduction in homogeneous material is described with Fourier equation:

$$qT = -\lambda(dt/dx), \quad (1)$$

where: qT – surface density of heat flux, passing through the plane, perpendicular to the heat flux, W/m^2; λ – heat conductivity coefficient of the material, W/m·°C; t – temperature, which alters along the x axis of heat flux, °C.

Heat conductivity coefficient λ is one of the basic thermal characteristics of any heat-insulating materials. As follows from the equation, the material's heat conductivity is the measure of heat, conducted by the material, which is numerically equal to the heat flux, passing through 1 m^2 area of the material, perpendicular to the heat flux direction, at temperature difference between the inner and outer surface equal to 1°C. The higher is the λ value, the more intensive is the heat conduction process. That's why the special heat-conductivity requirements are specified for heat-insulating materials; λ for these materials should be below 0.3 W/m·°C. The heat conductivity of a material at its constant density and structure can be influenced by a number of physical and chemical factors and phenomena. The most significant of them are humidity and rise of temperature [20].

Another way of heat exchange in enclosing structures is convection. At different temperature values of the inner and outer surface of the wall and the circumambient air, convective heat exchange takes place – the combined convection and heat conductivity process at the surface of the wall [21]. Convection can be natural and forced. Natural convection occurs due to the temperature difference between the wall surface and the circumambient air. Forced convection is created by means of various ventilation systems. Convection plays an important role in creating the comfortable indoor climate [22]. The value of convective heat flux is determined through the Newton equation:

$$q_k = \alpha_k (t_a - \tau), \tag{2}$$

where: q_k – heat flux, W, transferred by convection from the flowing air to the surface or vice versa; t_a– air temperature at the wall surface, °C; τ – temperature of the wall surface, °C; α_k – convective heat-transfer coefficient on the wall surface, W/m^2·°C.

The convective heat-transfer coefficient is a physical value, numerically equal to the amount of heat, transferred from the ambient air to the wall surface by means of convective heat exchange at temperature difference between air and surface equal to 1°C.

The third way of heat transfer is radiation or radiative heat exchange, which plays an important role in heat exchange between the inner surfaces of enclosing structures and the heating equipment. The intensity of heat transfer between surfaces by means of radiative heat exchange can be determined in simplified form by the temperature difference between these surfaces:

$$q_l = \alpha_l (T_1 - T_2), \tag{3}$$

where: T_1 and T_2 – temperature of surfaces, participating in radiative heat exchange, °C; α_l – radiant heat transfer coefficient on the wall surface, W/m^2·°C.

The characteristic of heat transfer process concerning the enclosing structures of buildings and constructions is a heating-performance coefficient, known as heat-transfer resistance of an enclosing structure and characterizing the level of heat-insulating properties of enclosing structures [23]. The higher is the heat-transfer resistance of an outer wall, the better is its heat-insulating ability, i.e. the lesser is the amount of heat transferred through it, and, consequently, the lesser are the heat losses. Heat-transfer resistance is measured in m^2·°C/W and is denoted as R. The physical significance of its numerical value can be explained from two perspectives. In the first case the numerical value of heat-transfer resistance shows the surface area in m^2, through which the heat flux of 1 W passes at temperature gradient at the surface equal to 1 °C. In the second case the numerical expression of R shows which temperature gradient in °C at the inner and outer surface of the enclosing structure would provide its thermal protection at the heat flux 1 W, passing through 1 m^2 of this structure's surface. Speaking about the heat-transfer resistance it's necessary to understand two of its elements: total heat-transfer resistance of the structure R_o and thermal resistance of the structure R_k.

R_o is determined with the sum of the thermal resistance of the structure R_k, equal to the sum of its layers' thermal resistances ($R_1 + R_2 + R_3 + \cdots + R_n$), and heat-transfer resistances of the near-wall air layers at the inner and outer surfaces of the structure R_v and R_n:

$$R_o = R_k + R_v + R_n = (R_1 + R_2 + R_3 + ... + R_n) + R_v + R_n \text{, m}^2\cdot°\text{C/W,} \tag{4}$$

The thermal resistance of homogeneous layers is determined by the formula:

$$R_n = B_n / \lambda_n \text{ ,m}^2\cdot°\text{C/W,} \tag{5}$$

where: B_n – thickness of n-layer in m; λ_n – heat conductivity coefficient of the material of n-layer, W/m·°C.

The heat-transfer resistance of near-wall air layers is determined by the formulas:

$$R_v = 1/\alpha_v \quad \text{and} \quad R_n = 1/\alpha_n \text{ ,} \tag{6}$$

where: α_v and α_n – heat conductivity coefficients of the inner and outer near-wall air layers, respectively, W/m^2·°C.

Due to the fact that the surface of real enclosing structures can have dissimilar temperatures because of their inhomogeneous structure, for practical calculations a concept of the «corrected value of heat-transfer resistance of the enclosing structure» – R_{pr} was introduced, which denotes the heat-transfer resistance of a one-layer enclosing of the same area, which the heat flux, similar to the one in the real structure, of the same temperature difference between the internal and external air, passes through [24]. It is determined as the weighted-average in area R_o value, with account of R_o of all the nonuniform parts of the structure:

$$R_{pr} = (R_{o1} \cdot F_1 + R_{o2} \cdot F_2 + R_{o3} \cdot F_3 + ... + R_{on} \cdot F_n) / (F_1 + F_2 + F_3 + ... + F_n). \tag{7}$$

or

$$R_{pr} = (t_v - t_n) / (Q/A), \tag{8}$$

where: Q – heat flux, passing through the structure (or its part), W; A – area of the structure (or its part), m^2.

Expression Q/A is the averaged density of the heat flux, passing through the averaged are of an enclosing structure, i.e.:

$$q^{pr} = \frac{Q}{A} \quad \text{or} \quad q^{pr} = \frac{(t_v - t_n)}{R_o^{pr}} \tag{9}$$

At making facade heat insulation with the use of efficient insulating materials the role of heat-conducting inclusions of the enclosing structure is minimized, and can be neglected [25]. Such part of the outer wall is characterized with the relative heat-transfer resistance R_o^{usl}. By its value we can judge about the heat transfer-performance uniformity of the enclosing structure through heat transfer performance uniformity factor r. It is the ratio of the corrected heat-transfer resistance of the structure to the relative heat-transfer resistance of its certain part:

$$r = \frac{R_o^{pr}}{R_o^{usl}} \tag{10}$$

This coefficient is always less than 1. In the practice of using facade heat insulation it is necessary to try to use its functional capacities in full. With that, the technological uniformity coefficient should tend to 1. Apart from solving the problem of heat-protective indices, it is of significant economic importance at determining the cost of heat insulation.

An important measure, associated with the corrected heat-transfer resistance is the heat-conductivity coefficient of the outer enclosing structure – the value, reciprocal to:

$$K = \frac{1}{R_o^{pr}}, \tag{11}$$

which is equal to the heat flux density, passing through the enclosing structure, at the temperature difference on its surfaces of 1°C. Hereof it follows, that the heat flux q, W/m^2, passing through the enclosing structure by means of heat transfer, is equal to:

$$q = K \cdot (t_v - t_n). \tag{12}$$

According to the requirements of SNiP 23-02-2003 «Buildings Heat Insulation» [26] the actual value of R_{pr} of the structure should not be less than the heat-transfer resistance, which provides the standard value of temperature gradient ΔT between the indoor air temperature t_v and the temperature of the inner surface of the enclosing structure τ_v. This sanitary requirement is expressed through condition:

$$R_{pr} \geq n \cdot \frac{t_v - t_n}{\Delta T} \cdot \alpha_v, \tag{13}$$

where: n – coefficient, taking into account the arrangement of the enclosing structures' outer surface towards the outdoor air; t_n – design outdoor air temperature in the cold season, °C.

At selecting material for heat-insulating and protective layer it is necessary to keep in mind that the proceeding of the above-mentioned heat-transfer processes is greatly influenced by the humidity conditions of the internal volume of the building and of the enclosing structures material. Ignoring this parameter at the design stage is a widespread mistake, which can reduce the functional and economic efficiency of façade systems to minimum [27]. The increase of humidity deteriorates the heat-saving properties of enclosing structures due to the increase of materials' heat conductivity coefficient, which in its turn results in the increase of heat losses and power consumption for heating. In this regard it is essential to determine in every particular case the ways of moisture penetration into the enclosing structures and to take measures for their waterproofing. This is especially important at the reconstruction of old housing stock with various extent of wear. The errors or over-optimism in the use of waterproof and water-repellent plastering or protective coatings at façade repairs have in many cases resulted in the aggravation of the already adverse conditions of the indoor climate [28].

The humidification of the indoor air and of the enclosing structures' materials is inevitable and has an adverse effect on sanitary and performance characteristics. Firstly, there are no absolutely solid and waterproof natural materials. Secondly, the outer enclosing structures should «breathe», so they should be vapor-permeable. As for such apparent sources of moisture as service, telluric and atmospheric moisture, the enclosing structures can be protected from them at the work execution stage. But in a structure in service the inadvertent natural processes take place, which are accompanied with humidification and can't be avoided. It is service moisture and hydroscopic moisture. The saturation of enclosing structures with service moisture is caused by internal sources, which can't be removed at the normal building operation. It is not only household and industrial moisture and moisture from environment control systems. Moisture is contained in the exhaled air as well. Under the manometric pressure the vaporous moisture moves outward, filling the pores of material. The ability of materials to adsorb moisture from the air results in the accumulation of hydroscopic moisture in its pores. Depending on the type and thickness of the outer wall it condensates on its inner surface or inside the wall, turning into condensed moisture.

Temperature drops cause the phase transition of water to the ice with the considerable increase of volume. This, in its turn, results in the formation of internal stresses in the material of the enclosing structure, which cause mechanical corrosion. At the rise of temperature the reverse phase transition of ice to water takes place. The internal stresses are released, leaving a network of microfractures from the previous freezing in the bulk of the structure. Then the process is repeated, and each new cycle contributes to the deeper penetration of moisture, and consequently, to even stronger deterioration. The larger is the difference between the seasonal and daily temperature fluctuations, the more frequent and intensive these processes are (the frost deterioration mechanism). The permanent action of alternating load on the dampened wall results in its intensive destruction. So, at selecting materials for façades heat insulation it is necessary to take into account their vapor permeabilty – i.e. the ability of materials and the structure, made of them, to let water vapor through it, and the vapor penetration – the process of vapor passing though material or enclosing structure.

Vapor permeabilty μ depends on the physical properties of material and reflects its ability to let the diffusing water vapor through it. It is numerically equal to the diffusion water vapor flux, mg/h, passing through m^2 of the area, perpendicular to the flux, at the partial pressure gradient of water vapor along the flux, equal to 1 Pa/m.

3. Conclusions

So, all the above-mentioned details allow us making a conclusion that taking into account the heat-transfer processes in enclosing structures and the conditions of their proceeding would allow efficiently designing heat insulation with the least possible costs. The real improvement of civic buildings' energy performance at installing façades heat insulation is possible only by using scientific approach in solving these problems. This would allow the economically and technologically viable selection of the type of heat insulation depending on the climatic location of an object, as well as of the materials and the installation technology. The most important thing in solving these problems is to ensure that instead of thermotechnical processes and phenomena influencing the operation of outer enclosing structures, an efficiently designed structure would regulate these processes by itself.

Acknowledgments
The article was prepared within development program of the Flagship Regional University on the basis of Belgorod State Technological University named after V.G. Shukhov.

References

[1] Kosukhin M M 2016 Ot istokov k sovremennomu sostoyaniyu i perspektivam razvitiya ZHKKH v Rossii [From the origin to the current state and development prospects of the housing and utilities sector in Russia] *Vestnik BGTU im. V.G. Shuxova* **12** pp 48-54 [In Russian]
[2] 2018 Udobnaya usad'ba [Comfortable manor] URL http://cozyhomestead.ru/Zhivotnie_34346.html
[3] Medvedev D 2013 Boleye 60% zhilishchno-kommunal'nykh sredstv iznosheny [Over 60 % of housing-and-utility funds are worn-out] URL https://novostienergetiki.ru/medvedev-fond-zhkx-strany-iznoshen-pochti-na-60/ [In Russian]
[4] Strategiya razvitiya ZHKKH do 2020 goda [Housing and utilities development strategy until 2020] URL http://www.minstroyrf.ru/press/opublikovana-strategiya-razvitiya-zhkkh-do-2020-goda/
[5] Ivanov G S 1996 Errors in norming a thermal protection level of enclosing structures *House construction* **9** pp 11-3
[6] Govorenkova T M and Savin D A 1999 Residential lease cooperation. Experience of new economic policy and possibility of its application in the modern-day Russia *Housing Yearbook* (Moscow)
[7] Korol E A 2001 The efficient enclosing structures with high thermal protection level *Industrial and civil construction* **9** pp 24-5
[8] Sozinov P 2016 Kapital'nyy remont ostalsya bez professionalov [Capital repairs are left without

professionals] URL https://ok-inform.ru/stroitelstvo/zhkkh/53919-kapitalnyj-remont-ostalsya-bez-profes-sionalov.html [In Russian]

[9] GOST3 1937-2011 Zdaniya i sooruzheniya. Reglament inzhenernykh izyskaniy i monitoringa [Buildings and constructions. Regulations of engineering survey and monitoring] (Moscow: Standartinform) [In Russian]

[10] Kosukhin M M, Kosukhin A M and Sokolovskaya D M 2017 Energosberegayushchiye tekhnologii i metody v stroitel'stve i ZHKKH [The energy-saving technologies and methods in construction and housing and utilities infrastructure] *Nauka i innovatsii v stroitel'stve: sbornik statey Mezhdunarodnoy nauchno-prakticheskoy konferentsii* (Belgorod: BGTU im. V.G. Shukhova) pp 237-47 [In Russian]

[11] Kosukhin M M, Sharapov O N, Bogacheva M A and Kosukhin A M 2016 The problems of energy saving in conditons of stable functioning, modernization and development of housing stock *Bulletin of BSTU named after V.G.Shukhov* **10** pp 51-61

[12] Gagarin V G 2009 The economic analysis methods of improving the heat insulation level of enclosing structures of buildings *AVOK* **1-3** pp 50-3

[13] Federal'nyy zakon ot 23.11.2009 №261 «Ob energosberezhenii i povyshenii energoeffektivnosti i vnesenii izmeneniy v nekotoryye zakonodatel'nyye akty Rossiyskoy Federatsii» [Federal Law of 23.11.2009 №261 «About energy saving and energy efficiency improvement, and modifications to certain legal acts of the Russian Federation»] [In Russian]

[14] Kosukhin M M , Kosukhin A M and Komarova K S 2018 Increasing of the Energy Efficiency in Civilian Buildings Applying Fixed Type of Facade System *Journal of Physics: Conference Series* 1066(1) 012012

[15] Doklad RAN Prezidentu o sostoyanii fundamental'nykh nauk v Rossiyskoy Federatsii i vazhneyshikh nauchnykh dostizheniyakh rossiyskikh uchenykh [Report of RAS to the President about the state of fundamental sciences in the Russian Federation and the most important scientific achievements of the Russian scientists] URL www.ras.ru/FStorage/Download.aspx?id=32f7083e-46e9-45ff-8a79-da17c196f507 [In Russian]

[16] Kosukhin M M, Kosukhin A M and Komarova K S 2018 Increasing of the energy efficiency in civilian buildings applying fixed type of facade systems *Journal of Physics: Conference Series* **1066(1)**

[17] Zabelskaya M 2003 Façade systems: problems and their solutions *Budmeister* **8** pp 26-9

[18] Kosukhin M M and Kosukhin A M 2017 The role of heat-transfer processes in improving the efficiency of thermal protection of civic buildings façade insulation *Vestnik BGTU im. V.G. Shuxova* **12** pp 21-6

[19] Tolstova Y I and Shumilov R N 2014 *Osnovy` stroitel`noj teplofiziki : uchebnoe posobie* [Basics of building thermal physics: Study guide] (Ekaterinburg: Izdatel'stvo Ural'skogo universiteta) p 104 [In Russian]

[20] Shoykhet B M 2007 Structure and permeability of fibrous heat-insulating materials *Energy saving* **7**

[21] SNiP 23-02-2003 2013 CR 50. 13330.2012 Teploizolyatsiya zdaniy. Peresmotrennoye izdaniye [Buildings Heat Insulation. Revised edition] (Moscow: Gosstroy Rossii) [In Russian]

[22] Vatin N I, Gorshkov A S and Nemova D V 2013 The energy efficiency of enclosing structures at capital repairs *Construction of unique buildings and structures* **3(8)**

[23] Glikin S M 2004 Designing and improvement of efficient enclosing structures *Industrial and civil construction* **6** pp 20-1

[24] Akhmyarov T A, Spiridonov A V and Shubin I L 2014 Making the exterior enclosing structures with the increased heat insulation level *Energy saving* **6**

[25] Drizhuk D and Flenkin M 2014 Problems of choosing a façade system *Construction technologies* **6** pp 34-7

[26] SNiP 23-02-2003 2014 Teploizolyatsiya zdaniy [Buildings Heat Insulation] (Moscow: Gosstroj Rossii FGUP CzPP) [In Russian]
[27] Biryukova T P 2005 The enclosing structures of walls with the use of up-to-date façade systems *Industrial and civil construction* **1** pp 50-1
[28] Kosukhin M M, Semak A V and Kosukhin A M 2016 Otsenka potentsiala energosberezheniya na osnove energoaudita [Evaluating the energy-saving potential on the basis of energy audit] *Vestnik BGTU im. V.G. Shuxova* **12** pp 89-94 [In Russian]

Improving the efficiency of flue gas desulphurization of TPS

Yu O Rikker[1], M V Kobylkin[2] and A G Batukhtin[3]

[1]Scientific Research Department, Transbaikal State University, Aleksandro-Zavodskaya St., 30, Chita, 672039, Russia
[2]Technopark, Transbaikal State University, Aleksandro-Zavodskaya St., 30, Chita, 672039, Russia
[3]Department of Thermal Power Plant, Transbaikal State University, Aleksandro-Zavodskaya St., 30, Chita, 672039, Russia

E-mail: yrikker@mail.ru

Abstract. The effects of sulfur oxides on the environment and human health have negative consequences. A significant amount of sulfur oxides is emitted into the air with flue gases from the TPS. Modern Russian TPS are not equipped with sulfur-trapping units due to their significant cost. The study is devoted to the research of low-cost methods that increase the efficiency of the flue gas desulphurisation process at coal-fired TPS. Existing methods are considered from the perspective of the presence of advantages and disadvantages. The potential of the use of purge water from boilers for flue gas desulfurization of TPS.

Key word. TPP flue gases, desulfurization, seawater, coal-fired thermal power plants, benz(a)pyrene, coal, flue gases, sulfur oxides, ecology, TPP.

1. Introduction

At the present stage of development of the fuel and energy complex, the problem becomes very urgent, consumption of energy resources, but also with the results of the impact of energy facilities (in particular, coal-fired power plants) on the environment. In the conditions of constantly growing energy consumption on the background of an increase in the amount of solid fuel burned, the lack of modern technologies and equipment for flue gas cleaning and recycling of man-made wastes to be introduced to the environment becoming a serious problem.

Energy development in any country can not do without taking into account the environmental situation: both the present and projected for the future. At the same time, requirements for development activities should be developed in accordance with the recommendations of WHO and environmental organizations.

Production of heat and electricity power, combined with a mass fuel combustion, leading to an increase in environmental pollution, energy complex is one of the major sources of air pollution. Prospects for changes and increases in the fuel and energy complex should be evaluated in terms of possible harm to public health and be accompanied by a set of measures to reduce it.

With the increase in energy production and the volume of fuel burned, the problems of human health and pollution of the biosphere can be significantly aggravated. Modern TPS plants, consuming up to 20,000 tons of coal per day, instead emit a quarter of the total coal dust of industrial enterprises, two-thirds of sulfur dioxide, over 90 tons of ash and soot, over 600 tons of sulfur oxides and nitrogen, which leads to persistent pollution of water and air basins of cities. One of the consequences of these emissions is acid rain.

Today, the majority of TPPs in Russia are equipped only with ash catchers, the remaining harmful emissions, including sulfur and nitrogen oxides, are emitted into the atmosphere without being captured, and therefore the share of emissions into the atmosphere of energy enterprises is about one fourth of the emissions of other industrial enterprises in the Russian Federation.

Abroad, beginning in the second half of the 20th century, industrial sulfur traps for flue gas cleaning are successfully introduced and operated. In the USA, Germany, Japan, Austria, the commissioning of TPS without flue gas desulfurization module is impossible. Foreign suppliers offer sets of flue gas cleaning systems, successfully implemented at foreign energy enterprises. In addition to the high cost of the installation itself, they are quite expensive to operate, while fines for emissions into the atmosphere in Russia are much lower. Implementation of projects for the installation of desulfurization facilities at existing TPPs requires the reconstruction of power units, and the construction of new stations increases the project cost by 10-20%, and therefore, Russian sewerage plants do not find use of sulfur-trapping plants, with the exception of pilot plants. [1]

2. Materials and Methods

One of the main pollutants emitted with flue gases TPPs are sulfur oxides. The emission of flue gases at thermal power plants reaches, on average, several million cubic meters per year. Sulfur dioxide adversely affects human health, inhibits the animal and plant world, increases the corrosion of metal structures, and under the action of ozone in the atmosphere oxidizes to trioxide and combines with water vapor and forms sulfuric acid vapors that precipitate with the sediments, acidifying the soil and water resources. [2]

When burning solid fuel, which is the predominant energy from the pyrite and organic sulfur dioxide formed and sulfur trioxide:

$S + O_2 = SO_2$;

$2FeS_2 + 5O_2 = 2FeO + 4SO_2$;

$SO_2 + \frac{1}{2} O_2 = SO_3$. [1]

According to G.F. Bystritsky, «... in natural types of fossil solid fuels, sulfur is found in three varieties:

1. organic SO, associated with other elements of fuel C, H, N and O in the form of complex organic compounds;
2. pyrite SC in the form of pyrite, FeS2 pyrite;
3. sulfate Sculfate in the form of sulfuric acid salts (gypsum, FeSO4, etc.)». [1]

Only the first and second types of sulfur are subject to burning, sulphate sulfur does not burn. So, G.F. Bystritsky writes: "Sulfates are oxides of sulfur, therefore the sulfur in them cannot burn. Organic and pyritic sulfur present in the fuel is burned to form toxic sulfur dioxide SO2 and (in small quantities) even more toxic sulfur dioxide SO3. Emission from combustion products causes air pollution. Organic and pyritic sulfur together form volatile combustible sulfur Sl. [1]

Flue gas desulphurisation methods, despite their diversity, can be divided into three large groups: dry, wet (regenerative), semi-dry or wet-dry (non-regenerative). These methods can be classified as follows: absorption (absorption of the removed components by the liquid), adsorption (methods based on the ability of certain solids to selectively absorb gaseous components), chemisorption (a group of methods based on the chemical interaction of gases and vapors with solid or liquid absorbents with the formation of poorly soluble chemical compounds). [2]

3. Results of modern research

The classical methods of flue gas desulfurization include various methods of using lime and milk of lime. Desulfurization methods, which are based on wet technologies, are characterized by intensive irrigation of flue gases with lime milk, which is a solution of water and lime, and the lime in this solution is divided into ions during the dissociation process. The contact of this solution with flue gases

IOP Conf. Series: Materials Science and Engineering **552** (2019) 012024 doi:10.1088/1757-899X/552/1/012024

leads to the fact that sulfur dioxide also passes into the ionic form and binds to lime ions, after which it is removed from the flue gases along with the solution.

To dry methods cleaning can be related:

- Dry additive method (chemisorption), alkaline-earth compounds (mainly limestone) are directly blown into the furnace or fed there with fuel. The method has a relatively low efficiency, processes «bergbau frschung», Germany; HOKCO, USA, etc.;
- SO_2 chemisorption with the use of copper oxide ("wop-shell" method);
- Catalytic oxidation of SO_2 to SO_3, to produce sulfuric acid as a result of the process, the «WSA» process developed by Haldor Topse Denmark;
- SO_2 adsorption using activated carbon or coke, to obtain the final product of dilute sulfuric acid or gypsum;
- Radiation-chemical cleaning of flue gases from sulfur oxides and nitrogen (radiolysis), developed by Steinmuller, Germany, also in Russia, Institute of Nuclear Physics. The flow of flue gases after the introduction of ammonia into it is irradiated by a beam of accelerated electrons. The final product is ammonium sulphate and ammonium nitrate.The classical methods of flue gas desulfurization include various methods of using lime and milk of lime. Desulfurization methods, which are based on wet technologies, are characterized by intensive irrigation of flue gases with lime milk, which is a solution of water and lime, and the lime in this solution is divided into ions during the dissociation process. The contact of this solution with flue gases leads to the fact that sulfur dioxide also passes into the ionic form and binds to lime ions, after which it is removed from the flue gases along with the solution.

Due to the availability of this material and the possibility of using the resulting semi-finished product (gypsum) [3] in construction, methods based on the use of lime are widespread, however, are not without drawbacks. Thus, the use of dry limestone gas cleaning method, despite its simplicity and accessibility, worsens the performance of boilers and has low efficiency. Wet limestone methods can impair the operation of electrostatic precipitators and increase abrasive wear of surfaces. At the same time, the potential in the field of desulphurisation suggests a possible modernization and maintenance of existing methods, despite the shortcomings described.

One of the ways to improve the wet methods of flue gas desulphurisation is ash leaching technology, which is relevant for coal-fired power plants with a high content of calcium oxides in ash emissions. Such TPPs in irrigation water are characterized by an increase in the content of calcium ions due to the additional leaching of ash trapped in the ash collector.

Free lime is converted to bicarbonate $Ca(HCO_3)$ with greater solubility than lime. The separation of the flow from the lixiviant is divided by means of a cyclone into condensed pulp and clarified water. Clarified water is used in scrubber irrigation to trap sulfur oxides. The pulp is discharged to the ash dump, compacting and sealing its bed and preventing the ingress of contaminated water into groundwater and water bodies. [4] Such use of a wet ash collector allows to catch about a third of the total flow of sulfur oxides and almost complete trapping of fly ash.

A more complete trapping of sulfur oxides is shown by a double alkaline technology, which allows scrubbing flue gases from sulfur oxides by 50-60% due to the use of soda Na_2CO_3 as a reagent. When using this purification technology, sulfur dioxide is neutralized by soda, resulting in the formation of sodium sulfites and sodium bisulfites, which, in turn, are treated with lime used as the second alkaline reagent. The combination of sodium sulfites and sodium bisulfites with lime forms gypsum and lye, which can be reused in the irrigation of the ash collector. [5]

Another possible way to improve the methods of wet gas cleaning is the use of boiler blowdown water that has not been subjected to a decrease in alkali. Ph quicklime is 12, whereas the pH of the boiler water is between 7.1 and 14.0, opening up an additional way to dispose of the purge water - mixing it with a solution of water and lime. Increasing the alkalinity of the solution accelerates the binding of sulfur dioxide to it. Thus, the use of alkaline purge water boilers for irrigation of flue gases can increase the efficiency of desulfurization for any of the methods described above at minimal cost.

However, in addition to the low cost of this method, there is a danger of deterioration of the internal surfaces of the boiler from the use of alkaline water and an increase in the brittleness of the metal, therefore, this approach requires further detailed study.

A separate direction can be distinguished desulfurization of fuel.

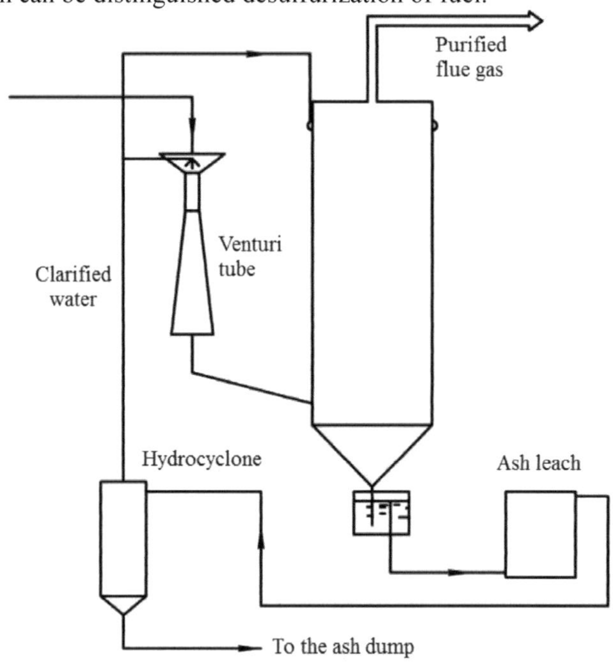

Figure 1. Schematic of a lixiviant.

Before any burning, any significant volumes of desulfurization can be discussed only with respect to fuel oils, which are desulfurized by hydrocracking, that is, when the fuel oil is saturated with hydrogen at high temperature, about 350-430°C, and high pressure, about 10-15 MPa, as which cobalt, nickel and molybdenum sulfides can be used. Hydrogen sulfide formed in the hydrocracking process is removed from the fuel oil, being separated by the adsorption method from the residual hydrogen.

Due to the high cost of desulfurizing solid fuel, it is practically not used in modern energy, except in cases where the processing of coal is necessary. [5]

More than half of the sulfur can be removed by physical cleaning (crushing and grinding). Technologically justified use of flue gas cleaning from sulfur oxides, but two-stage cleaning (fuel and flue gases burning it) significantly increases the cost of fuel in comparison with single-stage cleaning (only flue gases).

In addition to physical cleaning, coal is cleaned in other ways:

- adding limestone during coal gasification in a fluidized bed and into the lower vortex zone of the combustion chamber of a low-temperature vortex furnace (allows you to bind up to 80% of sulfur);
- processing of coal into gaseous / liquid hydrocarbons.

However, in spite of the potential of the methods described above, their practical value is low, which is due to the high cost of the final product, which is commensurate in value with oil. [5]

Desulfurization with seawater, the use of activated carbon, and the use of zeolites can be distinguished among the less common methods of flue gas desulfurization.

In the process of using seawater, its natural alkalinity is used to remove sulfur dioxide from flue gases [6, 7]. This method compares favorably with the use of lime, firstly, the use of water at much lower temperatures (30 - 40 ° C), and secondly, the lack of use of the suspension. However, seawater increases the corrosive wear of surfaces, which requires additional protection with a spray-coated flake coating or rubber lining, which increases installation costs.

Sea water is highly mineralized and contains various soluble compounds of many, including alkaline, elements with a total average concentration of about 35 g / kg sodium - 10.76 potassium - 0.38 calcium - 0.40 magnesium - 1.29 bicarbonate - 0 , 14, etc. The presence of the listed ions allows the use of seawater to neutralize the sulfur dioxide of flue gases. [8, p.82]

The principle of operation of desulfurization plants using seawater is similar to the principle of operation of wet limestone desulfurization, however, given the absence of reagent-preparing units and units of gypsum production, the installation cost is reduced by simplifying the technological scheme. [5, p.83]

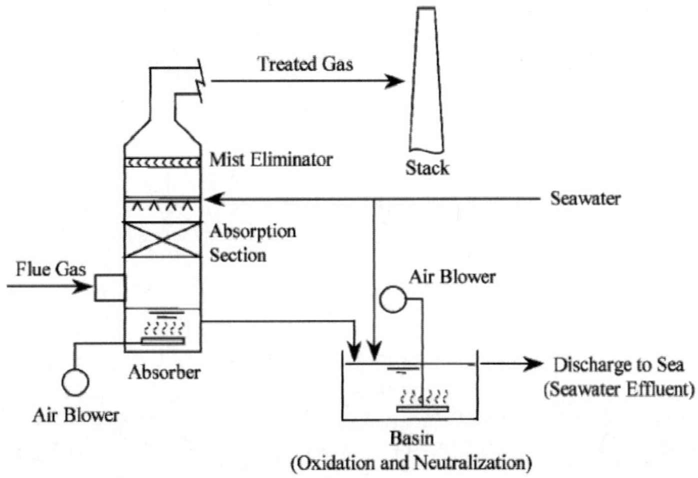

Figure 2. Simplified schematic diagram of desulfurization with seawater.

The simplified technological scheme reduces the cost of operation, including for additional wastewater treatment, due to the fact that the pulp obtained at the outlet of the sulfur treatment plant does not require treatment before discharge into the sea, since its composition practically does not differ from the composition of the marine water. [8, p.83]

For the desulfurization of gas is often used activated carbon [8] as an adsorbent. In particular, such purification is carried out by coke oven gas, from which various sulfur compounds are removed. The technology allows to achieve 75-80% gas purification. [9]

In addition to gas purification, there is a method of dust and gas cleaning of flue gases using the «Dusseldorf system» and activated carbon.

This cleaning system belongs to the class of wet-dry technologies based on the injection of lime mortar into the gas flow to suppress SOx. After that, the stream is dedusted in the electrostatic precipitator, thereby, in addition to ash, calcium compounds are also removed from the stream. The next stage of this system is cleaning of heavy metals, chlorine and fluorine derivatives of hydrocarbons, including dioxins and furans, as well as SOx and HCl breakthroughs. The final stage of purification is the removal of nitrogen oxides in a DeNOx reactor in a moving layer of activated carbon at a temperature of 130 ° C. [10]

The potential of this method is not studied sufficiently directly in the desulfurization of flue gases. It is known that the adsorption activity of activated carbon is maintained at a high level even after repeated regeneration cycles, in connection with which the desulfurization process can be cyclical in nature with regeneration, which can significantly affect resource saving. [5, p.44]

So, I. N. Shmigol notes that «in recent years, these methods have been increasingly used for cleaning air-gas mixtures from mercury vapor and organo-mercury compounds. Previously it was believed that activated carbon has a small adsorption capacity for mercury vapor and therefore it is rarely used in its pure form to remove mercury vapor from ventilating emissions. However, if activated carbon is pre-treated with chlorine, iodine, potassium permanganate, hydrogen sulfide or some other substances, its adsorption capacity increases dramatically, and when it comes into contact with mercury-contaminated gas, it is almost completely purified from mercury» [8]

Also of interest is the method of using zeolites as fuel additives. This method stands out favorably among others for its versatility, allowing for complex desulfurization of flue gases, using natural zeolite-containing raw materials as a reagent. [11, 12, 13]

This technology is relatively simple in comparison with alternative ones and does not require the construction of additional technological equipment. A feature of this approach is the continuous supply of zeolites to the flameless zone of the boiler directly during its operation to absorb sulfur oxides. [14, 15]

A. G. Batukhtin and V. V. Pinigin summarize: «this technology of adsorption cleaning of flue gases with the help of natural zeolites allows simultaneously with the removal of harmful gas emissions from products of combustion of fuel to improve the economic performance of heat-generating equipment. [16] So, for example, wiping natural zeolites in the convective shaft of the boiler in the amount of 15% in relation to fuel consumption at the nominal mode of operation allows reducing the emission of sulfur oxides by 70% and increasing the efficiency of the boiler gross by 1.6% by reducing the outgoing dew temperature gases». [17, 18, 19]

4. Discussion

The selection of an effective method of desulfurization, as a rule, is guided by the following criteria:

1. Baseline data: sulfur content in fuel, volume and temperature of flue gases;
2. Factor availability and prevalence of the sorbent;
3. Disposal of the final product of desulfurization: the possibility of its industrial use or the need to transport to the dump;
4. Installation of desulfurization equipment: the amount of investment in installation and maintenance;
5. The payback period. [20]

In practice, the following ways to reduce sulfur emissions are used:

1. Removal of sulfur directly from the fuel by thermal, mechanical or biological cleaning.
2. Regime-technological measures: the use of the fluidized bed, mixing of fuel with limestone before burning, elimination of fuel with high sulfur content from the fuel balance, reduction of equipment load, gasification, etc.
3. Flue gas desulfurization: dry, wet, semi-dry (wet or mixed). [21]

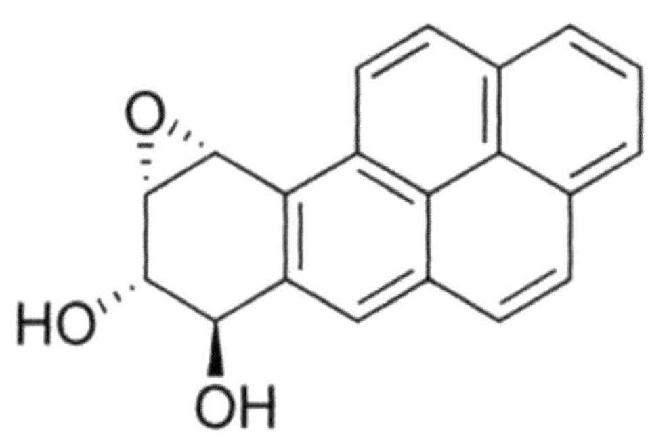

Figure 3. Chemical formula of benz (a) pyrene.

IOP Conf. Series: Materials Science and Engineering **552** (2019) 012024 doi:10.1088/1757-899X/552/1/012024

However, it should be borne in mind that the use of various technological measures related to the suppression of NOx and SO_2, as a result, leads to an increase in the amount of benzo (a) pyrene in the flue gases, which poses a rather serious complex problem to minimize harmful emissions.

Benz (a) pyrene - an aromatic compound, a member of the family of polycyclic hydrocarbons, formed during the combustion of hydrocarbon fuels - causes a high level of air pollution. [22]

Benz (a) pyrene is one of the most toxic combustion products belonging to the group of polycyclic aromatic hydrocarbons. It is a substance of the first class of hazard, with an extremely high level of dangerous effects on the environment and human health (carcinogenicity and blastomogenicity are noted upon contact with benzo (a) pyrene). At the same time, the mechanism of the occurrence of polycyclic aromatic hydrocarbons during fuel combustion has not been sufficiently studied today. [23]

In addition to the special danger, benzo (a) pyrene is extremely widespread, chemically and thermally stable, and when it enters the body it can accumulate. [24]

The relationship between the decrease in the sulfur content in the fuel and the increase in Benz (a) pyrene emissions is most pronounced. An example of this relationship for the BKZ-220 boiler with various sulfur contents in the fuel is shown in figure.

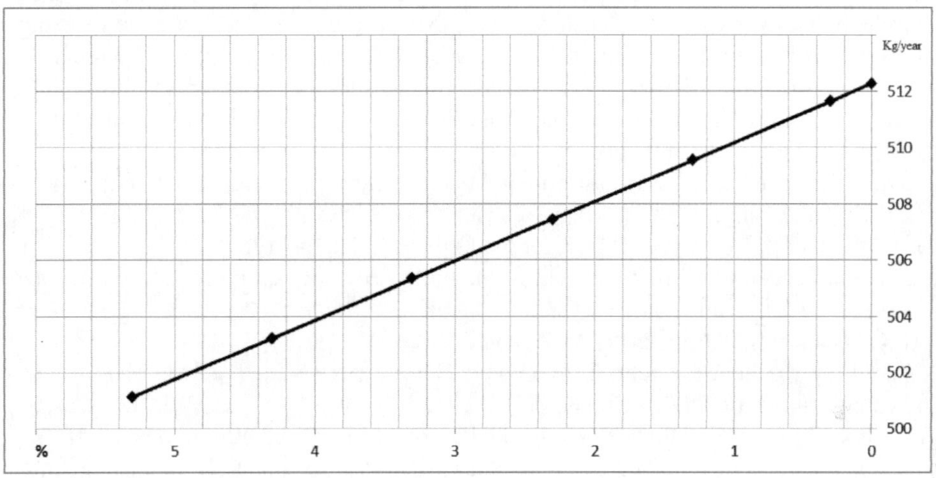

Figure 4. The dependence of emissions of benzo (a) pyrene on the sulfur content in the fuel.

5. Conclusion

Thus, the relationship between a decrease in the content of sulfur oxides and nitrogen and an increase in the content of benzo(a) pyrene poses a rather serious complex problem of minimizing harmful emissions in exhaust gases, among which solutions you can specify:

1. Introduction of energy-saving technologies for heat consumers (including control and metering devices);
2. Reduction of heat loss in networks;
3. Optimization of boiler operation modes; [25]
4. Improving the efficiency of fly ash trapping (including fine fractions).

In practice, an extremely unfavorable ecological situation has developed in Chita today. So, according to the results of the first half of 2018, Chita entered the list of Russian cities with the most polluted air. In addition to Chita, Petrovsk-Zabaykalsky was included in this list. Information about this was published on the website of Rosstat. Based on these data, 8 facts of air pollution were recorded, with benz (a) pyrene exceeding the maximum permissible concentration by 57 times. [6]

The President of the Russian Federation, V. V. Putin, has set the task of reducing pollutant emissions by 20% by 2024. This task has been set for 12 Russian cities, including Chita. The main air emissions in the city of Chita are accounted for by energy enterprises (CHP-1, CHP-2, numerous boiler houses), a dense stream of city transport and seasonal forest fires. [2]

References

[1] Bystrickij G F 2012 *Energy Basics* (Moscow: KnoRus)
[2] Kalinina A S, Romanovich D A and Trinchenko A A 2013 Sovremennoe sostoyanie problemy seroochistki dymovyh gazov teplovyh ehlektricheskih stancij [The current state of the problem of flue gas desulfurization of thermal power plants] *Trudy Sankt-Peterburgskogo politekhnicheskogo universiteta Petra Velikogo* **515** pp. 45 – 50 [In Russian]
[3] Rikker Y O, Kobylkin M V and Strelnikov A S 2017 *Kulaginskie chteniya: tekhnika i tekhnologii proizvodstvennyh processov* [Kulagin Readings: Technique and Technology of Production Processes] vol 17 (Chita: TbSU) pp 138-143 [In Russian]
[4] Rose L, Mary X A, Issac J S and Thomas B 2018 Power plant optimization for energy efficiency enhancement *International Journal of Mechanical Engineering and Technology* 9.700-707
[5] Rikker Y O, Batuhtin A G and Kobylkin M V 2018 Potencial ispol'zovaniya shchelochnoj produvochnoj vody kotlov v desul'furizacii dymovyh gazov [The potential of using alkaline boiler blowdown water in flue gas desulphurisation] *Veles* **5-1** pp 75-9 [In Russian]
[6] Oikawa Katsuo, Yongsiri Chaturong, Takeda Kazuo and Harimoto Takayoshi. 2003 Seawater flue gas desulfurization: Its technical implications and performance results *Environmental Progress* 22.67-73.10.1002/ep.670220118
[7] Caiazzo Giuseppe, Langella Giuseppe, Miccio Francesco and Scala Fabrizio 2013 An Experimental Investigation on Seawater SO2 Scrubbing for Marine Application *Environmental Progress and Sustainable Energy* 32.10.1002/ep.11723
[8] Bing Li and Chun Yuan Ma 2018 SO2 Removal by Powder Activated Carbon in a Drop Tube Furnace Experimental System *Key Engineering Materials* 783.34-40. 10.4028/www.scientific.net/KEM.783.34
[9] 2018 Metody rascheta toksichnyh vybrosov: Benz(a)piren C20H12 [Methods for calculating toxic emissions: Benz(a)pyrene K20X12] URL https://clck.ru/ES3jH [In Russian]
[10] 2018 Klassifikaciya sposobov seroochistki [Classification methods of desulfurization] URL https://studwood.ru/1319443/ekologiya/klassifikatsiya_sposobov_seroochistki [In Russian]
[11] Batuhtin A G 2016 Primenenie teplovyh nasosov dlya razvitiya teplofikacii [The use of heat pumps for the development of heating] *Nauchno-tekhnicheskie vedomosti Sankt-Peterburgskogo gosudarstvennogo politekhnicheskogo universiteta* **1 (238)** pp 28-36 [In Russian]
[12] Pinigin V V 2014 *Povyshenie kachestva prognozirovaniya vrednyh vybrosov ot kotel'nyh ustanovok TES* [Improving the quality of forecasting emissions from boilers of thermal power plants] (Chita: TbSU) [In Russian]
[13] Shelepova V S 2017 Benzapiren - himiko-biologicheskaya problemy sovremennosti (C20H12) [Benzapiren - chemical and biological problems of the present (C20H12)] *Pozharnaya bezopasnost': problemy i perspektivy* **8** pp 477-80[In Russian]
[14] 1998 Instrukciya po normirovaniyu vybrosov zagryaznyayushchih veshchestv v atmosferu dlya teplovyh ehlektrostancij i kotel'nyh RD 153-34.0-02.303-98 [Instructions on the regulation of emissions of pollutants into the atmosphere for thermal power plants and boiler RD 153-34.0-02.303-98] (Moscow: Sluzhba peredovogo opyta ORGREHS) [In Russian]
[15] Zvereva E R, Farahov T M and Iskhakov A R 2011 Snizhenie vrednyh vybrosov teplovyh ehlektrostancij [Reducing harmful emissions from thermal power plants]. *Vestnik KGEHU* **1** pp 39-44 [In Russian]
[16] Batuhtin A G and Pinigin V V 2012 Optimizaciya rezhimov raboty teplogeneriruyushchego oborudovaniya pri ispol'zovanii prirodnyh ceolitov dlya snizheniya vrednyh vybrosov [Optimization of operating modes of heat-generating equipment when using natural zeolites

to reduce harmful emissions] *Vestnik Chitinskogo Gosudarstvennogo Universiteta* **7(86)** pp 19 – 24 [In Russian]

[17] 2018 Snizhenie vybrosov dioksidov sery [Reducing sulfur dioxide emissions] URL http://portal.tpu.ru:7777/SHARED/m/MAR/education/Tab2/%D1%81%D0%B5%D1%80%D0%B011.pptx [In Russian]

[18] 2018 Sistemy desul'furizacii dymovyh gazov na morskoj vode [Seawater flue gas desulphurisation systems] URL http://www.antikorprom.com.ua/index.php/ru/seroochystka/systemy-desulfuryzatsyy/na-morskoi-vode [In Russian]

[19] Shmigol' I N 2018 Seroochistka dymovyh gazov dlya teplovyh ehlektrostancij Rossii [Desulfurization of flue gases for thermal power plants in Russia] URL http://ir.nmu.org.ua/bitstream/handle/123456789/577/35.pdf;jsessionid=72AB8D5C05A4998540B5D024DCDE0348?sequence=1 [In Russian]

[20] SHmigol' I and Ol'shanskij Yu P 2006 *Seroochistka dymovyh gazov teplovyh ehlektrostancij* [Desulfurization of flue gases of thermal power plants] (Moscow: IPKgossluzhby) p 56 [In Russian]

[21] 2018 Energetika: TES i AES. Vse o teplovoj i atomnoj ehnergetike [Energy: thermal power plants and nuclear power plants. All about thermal and nuclear energy] URL http://tesiaes.ru/?p=12673 [In Russian]

[22] 2018 Chita snova vozglavila spisok gorodov Rossii s samym gryaznym vozduhom [Chita again headed the list of cities in Russia with the dirtiest air] URL http://chita.bezformata.ru/listnews/rossii-s-samim-gryaznim-vozduhom/68685162/ [In Russian]

[23] Zhang Zhillong and Tang Mao-Ping 2002 Applications of scrubber desulphurisation of dim gases using the required sea water for six power units with a capacity of 600 MW el. stancii zhangzhou houshi *Zhongguo dianli-Elec. Power* **35(7)** pp 69-71

[24] 2018 5 gorodov Rossii s samym gryaznym vozduhom [5 cities of Russia with the dirtiest air] https://news.rambler.ru/ecology/37554801/?utm_content=rnews&utm_medium=read_more&utm_source=copylink [In Russian]

[25] Roeva N N, Voronich S S, Khlopaev A G, Zaitsev D A and Voronich N S 2018 Investigation of the Dynamics of Accumulation of Mobile Forms of Heavy Metals, Benz-(a)-Pyrene and Petroleum Products in Soils of Urbanized Territories *Ecology and Industry of Russia* 22.39-43.10.18412/1816-0395-2018-8-39-43

Application of thermohydraulic dispatcher for existing and new district heating systems

D O Romanov[1], Y V Yavorovsky[1], V V Smirnov[2]

[1]Department of Industrial heat Engineering Systems, National Research University MPEI, Krasnokazarmennaya St., 14, Moscow, 111250, Russia
[2]Department of Industrial Heat and Power Engineering, Ivanovo State Power Engineering University, Rabfakovskaya St., 34, Ivanovo, 153003, Russia

E-mail: dimrom09@yandex.ru

Abstract. Electricity consumption represents high share of operational costs in district heating (DH) systems. Recent attempts of decreasing these costs have led to appearing the idea of systems with distributed variable speed pumps (DVSP). One of the parts of such systems is thermohydraulic dispatcher (THD). This article is concerned with application of THD for DH systems. Existing DH systems with low available hydraulic head are suitable for installing THD instead of booster pumps. Electricity consumption and cost of equipment for several schemes of heat substation are estimated. Application for low temperature district heating (LTDH) systems is considered. Further ideas of development DVSP systems are proposed. Schemes of cascade systems based on THD and hydraulic profiles are presented. In result, cascade systems can reach highest possible electricity saving effect due to eliminating of throttle valves at heat source, heat substations and heat consumers.

1. Introduction

At the present time problems of energy saving are of much importance worldwide. Particularly, in Russia several federal laws were passed recently in order to stimulate energy saving in industry and public service [1-3]. One of the areas of focus is heat supply. Heat supply systems in Russia have following problems: outdated equipment, considerable transportation energy costs, critical condition of heat supply network, ineffective building envelope, insufficient automatization, all of them leading to considerable energy loses [4-7]. Mentioned problems can be solved by large-scale refurbishment of equipment, network and buildings. During the refurbishment it is important to pay attention to modern and perspective technologies in heat supply systems.

One of devices that can be used in DH systems for increasing its effectiveness is thermohydraulic dispatcher (THD). THD is vertical or horizontal pipe manifold of large diameter with low hydraulic resistance. It is a part of hydraulic system consisting of source circuit (circuits) and consumer circuit (circuits). THD allows making circuits hydraulically independent from each other. For example, increasing mass flow in consumer circuit doesn't affect mass flow in source circuit. Similarly, a consumer circuit doesn't influence another consumer circuit in case of using THD. This effect can be used to avoid mutual influence of various consumers (heating, hot water, ventilation consumers) on each other [8-9].

IOP Conf. Series: Materials Science and Engineering **552** (2019) 012025 doi:10.1088/1757-899X/552/1/012025

2. Possible applications of THD

2.1. Heat source
Since THD separates circuits, it is reasonable to install THD in boiler room to separate source circuit and network. Moreover, several boilers working for the same DH system can be separated as well. Another advantage of connection to THD is constant flow rate through boiler, which helps avoid minimal flow rate regimes. Recirculation line is usually installed to increase boiler inlet temperature and prevent thermal shock and risk of sulphuric corrosion. THD installed at boiler room plays the role of recirculation line.

2.2. Heat substation
Installation of THD at heat substation allows separating main network from consumers. Besides, space heating (SH) and hot water supply systems (HWS) are separated as well, so their mutual influence, which usually leads to hydraulic and thermal disbalance, is disposed. There are several ways of connecting consumers to heat substation [10]. So, THD represents the connection similar to indirect connection, where THD replaces heat exchanger. But static pressure of primary and secondary circuits for THD connection is the same unlike with indirect connection. Since direct connection (open HWS) of HWS system is to be forbidden in 2022 in Russia [2], only indirect connection (closed HWS) is considered in this article. At the same time either direct or indirect connection of SH system is possible.

2.3. Heat consumers
SH systems are characterised by distributed network inside a building and adjusted hydraulically by balancing valves which throttle redundant hydraulic head. Although this system is common, it is not effective due to energy loss during throttling. THD can be used for SH systems to separate consumers form each other and to avoid throttling losses by using variable speed pumps for every consumer instead of balancing valves and central pump.

The same solution can be adopted for HWS systems, where THD can provide hydraulic independence to consumers. Now HWS systems are not designed properly in some cases, which results in hydraulic disbalances while several consumers are turned on. For example, having adjusted temperature of water for shower bath, consumer is likely to have to readjust it after another water consuming unit is turned on. So, application of THD for HWS systems will bring more comfort to dwellers of buildings.

2.4. Application of THD for LTDH systems
Low temperature district heating (LTDH) is promising technology which is actively being researched in Europe [11-15]. It is well-known fact that electricity consumption in DH system can be decreased by increasing supply temperature and decreasing return temperature. So, cases of low available hydraulic head may occur during transition from existing DH systems to LTDH systems. Since existing DH systems were designed for particular temperature difference, decreasing of temperature in supply line can lead to lower temperature difference and result in higher flow rate causing bottlenecks [16]. Application of THD for such cases can be a good solution for tackling bottlenecks problem. Moreover, THD allows systems to operate under lower differential pressure, which can be used for creating low temperature and pressure district heating systems.

3. Methods
THD suits well for distant consumers with low hydraulic head. So, abstract heat substation with available hydraulic head of 10 kPa was analysed. Three schemes of heat substation were considered: indirect connection, direct connection, and THD connection. Design temperatures of supply and return lines for primary circuit are 130°C and 70°C accordingly; for secondary circuit those temperatures are 95°C and 70°C. Different heat outputs were considered: 800 kW (300 kW for SH and 500 kW for HWS), 2400 kW (900 kW for SH and 1500 kW for HWS), and 4800 kW (1800 kW for SH and 3000

kW for HWS). The equipment was chosen and prices for heat exchangers, pumps, pipes, valves, controls and instruments were received from producing companies. Then electricity consumption during heating period in Moscow was estimated. tables 1, 2 and 3 show input data and the results of calculation of electricity consumption during heating period for different connection schemes and different heat output. Electricity consumption was estimated under condition that all pumps work in nominal mode. But HWS circulating pumps may require less energy during night time which is not considered in this paper. Also electricity consumption of make-up pumps for indirect scheme was neglected. Although coefficient of performance (COP) of pump depends on its construction, the same value (0.6) was taken for every considered pump. Power of pump was calculated as

$$N = \frac{Q * \Delta P}{3600 * \eta}$$

N is power of pump, kW.
Q is volume flow rate, m^3/hr.
ΔP is pump head, kPa.
η is COP of pump.
Electricity consumption was calculated as

$$E = N_{\Sigma} * \tau * 24$$

E is electricity consumption, kW*hr,
N_{Σ} is total power of pumps, kW,
τ is duration of heating period, which is 205 days for Moscow [17].

4. Results

Figure 1 shows estimated cost of equipment for mentioned schemes of heat substation in dependence of heat output. While cost of equipment for schemes with direct connection and THD connection is almost identical for considered range of heat output, indirect connection scheme costs much more because of high heat exchangers price.

Table 1. Electricity consumption of different heat substation schemes. Total heat output is 800 kW.

	Indirect				Direct				THD			
	Volume flow rate (m3/hr)	Pump head (kPa)	COP of pump	Power of pump (kW)	Volume flow rate (m3/hr)	Pump head (kPa)	COP of pump	Power of pump (kW)	Volume flow rate (m3/hr)	Pump head (kPa)	COP of pump	Power of pump (kW)
Booster pump (primary circuit)	11.5	100	0.6	0.532	11.5	100	0.6	0.532	_	_	_	_
HWS circulating pump (primary circuit)	_	_	_	_	_	_	_	_	7.2	50	0.6	0.167
DH circulating pump (secondary circuit)	8.6	100	0.6	0.398	_	_	_	_	10.3	50	0.6	0.238
Mixing pump	_	_	_	_	6	50	0.6	0.139	_	_	_	_
HWS circulating pump (secondary circuit)	2.2	100	0.6	0.102	2.2	100	0.6	0.102	2.2	100	0.6	0.102
Total power	_	_	_	1.032	_	_	_	0.773	_	_	_	0.507
Electrictiy consumpiton during heating period, kW*hr	_	_	_	5079	_	_	_	3804	_	_	_	2494

IOP Conf. Series: Materials Science and Engineering **552** (2019) 012025 doi:10.1088/1757-899X/552/1/012025

Table 2. Electricity consumption of different heat substation schemes. Total heat output is 2400 kW.

	Indirect				Direct				THD			
	Volume flow rate (m3/hr)	Pump head (kPa)	COP of pump	Power of pump (kW)	Volume flow rate (m3/hr)	Pump head (kPa)	COP of pump	Power of pump (kW)	Volume flow rate (m3/hr)	Pump head (kPa)	COP of pump	Power of pump (kW)
Booster pump (primary circuit)	34.4	150	0.6	2.389	34.4	150	0.6	2.389	_	_	_	_
HWS circulating pump (primary circuit)	_	_	_	_	_	_	_	_	21.5	100	0.6	0.995
DH circulating pump (secondary circuit)	25.8	200	0.6	2.389	_	_	_	_	30.9	100	0.6	1.431
Mixing pump	_	_	_	_	18	100	0.6	0.833	_	_	_	_
HWS circulating pump (secondary circuit)	6.4	200	0.6	0.593	6.4	200	0.6	0.593	6.4	200	0.6	0.593
Total power	_	_	_	5.370	_	_	_	3.815	_	_	_	3.019
Electrictiy con-sumpiton during heating period, kW*hr	_	_	_	26422	_	_	_	18769	_	_	_	14851

Table 3. Electricity consumption of different heat substation schemes. Total heat output is 4800 kW.

	Indirect				Direct				THD			
	Volume flow rate (m3/hr)	Pump head (kPa)	COP of pump	Power of pump (kW)	Volume flow rate (m3/hr)	Pump head (kPa)	COP of pump	Power of pump (kW)	Volume flow rate (m3/hr)	Pump head (kPa)	COP of pump	Power of pump (kW)
Booster pump (primary circuit)	68.7	200	0.6	6.361	68.7	200	0.6	6.361	_	_	_	_
HWS circulating pump (primary circuit)	_	_	_	_	_	_	_	_	43	150	0.6	2.986
DH circulating pump (secondary circuit)	51.6	250	0.6	5.972	_	_	_	_	61.9	150	0.6	4.299
Mixing pump	_	_	_	_	36.1	150	0.6	2.507	_	_	_	_
HWS circulating pump (secondary circuit)	12.9	250	0.6	1.493	12.9	250	0.6	1.493	12.9	250	0.6	1.493
Total power	_	_	_	13.826	_	_	_	10.361	_	_	_	8.778
Electrictiy con-sumpiton during heating period, kW*hr	_	_	_	68026	_	_	_	50977	_	_	_	43187

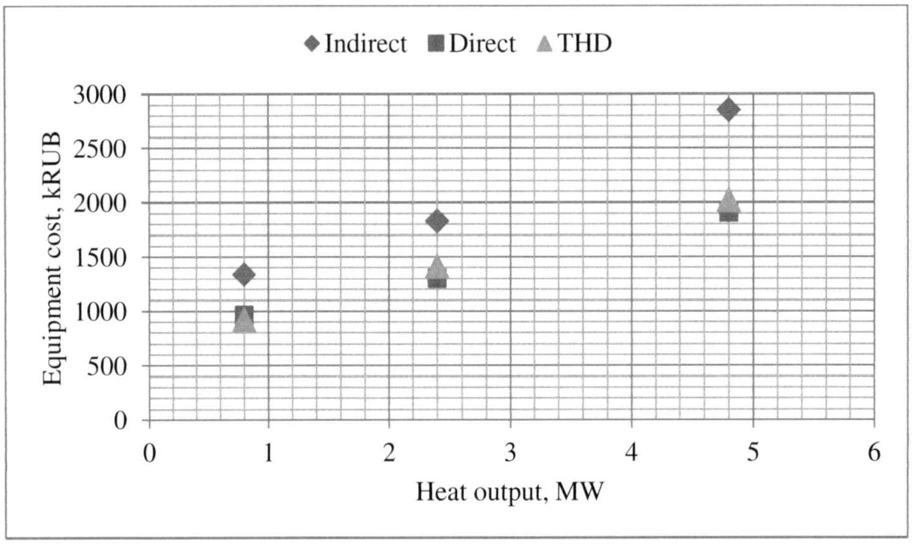

Figure 1. Dependence of equipment cost for different connection schemes from heat output.

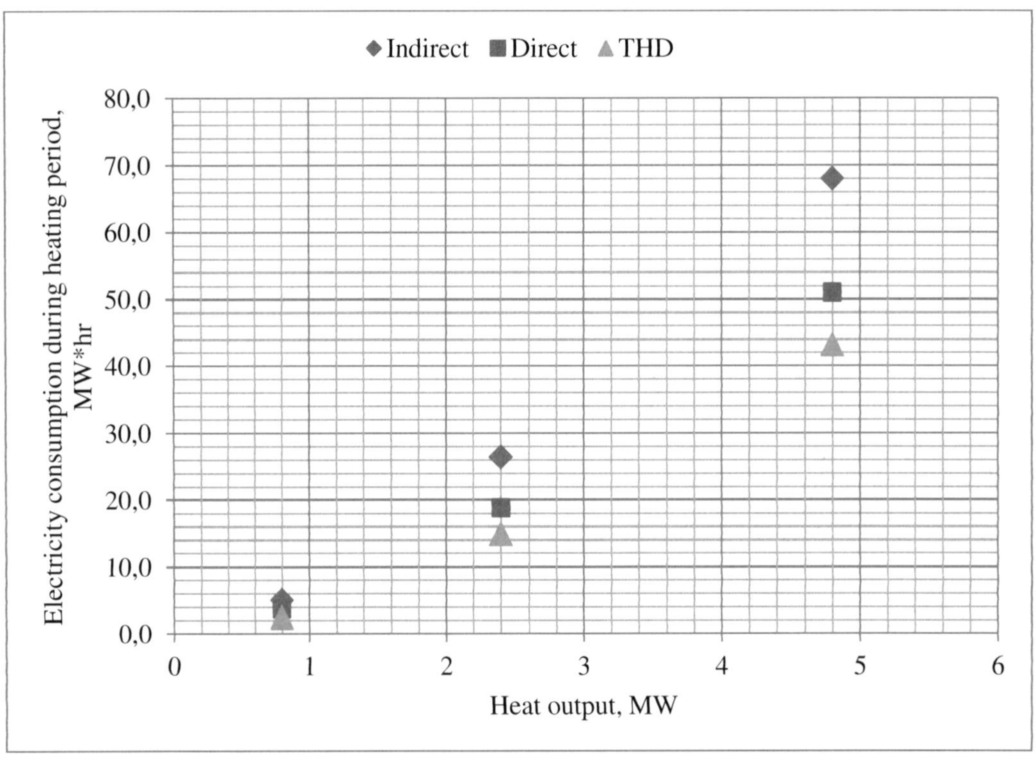

Figure 2. Dependence of electricity consumption during heating period for different connection schemes from heat output.

Figure 2 shows estimated electricity consumption during heating period for different connection schemes from heat output. Scheme with THD connection has less electricity consumption among all considered heat outputs. The difference between electricity consumption of different schemes increases with increase of heat output. Since available hydraulic head is low, it is necessary to apply booster pumps in schemes with both direct and indirect connections. The scheme with THD connection can work without booster pumps, because usually THD operates under low differential pressure. That is the explanation for the best performance of scheme with THD connection in terms of electricity consumption. The scheme with indirect connection has the highest energy consumption because of high hydraulic resistance of heat exchangers.

5. Discussion

The problem of electricity saving in DH systems is paid increased attention to. One of the ways to tackle this problem is to utilise redundant heat carrier pressure in district heating by using a pump with rotor wheel designed to work as turbine thus generating electricity instead of throttling redundant hydraulic head [18-19].

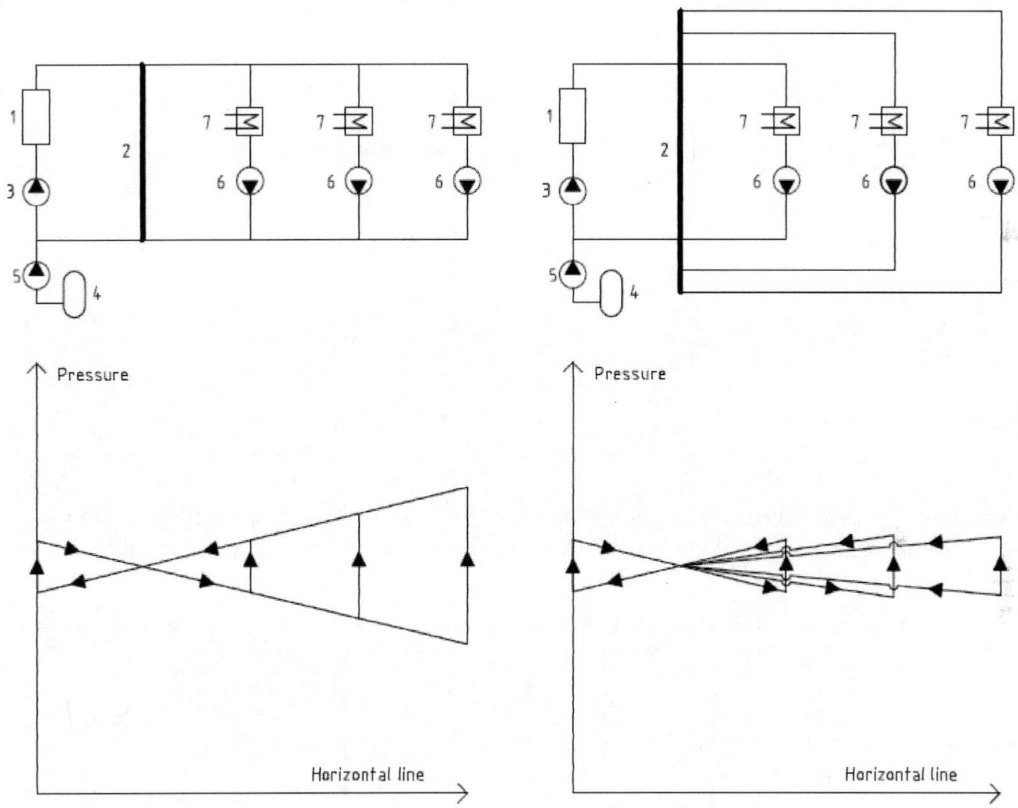

Figure 3. Scheme and hydraulic profile of DVSP system proposed in [20] (on the left) and scheme and hydraulic profile of modified DVSP system (on the right). 1 – heat source, 2 – THD, 3 – circulating pump, 4 – water tank, 5 – make-up pump, 6 – circulating pumps, 7 – heat consumers.

Another way is to install new DH systems with distributed variable speed pumps (DVSP), where THD is used. The concept of DVSP systems is to eliminate regulating valves and throttling losses and rely on variable speed pumps as devices for control and regulation [20-24]. In article [20] Chinese researchers proposed new DH system which is shown on the left side of figure 3. The device called hydraulic connector in the article seems to be the same as THD. Although proposed scheme has electricity saving effect, the effect is not as great as possible. Hydraulic profile tends to widen after zero pressure difference point, which contributes to more electricity consumption. This tendency can be decreased by either using main network pipes of larger diameter or connecting all consumers to THD

IOP Conf. Series: Materials Science and Engineering **552** (2019) 012025 doi:10.1088/1757-899X/552/1/012025

which is shown on the right side of figure 3. Nevertheless, both methods might have significant capital costs due to increased pipe diameters or number of pipes accordingly. On the other hand, increased number of pipes may result in increasing of overall heat supply reliability as every consumer has its own independent pipelines and breakdown of one consumer doesn't affect the others. So, these factors should be taken into account when estimating operational benefits and assessing overall efficiency of modifications to DVSP scheme.

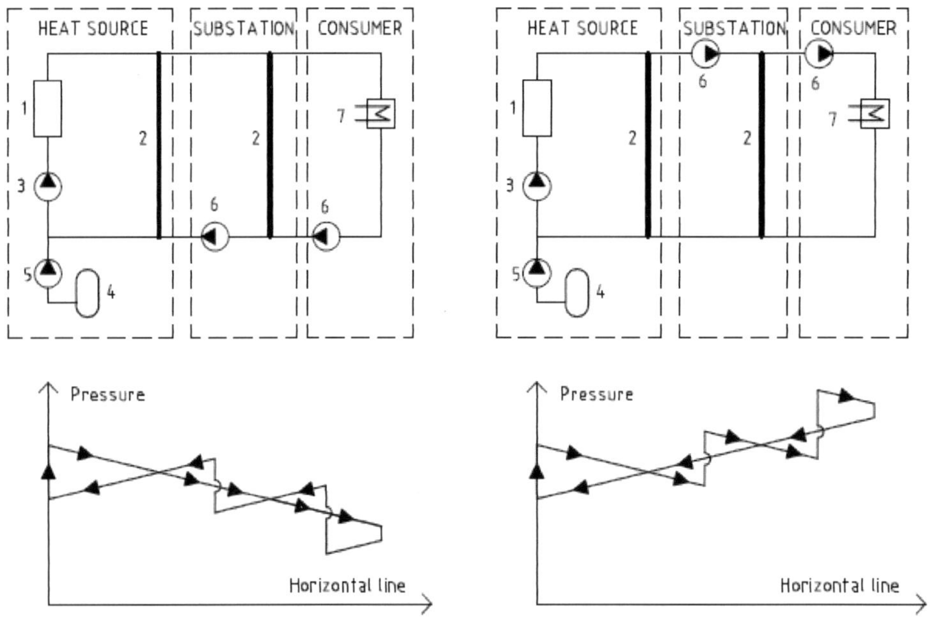

Figure 4. Cascade scheme with pumps on return line and its hydraulic profile (on the left) and cascade scheme with pumps on supply line and its hydraulic profile (on the right).

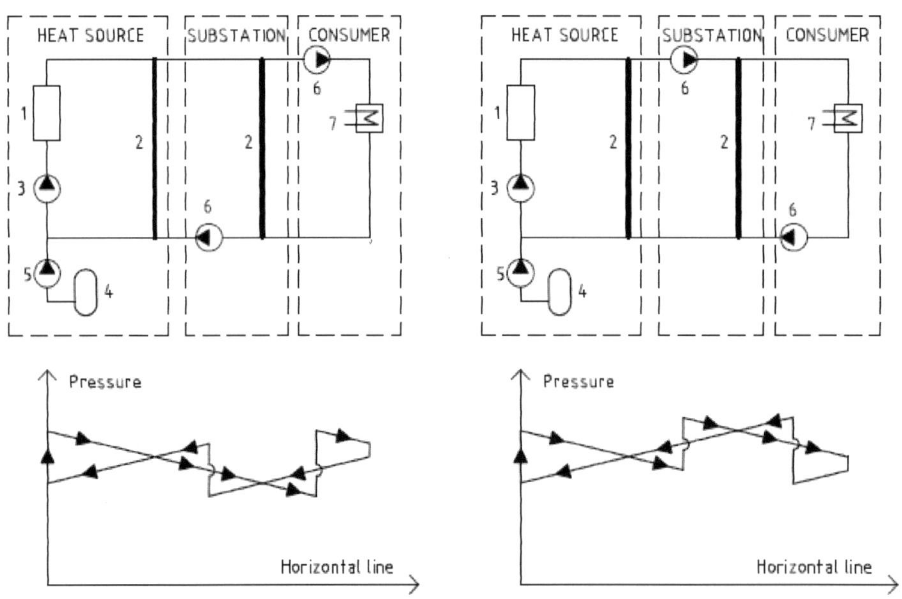

Figure 5. Cascade schemes with pumps on both return and supply lines and its hydraulic profiles.

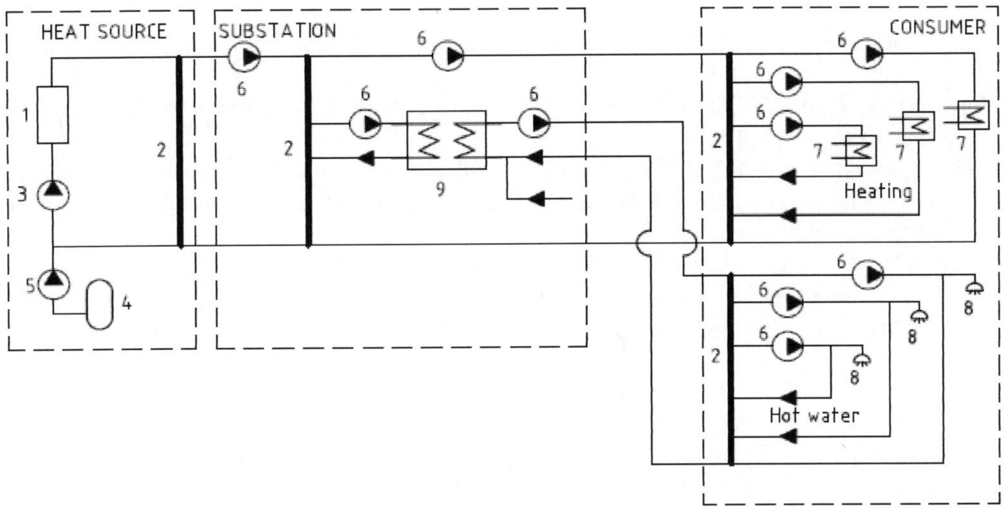

Figure 6. Principal cascade scheme of DH system. 8 – hot water consuming units, 9 - heat exchanger.

In articles [21-24] electricity consumption, economic and energy saving analyses of DVSP systems are considered as well as method of system regulation. But the articles' main focus is investigation of Source-Substation chain. One of the aims of this article is to propose Source-Substation-Consumer chain based on THD application. Such cascade scheme may have highest possible electricity saving effect due to avoiding of throttle valves in the whole DH system. Cascade schemes based on THD application are illustrated on figure 4. The scheme on the left side includes pumps installed on the return line after THD. It makes hydraulic profile go down and pressure in supply line declines which can cause evaporation of water. However, this is not going to be a problem for low temperature district heating systems. The scheme on the right side includes pumps installed on the supply line after THD. It makes hydraulic profile go up and pressure in return line increases which can cause breakdown of customer's heating units. In order to avoid these disadvantages cascade schemes with pumps on both return and supply lines are proposed and illustrated on figure 5. Hydraulic profile of such systems goes both up and down and pressure in supply and return lines is more stable. Figure 6 summarizes previous information. It describes SH and HWS systems connected to TDH at consumer's side, then there is another THD at heat substation, and one more THD at heat source, all of them installed in series.

6. Conclusion

THD can be installed at consumer's side, heat substation, and heat source separating hydraulically all of them from each other. At the same time it leads to electricity saving effect due to avoiding of throttle valves. Existing DH systems as well as LTDH systems with low available hydraulic head are suitable for installing THD instead of booster pumps. Result of estimation shows that it has significant electricity consumption reduction without increasing of capital costs. Cascade systems based on THD application were proposed. It seems to be perspective way of combining benefits of installing THD and distributed variable speed pumps in DH systems leading to the highest energy saving effect for the whole DH system.

Acknowledgements
The work was performed under support of state order № 13.9613.2017/8.9.

References

[1] Federal'nyy zakon №261 ot 23.11.2009 Ob energosberezhenii i povyshenii energoeffektivnosti [Federal law №261 23.11.2009 About energy saving and increasing energy efficiency] [in Russian]

[2] Federal'nyy zakon №190 ot 27.07.2010 O teplosnabzhenii [Federal law №190 27.07.2010 About heat supply] [in Russian]

[3] Federal'nyy zakon №382 ot 03.12.2011 O gosudarstvennoy informatsionnoy sisteme toplivno-energeticheskogo kompleksa [Federal law №382 03.12.2011 About state information system of fuel and energy sector] [inRussian]

[4] Schelokov A I and Filatova A I 2011 Sravnitel'nyy analiz effektivnosti teplosnabzheniya zhilishchno-kommunal'nogo khozyaystva [Comparative analysis of effectiveness of housing and public utilities heat supply] *Zhurnal Samarskogo gosudarstvennogo tekhnicheskogo universiteta* **2** [in Russian]

[5] Schelokov Y M 2009 Problemy teplosnabzheniya gorodov [Heat supply problems of cities] *Energetika: ekonomika, tekhnika i ekologiya* **5** [in Russian]

[6] Bashmakov I A 2010 Povysheniye effektivnosti sistem teplosnabzheniya [Increasing effectiveness of heat supply systems] *Energosberezheniye* **2** [in Russian]

[7] Panferov S V, Telegin A I and Panferov V I 2010 Nekotoryye problemy energosberezheniya i avtomatizatsii sistem teplosnabzheniya zdaniy [Some problems of energy saving and automation of buildings heat systems] *Zhurnal Yuzhno-Ural'skogo gosudarstvennogo universiteta* **22** [in Russian]

[8] Romanov D O, Yavorovsky Y V et al. 2017 Application of thermohydraulic dispatcher in low temperature district heating systems for decreasing heat carrier transportation energy cost and increasing reliability of heat supply *IOP Conf. Series: Journal of Physics: Conf. Series* **891**
Romanov D O, Yavorovsky Y V, et al. 2018 Experimental research of thermo-hydraulic separators and dispatchers in heat supply systems *Solid State Phenomena* **284** pp 1385-9

[9] Frederiksen S and Werner S 2013 District Heating and Cooling *Studentlitteratur Ab*

[10] Lund H, Werner S et al. 2014 4th Generation District Heating (4GDH). Integrating smart thermal grids into future sustainable energy systems *Energy* **68** pp 1-11

[11] Lund H, Duic N et al. 2016 Smart energy systems and 4th generation district heating *Energy* **110** pp 1-4

[12] Ziemele J, Kubule A et al. 2016 Multi-perspective methodology to assess the transition to 4th generation district heating systems *Energy Procedia* **113** pp 17-21

[13] Paiho S and Reda F 2016 Towards next generation district heating in Finland *Renewable and Sustainable Energy Reviews* **65** pp 915-24

[14] Schmidt D et al. 2017 Low temperature district heating for future energy systems *Energy Procedia* **116** pp 26-38

[15] Brange L, Lauenburg P et al. 2017 Bottlenecks in district heating networks and how to eliminate them. A simulation and cost study *Energy* **137** pp 607-16

[16] SP 131.13330.2012 Stroitel`naya klimatologiya [SP 131.13330.2012 Building climatology] [in Russian]

[17] Volkova T A 2014 Ispol'zovaniye izbytochnogo davleniya magistral'nogo setevogo teplonositelya dlya povysheniya nadezhnosti i effektivnosti sistemy tsentralizovannogo teplosnabzheniya [Utilisation of redundant main network heat carrier pressure for increasing reliability and efficiency of district heating system] *PhD thesis* [in Russian]

[18] Chernenkov V et al 2018 Increasing the energy efficiency of the heat supply system with the help of the device for recuperation of throttling head *FEFU: School of Engineering Bulletin* **2/35** (in Russian)

[19] Yan A Zhao J et al 2013 Hydraulic performance of a new district heating systems with distributed variable speed pumps *Applied Energy* **112** pp 876-85

[20] Sheng X and Duanmu L 2016 Electricity consumption and economic analyses of district heating system with distributed variable speed pumps *Energy and buildings* **118** pp 291-300

[21] Sheng X and Duanmu L 2016 Energy saving analyses on the reconstruction project in district heating system with distributed variable speed pumps *Applied Thermal Engineering* **101** pp 432-45

[22] Sheng X and Duanmu L 2017 Energy saving factors affecting analysis on district heating system with distributed variable frequency speed pumps *Applied Thermal Engineering* **121** pp 779-90

[23] Wang H, Wang H and Zhu T 2017 A new hydraulic regulation method on district heating system with distributed variable speed pumps *Energy Conversion and Management* **147** pp 174-89

Coagulant on the base of industrial ferriferous waste for the preliminary treatment of multi-component sewage

S V Sverguzova[1], Zh A Sapronova[1], A V Svyatchenko[1], E V Fomina[1] and A A Malysheva[2]

[1]Department of Industrial Ecology, Belgorod State Technological University of V.G.Shukhov, Kostyukov St., 46, Belgorod, 308012, Russia
[2]Department of Heat and Gas Supply and Ventilation, National Research University Moscow State University of Civil Engineering, Yaroslavskaya road, 26, Moscow, 129337, Russia

E-mail: sapronova.2016@yandex.ru

Abstract. The paper gives the results of a study on the processing of model systems by a coagulating suspension made on the basis of dust from electric arc furnaces (EAF). Dust of electric furnace steelmaking is a large-tonnage waste containing in its composition a large number of iron compounds. When treating EAF with hydrochloric acid, formed ions $FeOH^+$, $FeOH^{2+}$, $[Fe(OH)_2]^+$ in this process provoke coagulation, which leads to the adhesion of particles in the system and their sedimentation. It is shown that the suspension is effective for clarifying aqueous media from suspended particles (kaolin clay). It has been established that precipitation occurs more effectively in a neutral and alkaline environment than in an acidic one. When processing multicomponent model systems (industrial oil stabilized with sodium lauryl sulfate and clay particles), the efficiency of clarification decreases, but remains high enough (65%). Taking into account the research results, the coagulating suspension can be used at the stage of pretreatment of highly contaminated sewage of mixed composition to facilitate and reduce the cost of subsequent stages of water treatment.

Keywords. Water treatment, coagulation, industrial waste, wastewater, oil products, multicomponent sewage, colloidal systems, clay suspension

1. Introduction

The global environmental crisis is the result of the cumulative anthropogenic impact of all states and peoples of the world on nature [1]. Water consumption has now reached huge volumes. Table 1 presents data on groundwater and surface water abstraction in some countries, 2005-2015 (million m^3).

Sewage resulting from economic and industrial activities is a big problem for all countries of the world, and they require new solutions.

Most sewage is a multicomponent polydisperse system, containing chemicals of various nature and inclusions of varying sizes. This introduces difficulties in the process of water treatment, requiring the use of devices and methods in various combinations. So, one of the most toxic, and at the same time, widespread pollutants of our time are petroleum products [3-5]. Sources of their inflows into natural ecosystems are multiple and the composition of the resulting sewage is also different. In addition to the petroleum products themselves, drainages in most cases contain suspended substances, synthetic surfactants, dyes, etc. These components lead to the stabilization of oil emulsions, and, as a result, make water treatment more difficult [6, 7]. Therefore, the pre-cleaning step is necessary in many waste treatment plants.

Table 1. Groundwater and surface water abstraction data

Country	Groundwater abstraction			Surface water abstraction		
	2005	2010	2015	2005	2010	2015
Belgium	636	612	632	5753	5341	no data
Bulgaria	597	557	558	5439	5403	5071
Czech Republic	385	377	366	1564	1573	1237
Estonia	274	296	199	1304	1546	1525
Greece	3772	5615	5611	5882	4319	4297
Spain	6387	6601	6304	31643	29009	26613
Poland	2633	2722	2608	8889	8923	8486
United Kingdom	2336	2152	2053	7988	6111	5232

Coagulation is one of the most effective and easy in using methods for extracting suspended substances from aqueous systems [8-11].

Coagulation cleaning can be divided into two main stages:

- Coagulation itself, i.e. the process of destabilizing the colloid system, achieved by the use of coagulant substances having a short duration in time;
- Flocculation this is inducing the approach of destabilized particles to form agglomerates which are capable to precipitate itself under the influence of gravity, this stage can last more than half an hour [12, 13].

Depending on the size of particles dispersed in water, the time of their precipitation may vary from 0.1. s (in the case of coarse systems) to decades (fine systems), or the absence of precipitation at all. Due to the very small size, the only way to precipitate, and, subsequently, isolate such particles, is to cause their aggregation [14, 15]. However, the spontaneous occurrence of such a process is impossible due to the action of repulsive forces of like charges of particles, which determines the stability of the system to precipitation [13, 17, 18].

The coagulant substances added to the water provokes a decrease in the surface charge, after which the particles can become close and Van der Waals forces start acting (figure 1) [19].

Most of all aluminum and iron salts are used as coagulants, the hydrolysis of which is accompanied by the formation of large flakes [19–21].

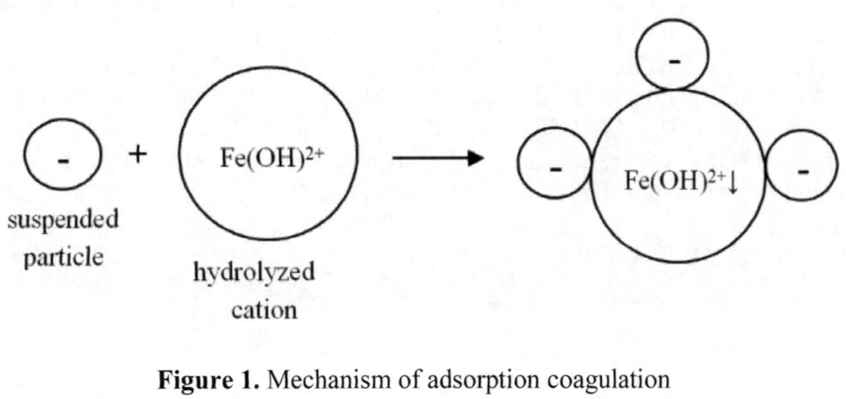

Figure 1. Mechanism of adsorption coagulation

IOP Conf. Series: Materials Science and Engineering **552** (2019) 012026 doi:10.1088/1757-899X/552/1/012026

Various parts of plants can be used as coagulants. There are known works on the application of Okra (Abelmoschus esculentus) [22], Moringa oleifera, Dolichos lab lab and Cicer arietinum [23].

In work [12], KlarAid CDP1326 coagulants (contains ferric sulfate), KlarAid PC1195 (1, 2 – ethanediamine) and KlarAid IC1176 (contains aluminum chloride) were used to purify sewage contaminated with oil products and suspended substances.

Thus, coagulation cleaning is an important stage of water conditioning and purification in the treatment of multi-component sewage, which permit to isolate a significant part of the dispersed pollutants.

Dust of electric furnace steelmaking is a large-tonnage waste containing in its composition a large number of iron compounds (figure 2) [24, 25].

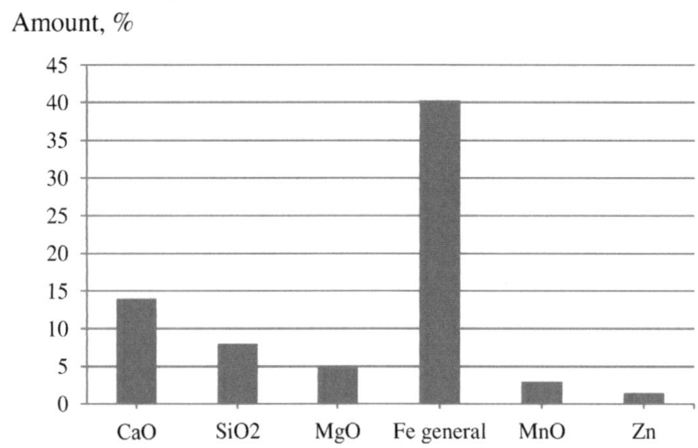

Figure 2. Element-oxide composition of EAF dust

According to the well-known scientific data [26], when treating EAF with hydrochloric acid, the following reactions will occur:

$FeO + 2HCl \rightarrow FeCl_2 + H_2O$;

$Fe_2O_3 + 6HCl \rightarrow 2FeCl_3 + 3H_2O$;

$Fe + 2HCl \rightarrow FeCl_2 + H_2$;

$Fe_3O_4 + 8HCl \rightarrow FeCl_2 + 2FeCl_3 + 4H_2O$

$CaSiO_4 + 2H_2O \rightarrow 2Ca\,(OH)_2 + H4SiO_4$

The formed ions $FeOH^+$, $FeOH^{2+}$, $[Fe\,(OH)_2]^+$ in this process provoke coagulation, which leads to the adhesion of particles in the system and their sedimentation [27].

Since a significant amount of calcium oxide is present in the form of compounds in the dust composition of EAF, then when the pH of the medium is increased due to their dissolution, the following ions can be formed: $[Fe(OH)_4]^-$, $[Fe(OH)_5]^{2-}$, $[Fe(OH)_6]^{3-}$ and others [21, 28].

Silicates present in the dust are able to dissolve in an acid medium with the formation of silicic acids, and then polysilicon, both linear and volumetric types [25]:

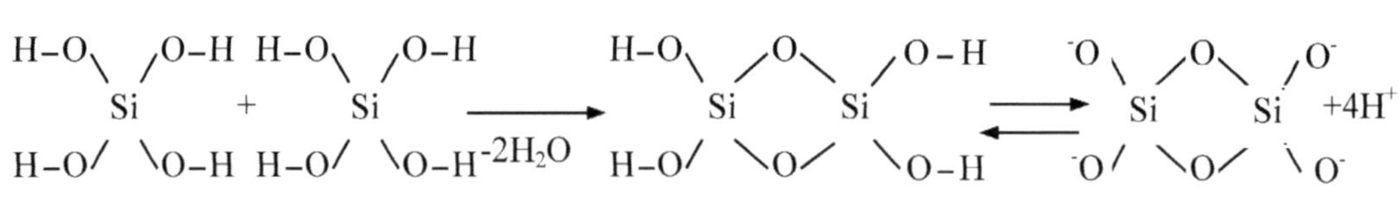

These compounds have the properties of flocculants, which intensifies water treatment.

It was previously established that the coagulant suspension (CS) is effective for the precipitation of suspensions due to the presence of natural clay particles, as well as organic emulsions formed from components of cow and soy milk [29, 30]. It was of scientific interest to find out the effectiveness of the release of suspended substances from model systems of mixed composition: oil product emulsions and inorganic suspended substances.

2. Materials and methods

The test involved a coagulating suspension produced on the basis of EAF dust treated with HCl (1N solution), and model wastewater. For making efficient coagulating suspension, according to the preliminary studies, the proportion of 6.5 ml HCl / 1 g of dust is necessary.

For making suspended particles in a modeled wastewaters kaolin clay was used.

To prepare the model emulsion, an industrial oil of general purpose I-20A (GOST 20799-88) [31] in the amount of 1 g / dm^3 and sodium lauryl sulfate (0.005 g / dm^3) was added to the tank with tap water to stabilize the water-oil system. The emulsion obtained after stirring on an automatic mixer for 24 hours was characterized by high stability and did not break down for several days.

The clarification of the emulsions was determined by the turbidity (NTU) using the HI 98703 Portable Turbidimeter (Hanna Instruments, USA)

3. Results and discussion

Some studies have been carried out to determine the applicability of the obtained CS for the purification of various aqueous media containing inorganic and organic pollutants.

Figure 3 shows the graphs of clay precipitation in model sewage without the addition and with the addition of CS (1 cm^3 / dm^3).

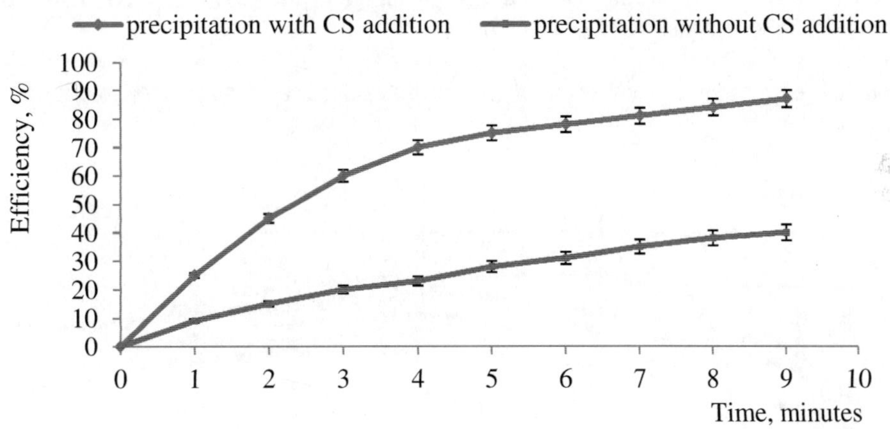

Figure 3. Sedimentation curves of precipitation of suspended substances in clay suspension

It is obvious that the addition of a CS to the water system intensifies the precipitation of suspended particles, which confirms the previously expressed assumptions.

Some researches were carried out to study the effect of the amount of CS added to the model systems on the efficiency of particle deposition (figure 4).

From the data presented in the last figures, it is seen that the addition of a suspension in amounts of 1 and 2 ml / dm^3 allows achieving high cleaning efficiency in the first 30 minutes of interaction; the greatest effect is achieved during 2 hours. Adding more amount of CS does not lead to an increase of results, on the contrary, efficiency decreases, especially it is seen at prolonged defecate. This is probably due to the fact that the coagulation limit for this system is in the range of 1.5-2 ml / dm^3. When the concentration of CS is increased in the system, the particles are recharged in the suspension, as a result of which the cleaning efficiency decreases.

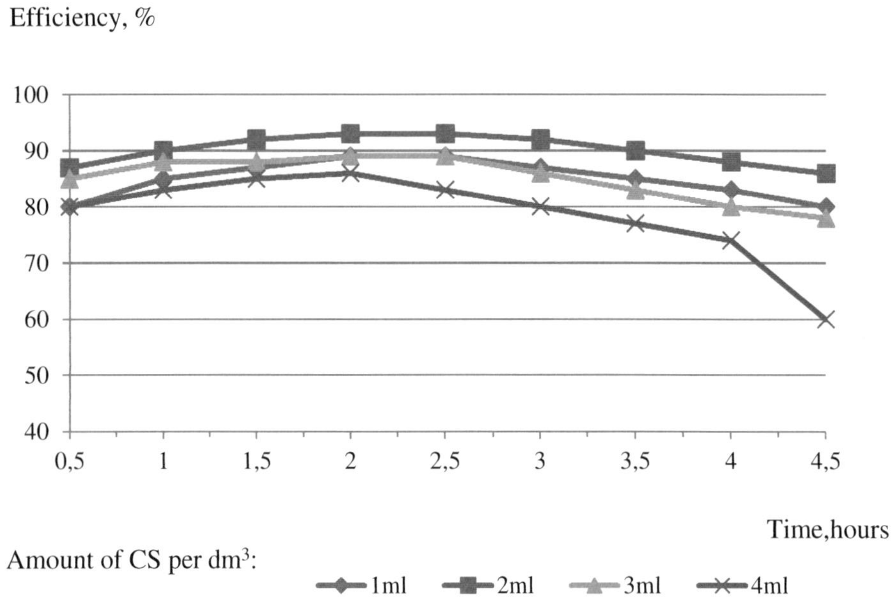

Figure 4. Influence of the amount of added CS and defecate time on the efficiency of cleaning model clay suspensions ($C_{initial}$ 1 g/ dm^3)

With an increase in the concentration of clay particles in the system, the coagulation limit decreases (figure 5).

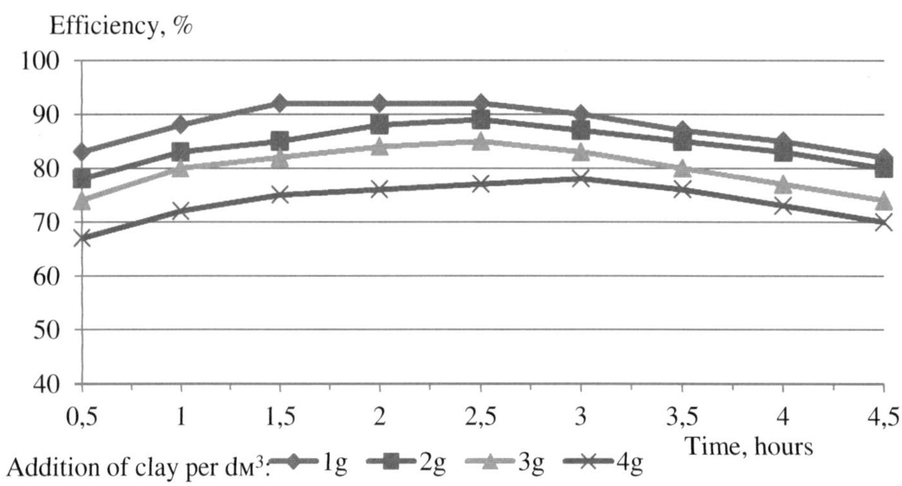

Figure 5. Influence of the efficiency of cleaning model suspensions on the initial clay concentration

Thus, it can be seen from figure 4 that the greatest efficiency of clarification of the emulsion at the initial clay concentration of 1 g/dm^3 is achieved by the addition of 1.5 ml of CS, while for the clay concentration of 3 g/dm^3 the required dose of CS is 2.5-3 ml/dm^3.

Thus, preliminary tests are necessary when using CS in industrial conditions to determine the optimum amount of coagulant added, based on the concentration of suspended substances in the sewage.

To clarify the effect of pH on the precipitation efficiency of suspended substances using CS, further studies were carried out.

The influence of the initial pH of the aqueous medium on the efficiency of clarification of model waters at a clay concentration of 1 g/dm^3 and the amount of the CS additive 0.5 g/dm^3 was investigated.

It has been established that precipitation occurs more effectively in neutral and alkaline environments than in acidic one (figure 6).

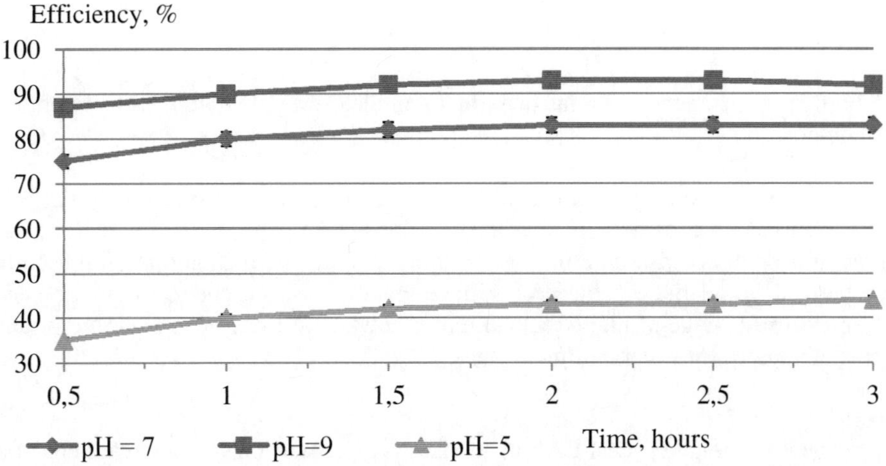

Figure 6. The effect of pH of the aquatic environment on the effectiveness of the cleaning of clay suspensions

The further researches were carried out to assess the effectiveness of clarification of multi-component sewage containing petroleum products. The emulsion with the initial content of industrial oil 1 g/dm^3 and suspended substances (clay) 1 g/dm^3 undergoes the contact with coagulating suspension in an amount of 2.5 ml/dm^3 for an hour and in this case stirring in the first 5 minutes was made for the better contact of the reagents, then the liquid was naturally defecated and the clarified layer of water was analyzed (figure 7).

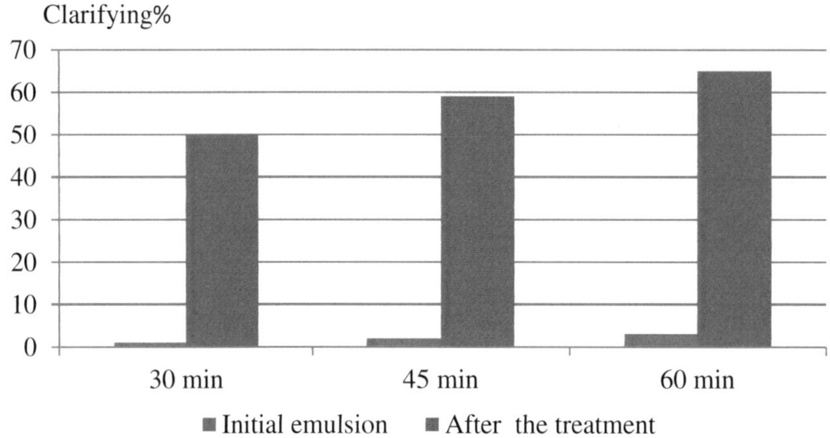

Figure 7. Efficiency of clarification of model sewage of multicomponent composition

From the results of the experiment it is evident that the coagulating suspension has an effective effect on waters of mixed composition. The stability of the system is partially violated and the effect of emulsion alteration is observed. Some of the contaminants are sedimentations on the bottom of the vessel, some part of the oil rises to the near-surface layer. Thus, the coagulating suspension can be used in the pre-treatment stage of highly contaminated sewage of mixed composition to facilitate and reduce the cost of subsequent water treatment steps.

4. Conclusion

Coagulating suspension based on dust EAF is an effective preparation for the treatment of sewage of various compositions. High efficiency is achieved during clarifying of model waters made on the basis of kaolin clay, so the efficiency in 93% is achieved by adding a suspension in the amount of 2 ml / dm^3 to model waters contaminated with clay particles at a concentration of 1 g / dm^3. When processing complex systems containing industrial oil stabilized with sodium lauryl sulfate and clay particles, the efficiency of clarification reaches 65%. Thus, the coagulating suspension can be used in the pre-treatment stage of highly contaminated sewage of mixed composition to facilitate and reduce the cost of subsequent water treatment steps.

Acknowledgement
The work is realized in the framework of the Program of flagship university development on the base of the Belgorod State Technological University named after V.G. Shoukhov, using equipment of High Technology Center at BSTU named after V.G. Shoukhov.

References

[1] Rogozhina N G 1999 The threat of a global environmental crisis and national interests of developing countries. *Proc. Int. Conf. environment and development in the Arab world* (Moscow) pp 7-15
[2] Energy, transport and environment indicators 2017 *Statistical books* (Luxembourg: European Union) *Print: DOI:10.2785/889945* p 244
[3] Shaikhiev I G, Fazullin D D and Mavrin G V 2017 Modified PTFE-PANI membranes for the recovery of oil products from aqueous oil emulsions *Petroleum chemistry* vol 57 **2** pp 165–71
[4] Annunciado T R, Sydenstricker T H D and Amico S C 2005 Experimental investigation of

various vegetable fibers as sorbent materials for oil spills *Marine Pollution Bulletin* vol 50 pp 1340–6
[5] Balandina A G, Khangildin R I, Ibragimov I G and Martyasheva V A 2015 Analysis of the impact of enterprises of the petrochemical complex on the hydrosphere and ways to minimize their negative influence *Bashkirsky chemical J.* vol 22(1) pp 115–26
[6] Kuzubova L I and Morozov S V 1992 Purification of oily sewage: Analyt. review (Novosibirsk: SB RAS. SPSTL, NIOH) p 72
[7] Anapolsky V N, Oliferuk S V and Romanenko A P 2011 Purification of oily sewage *Plumbing, heating, air conditioning* **1(109)** pp 28–31
[8] Sahu O P and Chaudhari P K 2013 Review on chemical treatment of industrial waste water *J. Appl. Sci. Environ. Manage* vol 17(2) pp 241–57
[9] Postolachi L, Rusu V, Lupascu T and Maftuleac A 2015 Improvement of coagulation process for the prut river water treatment using aluminum sulphate *Moldova J. General, Industrial and Ecological Chemistry* vol 10(1) pp 25-32
[10] Ebeling J M 2004 Application of chemical coagulation aids for the removal of suspended solids (TSS) and phosphorus from the microscreen effluent discharge of an intensive recirculating aquaculture system *North American journal of aquaculture* vol 66 (3) pp 198–207
[11] Mousa K M, Hadi H J and Mousa K M 2016 Coagulation flocculation process for produced water treatment *International journal of current engineering and technology* vol 6(2) pp 2347–5161.
[12] Khalid M M and Hind J H 2016 Coagulation Flocculation process for produced water treatment *International Journal of Current Eng. and Tech* vol. 6(2) (*E-ISSN 2277–4106, P-ISSN 2347–5161*)
[13] Tzoupanos N D and Zouboulis A I 2008 Coagulation-flocculation processes in water wastewater treatment: the application of new generation of chemical reagents 6th IASME *WSEAS Inter. Conf. on heat transfer, thermal engineering and environment* (Greece) pp 309–17
[14] Starostina I V, Antipova A N, Ovcharova I V and Starostina Yu L 2018 Porous materials based on foaming solutions obtained from industrial waste *IOP Conference Series: Materials Science and Engineering* **327(3)** 032052
[15] Starostina I V, Stolyarov D V, Anichina Y N and Porozhnyuk E V 2018 The carbonaceous sorbent based on the secondary silica-containing material from oil extraction industry *IOP Conference Series: Earth and Environmental Science* **107(1)** 012075
[16] Nyklema J 1995 Fundamentals of interface and colloid science *Solid – Liquid interfaces* vol 2 p 384–558
[17] Ahmad H A 2016 Quantitative comparison between chemical coagulation and biological treatment of municipal wastewater *International journal of applied engineering research* vol 11(18) pp 9424 – 29
[18] Borchate S S, Kulkarni G S and Kore V S 2014 A Review on Applications of Coagulation-Flocculation and Ballast Flocculation for Water and Wastewater *International Journal of Innovations in Engineering and Technology (IJIET)* vol 4(4) pp 216-23
[19] Ebeling J, Ogden S, Sibrell P and Rishel K 2004 Application of chemical coagulation aids for the removal of suspended solids (TSS) and Phosphorus from the Microscreen Effluent Discharge of an Intensive Recirculating Aquaculture System *North American Journal of Aquaculture* vol 66 pp198–207
[20] Sahu O P and Chaudhari P K 2013 Review on Chemical treatment of Industrial Waste Water *J. Appl. Sci. Environ. Manage* vol 17(2) pp 241–57
[21] Friedrichsberg D A 1984 Course of colloid chemistry (Leningrad: Chemistry) p 368
[22] RAJI Yu O, Abubakar L, Giwa S O and Giwa A 2015 Assessment of Coagulation Efficiency of Okra SeedExtract for Surface Water Treatment *International Journal of Scientific & Engineering Research* vol 6(2) pp 719–25

[23] Jisha T J and Chinnamma M A 2017 Effect of Natural Coagulants on the Treatment of Automobile Service Station Waste Water *International Journal of Engineering Development and Research* vol 5(2) pp 358–64

[24] Sukhanov E V 2015 Physical and chemical properties of the dust of the electrical steel works (ESW) / Fundamental and applied research in the field of chemistry and ecology: materials of the Int. Scien. and Prac. Conf. of Students, Postgraduates and Young Scientists (Kursk: Yugo-Zap. state University, CJC «University Book») pp 154–6

[25] Sverguzova S V 2008 Complex sewage neutralization, reclamation of sewage sludge and secondary use of gypsum and metal containing industrial wastes Dissertation for the degree of Doctor of Technical Sciences (Kazan) p 514

[26] Svergusova S V, Starostina I V, Sukhanov E V and Sapronov D V 2015 EASSF dust-based coagulant *Bulletin of BSTU named after V G Shoukhov* **10** pp 202–5

[27] Sukhanov E V 2016 Colloid-chemical aspects of obtaining iron-containing coagulant-flocculant based on dust of electric steelmaking production Thesis for the degree of candidate of technical sciences (Belgorod) p 160

[28] Adam N K, Tolstoy D M and Akhmatova A S 1947 Physics and chemistry of surfaces (Moscow: Gostekhizdat) p 536

[29] Sverguzova S V, Sapronova Zh A, Porozhnyuk L A and Svyatchenko A V 2016 Production of ironcontaining coagulant for wastewater purification from organic suspensions *Bulletin of Samara Scientific Center of RAS* vol 18 **5(2)** pp 362–6

[30] Svergusova S V and Svtatchenko A V 2018 Electrokinetic phenomena accompanying treatment of soy milk production wastewater by agent based on electric arc furnace dust *IOP Conf. Series: Materials Science and Engineering* vol 327 042096

[31] GOST 20799-88 2005 Masla industrial'nyye. Tekhnicheskiye kharakteristiki [Industrial oils. Specifications] (Moscow: Standardinform) p 7 [In Russian]

Parametric optimization of an architectural object's form as a method to improve its energy efficiency

N A Vasilenko

Department of Architectural Designs, Belgorod State Technological University of V.G. Shukhov, Kostyukov St., 46, Belgorod, 308012, Russia

E-mail: vasilenko.na@bstu.ru

Abstract. Parametric optimization of an architectural object's form is one of the design tasks, which architects and project designers face at the stage of structural concept development and design. Improving the performance qualities of a design solution, while providing its energy efficiency, is the purpose of system research in the sphere of architecture and urban planning. The article gives the comparison of calculated compactness ratios of the external envelope for one-, two- and three-, four-, five-, six-, seven-storey 6-apartment compact-form residential houses, comparable in the floor surface area criterion. The application of BEM software packages on the basis of BIM-modeling of buildings is reasonable to substantiate with the optimal functional (consumer) properties of an architectural object. Systematic methodology allows considering the architectural and design problems in the interdisciplinary way. Computer modeling of architectural objects should solve the optimization tasks on the basis of version design and contain an optimization module, which would allow taking the optimal decision at the specified design constraints (conditions) and normative parameters. Computer modeling of architectural objects with the use of structural optimization methods for the version design is a promising area, which requires the formalization of project procedures and developing automated design system tools for designing buildings of various purposes.

Key words. Parametric optimization of an architectural object's form, form making, energy efficiency, compactness principle, compactness criterion of a building form, modeling methods, computer modeling of architectural objects, optimization problem, optimization module, version design, transit area method.

1. Introduction

Optimization of space-planning solutions of a building is one of the main energy-saving measures at the stage of structural concept development and design. The theoretical insights in the form-making of energy-efficient residential buildings, passive houses and the possibilities of their implementation are considered in research works [1–8]. The questions of selecting an optimal space-planning solution for multi-storey residential buildings are dealt with in works [9, 10]. In modern science the research works, dealing with BIM-technologies are carried out in a number of areas [11–15]. The special attention is paid to using Building Energy Modeling (BEM) on the basis of Building Information Modeling (BIM) of buildings and constructions [16, 17]. The analysis of experience of designing energy-efficient buildings and applying the space-planning methods of their form-making with account of using non-conventional energy sources is considered in [18–20]. The formation of environmentally-friendly and energy-saving low-rise housing development at the territory of Russia is considered in research works [21–24]. To systematic methods of architectural solutions' quality assessment a number of works [25–27] are devoted. The theoretical questions of decision-making and optimization are considered in works [28–30]; the methodology of system synthesis of decision-making techniques is dealt with in [31].

2. Materials and Methods

The object of this research is the integral architectural objects as systems (subsystems) of various hierarchy levels and functional complexity through the example of multi-dwelling medium-rise residential buildings. The subject of research is the optimization procedures of space-planning solutions and the geometrical parameters of architectural objects' forms, aimed at improving their energy efficiency.

The methodology of research is based on methodological provisions of general systems theory in relation to architectural objects as the integral systems (subsystems) of a certain hierarchical level, as well as on the version design and the comparative analysis of space-planning solutions and engineering-and-technical solutions of buildings through the example of multi-dwelling medium-rise residential buildings.

The purpose of research is to determine and substantiate the essence of parametrical optimization of an architectural object's form as a component of a system procedure of optimizing space-planning solutions of architectural objects, aimed at improving their energy efficiency (through the example of multi-dwelling medium-rise residential buildings).

3. The theoretical part

In this research work the energy efficiency of a building is understood as providing measures for saving energy for building's climatisation. In 2011–2012, due to the demand for green construction and buildings and constructions certification in Russia, a number of standards, including the principal provisions of rating systems LEED, BREEAM, DGNB, HQE and normative requirements of the Russian GOSTs and SP [32–33], have been developed.

The selection of the shape of building, its number of storeys and configuration depend on a system of internal and external factors [34], which influence the form-making and functioning of architectural objects [25]:

– group of factors of the «Population» subsystem, which reflects demographical, social and economical aspects of environmental conditions;

– group of factors of the «Artificial environment» subsystem, which includes the scientific and technical aspects and production and consumption sphere;

– group of factors of the «Natural environment» subsystem, which reflects natural and climatic conditions, topographical and geological features of terrain, the state and potential of natural resources.

The development of computer engineering means has resulted in the increase of decision-making speed in most cases due to automation of algorithmic operations [31]. Determining an optimal design solution consists of the following basic procedures:

1) modeling (building a system's model);
2) determining the efficiency criteria of solutions;
3) optimization (comparing the efficiency of acceptable solutions);
4) selection of the optimal solution from comparable variants.

The modeling functions are [28–30]:

– descriptive function;
– forecasting function;
– normative function.

In the present-day computational design of buildings and city-planning objects we can single out two basic tendencies: parametric (generative) and kinetic (robotic). Figure 1 presents a scheme, which shows the main approaches in modern computational design of buildings and urban-planning objects.

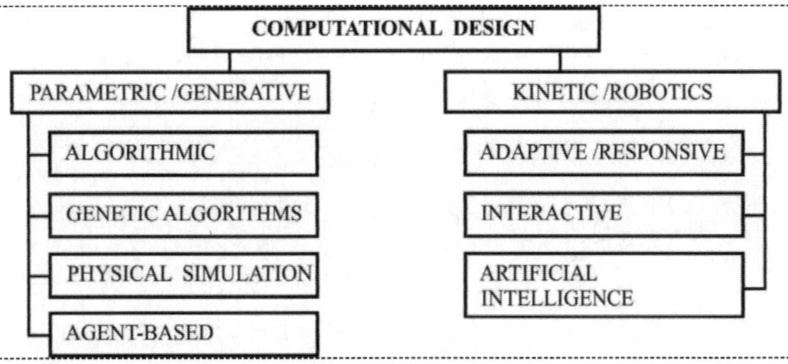

Figure 1. The main approaches in present-day computational design of buildings and city-planning objects.

Parametric optimization of a building's form is a multicriteria problem, which should meet, first of all, functional requirements. Applying the general compactness (energy minimization) principle to architectural objects of any hierarchy level reflects the striving to minimize various system communication (linking) values. Achieving the compactness of the design solution in each specific case is possible only at achieving the conditions [25]:

$$P_i \to Pn_i;\ Pn_i \to const;\ C_j \to \min \tag{1}$$

where P_i – actual (design target) values of the i–th type of works of the first group (normalizable processes);

Pn_i – normative parameters of the i–th type of works of the first group (normalizable processes);

C_j – parameters of the j–th type of works of the second (communicational) group.

So, the closer are the target values of the processes P_i to the norm and the lower are expenditures for communications, the more compact is the solution, i.e., the more economically efficient are the costs for power supply. In figure 2, 3 the diagrams of normalizable processes (P) and communicational processes (C) alteration in architectural systems are presented [35]. Optimization of functional and planning solutions through the example of residential housing architecture is reflected in transit area method, described in works [26].

Among the space-planning solutions, which influence the energy efficiency of buildings, we can single out the following:

– locating on the lot (with account of the land topography and nearby buildings);
– building orientation with respect to the cardinal directions;
– number of storeys;
– form;
– floor plans configuration;
– storey height, determined with sanitary-hygienic and functional requirements;
– the presence of through openings between different parts of buildings;
– unevenness of the building's facades with bay-windows, balconies, protruding or sinking articulations (the number of exterior angles, which reduce the heat-saving properties of the outer envelope structures);
– the area of glassed façade surfaces with account of illumination norms;
– correlation of depth and width of rooms.

The thermal efficiency of buildings is considerably influenced by their facades' roughness. Depending on the degree of the façade's unevenness with bay-windows, balconies, protruding of sinking

IOP Conf. Series: Materials Science and Engineering **552** (2019) 012027 doi:10.1088/1757-899X/552/1/012027

partitions, the number of exterior angles, which reduce the heat-saving properties of the outer envelope structures, the thermal efficiency of a building can be reduced by 12–15 % [1].

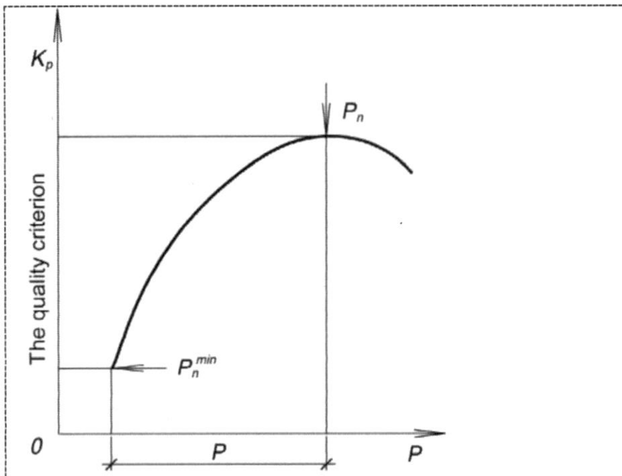

Figure 2. Diagram of normalizable processes alteration in architectural systems: K_p – quality criterion of normalized processes; *Pn* – normative parameter of normalizable processes; $Pn^{\min}$ – minimal normative parameter of normalizable processes; *P* – normalizable processes in architectural systems.

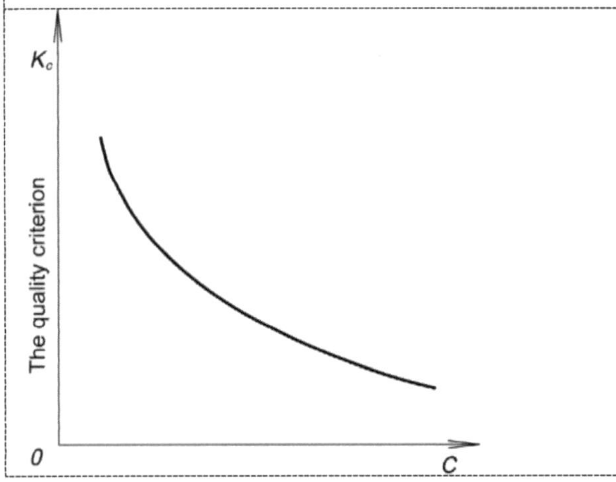

Figure 3. Diagram of non-normalizable (communicational) processes alteration in architectural systems: K_c – quality criterion of communicational processes; *C* – communicational (non-normalizable) processes in architectural systems.

The calculated building compactness ratio is the energy-efficiency criterion of the building's form [1, 9]:

$$k_e^{des} = A_e^{sum} / V_n \tag{2}$$

A_e^{sum} – the total area of all the internal surfaces of exterior walls and attic floorings, including the ceiling of the upper storey and flooring of the lower heated space, m^2;

V_n –the heated volume of the building, confined with the internal surface of the building envelope, m^3.

For low-rise buildings the ratio of enclosing structures area to the volume increases in comparison with multistorey buildings, so, the compactness ratio increases with the decrease of the number of storeys. The building of the most compact shape has the highest thermal-efficiency (energy-efficiency) properties.

According to the research [26] the most compact solution for one-room apartments at the centrical grouping is a 6-apartment section. Through the example of planning solutions of a 6-apartment medium-rise residential building (up to 7 storeys) of various plan form let us determine a calculated compactness ratio for several variants. In figure 4 a planning solution for a residential section of a 6-apartment medium-rise meridional-oriented residential building (variant № 1) is presented. Such scheme of planning solution assumes the blocking of residential sections in the narrow sides. The floor area by the inner surface of the building's outer walls is 285.33 m^2.

To increase the compactness of form of planning solutions and their better comparability let us reduce the unevenness of facades and the number of exterior angles. Let us reduce the compared planning solutions to comparability upon the criterion of area, limited by the inner surface of outer enclosing structures of the building. In figures 5–8 there are schemes of planning solutions for a residential section of a 6-apartment medium-rise tower residential house (variants № 2–5). The compared planning solutions have a corridor system of blocking one-room apartments. For comparison in planning solutions schemes let us take the general dimensions by the internal perimeter of outer walls. The presented variants № 2–5 of the planning solutions suppose the arrangement of sections as stand-alone, without blocking.

The recommended room height with account of the residual air pollution zone in its upper part according to the research [21] makes up over 3.0 m from floor to ceiling. In accordance with this criterion, in our calculations we take the indoor height from floor to ceiling 3.0 m, floor thickness 0.3 m. Let us determine for each of the five variants the calculated compactness ratios k_e^{des} for one-, two- three-, four-, five-, six-, seven-storey residential buildings (table 1).

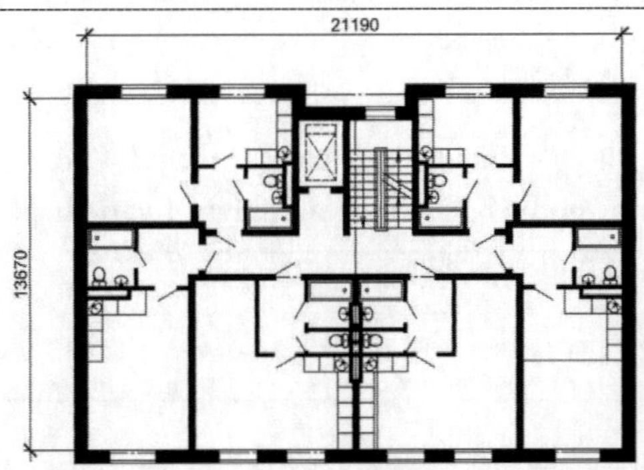

Figure 4. Scheme of the planning solution for a residential section of a 6-apartment medium-rise meridional-oriented residential building (variant № 1).

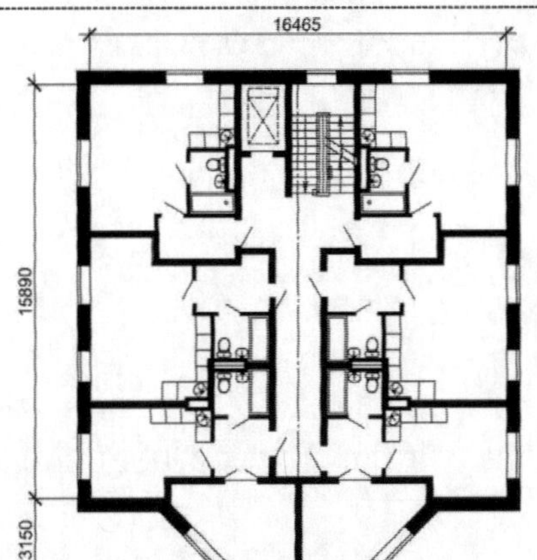

Figure 5. Scheme of the planning solution for a 6-apartment medium-rise tower residential house (variant № 2).

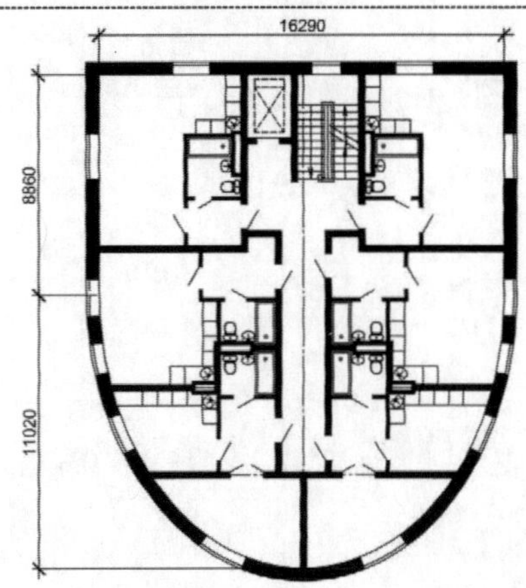

Figure 6. Scheme of the planning solution for a 6-apartment medium-rise tower residential house (variant № 3).

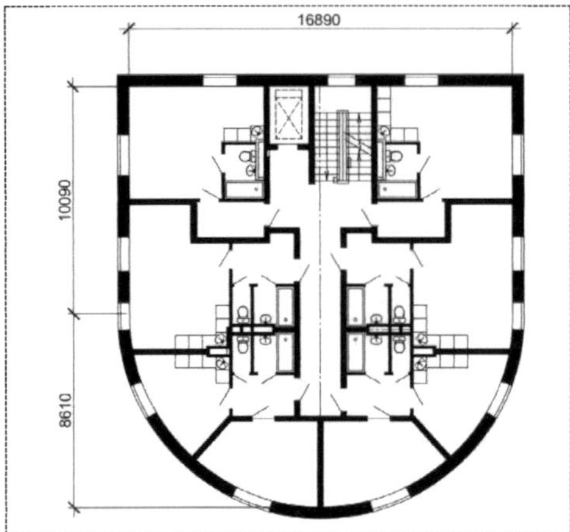

Figure 7. Scheme of the planning solution for a 6-apartment medium-rise tower residential house (variant № 4).

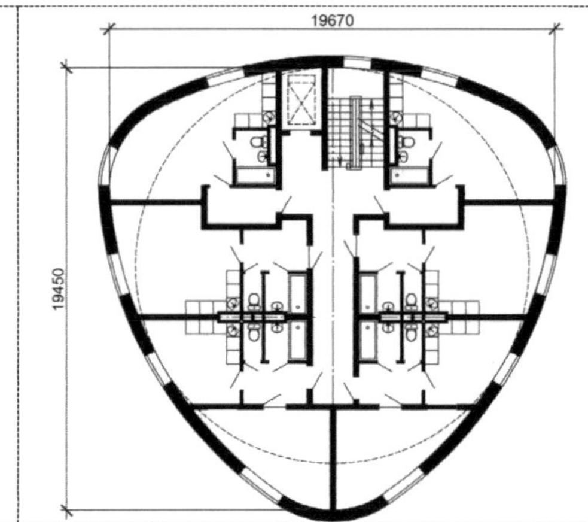

Figure 8. Scheme of the planning solution for a 6-apartment medium-rise tower residential house (variant № 5).

Table 1. Comparison of the calculated compactness ratios k_e^{des} for one-, two- three-, four-, five-, six-, seven-storey residential buildings of various plan form.

Plan form (m)	The area of one storey (m²)	A_e^{sum} for one storey (m²)	The calculated compactness ratio k_e^{des} (m⁻¹) for a building with						
			1 storey	2 storeys	3 storeys	4 storeys	5 storeys	6 storeys	7 storeys
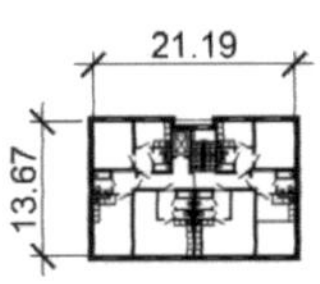	285.33	785.03	0.917	0.568	0.459	0.405	0.374	0.353	0.338
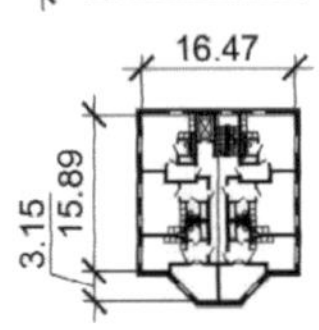	285.33	772.63	0.903	0.553	0.444	0.391	0.359	0.339	0.324
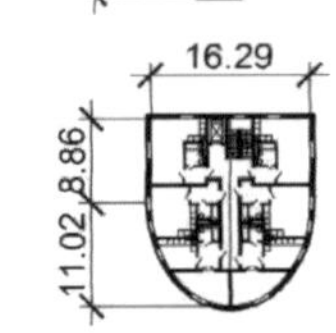	285.33	763.51	0.892	0.543	0.434	0.380	0.349	0.328	0.313
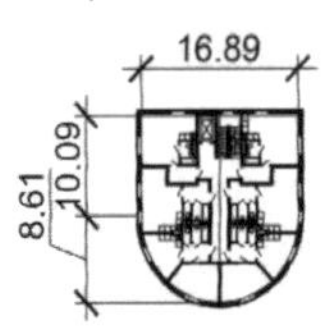	285.33	762.48	0.891	0.542	0.432	0.379	0.348	0.327	0.312
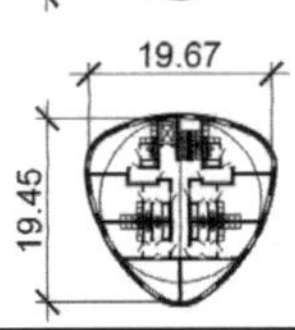	285.33	754.76	0.882	0.533	0.423	0.370	0.339	0.318	0.303

4. Results

The comparative analysis of the calculated compactness ratios k_e^{des} for one-, two- three-, four-, five-, six-, seven-storey multi-dwelling residential buildings of the centrical grouping and various plan forms allows us making the following conclusions. According to the calculation, the most energy-efficient is the streamlined plan form (variant № 5), presented in figure 8. At the increase of the plan form compactness, the «opening» of the apartments to the heated external perimeter of walls increases as well, the rooms become longer and less deep.

In the research work [9] it is described that at the increase of compactness ratio of the form by 0.01 energy consumption increases by 2.6%. Compactness ratio of the form k_e^{des} for one-storey building from variant №5 as compared to variant № 2 decreases by 0.01 m^{-1}, which corresponds to the decrease of expenditures for heating by 2.6%; for two-, three-, four-, five- and six- and seven-storey buildings of the comparable variants (variant № 5 and № 2) k_e^{des} decreases by 0.02 m^{-1}, which corresponds to the decrease of expenditures for heating by 5.2 %. The calculated compactness ratios of the form k_e^{des} for one-, two-, three-, four-, five-, six- and seven-storey buildings from variant № 5 as compared to variant № 1 decrease in each case by 0.035 m^{-1}, which corresponds to the decrease of expenditures for heating by 9.1%.

Parametric optimization of an architectural object's form is determined not only by achieving the most compact shape of object, floor plans, minimization of facades unevenness etc. This procedure is complex, multilayered and requires taking into account the functional (consumer) properties of the accepted design-planning solutions. For example, at achieving the most compact round or spherical plan form, one could miss the other important functional-planning requirements in the designed object.

The up-to-date methods of computer modeling allow, on the basis of optimization task setting, developing an optimization module, based on the system principle of compactness (minimization of energy), which would at the same time take into account the functional and planning aspects of the designed object.

5. Discussion

Parametric optimization of buildings of any functional purposes as means of improving their energy efficiency requires system approach with the use of computer modeling methods. The object of optimization is searching the optimal space-planning and functional solution of an architectural structure. To find the optimal solution it is necessary to determine the efficiency criterion of architectural objects as integral systems (subsystems) of a certain hierarchical level.

6. Conclusion

Parametric optimization of buildings of various functional purposes with the subsequent development of norms and standards for making design solutions and their large-scale implementation in housing construction is a priority interdisciplinary task. Through the example of one-, two- and three-, four-, five-, six-, seven-storey 6-apartment residential buildings of the compact plan form, comparable in the floor area criterion, the comparison of calculated compactness ratios of their external envelopes has been performed. The further research in optimization of buildings' form is reasonable to carry out with account of the object's functional properties optimization and the economic feasibility of architectural and structural solutions. So, the parametric optimization of a building's form is a task of optimizing the quantitative and qualitative characteristics of an architectural solution at the stage of structural concept development and design. Development of algorithms in computer modeling is possible at setting an optimization task and determining an efficiency criterion of solutions. Optimization becomes possible at the comparison (commensuration) of the efficiency of acceptable solutions.

Acknowledgments
The article was prepared within development program of the Flagship Regional University on the basis of Belgorod State Technological University named after V.G. Shukhov.

References

[1] Belyaev V S, Granik Yu G and Matrosov Yu A 2016 *Energoeffektivnost' i teplozazhita zdanyi* [The energy efficiency and heat insulation of buildings] (Moscow: ASV) p 396 [In Russian]
[2] Hossain F 2018 *Sustainable Design and Build* (Butterworth-Heinemann Publisher) p 432
[3] Smirnova S N 2009 *Teoreticheskaya model' energoeffektivnogo zhilogo zdaniya* [A theoretical model of an energy efficient residential building] *Privolzhsky scientific journal* (Nizhny Novgorod: NNGASU) **2** p 86–91
[4] Kubba S 2017 Handbook of Green Building Design and Construction: *LEED, BREEAM, and Green Globes* (Butterworth-Heinemann Publisher) p 1064
[5] Givoni B 1992 Comfort Climate Analysis and Building Design Guidelines Comfort, climate analysis and building design guidelines *Energy and Buildings* **18** pp 11–23
[6] Clarke J A 2001 *Energy Simulation in Building Design, 2nd edition* (Butterworth-Heinemann, Oxford) p 362
[7] Attia S, Gratia E, De Herde A and Hensen J L M 2012 Simulation-based decision support tool for early stages of zero-energy building design *Energy and Building* **49** pp 2–15.
[8] Attia S, De Herde A, Gratia E and Hensen J L M 2013 *Achieving informed decision-making for net zero energy buildings design using building performance simulation tools* (Springer International Publishing AG:Building Simulation) vol 6(1) pp 3–21
[9] Li Ruixin and Bantserova O L 2015 *Issledovanie voprosov energoeffektivnosti mnogoetazhnyh zhilyh zdanyi v Kitae* [Research on energy efficiency of multi-storey residential buildings in China] (Promyshlennoe i grazhdanskoe stroitel'stvo) **11** pp 52–56 [In Russian]
[10] Kologrivova L B and Molodkin S A 2006 Kompleks energosberegauzhih reshenyi pri proektirovanii mnogoetazhnyh zhilyh zdanyi [A set of power-saving solutions at designing multi-storey residential buildings] *Promyshlennoe i grazhdanskoe stroitel'stvo* **10** pp 51–3 [In Russian]
[11] Volkov A Chelyshkov P and Khromenkova A 2017 BIM cost analysis of transport infrastructure projects *IOP Conference Series: Managemant Earth and Environment Science 19 Ser Energy Management of Municipal Transportation Facilities and Transport EMMFT* 012203
[12] Volkov A A and Kuzina O N 2018 Funkcional'no-komplementarnaya teoriya organizatzii upravleniya ekspluatatziey kompleksa zdanyi i infrastruktury [Functional-Complementary Theory of Organization of Operation Management for Complex of Buildings and Infrastructure] *Promyshlennoe i grazhdanskoe stroitel'stvo* **9** pp 61–67 [In Russian]
[13] Vasilkin A A 2018 K integrirovaniu instrumentov strukturnoy optimizatzyi v SAPR [To Intagration of Tools of Structural Organization Into CAD] *Promyshlennoe i grazhdanskoe stroitel'stvo* **9** pp 52–60 [In Russian]
[14] Chelyshkov P D, Grossman Ya E and Khromenkova A A 2018 Podhody k analizu informatzionnyh modelei zdanyi i kompleksov [Approaches to the Analysis of Information Models of Buildings and Complexes] *Promyshlennoe i grazhdanskoe stroitel'stvo* **2** pp 72–5 [In Russian]
[15] Danilina N V 2018 Primenenie BIM-tekhnologiy na stadyi gradostroitel'nogo proektirovaniya [Application of BIM-Technologies at the Stage of Urban Planning Design] *Promyshlennoe i grazhdanskoe stroitel'stvo* **9** pp 48–51 [In Russian]
[16] Chelyshkov P D, Grossman Ya E and Khromenkova A A 2017 Analiz programmnyh kompleksov dlya energomodelirovaniya [Analysis of Software Products for Energy Simulation] *Promyshlennoe i grazhdanskoe stroitel'stvo* **7** pp 79–84 [In Russian]

[17] Garg V, Mathur J, Tetali S and Bhatia A 2017 *Building Energy Simulation: A Workbook Using Design Builder*™ (Boca Ration: CRC Press. Taylor and Francis Group) p 464

[18] McArthur J, Herrera N and Mantha P 2014 International Sustainability Systems Comparison – Key International Sustainability Systems: Energy and Water Conservation Requirements (Ove Arup & Partners Ltd) p 65

[19] Shamseldin A K M 2018 Unifying some variables' effects among different environmental assessment methods of buildings *HBRC Journal* **14** p 93–103
Dirlich S 2011 A Comparison of Assessment and Certification Schemes for Sustainable Building and Suggestions for an International Standard System *The IMRE Journal (Germany: TU Bergakademie Freiberg)* **5(1)** URL http://tu-freiberg.de/sites/default/files/media/imre-2221/IMREJOURNAL/2-dirlich.pdf

[20] Petrova Z K 2016 Organizatziya maloetazhnoyi zhiloyi zastroyki v sisteme rasseleniya Rossyi [The organization of low-rise residential areas in the settlement system of Russia]: doctoral thesis: 05.23.22: defended 26.04.2016. (Moscow: Moskovskyi arkhitekturnyi institut) p 345 URL http://www.marhi.ru/sciense/author/petrova/1_petrova_diss.pdf

[21] Tabunshikov Yu A 2014 *Energoeffektivnye poselki i zhylye raiony* [The Energy Efficient Settlements and Residential Areas] *Sustainable building technologies. Zvt.avoc.ru.* pp 30–39 [In Russian]

[22] Sayanov A A, Golubeva E I and Korol T O 2018 *Kottedzhnye poselki kak novyi element sistemy rasseleniya v Rossyi* [Cottage Estates as a New Element of Settlement System in Russia] (Moscow: Lenand) p 208 [In Russian]

[23] Lavrik G I 2007 *Metodologicheskie osnovy raionnoy planirovki: Vvedenie v demoekologiu* [The Methodological Basis of Settlement Planning. Introduction to Demoecology] (Belgorod: BSTU) p 117 [In Russian]

[24] Zagorodnyuk L H, Koryakina A A, Sevostyanova K I and Khaheleva A A 2018 Heat Insulating Composite Mixtures with Technogenic Materials *J. Phys.: Conf. Ser.* **1066** 012011

[25] Lavrik G I 2007 *Metody otzenki kachestva zhylizha: Issledovanie. Proektirovanie. Ekspertiza* [Methods for Assessing the Quality of Housing. Research, Design, Expertise] (Belgorod: BGTU) p 100 [In Russian]

[26] Zaharov Y I and Derevyanko V V 2014 *Metody analiza architekturno-kontzeptual'nyh reshenyi* [Methods of Architectural and Conceptual Solutions Analysis] *Vestnik PDABA* **9** pp 54–58

[27] Larichev O I 1979 *Nauka i iskusstvo prinyatiya reshenyi* [Science and the Art of Decision Making] (Moscow: Nauka) p 200 [In Russian]

[28] Peregudov F I and Tarasenko F P 1989 *Vvedenie v sistemnyi analiz* [Introduction to System Analysis] (Moscow: Vysshaya Shkola) p 367 [In Russian]

[29] Rykov A S 2005 *Modeli i metody sistemnogo analiza: prinyatie reshenyi i optimizatziya* [The Models and Methods of System Analysis: Decision - Making and Optimization] (Moscow: MISiS) p 367 [In Russian]

[30] Oleynikov D P, Butenko L N and Oleynikov S P 2018 *Sozdanie metodologii systemnogo sinteza metidov prinyatiya resheniy* [The Creation of a Methodology for Systematic Synthesis Methods in Decision - Making] (Moscow: Lenand) p 112 [In Russian]

[31] STO NOSTROY 2.35.4-2011. Zelenoe stroitel'stvo. Zdaniya zhulye i obzhestvennye. Reytingovaya systema otzenki ustoychivosty sredy obitaniya [STO NOSTROY 2.35.4-2011. Green building. Residential and Public Buildings. Rating System of the Sustainability of the Environment] (NP AVOK, OAO CNIIPromzdanyi, OAO NPO TERMEK) p 57 [In Russian]

[32] STO NOSTROY 2.35.68-2012. Zelenoe stroitel'stvo. Zdaniya zhulye i obzhestvennye [Green building. Residential and Public Buildings. Taking into account regional features in the rating system for assessing the sustainability of the habitat] (NP AVOK, OAO CNIIPromzdanyi, OAO NPO TERMEK) p 45 [In Russian]

[33] GOST P 54964 – 2012. Otzenka sootvetstviya. Ecologicheskie trebovaniya k ob'ektam nedvizhimosti [GOST P 54964 – 2012. Conformity assessment. Environmental

requirements for real estate] (NP AVOK, OAO CNIIPromzdanyi, OAO NPO TERMEK) p 48 [In Russian]

[34] Il′ina T N and Isaeva V V 2018 Economic Feasibility of Ventilation Cleaning Emissions in Underground Parkings *J. Phys.: Conf. Ser.* **1066** 012007

[35] Kosukhin M M, Kosukhin A M and Komarova K S 2018 Increasing of the Energy Efficiency in Civilian Buildings Applying Fixed Type of Facade Systems *J. Phys.: Conf. Ser.* **1066** 012012

The use of modern highly precise ultrasonic gas flowmeters and etalon stands to enlarge industrial resource efficiency

M A Velichko[1], I B Kostina and Yu P Gladkikh

Department of Informatics, Natural Sciences and Teaching Methods, Belgorod State National Research University, Pobeda St., Belgorod, 85, 308015, Russia

E-mail: gladkikh@bsu.edu.ru

Abstract. In this article we have described our new development that is the ultrasonic flowmeter with high dynamic range of 1 to 2000 for industrial applications. Its high accuracy of measurement can help gas consuming enterprises to save resources. Its software and the software of corresponding computing device is able to avoid gas leakage, to minimize energy consumption and to save human resources while maintaining metrological data. Shown are the new anti-adhesion technologies that can help to increase measurement accuracy and enlarge interval between counters verifications. Described is the low power consumption that makes it possible to use this ultrasonic flowmeter in hard remote environment without direct management for a period of several months. Experiments held in Turkmenistan have shown that device indications didn't drift and remained stable during the year that is a great advantage in comparison to rotary and turbine flowmeters. Also described is the mobile ultrasonic calibration stand that uses the same physical principles and similar software. Outlined is the usage of modern wireless technologies to collect and transmit metrological data.

Key words: ultrasonic flowmeter, verification stand, etalon, WiFi, gas consumption, energy saving.

1. Introduction

Gas is the main fuel in Russia and it shares more than 50% of primary energy consumption, that is quite a lot according to the world standards. No country with advanced economy has such a high share of gas in the fuel balance. As a comparison, the UK's gas ratio in energy consumption is 40%, for Netherlands this figure is 38%, for Canada – 27%, for the US – 26%. Hydropower predominates in Norway and its share of gas is only 9%. However, at the background of such countries as Iran, where gas also gives 55% of all primary energy, or Algeria, where its share is 60%, Russia looks quite organically.

Nevertheless, gas consumption in Russia is huge. It is reasonable to mention that it is equal to the combined consumption of Germany, France, Italy, Japan, China and India. Annually Russia burns and processes more than 420 billion cubic meters of gas, yielding only the USA on this indicator. In Russia the share of exported blue fuel reaches only 30%, while the rest of it is consumed domestically. In order to maintain its position in foreign gas markets or even win new ones, Russia needs to reduce its domestic consumption.

2. Materials and Methods

Measurement and comparison of gas flow rates at all stages from gas companies to gas consumers is a rather complicated job. In particular, it includes efficient gas metering and accounting, gas flow monitoring and telemetry [1].

Gas flow rates have to be measured under various temperature, pressure and humidity conditions. Different gas composition, changing dynamic range is also a great challenge. There is no uniform solution for all circumstances, multiple physical principles are used to measure gas flow.

The gas flowmeter usually consists of three main parts, such as sealed case, measuring converter and counting device. The measuring mechanism is the main element, and the principle of converting the flow into mechanical, electrical or other impact on the counting device determines the type of flowmeter. Mechanical converters use different properties of gases to move the lever acting on the counting device. Electronic sensors convert the flow properties into an electrical pulse that increases the indication on the display.

3. Results

The measuring convertor uses the physical properties of the gaseous substance flow. According to the principle of its operation, there are several types of counters, particularly:

1) membrane or diaphragm meter, in which gas alternately fills the two measuring chambers separated by a membrane. Each camera has a fixed volume. The input and output valves are asymmetrically synchronized and alternately inlet and outlet gas. The movements of the membrane are converted by the lever into a turn of the counting device drive, which shows the volume of fuel passed through the device;

2) rotary gas meter, that uses the gas pressure in special tanks to drive two eight-shaped shafts (rotors), one of which is connected to the drive of the counting mechanism. Counter translates the number of revolutions in the amount of fuel;

3) turbine meter principle is based on flow rate measuring by counting the number of revolutions of turbines that rotate under the flow pressure. Such metering devices are used only in municipal and industrial enterprises to account for large volumes;

4) vortex flowmeter uses the dependence of the frequency of pressure fluctuations that occur after the jet of a gaseous substance is given a vortex shape. The mechanism of accounting for these oscillations is complicated, requires the use of complex microprocessor technology, therefore, such devices are expensive and are used only in large gas distribution enterprises;

5) ultrasonic flowmeter, that usually uses the difference of ultrasound waves propagation times along and against the flow to estimate the speed of flow and then convert it to gas consumption. Ultrasonic flowmeters also need high-accuracy electronics to convert nano- and pico-second pulses into digital signals [2, 3, 4].

Ultrasonic flowmeters are increasingly used in various industries. Combining the advantages of vortex and jet flowmeters, such as high measurement accuracy, wide dynamic range, low inertia, no additional pressure losses, simple design – ultrasonic flowmeters in most tasks successfully replace rotary and turbine flowmeters.

4. Discussion

Industrial ultrasonic flowmeter. We have recently developed the new industrial time-of-flight ultrasonic gas flow meter Irga RU that has already been entered into the State register of measuring instruments.

Its operation principle is based on the time-to-pulse method of gas flow measuring. It consists in measuring the time of ultrasonic pulses passage in the direction of the gas flow in the pipeline and against it. Excitation and reception of pulses is carried out by piezoelectric transducers, which are installed in an all-metal case of the flowmeter at an angle (from 30° to 45°, depending on the version) to the flow direction [5, 6]. The speed of ultrasound in the medium depends on the physical and chemical properties of this medium: temperature, pressure, etc. At the same time, it is much greater than the speed of the medium, so that the actual speed of ultrasound in a moving medium is not much different from its speed in a stationary medium. The difference in transit times Δt even at flow rates of about 10 m/s is a fraction of a microsecond, while the measurement error should not exceed a few nanoseconds [7].

These circumstances necessitate the use of complex electronic circuits in combination with microprocessor technology, providing compensation for the influence of these factors.

Structurally, the flow meter consists of three blocks:

1) primary flow converter, which is a housing with built-in ultrasonic transceivers;

2) electronic unit, which carries out the reception and transmission of signals through ultrasonic transceivers, their conversion, processing and calculation of the volume flow of gas under operating conditions, followed by the formation of the output signal;

3) power supply unit with built-in spark protection barrier to provide explosion protection circuits if necessary.

The electronic unit controls ultrasonic transceivers, performs receiving, processing, converting and transmitting signals to the computing device. These signals contain, in particular, information about the propagation time of the ultrasonic pulses, that is required to calculate the volumetric gas flow rate in operating conditions:

$$v = \frac{L \cdot (t_2 - t_1)}{2 \cdot t_1 \cdot t_2 \cdot \cos\alpha},$$

where: v is velocity of flow in the pipeline, L is the distance between the transceivers, α is the angle between the axis of the sensors installation and the axis of the conduit, $t1$ and $t2$ are times of ultrasonic pulses propagation along the flow and against it.

The design of flow meter eliminates the possibility of unauthorized influence on the flow meter software and measurement information. We have developed and successfully tested our own control circuits for piezoelectric transceivers, and also created new original algorithms for processing received signals. It allowed us to:

– significantly expand the dynamic range of gas flow measurement to values of the order of 1:2000;

– increase the accuracy of measurements to 1% in most tasks;

– use our flowmeter on pipelines of almost any diameter;

– measure the flow rate of various physical and chemical properties of gas media;

– carry out measurements at high speed, in a wide range of temperatures and pressures;

– not to introduce additional pressure losses into the pipeline.

At the same time, our flow meter has a fairly simple design and does not require significant lengths of straight sections before and after its installation.

We have also mastered and successfully tested the technology of applying anti-adhesive coating on the inner walls of the flowmeter to avoid sticking of the solid fraction of the measured medium. This helped us to reduce the probability of non-reproducibility of measurement results by 0.2%. In future it will also lead to calibration interval decrease [8, 9, 10].

In our flowmeter design we have used modern time-to-digital microcontroller convertors, that apply energy saving technologies. We have also developed energy efficient algorithms of transmission, receiving and processing of ultrasonic pulses. Power consumption has decreased to several μW. It helped us to implement autonomous power supply from rechargeable batteries [11, 12, 13, 14].

Our colleagues from Turkmenistan gave us a possibility to test our devise powered by solar panels in desert conditions with almost cloudless weather and periodic changes in energy supply. Experiment has shown that our flowmeter is able to operate without direct maintenance for months.

In this case the only challenge is to receive information from the remote flowmeter, so we had to organize telemetry system and to develop additional software. We have tried several solutions, while transmitting metrological information from the flowmeter, such as radio modems, GSM/GPRS or 4G/3G modems. The most efficient solution for the desert was the radio modem with code modulation, that transmitted data to more than 7 km to the central node. 3G signal was high enough only in the central node, where we created a solar powered microcontroller based HTTP client that periodically posted data to the web server located in Ashgabat [15, 16,17].

Mobile ultrasonic calibration stand.

Though ultrasonic flowmeters are very effective in a high dynamic range, they also have to be calibrated, that means that their indications have to be synchronized with the etalons. This is a standard procedure for all metrological devices despite of their prices, conditions used, dynamic ranges, types, sizes and other properties.

In accordance with the Federal law of Russian Federation that deals with "Ensuring the unity of measurements" it is necessary to verify the gas meters. Only in the course of such verification it becomes clear whether the equipment meets the metrological requirements. Verification is carried out before putting the devices into operation and periodically during their exploitation [18, 19, 20]. This is associated with certain difficulties both for industrial enterprises, that have to either stop the section of gas consumption, or install a similar flowmeter for the time of verification of the main one, and also for private consumers, who have to remain without gas for the period of verification. In any case, verification takes time and money.

Verification can produce only specialized organizations with high-precision stands, that have to be included into the State register of measuring instruments. Usually these stands are bulky systems that require a lot of free space and a highly skilled operator. In addition, such stands or their parts (e.g. nozzles, in the case of stands on critical nozzles) themselves require periodic recalibration in standardization organizations. Such stands cost about several million rubles [21].

We have applied similar to Irga-RU principles of flow measurement, electronic circuits and signal processing algorithms while developing and creating ultrasonic testing facility, which we have called KRAB-UM (Its appearance in shown in figure 1.). The principle of its operation lies in automatic selection of verification points and a set of conditions for measurements, monitoring the readings of temperature and pressure sensors, automatic flow control, comparison of the volumes of gas passed through the household meter and through a high-precision ultrasonic reference system, analysis of the measurements and decision-making on the success of verification.

It is the first ultrasonic mobile calibration stand, which among other things uses wireless technology for transmission, storage and display of metrological and other data, and is able to work on batteries for 8 hours. The picosecond accuracy of electronic calculation of the time difference between the ultrasonic pulses passing between the receiver and the transmitter, which was impossible until recently, as well as the improved design of the reference device hydraulic part, made it possible to develop a compact high-precision reliable stand [22, 23].

ESP8266 board was chosen as the main module of the stand control (figure 1). It has accumulated data from:

– ultrasonic sensor control module;
– power management module;
– digital thermometer, digital pressure sensor;
– reed switch connected to the verifiable domestic meter.

Based on the results of data processing obtained from various modules, the ESP8266 controller generated information about the reference gas flow rate in operating conditions. The reference flow rate was calculated from the calibration table recorded in the ESP8266 EEPROM memory (EEPROM-Electrically Erasable Programmable Read — Only Memory is electrically erasable reprogrammable memory, a type of nonvolatile memory used in computers and other electronic devices to store relatively small amounts of data but with the ability to read, delete, or write bytes individually) [24].

For domestic gas meters equipped with a slot for connecting an electromagnetic sensor (reed switch), an output for reading signals from such a sensor was provided in the ultrasonic calibration unit. Interfacing with household meters with the help of reed switch made it possible to significantly simplify the process of verification and improve its accuracy, reducing to zero the operator's error. For meters that do not provide for the connection of the reed switch, the operator had to visually monitor the passage of a certain volume of gas and then manually record it in the inspection program. The mobile unit allowed to check up to 4 domestic meters at the same time, but it made sense only when using reed switches, otherwise 4 operators would be needed for accurate visual control of household meters.

The flow rate generated in ESP8266 took into account the difference in gas temperature during verification and during the initial calibration of the installation itself. The instantaneous temperature value was also displayed as reference information in the verification program. A platinum resistance thermometer mounted near the location of one of the ultrasonic sensors was used to measure the temperature. To digitize the thermometer signal, a 24-bit analog-to-digital converter was used, which made it possible to measure the temperature value up to a fraction of a degree. The inertia of the thermometer is high enough, but the temperature of the gas in the pipeline does not undergo sudden changes, so this fact does not significantly affect the accuracy of measurements [25, 26, 27].

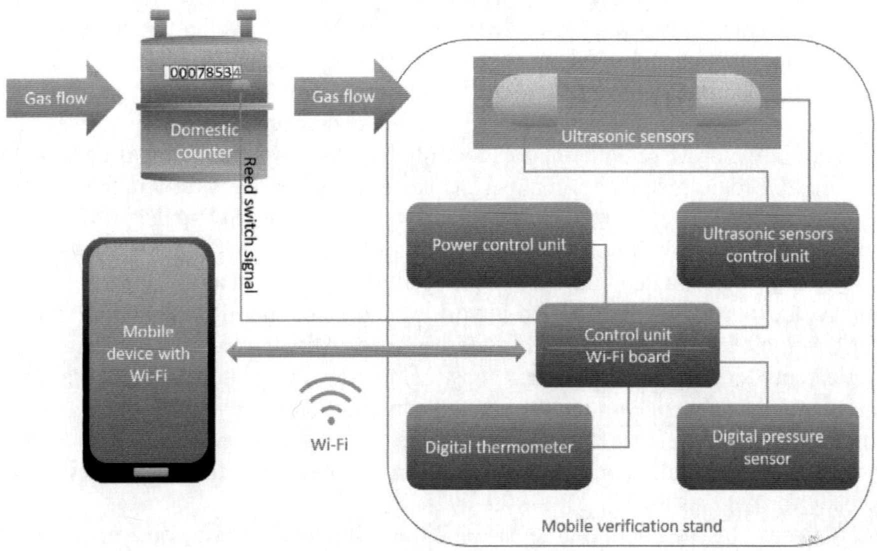

Figure 1. Mobile verification stand general scheme

When carrying out verification without removing the domestic counter seal on it should remain intact, so to account for the pressure drop, we were forced to measure the pressure at least twice at each point of verification: with the inlet valve closed and open. The digital pressure sensor used data transmitted once a minute to the ESP8266 controller via the UART protocol. It did not need additional analog-to-digital converters. But it was necessary to use a software UART, as the built-in UART in ESP8266 was already reserved. Opening / closing of the valve was carried out by the operator manually, the information about the measured pressure was stored already in the verification program, so the pressure drop was taken into account in the flow value outside the ESP8266 controller.

Power supply of the mobile unit in the absence of a 220V network was provided by a 5V battery operated by the battery shield controller. It has been technically and programmatically modified for optimal operation as part of the mobile calibration facility. In particular, the indicator diodes were transferred to the unit housing, and information about the battery charge and other power parameters was transmitted to the ESP8266 controller via the I2C Protocol (its software implementation). Information about the battery charge was also displayed in the verification program.

The most important information for verification – the time difference between the propagation of ultrasonic pulses – was generated in ultrasonic transceivers control board. This information was transmitted via the UART Protocol and converted to a flow rate according to the calibration table. Then the temperature difference was corrected and in this form the information was received in the verification program and in the "Current flow" field was displayed on the screen of a mobile or stationary device. The program for the AVR controller that controled this board was written in Assembler.

Programming ESP8266 was carried out in C++ in the Arduino IDE. The program blocks for each of the components described above were contained in separate files. The program also included a file of the real-time operating system functions, procedures to work with Wi-Fi, functions for the HTTP, DNS server, etc. The interface was written in JavaScript in NetBeans environment. HTML-pages, CSS and JavaScript files were entered into the memory of the ESP8266 controller using a special script.

All the information described above (service and necessary for verification) was transmitted via Wi-Fi network to a computing device (smartphone, tablet, desktop computer or laptop) for further processing and storage. To do this, the ESP8266 module was set to the access point mode. The name of the Wi-Fi network (ESPWiFi) and the password to access it, as well as the IP address of the server (which was also the ESP8266 module) were recorded in the program of the ESP8266 controller. Network ESPWiFi automatically became viewable for all mobile devices. After connecting to the network in the command line of the mobile device browser, it was enough to dial the IP address of the ESP8266 module to load the page with the verification program. In this case, the computing device could be under control of any operating system. The main condition of the program was installed on a computer device modern browser with support for HTML5. (The application was designed to work in Chrome and Yandex Browser, so its optimal performance was guaranteed only in these browsers.)

Information on the Wi-Fi network was transmitted in the form of JSON-files. ESP8266 acted as an HTTP server responding to continuous requests from an HTTP client (modern browser). On HTML pages, the information was updated without reloading all the content, using AJAX technology.

Thus, we have developed and assembled a working model of mobile ultrasonic reference unit for automatic verification of household meters. In the case of verification of counters with the support of electromagnetic sensors, the process can be almost completely automated and simultaneously carried out for 4 counters. The only concern of the operator, in addition to switching the verification points, will be the measurement of pressure with the inlet valve closed and open. This can be corrected if necessary by installing electronically controlled valves. The unit has been successfully tested in the laboratory and is ready for field testing.

The KRAB-UM can be used both on the gas pipeline that allows to carry out verification of domestic and communal gas counters without removal, and as stationary reference installation (in this case it needs a flow compressor).

Thus, the mobile calibration using our ultrasonic stand can greatly facilitate life for private gas consumers, saving them from necessity to remove the counters and take them for testing, as well as for industrial enterprises. It is quite easy to work with our stand, because the whole process is automated and requires only to move between pages in a Web browser, and pressing one button. The information on the performed verification is stored and encrypted, which ensures the reliability of the results and the ability to trace the process of all measurements.

Figure 2. Mobile ultrasonic calibration stand

5. Conclusion

So we have developed and created the modern industrial ultrasonic flowmeter with great dynamic range. Its high accuracy of measurement can help gas consuming enterprises to save resources. Its software and the software of corresponding computing device is able to avoid gas leakage, to minimize energy consumption and to save human resources while maintaining metrological data. Our flowmeters can be organized in the network operated from single remote central unit. New anti-adhesion technologies can help to increase accuracy of measurement and enlarge interval between verifications. Low power consumption and high reliability makes it possible to use our ultrasonic flowmeter in hard remote environment without direct management for a period of several months. Experiments held in Turkmenistan have shown that device indications didn't drift and remained stable during the year, that is a great advantage in comparison to rotary and turbine flowmeters.

Using the same physical principles and similar to our ultrasonic flowmeter software we have also created the mobile ultrasonic calibration stand. It uses modern wireless technologies to collect and transmit metrological data. It is compact, energy effective and easy-to-use. Almost any mobile device can operate as an interface for verification computer, receiving metrological data from both etalon and counter under calibration over local WiFi network. It is the first mobile ultrasonic etalon device.

Both ultrasonic flowmeter and mobile ultrasonic etalon stand use modern high accuracy time-to-digital technologies, that are at the same time very energy effective. Low power consumption makes it possible to operate from batteries of from solar panels. It can become a very interesting option for industrial measuring devices. Besides our devices can be operated by non-specialists, as they are almost fully automated. It can save time and financial resources. Their most significant features is high precision and high dynamic range, as the greater accuracy leads to better effectiveness of gas resources saving.

References

[1] Velichko M, Satler O, Krasovskaya L, Bronnikova M, Belyaeva I, Kostina I and Gladkikh Yu 2018 Calibration of Ultrasonic Flow Meter on Wi-Fi Network Using a Web Browser. *J. Advanced Research in Dynamical and Control Systems (Special Issue)* vol 8 pp 1593-96

[2] Velichko M, Satler O, Krasovskaya L, Erina T, Belyaeva I and Gladkikh Yu 2017 Using Drone as WiFi Access Point During Infrared Thermography for Subsidiary Data Acquisition. *J. Fundam. Appl. Sci.*, vol 9 pp 1279-88

[3] Hamidullin V, Malakhanov R and Hamidullina E 2001 Statics and Dynamics of Ultrasonic Flowmeters as Sensing Elements for Power Control Systems *IEEE Conf. on Control Applications - Proceedings Proceedings of the 2001* pp 680-5

[4] Fraṇa D, Ono Y and Jen C 2011 Contrapropagating Ultrasonic Flowmeter Using Clad Buffer Rods for High Temperature Measurements *Journal of Dynamic Systems, Measurement and Control, Transactions of the ASME* pp 137-44

[5] Hu L, Qin L, Mao K, Chen W and Fu X 2016 Optimization of Neural Network by Genetic Algorithm for Flowrate Determination in Multipath Ultrasonic Gas Flowmeter *IEEE Sensors Journal*) vol 16 pp 1158-67

[6] Luca A, Marchiano R and Chassaing J 2016 Numerical Simulation of Transit-time Ultrasonic Flowmeters by a Direct Approach *IEEE Transactions on Ultrasonics, Ferroelectrics, and Frequency Control* vol 63(6) pp 886-97

[7] Gryshanova I 2007 The Effect of Duct Cross-sectional Shapes on Metrological Performance of Ultrasonic Flowmeters. *Digest Proceedings of the 5th Joint ASME/JSME Fluids Engineering Summer Conference, FEDSM* pp 27-9

[8] Hamidullin V K 1995 Dynamics of Ultrasonic Flowmeters *Digest: Proceedings of the IEEE*

Ultrasonics Symposium pp 1109-13

[9] Gerasimov S, Glushnev V and Serov N 2018 Analysis of Calibration and Verification Indirect Methods of Ultrasonic Flowmeters International Conference on Advanced Technologies, Systems and Services in Telecommunications, *TELSIKS – Proceeding* vol 13 pp 451-54

[10] Bates R 2013 Development of a Custom On-line Ultrasonic Vapour Analyzer/Flowmeter For the Atlas Inner Detector, with Application to Gaseous Tracking and Cherenkov Detectors *Journal of Instrumentation* vol 8 pp 1002-04.

[11] Gershman E M, Kruglo D S, Fafurin V A, Borisov A M and Sabirzyanov A N 2017 Flow measurement of two-phase flow by ultrasonic flow meters *Transactions of Academenergo* vol 1 pp 31-7

[12] Mansfeld D A, Sanin A G, Volkov G P, Belyaev R V and Morozkin D V 2017 Ultrasonic flow of gas from overhead sensors *Scientific notes of the physics faculty of Moscow University* vol 5 pp 175-80

[13] Kuranov K N and Sergeev S S 2016 Automated calculation of parameters of ultrasonic flowmeters. *In the book: the Creation of resource-saving machines and technologies Abstracts of the Republican scientific and technical conference* (Ministry of education and science of the Republic of Belarus, Mogilev machine-building Institute) pp 100-05

[14] Strelnikov E A and Eremin I Yu 2016 The issue of improving the accuracy of ultrasonic flowmeters *Legislative and applied Metrology* vol 1(140) pp 41-4

[15] Abzalilova E R and Galimullina E 2017 Reducing the influence of vibration on the measurement accuracy of ultrasonic flowmeters *Bulletin of science and education* vol 1(25) pp 33-5

[16] Tyryshkin R A., Sabirzyanov A N and Fafurin V A 2016 Numerical simulation of the correction coefficient of ultrasonic flowmeters *Mathematical methods in technics and technologies – mmtt* vol 9(91) pp 17-23

[17] Biryukov E A 2018 The transducers of ultrasonic flowmeters *In the book: traditional and innovative science: history, modern condition and prospects collection of papers of International scientific-practical conference* pp 10-2

[18] Borisov V A, Gershman E M, Pruglo D S and Fafurin V 2017 Experimental study of the influence of heterogeneity of the temperature field on metrological characteristics of the ultrasonic flowmeter *Transactions of Academenergo* vol 2 pp 7-18

[19] Vasilyev T R and Kokuev A G 2018 Oil products export control based on artificial neural network *Bulletin of Astrakhan state technical University. Series: Management, computer engineering and Informatics* vol 2 pp 43-52

[20] Abzalilova Y R, Vydrin D F, Galimullina E E and Vdovin A K. 2017 Methods of flow measurement *Scientific research* vol 2(13) pp 20-1

[21] Terekhin A and Pashnina N 2016 Ah. Evaluation of the effect of the number of beams of ultrasonic flow transducers on the quality characteristics of measurement *Digest: Prom-Engineering proceedings of the II international scientific and technical conference. FGBOU VPO "South Ural state University"* pp 404-07

[22] Polovneva S I, Yelshin V V and Tolstoy M Y 2009 Flow measurement of gases and liquids *Irkutsk: Teaching aid national research Irkutsk state technical University* pp 173-6.

[23] Kositsyn N In, Petrov V V and Petrov A V 2016 Calculation of Field Amplitude Distribution of Ultrasonic Wave Beam in Moving Media. *News of Saratov University. New series. Series: Physics* vol 16 (1) pp 27-33

[24] Sheshukov I Y 2015 Metering devices, errors of measurement and calculation algorithms *Legislative and applied Metrolog* vol 3(136) pp 24-40

[25] Isaev I A, Khakimov D R, Garchev A I and Fafurin V A 2013 Investigation of the metrological characteristics of the ultrasonic transducer gas flow *World of measurements* vol 3 pp 8-15

[26] Bogush M V 2007 Design of piezoelectric transducers for ultrasonic flowmeters of gas *Sensors and systems* vol 8 pp 8-11

IOP Conf. Series: Materials Science and Engineering **552** (2019) 012029 doi:10.1088/1757-899X/552/1/012029

Method development for the thermotechnical calculations utilizing reduced characteristics of the RDF fuels

N V Kornilova[1] **and P A Trubaev**[2]

[1]Department of Electric Power Engineering and Automation, Belgorod State Technological University of V.G. Shukhov, Kostyukov St., 46, Belgorod, 308012, Russia
[2]Department of Energy Engineering of Heat Technologie, Belgorod State Technological University of V.G. Shukhov, Kostyukov St., 46, Belgorod, 308012, Russia

E-mail: trubaev@gmail.com

Abstract. The paper presents method of calculation of the heat leakage and efficiency by using method of indirect heat balance, utilizing data from gas analysis, which can be employed for operational evaluation plant operation during burning of different types of waste materials. Taking into account that practically only oxygen and carbon dioxide is being measured in the exhaust gases, calculation method for heat leakage with exhaust gases, using measurements of only this component, without utilizing carbon dioxide content in combustion products, was developed. Carbon dioxide and oxygen calculated percentage content in dry complete combustion products in actual air consumption was measured with consideration of incomplete combustion. Heat leakage for low power boilers and furnaces was analyzed, heat leakage calculation method for this kind of plants was introduced. Using design factors analysis of the solid fuel boilers for burning low-calorie fuel and wood waste, was developed dependence of the outside walls specific area of the boiler on the boiler's efficiency. Based on the developed method for seven types of the RDF fuels, tabular and graphic dependences for reduced factors, characterizing heat leakage to the atmosphere and incomplete combustion, were introduced. Suggested method was tested for calculating losses in solid fuel water heating boiler with grate stoker and designed output power of 200 kW during burning of the briquetted waste materials.

1. Introduction

The most efficient way to recycle solid municipal waste (MSW) is a thermal utilization with energy production [1]. The paper [2] shows, that from economic point of view, thermal utilization of MSW is more efficient compare to the other methods. Usage of the alternative fuel, in comparison to traditional energy plants on organic fuel, has similar impact on the environment [1]. Partial presence of RDF in the fuel mixtures decreased amount of SO_2 emission and didn't change the amount of NO_x emission [3]. Analysis of RDF and coal simultaneous combustion discovered that presence of RDF doesn't affect SO* emission, emission of NO, HCI, N20 slightly increases, and dioxin amount in slag grows in direct ratio to RDF percentage in fuel [4]. Findings in the paper [5] indicate that addition of the solid municipal waste to the bio fuel improves its energetic characteristics.

Additional advantage of the energy recuperation from combustion method is that for energy production can be used MSW with wide range qualities [6].

Use of different types of bio fuels and RDF with traditional organic fuels simultaneously requires information about combustion process characteristics and methods for evaluation of efficiency of the waste combustion plants for firing fuel with different component proportions [7, 8].

IOP Conf. Series: Materials Science and Engineering **552** (2019) 012029 doi:10.1088/1757-899X/552/1/012029

Bio fuel has many different components [9], therefore its physical and chemical properties are varying in a big way and that takes effect on the combustion characteristics [10]. RDF presence has significant effect on the combustion characteristics, when RDF is over 10% effectiveness of combustion decreases [3]. The paper [11] explored combustion of different types of waste. Emissions of CO in flue gases, for the most conditions, have met the ecological requirements. Combustion thermal effectiveness was changing in a wide range and heat efficiency was as such:

Wood	65,0...98,9%
Paper	74,8...99,7%
Cellulose	63,3...99,6%
RDF	50,1...91,8%

Traditional direct balance method for efficiency test during fuel combustion at the waste firing plants has low accuracy because physical and chemical qualities of bio fuel varies in a wide range [2, 10, 12], in addition to that, waste management effectiveness in different regions is uneven [13]. As paper [14] states, carbon content in MSW fluctuated from 40 to 99 %. On an average organic part in MSW can be described by the empirical equation $C_{32}NH_{55}O_3$.

During calorimetric properties of waste measurements determination, that defines precision of the calculation results for direct balance method, great difficulties emerges. Many publications show that MSW properties are much different from theoretical or reported values [15, 16] as well as thermal properties, calculated by experimental approach and by heat balance analysis [17].

The paper [18] presents research study of MSW characteristics by calorimetric and TGA methods that led to the empirical model for combustion value calculations, which demonstrated that combustion value of RDF fuel is within a broad range. The most significant differences are in the combustion heat and heat values that were calculated experimentally. This distinction is near 5% on average, as the paper [5] shows. The paper [19] made conclusion that method of determination of the combustion heat by calorimeters gives inaccurate information for RDF.

On the other hand, research of the composted RDF firing determined that self-supporting combustion process lasts a long time [20], that makes it difficult to find the exact amount of released heat energy for specified time length. It complicates application for calculation of burning of RDF methods of numerical and computer modeling [21, 22].

As a result of the performed analysis one can arrive at conclusion that the most precise method to evaluate operational effectiveness of the plants for waste combustion is the indirect balance method.

2. Boiler efficiency test utilizing indirect heat balance (M V Ravich method) with fuel reduced characteristics

It is difficult to determine exact characteristics and amount of the RDF being burnt during waste combustion. Also, this plant work in transient mode with RDF batch loading. In such conditions, calculations of the heat balance and efficiency by direct method have low accuracy and it is necessary to utilize indirect balance method calculations. During service conditions, method of indirect balance provides much more accuracy, having the same precision of measurements, for efficiency test against direct balance method and demands smaller number of the measured variables. In the indirect balance method heat leakage is being calculated using temperature and composition of the exhaust gases.

For the industrial plants, operating on the specific fuels, M V Ravich developed method of thermotechnical calculations utilizing fuel reduced characteristics, which gives options for generalization, simplification and measurements accuracy increase [23].

Efficiencycan be determined by utilizing specific heat losses relative to fuel combustion value:

$$\eta = 100 - (q_2 + q_3 + q_4 + q_5 + q_6), \%, \qquad (1)$$

where q_2 – losses due to exhaust gases; q_3 – losses due to chemical incomplete combustion (emitted combustibles); q_4 – losses due to mechanical incomplete fuel combustion; q_5 - losses due to thermal radiation from the surface of the boiler; q_6 - losses due to sensible heat of ash.

This section offers development of the method for calculating heat losses and efficiency of the plants for burning waste materials using indirect balance method based on gas analysis data. Taking in consideration that in practice, content of the exhaust gases is based on the oxygen and carbon dioxide content, method of calculation heat loss with exhaust gases q_5, using measurements of only these components, was developed. Heat loss for low power boilers and furnaces was analyzed, then method for heat loss q_5 calculation for those plants was suggested. Let's examine specific heat loss calculation (parameters of the heat balance).

2.1. *Losses due to exhaust gases q_2*

Excess air coefficient α in the exhaust gases content can be determined by the following formula:

$$\alpha = 21/(21 - O_2 - 0{,}5CO)$$

During fuel combustion with excess air ($\alpha > 1$) specific heat losses due to exhaust gases q_2 determined by the equation

$$q_2 = 100\,\frac{t_{fg} - t_a}{t_{max}}\left(c' + (h - 1)\cdot b\cdot k\right) = (t_{fg} - t_a)\,Z, \qquad (2)$$

where t_{eg} - flue gases temperature, °C; t_a - temperature of the air being used for combustion, °C; $t_{\max}$ - heat capacity of the fuel or maximum temperature that can be achieved during complete combustion of the moist fuel in theoretically necessary volume of the dry air at a temperature of 0°C and absence of heat leakage, determined by calculation, °C; Z - reduced factor, characterizing heat losses into the outside atmosphere.

Coefficients in the equation (2) are determined in the following way.

Coefficient c' – relation of the mean specific volume heat capacity of the combustion products c_{cp} and $c_{cp\,\max}$ determined for theoretical conditions at given $\alpha = 1$) in temperature range from 0°C to t_{cp} and $t_{\max}$ respectively:

$$c' = c_{cp}/c_{cp\,\max}.$$

Coefficient h - variation of the dry combustion products volume in comparison with theoretical one due to dilution with air. During solid fuel combustion (without considering CH_4 content in the incomplete combustion products) coefficient can be determined by the following formula

$$h = CO_2^{\max}/\,(CO_2' + CO'),$$

where $CO_2^{\max}$ - content of the CO_2 in dry combustion products in theoretical conditions (at given $\alpha=1$),%; CO_2', CO' – gases content in dry combustion products, %.

Usually, content of the CO_2' in combustion products is not measured. Part of the CO_2 amount, during incomplete combustion, is being substituted for an equal amount of the CO, at the same time, in the combustion products and exhaust gases amount of the O_2 increases by amount equal 0,5 of the C contents. Therefore, amount of the dry combustion products $V'_{dry.cp}$, during incomplete combustion, can be determined from the dry complete combustion products volume $V_{dry.cp}$ in the following way:

$$V'_{dry.cp} = V_{dry.cp} + 0{,}5\frac{CO'}{100}V'_{dry.cp}, \text{ или } V'_{dry.cp} = \frac{V_{dry.cp}}{1 - 0{,}005CO'}.$$

Content of the CO_2', % in dry incomplete combustion products $V'_{сух.пг}$ can be determined using content of the carbon dioxide in complete fuel combustion by the following formula:

$$CO'_2 = \frac{V_{CO_2} - 0{,}01CO'V'_{dry.cp}}{V'_{dry.cp}} = CO_{2dry}(1 - 0{,}005CO') - CO',$$

where CO_2 - calculated percentage content of the carbon dioxide in dry complete combustion products in actual air consumption; CO' - content of the carbon monoxide in the exhaust gases, determined by the gas analysis data, %.
Therefore

$$h = CO_2^{max} / (CO_{2dry}(1 - 0{,}005CO')).$$

In the absence of the CO data content, coefficient h can be determined by the fuel combustion calculation.

$$h = CO_2^{max} / CO_2^{calc} = V_{dry.cp} / V_{dry.cp\ theor},$$

where CO_2^{calc} – calculated content of the CO_2, %, in dry combustion products at given α; $V_{dry.cp\ theor}$ – volume of the dry combustion products at given α = 1.

Coefficient b – volume to volume ratio of dry and moist combustion products in theoretical conditions (at given α = 1):

$$b = (V_{CO_2} + V_{N_2}) / (V_{CO_2} + V_{N_2} + V_{H_2O}),$$

where V_{CO_2}, V_{N_2}, V_{H_2O} – volume of CO_2, N_2, H_2O, м^3 in combustion products.

Coefficient k – relation of the mean specific volume heat capacity of the air $c_{a\ cp}$ in temperature range from 0°C to t_{cp} and combustion products $c_{cp\ max}$ (calculated at given α = 1) in temperature range from 0° to t_{max}

$$k = c_{a\ cp} / c_{cp\ max}.$$

2.2. ***Losses due to chemical incomplete combustion q_3***

Losses due to chemical incomplete combustion q_3 were determined by the formula:

$$q_3 = 100\ Q_{und} h/p,$$

where Q_{und} – heat value of the incomplete combustion products, that are contained in the combustion products, kJ/m^3;

$$Q_{und} = 126{,}4CO,$$

where CO – content of the gases in the combustion products, %; p – low operational heat value of the fuel, based on 1 m^3 of dry combustion products, generated during combustion in theoretical conditions (at given α = 1).

2.3. ***Losses due to mechanical incomplete combustion of the fuel q_4***

Heat loss due to mechanical incomplete combustion of the fuel q_4 can be omitted, because in practice, content and consumption of the burning waste determined with low accuracy, and due to big size of the briquettes, pieces of unburnt fuel stays in the boiler.

2.4. ***Losses due to thermal radiation from the wall surface of the boiler q_5***

As opposed to the rest of the overall efficiency components η to calculate specific heat loss q_5 it is necessary to know heat value Q, kW, that is entering the boiler. Heat loss from the wall surface of the boiler q_5, %, is written as:

$$q_5 = \frac{Q_{wall}}{Q} 100\%,$$

where Q_{wall} – heat loss from the wall surface of the boiler, kW.

Value Q_{wall} can be determined by utilizing value of the fuel consumption rate and its heat value or utilizing power and efficiency of the boiler or furnace:

$$q_5 = 0{,}01\eta \frac{Q_{wall}}{P} = \frac{Q_{wall}}{G_f Q_l^w} 100\%,$$

where G_f – fuel consumption rate, kg/s; Q_{wall} – heat loss from the wall surface of the boiler, kW; P – power, kW.

Heat loss from the wall surface of the boiler q_5 for small volume furnaces or boilers will be critical due to big value of the relation of the surface area to the inside volume. In general, taking into account value of the loss Q_{wall}, kW, can be written as:

$$Q_{wall} = \left(\frac{1}{\alpha_{inn}} + \sum \frac{x_i}{\lambda_i} + \frac{1}{\alpha_{out}}\right)^{-1} (F_{wall} + F_{arch}(t_c - t_a), \tag{3}$$

where α_{inn}, α_{out}– coefficients of the heat transfer from gases to the inner wall surface of the furnace (boiler) and from the outer wall surfaces and the arch into the outside atmosphere, W/(m^2·K); x_i, λ_i – thickness, м, and coefficient of conductivity, W/(m·K), of the layers in the wall design; F_{wall}, F_{arch} – surface area of the wall and arch of the furnace, m^2; t_c, t_a – mean temperature inside furnace (boiler) and outside temperature, °C.

Based on the design parameters analysis of the solid fuel boilers for combustion of the low calorie fuel and wood waste, dependence of the specific area of the boiler's outer surface f_{wall}, м^2/kW, on the power P was developed (figure 1):

$$f_{wall} = \frac{F_{wall}}{P} = 0{,}00656 + \frac{1{,}23}{P^{0{,}5}}. \tag{4}$$

In the equation (1) attempt to determine coefficient of the heat transfer α_{inn} from gases to the furnace's inner wall surface can be difficult, because in the standard literature [24, 25] dependences were obtained for high productivity boilers during conventional fuel burning, but calculating formulas for engineering calculations of the heat transfer processes, that are presented in the reference book [26] and similar ones, were obtained for ideal conditions, which are much different from presented here.

Let's accept an assumption, that inner temperature of the boiler's or furnace's wall equal to mean temperature of the combustion products t_c.

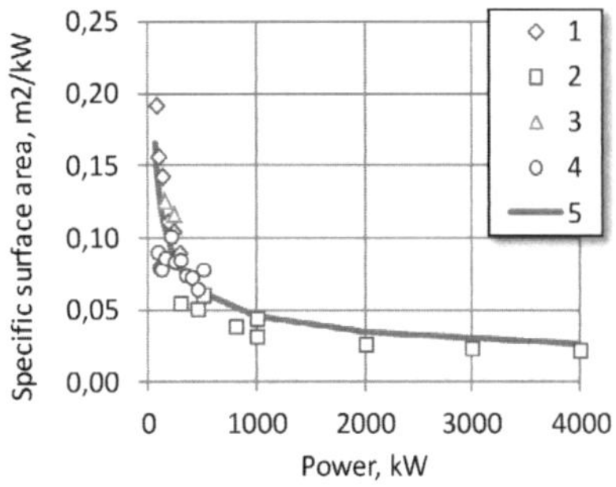

Figure 1. Dependence of the specific area of the solid fuel boilers for burning low calorie fuel on the power: 1 - Boilers AO "Motor Sich" (Ukraine), boilers NPP "Belkotlomash" (Belarus); 2 - series KB-PM2; 3 - series KB-Srm (wood waste); 4 - industrial boilers "TEPLOV T" (OOO "ZKO Teplov", Russia); 5 – approximating curve (1)

Using tubular's data, shown in the papers [25, 27], equations for calculation of the heat transfer coefficient from the boiler's outside walls into the outside atmosphere were obtained, that include heat transfer by means of convection and radiation, W/(m^2·K):

$$\alpha_{out} = 5{,}7 + 0{,}14 \cdot t_{out.wall} - 6{,}65 \cdot 10^{-4} \cdot t^2_{out.wall} + 1{,}7 \cdot 10^{-6} \cdot t^3_{out.wall}; \quad (5)$$

for which range of use: $t_{out.wall}$ = 25...200°C; mean error of the approximation 0,9%; maximum error of the approximation 3,2%, or

$$\alpha_{out} = 7 + 0{,}072 \cdot t_{out.wall}, \text{ at given } t_{out.wall} = 0...500°\text{C}. \quad (6)$$

for which range of use: $t_{\text{ст нар}}$ = 25...200°C; mean error of the approximation 4,2%; maximum error of the approximation 7,9%.

In accordance with technical standards outside surface temperature of the hot water boilers parts shouldn't be more than 45°C during making of the hot water with temperature up to 115°C and no more than 55°C for water with higher temperature.

Figure 2 presents obtained by the equations (5) and (6) dependences for determining losses q_5 utilizing known outside surface temperature of the boiler's or furnace's body, that were determined by the following formula

$$Q_{\text{CT}} = \alpha_{out} F_{wall}(t_c - t_a) = (7 + 0{,}072 t_{out.wall}) F_{wall}(t_c - t_a). \quad (7)$$

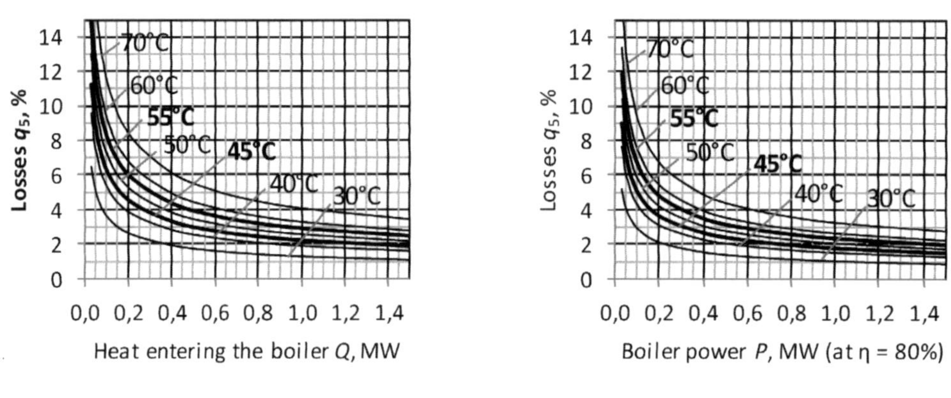

Figure 2. Heat losses into the outside atmosphere for different temperatures of the outside surface of the frame and outside atmosphere temperature at 0°C: **a** - in accordance with the amount of heat that entered boiler or furnace; **b** - in accordance with boiler's power

IOP Conf. Series: Materials Science and Engineering **552** (2019) 012029 doi:10.1088/1757-899X/552/1/012029

Disadvantage of utilizing formula (5) is in big error, related to the fact of averaging values of the surface temperatures and different factors having impact on the heat loss coefficient.

In such a manner, one can use simplified equation (1) with regard of equations (2) and (3) to calculate heat loss into the outside atmosphere using known outside walls surface temperature of the plant:

$$Q_{wall} = \left(\sum \frac{x_i}{\lambda_i} + \frac{1}{7 + 0{,}072 t_{out.wall}}\right)^{-1} F_{CT}(t_c - t_a),$$

or (8)

$$Q_{wall} = \left(\sum \frac{x_i}{\lambda_i} + \frac{1}{7 + 0{,}072 t_{out.wall}}\right)^{-1} (0{,}00656P + 1{,}23P^{0,5})(t_c - t_a).$$

2.5. ***Losses due to sensible heat of ash q_6***

On condition that carryover of the heat loss with ash, being removed from the boiler, is absent

$$q_6 = h_{ash} A^w / Q_l^W,$$

where h_{ash} – sensible heat of ash, kJ/kg, that can be determined by the following formula $h_{ash} = c_{ash} \cdot t_{ash}$. Heat capacity of the ash is presented in the paper [24], using data from this paper equation of linear regression $h_{ash} = 0{,}96 \cdot t_{ash}$ was developed. Utilizing this formula, losses due to sensible heat of ash can be as follows:

$$q_6 = 0{,}96 \cdot t_{ash} A^w / Q_l^W.$$

3. Reduced coefficient calculation for different types of RDF fuel

Objective of this section is to put in use simplified thermotechnical calculation methods (developed by M.B.Ravich) for boilers, firing wastes of different types. Components of waste presented in the table1.

Table 1. Contents of different types of waste.

Number of fuel	Name	Component content, wt.								
		Paper	Food waste	Tex-tiles	Wood	Leather, Rubber	Plastic	Flunko uts	Non-flamma-ble mass	Total
1	Paper	1	–	–	–	–	–	–	–	1
2	Wood		–	–	1	–	–	–	–	1
3	MSW	0,30	0,30	0,05	0,05	0,01	0,15	0,04	0,10	1
4	MSW 30% and wood 70%	0,09	0,09	0,015	0,715	0,003	0,045	0,012	0,03	1
5	MSW 50% and wood 50%	0,15	0,15	0,025	0,525	0,005	0,075	0,02	0,05	1
6	Plastic 30% wood 70%	–	–	–	0,7	–	0,3	–	–	1
7	Plastic 50% wood 50%	–	–	–	0,5	–	0,5	–	–	1

IOP Conf. Series: Materials Science and Engineering **552** (2019) 012029 doi:10.1088/1757-899X/552/1/012029

For all listed kinds of RDF fuel, for humidity levels 0, 10, 20, 30, 40, 50%, elemental operational content was determined. Based on section 2.1.2 coefficients Z and p were determined. Dependence of the coefficient Z on the CO content in the exhaust gases takes the linear form, therefore it can be taken into account utilizing correction coefficient k.

$$Z = Z_0(1 + k \cdot \mathrm{CO}),$$

where Z_0 - coefficient Z with amount of CO = 0% in combustion products; CO – amount of the carbon oxide in the exhaust gases, vol.%.

Coefficient k determined by the values of the coefficients Z.

$$k = \frac{\left(\sum\limits_{W=0,10,20,30,40,50\%} \frac{Z_{CO=2\%}}{Z_{CO=0\%}} \right) / 6 - 1}{2},$$

where $Z_{CO=2\%}$ and $Z_{CO=0\%}$ – coefficient Z with amount of the carbon oxide 2 and 0% in the combustion products.

Values of the coefficient Z and correction coefficient k for different types of RDF presented on the figure 3–11.

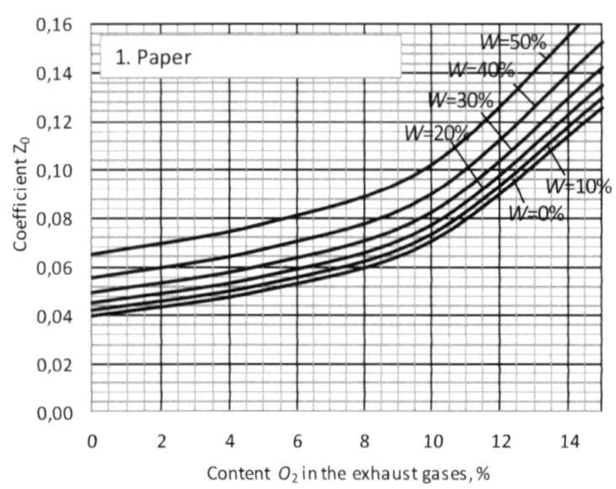

Figure 3. Coefficient Z (at given CO = 0%) for Fuel 1

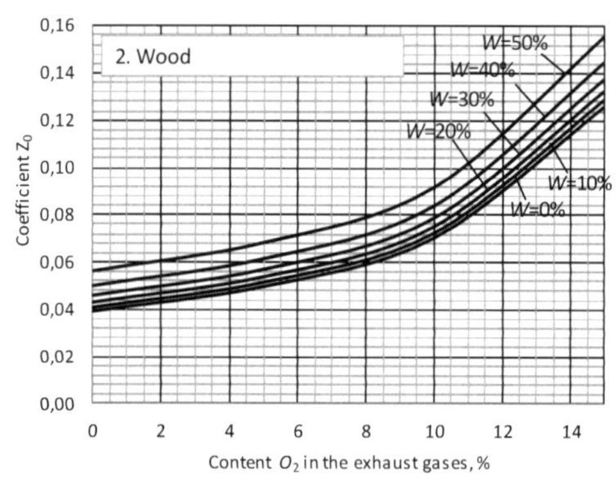

Figure 4. Coefficient Z (at given CO = 0%) for Fuel 2

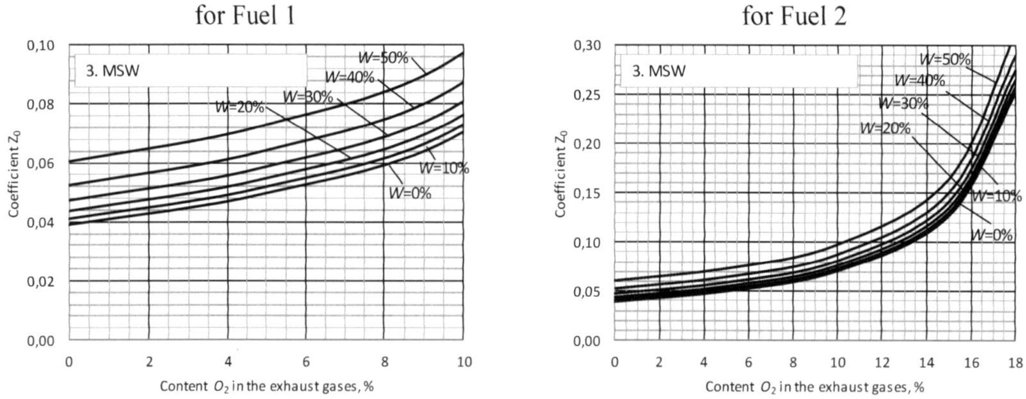

Figure 5. Coefficient Z (at given CO = 0%) for Fuel 3

IOP Conf. Series: Materials Science and Engineering **552** (2019) 012029 doi:10.1088/1757-899X/552/1/012029

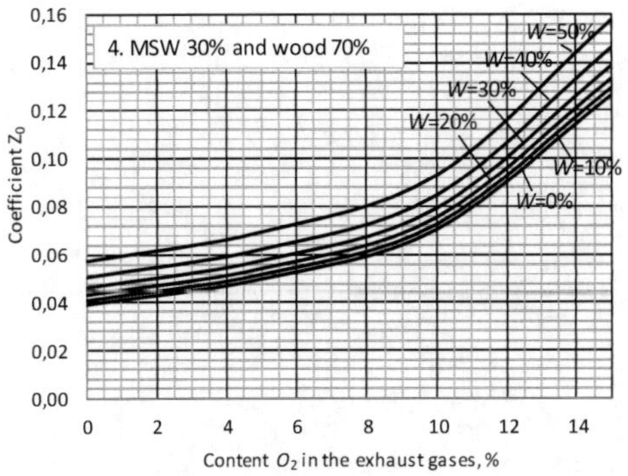

Figure 6. Coefficient *Z* (at given CO = 0%) for Fuel 4

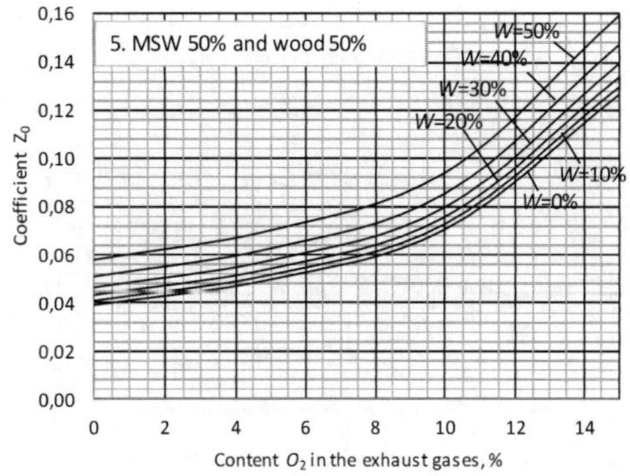

Figure 7. Coefficient *Z* (at given CO = 0%) for Fuel 5

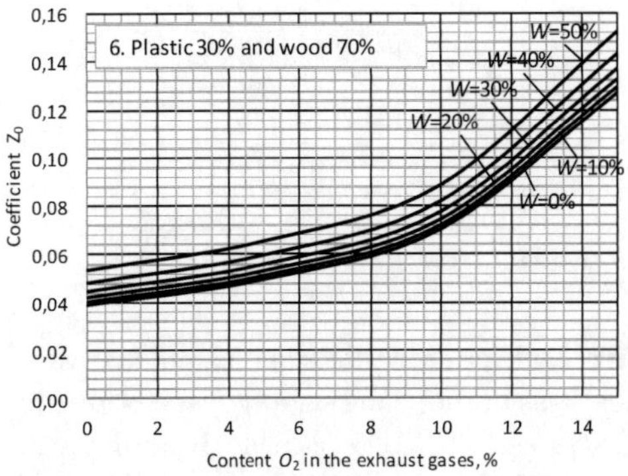

Figure 8. Coefficient *Z* (at given CO = 0%) for Fuel 6

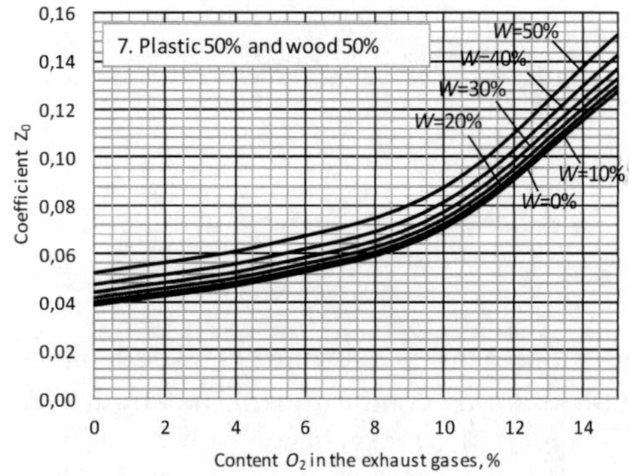

Figure 9. Coefficient *Z* (at given CO = 0%) for Fuel 7

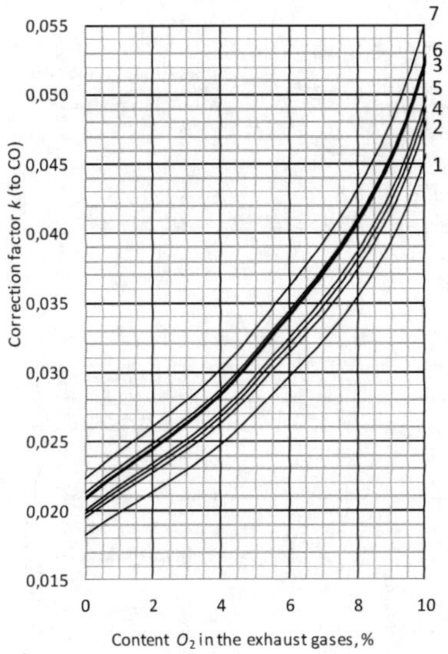

Figure 10. Correction coefficient *k* for RDF

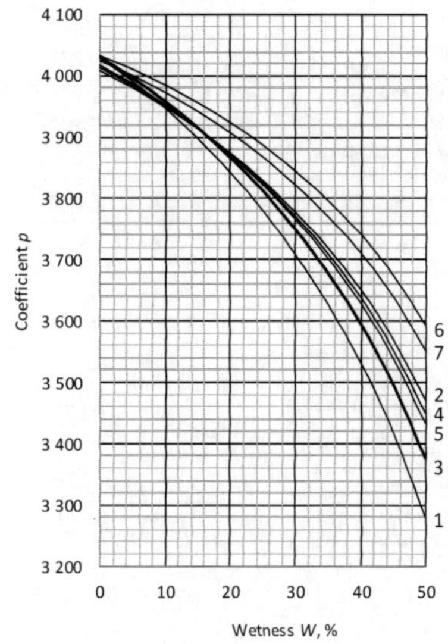

Figure 11. Coefficient *p* for RDF

4. Application of a method for the analysis of operation of the boiler

Utilizing proposed method, calculation of the losses in hot water solid fuel boiler with grate stoker and design power of 200 kWt during briquetted waste combustion was made. [28, 29]. Each measurement series correspond to single fuel stowing (figure 12-15).

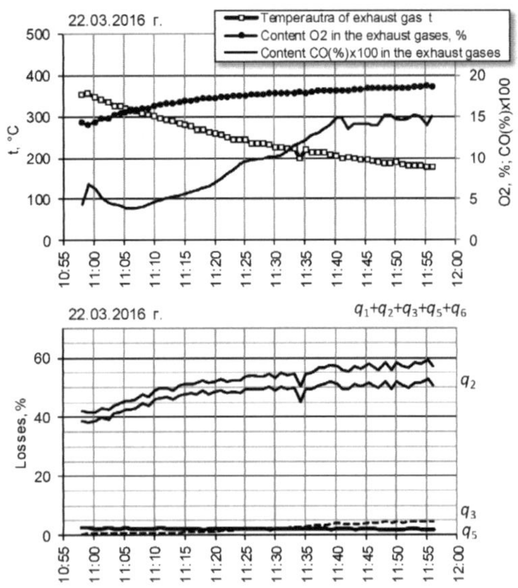

Figure 12. Measuring data and calculated values of heat loss during combustion of the briquettes, consisting of 70% of wood and 30% of plastic (Fuel 6)

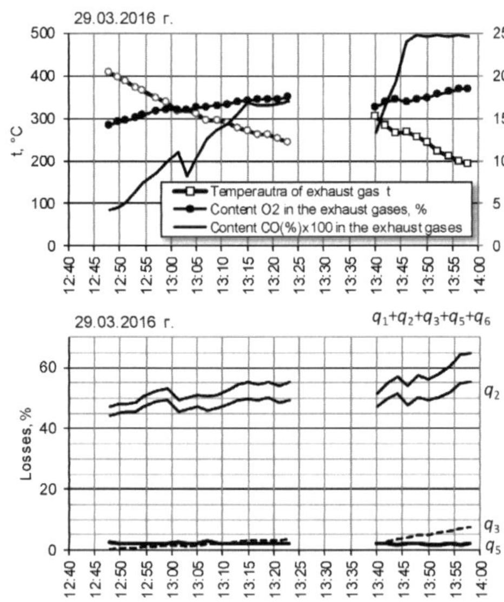

Figure 13. Measuring data and calculated values of heat loss during combustion of the briquettes, consisting of 70% of wood and 30% of plastic (Fuel 6). Boiler operated: 12:48…13:23 – with forced flow; 13:40…13:58 – without forced flow

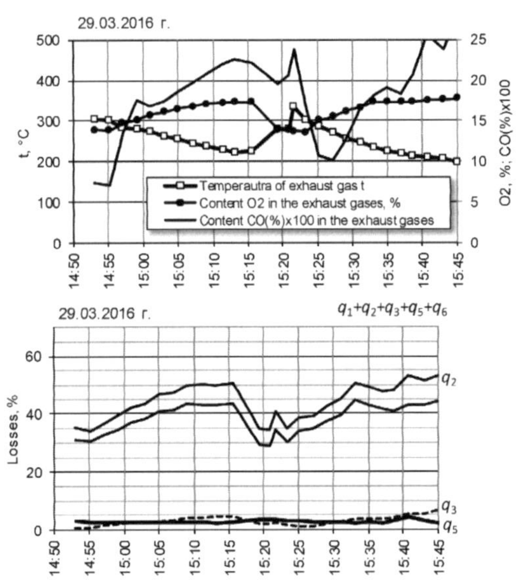

Figure 14. Measuring data and calculated values of heat loss during combustion of the briquettes, consisting of 70% of wood and 30% of plastic (Fuel 6)

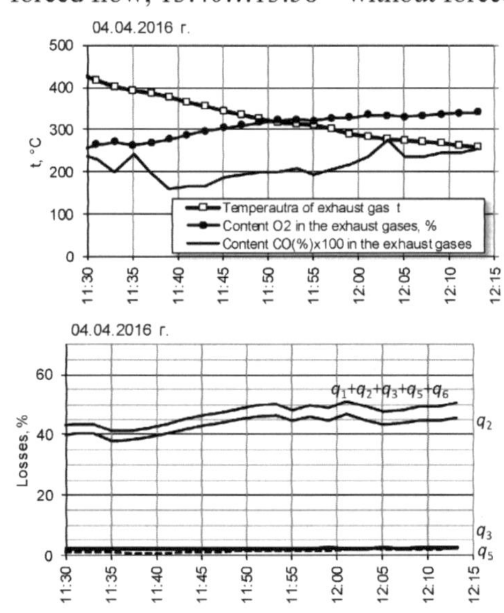

Figure 15. Measuring data and calculated values of heat loss during combustion of the briquettes, consisting of 50% of wood and 50% of plastic (Fuel 7)

Boilers operated with batch loading during the test, therefore listed data include not only active combustion period bat also fuel remains afterburning. This fact allowed to apply proposed method for the combustion mode with different efficiency.

Calculated losses were as such:

- with exhaust gases q_2..27,9…55,2%;
- with chemical incomplete combustion q_3........0,2…7,6%;
- into outside atmosphere q_5..............................1,6…4,3%;
- boiler's overall efficiency..............................35,2…67,7%.

Therefore, by using data presented in the tables and pictures, it is possible to calculate heat balance of boilers and furnaces utilizing simplified method for combustion of different types of waste.

The offered method can be also used when calculating burning of landfill gas which is characterized by non-constant composition [30].

5. Conclusions

Method of calculation of the heat leakage and efficiency of the plants for waste combustion, using method of indirect heat balance, based on data from gas analysis, was developed. Taking into account, that practically only oxygen and carbon dioxide is being measured in the exhaust gases, was developed calculation method for heat leakage with exhaust gases, using measurements of only this component, without utilizing carbon dioxide content in combustion products. Carbon dioxide and oxygen calculated percentage content in dry complete combustion products in actual air consumption was measured with consideration of incomplete combustion.

Heat leakage for low power boilers and furnaces was analyzed, heat leakage q_5 calculation method for this kind of plants was introduced. Based on design factors analysis of the solid fuel boilers for burning low-calorie fuel and wood waste, was developed dependence of the outside walls specific area of the boiler on the boiler's efficiency. This article introduced graphic dependences of the specific heat loss into the atmosphere, with different temperatures of the outside walls surface and atmosphere, on the amount of heat, that entered boiler or furnace and on the boiler's efficiency.

Based on the developed method for seven types of the RDF fuels, tabular and graphic dependences for reduced factors, characterizing heat leakage to the atmosphere and incomplete combustion, was introduced. It was discovered, that dependence of the coefficient Z on the CO content in the exhaust gases takes linear form, therefore to calculate amount of CO correction coefficient k should be used. Calculated tabular and graphic dependences can be used for plants operational evaluation during combustion of different types of waste.

Suggested method was utilized for calculating losses in solid fuel water heating boiler with grate stoker and designed output power of 200 kW during burning of the briquetted waste materials.

Acknowledgments

The article was prepared within development program of the Flagship Regional University on the basis of Belgorod State Technological University named after V.G. Shukhov.

Referens

[1] Pavlas M, Touš M, Bébar L and Stehlík P 2010 Waste to energy – An evaluation of the environmental impact *Applied Thermal Engineering* **30(16)** 2326-32

[2] Panepinto D, Blengini G A and Genon G 2015 Economic and environmental comparison between two scenarios of waste management: MBT vs thermal treatment *Resources, Conservation and Recycling* **97** 16-23

[3] Sever Akdağ A, Atımtay A and Sanin F D 2016 Comparison of fuel value and combustion characteristics of two different RDF samples *Waste Management* **47(B)** 217-24

[4] Dong C et al 2002 Experimental study on the co-firing of municipal refuse and coal in a circulating fluidization bed *Dongnan Daxue Xuebao, Ziran Kexueban* **32(1)** 95–9 [In Chinese]

[5] Boumanchar I, Chhiti Y, Alaoui F E M, Ouinani A E, Sahibed-Dine A, Bentiss F, Jama C, Bensitel M 2017 Effect of materials mixture on the higher heating value: Case of biomass, biochar and municipal solid waste *Waste Management* **61** 78-86

[6] Bessi C, Lombardi L, Meoni R, Canovai A and Corti A Solid recovered fuel: An experiment on classification and potential applications *Waste Management* **47(B)** 184-194

[7] Hupa M 2005 Interaction of fuels in co-firing in FBC *Fuel* **84(10)** 1312-9

[8] Kozhukhova N I, Lebedev M S, Vasilenko M I and Goncharova E N 2018 Fly Ash Impact from Thermal Power Stations on the Environment *J. Phys.: Conf. Ser.* **1066** 012010

[9] Suslov D Yu, Ramazanov R S and Shvydkaya M A 2018 To the Question of Biogas Combustion in Vortex Burners *J. Phys.: Conf. Ser.* **1066** 012028

[10] Yang Y B, Ryu C, Khor A, Yates N E, Sharifi V N and Swithenbank J 2005 Effect of fuel properties on biomass combustion. Part II. Modelling approach—identification of the controlling factors *Fuel* **84(16)** 2116-30

[11] Saxena S C 1998 Fluidized-bed incineration of solid pellets: Combustion and co-combustion *Energy Conversion and Management* **39(1–2)** 127-41

[12] Faitlia J, Magyara T, Erdélyib A and Murányic A Characterization of thermal properties of municipal solid waste landfills *Waste Management* **36** 213-21

[13] Malinauskaite J et al 2017 Municipal solid waste management and waste-to-energy in the context of a circular economy and energy recycling in Europe *Energy* **141** 2013-44

[14] Komilis D, Evangelou A, Giannakis G and Lymperis C 2012 Revisiting the elemental composition and the calorific value of the organic fraction of municipal solid wastes *Waste Management* **32(3)** 372-81

[15] Horttanainen M, Teirasvuo N, Kapustina V, Hupponen M, Luoranen M 2013 The composition, heating value and renewable share of the energy content of mixed municipal solid waste in Finland *Waste Management* **33(12)** 2680-6

[16] Vargas-Moreno J M, Callejón-Ferre A J, Pérez-Alonso J and Velázquez-Martí B 2012 A review of the mathematical models for predicting the heating value of biomass materials *Renewable and Sustainable Energy Reviews* **16(5)** 3065-83

[17] Savage G M 1989 Thermal conductivity and specific heat of densified refuse derived fuel *Waste Management & Research* **7(1)** 83-92

[18] Shi H, Mahinpey N, Aqsha A and Silbermann K 2016 Characterization, thermochemical conversion studies, and heating value modeling of municipal solid waste *Waste Management* **48** 34-47

[19] Vesilind P A, Martello W P and Gullett B 1981 Calorimetry of refuse derived fuels *Conservation & Recycling* **4(2)** 89-97

[20] Kuznetsov V A 2017 Mathematical Model of the Radiative Heat Exchange in the Selective Gases of a Diffusion Flame *Journal of Engineering Physics and Thermophysics* **90(2)** 357-65

[21] Suzuki T, Tsuruda T, Ogawa Y and Liao C 2005 A study on extinction of RDF (refuse derived fuel) pile *Fire Safety Science. In: Proceedings 8th International Symposium. International Association for Fire Safety Science (IAFSS), Beijing, 18–23 Sept 2005* pp 789-800

[22] Kuznetsov V A and Trubaev P A Resources and Problems of the Mathematical Simulating Thermo-Technological Processes *Journal of Physics: Conference Series* **1066(1)** 012024

[23] Ravich M B 1966 *Uproshhennaya metodika teplotexnicheskix raschyotov* [The simplified technique of heattechnical calculations] (Moscow: Izd-vo AN SSSR) [In Russian]

[24] Kuzneczov N V et all 1973 *Teplovoj raschet kotel`ny`x agregatov. Normativny`j metod* [Thermal calculation of boiler units. Standard method] (Moscow: E`nergiya) [In Russian]

[25] Kagan G M et al 1998 *Teplovoj raschet kotlov. Normativny`j metod* [Thermal calculation of boilers. Standard method] (Sankt Peterburg: Izdatel`stvo NPO CzKTI) [In Russian]

[26] Kutateladze S S 1990 *Teploperedacha i gidrodinamicheskoe soprotivlenie. Spravochnoe posobie* [Heat transfer and hydrodynamic resistance. Handbook] (Moscow: E`nergoatomizdat) [In Russian]

[27] Kacevich L S 1959 *Raschet i konstruirovanie e`lektricheskix pechej* [Calculation and designing of electric ovens] (Moscow: Gose`nergoizdat) [In Russian]

[28] Kornilova N V and Trubaev P A 2018 Analysis of MSW Combustion temperature in a hot water boiler with the low-capacity *Journal of Physics: Conference Series* **1066(1)** 012003

[29] Trubaev P, Kornilova N, Shein N and Verevkin O 2017 Study of refuse derived fuel use with pyrolysis furnace *Advances in Engineering Research: Actual Issues of Mechanical Engineering (AIME 2017)* vol 133 pp 866-71

[30] Trubaev P A, Verevkin O V, Grishko B M, Tarasyuk P N, Shchekin I I, Suslov D Yu and Ramazanov R S 2018 Investigation of Landfill Gas Output from Municipal Solid Waste at the Polygon *Journal of Physics: Conference Series* **1066(1)** 012015

IOP Conf. Series: Materials Science and Engineering **552** (2019) 012030 doi:10.1088/1757-899X/552/1/012030

Thermotechnical tests of the low power solid fuel heating boilers while incinerating briquetted RDF

N V Kornilova[1] and P A Trubaev[2]

[1]Department of Electric Power Engineering and Automation, Belgorod State Technological University of V.G. Shukhov, Kostyukov St., 46, Belgorod, 308012, Russia
[2]Department of Energy Engineering of Heat Technologie, Belgorod State Technological University of V.G. Shukhov, Kostyukov St., 46, Belgorod, 308012, Russia

E-mail: trubaev@gmail.com

Abstract. Objective of this paper is an analysis of briquetted waste combustion in the low power boilers. Research was conducted for the 100 kW pyrolysis heating boiler and 200 kW heating boiler with grate stoker. Briquettes were made by compacting milled and crushed materials into cylinders 50 mm in diameter, with a length of 100-200 mm, density of 800 kg/м3. It was established that process of combustion in low power boilers has low thermal efficiency, value of which fluctuated within 60...74% for wood combustion and 50...70% for RDF briquettes combustion. Burning process was stable and characteristics of the RDF combustion were similar to the wood combustion. Incineration of the fuel briquettes had better effectiveness than burning of solid wood. During RDF briquettes combustion high CO content (up to 0,2% or 2000 ppm) in flue gas was observed, which is an indication of insufficient burning time. It may be concluded that waste incineration in low-duty plants is viable, but to secure complete combustion and desired temperatures process of combustion and heat extraction should be separate in terms of design.

1. Introduction

Arranging for recycling of municipal solid waste (MSW) is a challenging technical problem, which consist of logistics management, environmental issues, energy and economics aspects [1]. In the paper [2] has been noted that MSW recycling is not just commercial investment project but also a need for society survival that cannot be overlooked.

Landfill is a main way of recycling of municipal solid waste (MSW) for most of the countries [3]. Ecological requirements, possibility of ground water contamination in particular, and landfill gas recycling [4] increases cost of waste disposal [5]. One of the most efficient ways of waste recycling is a thermal utilization with energy generation [5, 6], that allows to reduce costs for waste recycling when marketable energy is being produced in the process [7]. Table 1 presents comparison of the popular MSW recycling methods with energy production. Additional advantage of the recuperation of energy from waste by incinerating is a fact that for energy production can be used MSW with wide range of quality [8].

Table 1. Comparison of MSW recycling methods [9].

MSW recycling method	Anaerobic fermentation	Gasification	Incineration
Energy production per ton of waste	314 kWh of electricity	456 litre of synthetic gas (equivalent of 2676 kWh)	58,4 kWh of electricity
Type of waste	Food waste, green biomass	Wood, green biomass	All waste except metal, stone and glass particles
Necessityfor separation	Yes	Yes	Needs removal of non-combustible inclusions

Competitive advantage of electricity-producing waste treatment is determined by the following factors [2]:

- high cost of waste recycling on landfills;
- high cost of electricity in the region;
- low cost for collection and separation of waste;
- high effectiveness of waste incinerating plants.

From the economic point of view, in most cases, thermal utilization is more efficient in comparison to the other types of MSW recycling, which is stated in the paper [10]. Comparison of waste "life cycle" which includes economic and environmental aspects, that was carried out in the paper [11], has shown that biological way of recycling is more preferable from the environmental point of view, while incineration and gasification from the economical point of view. The paper [12] has stated that use of shredded RDF in conventional co-generating plants gives positive ecological and thermotechnical results.

Papers [5, 13] suggested criterions to conduct initial estimates for such projects. Decision for simultaneous production of electricity and heat is justified only if there is a large consumer of heat power available [14]. In the paper [15] analysis of RDF incineration at the power plant was performed using second law of thermodynamics. This analysis led to the conclusion that effectiveness of the system is very low and steam generation creates most of the energy losses. Similar conclusions were drawn in the paper [16], where energy analysis of a steam boiler incinerating waste presented. According to the analysis report, exergy efficiency of boiler units is a direct function of heated water and steam temperature in it. The paper [17] presents analysis of the other types of waste thermal utilization (gasification, pyrolysis, plasma technology, combinations), that suggests matching process and waste type, which makes those methods not versatile. Temperature of waste incineration in the heat-exchange unit entrance, not more than 950°C, guaranty that ash deposits, contaminating heating surfaces, will be non-coherent and can be easily cleaned out [7]. The highest efficiency of the plants for waste incineration is no more than 65-70% [7].

Use of alternative fuel has similar impact on the environment in comparison to traditional fuel in the power plants [6]. Partial presence of RDF in the fuel mixes decreased amount of SO_2 in the emissions and didn't change amount of NO_x [18]. Test of simultaneous burning of RDF and coal indicated that presence of RDF has no influence on SO* emissions, NO, HCI, N_20 emissions slightly increased, and amount of dioxin in ash and slag increased in sync with RDF content in the fuel [19]. Results in the paper [20] has shown that blending of solid municipal waste with biotype fuel improves its energy characteristics.

Waste recycling plants in USA produce electricity or work in co-generating mode, 5% of plants make steam for process purposes [21]. In addition, in spite of large number potential use of all waste corresponds to electricity generating output of more than 2800 MW, which is only 0,33% of the whole US generating market [22]. In Europe waste utilization structure differs from country to country [3]. In some countries (Greece, Hungary, Iceland, Ireland, Great Britain) most of waste goes to landfills; in Danmark, Switzerland, Luxemburg, Netherlands half of waste is being incinerated for energy production.

Waste incineration without preliminary treatment is being characterized by sufficient degree of reliability that is comparable to other types of heat energy and electricity production [7]. But as of now, incineration technologies developed for large scale production with waste utilization of 100 thousand tons per year [23, 24]. Some of the papers studied effectiveness of waste combustion for small towns. For example, based on analyses of cities with population of 100500 to 1 million people the paper [25] has shown increase of effectiveness of electricity production from waste instead of landfill utilization if city size increases. The paper [26] analyzed data for industrial combustion technologies and gasification for medium-sized municipals and indicated that the best economic results can be achieved by using unsorted municipal waste.

Existing situation overview suggests following conclusion that incineration is effective way of utilization if thermotechnical and ecological requirements are respected. But as of now industrial combustion technologies for MSW incineration has being developed only for utilization of 100 thousand tons per year (10 t per hour).

2. Materials and Methods

The papers objective was:

- analysis of briquettes combustion in the pyrolysis water heating boiler in comparison to solid wood, that by design supposed to be used in the boiler (Fuel 1 and Fuel 2);
- analysis of combustion of different composition briquettes in the water heating boiler with grate stoker (Fuel 3 and Fuel 4);
- analysis of flue gas content for different types of fuel combustion.

The paper examined three types of units that was used for waste incineration:

- 100 kW pyrolysis water heating boiler;
- 200 kW water heating boiler with grate stoker.

Boilers have been installed at the industrial site in Belgorod city (Russia) and designated to yield hot water for heating of two, 600 and 1900 m^2, administrative buildings. Thermal loads, taken from the building thermal design, are 32 and 86 kW respectively, that exceeds standard by approximately twofold (standard thermal loads for the mentioned above buildings are 13 and 41 kW). Heating system, in this example, is closed dependent type.

2.1. 100 kW pyrolysis boiler

Boiler "KO-100", made by "Boiler equipment factory in Kostroma city" (trademark "F.B.R.G.") (figure 1), is an all welded steel design that consists of two combustion chambers. At the lower chamber, which is a gasification chamber, where air flow is limited and temperature is high, fuel separates into solids and volatile fraction. Upper chamber is designed for afterburning of the released gas. Lower part of the gasification chamber has a nonadjustable window for an air intake and window with door to adjust oxygen intake, latter, according to the manual, should stay open only during boiler ignition. Smoke exhaust duct has damper.

Figure 1. Pyrolysis boiler KO-100

Pyrolysis boiler allows single loading of fuel to burn much longer and use as fuel large un-milled pieces of waste. According to the manual, boiler has following characteristics:

Heat power	100 kW;
Heated space area (floor to ceiling height 3 m)	900 m^2;
Efficiency	80-90%;
Operating temperature	90 °C;
Operating pressure	2-3 bar;
Furnace volume	0,65 m^3;
Flue gas duct diameter	250 mm;
Weight of the boiler	950 kg;
Fuel consumption	0,36 m^3/per day.

2.2. 200 kW water heating boiler with grate stoker

Solid fuel boiler made by OOO "TK Ecotrans" (figure 2) is an all welded steel design. Size: 1230 mm (width) x 1600 mm (depth) x 2170 mm (height). Fuel combustion is on the grate, air flow can be natural as well as by forced-flow fan. Boiler walls are insulated by the mineral wool Izovol CT-40

Figure 2. Solid fuel water heating boiler

2.3. Description of the incinerated samples

Solid wood and railway sleepers as well as cylindrical briquettes were used for measurements. Cylinder diameter is 50 mm, length 100-200 mm, density 800 kg/m^3, made by compacting grounded and crumbled wood. Materials were incinerated during tests and results presented in the Table 2.

Table 2. Incinerated waste characteristics.

Boiler type	Fuel number	Composition	Form
Pyrolysis boiler	1	Wood (pine tree)	Large ungrounded fragments
	2	Wood granules	Cylindrical briquettes
Boler with grate stoker	3	Wood 70%, plastic 30%	Cylindrical briquettes
	4	Wood 50%, plastic 50%	Cylindrical briquettes

For Fuel 3 and Fuel 4 humidity was measured by using method of drying out quantity of substance in the muffle furnace. Bulk humidity as follows: Fuel 3 - 13,6%, Fuel 4 - 13,5%.

2.4. Test schedule and instruments

Boilers has been tested between 2013 and 2016. Test schedule and measured parameters presented in the table 3. Table 4 presents instruments used for measurement.

Table 3. Measurements schedule.

Type of the boiler	Date of test	Time of test	Fuel number	Measurement of flue gas	Measurement of heat transfer agent characteristics	Measurement of temperature inside furnace
Pyrolysis	03.12.2013	13:00–14:00	1	X	–	–
	04.12.2013	13:40–15:50	2	X	X	–
With grate stoker	22.03.2016	10:48-11:56	3	X	X	–
	29.03.2016	12:48-13:58	3	X	X	X
		14:53-15:45	4	X	X	–
	04.04.2016	11:20-13:10	4	X	X	X

Table 4. Instruments used for measurement.

Name	Measured parameters
Gas analyzer Testo 330-1 LL	Gas temperature O2, CO content
Gas analyzer GANC-4	Mass concentration of hazardous substances (substance list is determined by availability of the cassettes)
Ultrasonic flowmeter "PanametricsPT878"	Volume flow rate of heat transfer agent
Infrared pyrometer Testo-845	Fuel surface layer temperature, temperature of pipe outer surface

2.5. Testing methods for boilers

Testing methods description:

1. Fuel load and setting boiler to the nominal operating mode.
2. Measurement of water temperature in supply and return pipelines and water flow in return pipeline with ultrasonic flowmeter Panametrics PT878 and pyrometer/contact thermometer Testo-845 (3 min interval).
3. Temperature and flue gas content (O_2, CO) measurement with gas analyzer "Testo 330-1 LL" (1 min interval).
4. Measurement of emissions in exhaust duct after the boiler (CO, NO_2, NH_3, mercaptans R-SH, H_2S, phenol) with portable gas analyzer GANC-4.
5. Measurement of fuel combustion temperature in the furnace with pyrometer Testo-845 (15-20 min interval).

Using obtained during tests data amount of produced heat energy, heat losses and boiler unit efficiency (by indirect heat balance method using simplified thermotecnical calculation technique) were calculated.

3. Results and Discussion

3.1. Effectiveness evaluation of waste incineration in pyrolysis boiler

It was established that during firewood combustion process of burning in the boiler is stable and disrupted only when firewood is being loaded. Closed flue damper doesn't interrupt combustion mode (no increase of CO or O_2 content), but flue gas temperature after the boiler decreases significantly. Efficiency of the boiler in stable operational mode is 67-74%, which is lower than numbers from the manual.

Two series of tests were conducted. First one evaluated operation of the boiler with closed door in order to stop air flow, during second test the door was partially opened in order to let air in. Tests results presented in the figures 3, 4. Briquettes loading happened at 13:42; 13:58 и 15:13

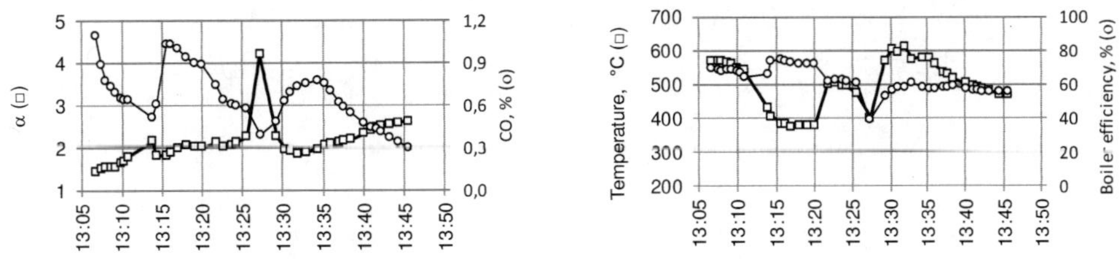

Figure 3. Measured results of wood incineration (fuel 1), 03.12.2013

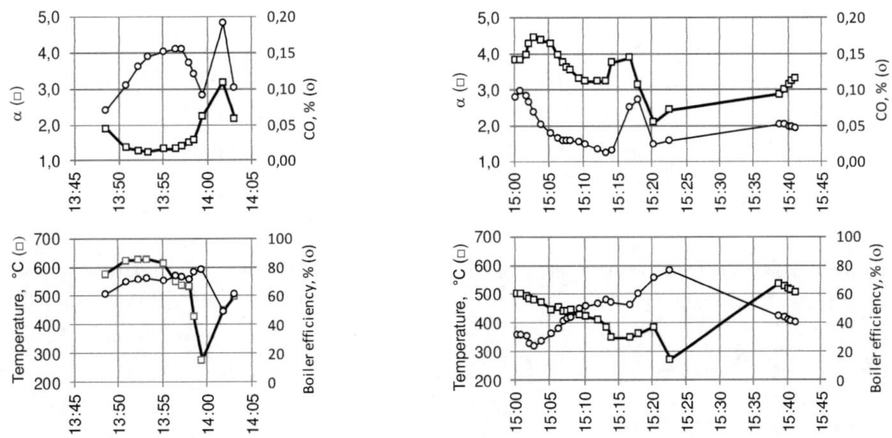

Figure 4. Measured results of wood briquettes incineration (fuel 2), 04.12.2013

3.2. Evaluation of boiler heat power

Since boiler operates in unstable mode, its capacity changes in the course of time. The following graph presents capacity that was received by calculation using measured temperature and flow rate of the heat transfer agent.

Capacity was determined by utilizing measured data of heat transfer agent flow rate V, m^3/sec, being heated in the boiler (measured with ultrasonic clamp-on flowmeter) and temperature of supply t_{sup} and return t_{ret} water, °C, measured with fixed temperature sensors, installed in supply and return water pipelines.

Boiler thermal output:

$$Q = 4{,}186 \cdot (t_{sup} - t_{ret}) V \rho, \text{ kW},$$

where t_{sup}, t_{ret} – supply and return water temperature, °C; V – volume flow rate, m^3/sec; ρ – water density, kg/m^3.

Heat generated from fuel incineration:

$$Q_f = Q / \eta,$$

where η – boiler efficiency, determined by data from gas analysis and indirect heat balance calculation utilizing simplified thermotechnical method.

Calculated capacity presented in figure 5.

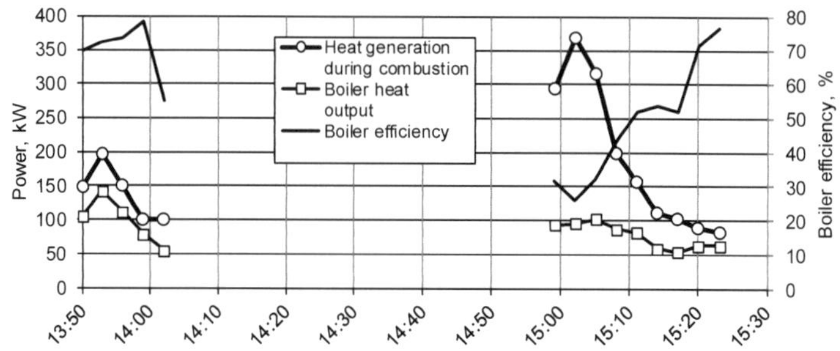

Figure 5. Boiler capacity for wood pellets incineration (Fuel 2) 04.12.2013

Conclusions after completed boiler thermotechnical tests for combustion of wood and briquettes (Fuel 1 and Fuel 2)

Combustion process in the boiler is stable and gets interrupted only at fuel loading.

Boiler mean hourly capacity for wood briquettes combustion is 95.6 kW, that is 5.6% more than boiler mean hourly capacity for firewood combustion.

Boiler efficiency doesn't exceed 74%, mean efficiency at stable operation mode – 60-74%, which is lower than indicated in the manual (84%). Fuel in the boiler burns out completely, indication of which is low CO content in flue gas.

3.3. Evaluation of effectiveness of briquetted waste incineration in the boiler with grate stoker

For measurements briquettes were incinerated. Briquettes had form of a cylinder with 50 mm in diameter, 100-200 mm in length, 800 kg/m^3 in density, made of compacted grated and crushed wood. Briquettes content: Fuel 3 (wood 70%, plastic 30%); Fuel 4 (wood 50%, plastic 50%). Tests continued for 3 days.

Table 5 presents average measured data from tests.

Table 5. Mean values of characteristics for each fuel incineration mode.

Date and time of measurement	Type of fuel	Flue gas temperature t_{fg}, °C	Content of CO, ppm	Excess air coefficient
22.03.16 10:48-11:56	Fuel 3	263	917	5,7
29.03.16 12:48-12:57		322	1012	4,3
29.03.16 13:03-13:13		291	1390	4,9
29.03.16 13:42-13:52		270	1823	5,4
29.03.16 14:53-15:03	Fuel 4	260	1668	4,9
29.03.16 15:31-15:43		279	1438	3,7
04.04.16 12:25-12:33		212	1832	10,6
04.04.16 12:37-12:47		235	1961	9,3
04.04.16 12:01-12:11		277	1305	5,1
04.04.16 11:41-11:51		283	1170	4,8

Figure 6 presents change in parameters of the boiler in the process of tests.

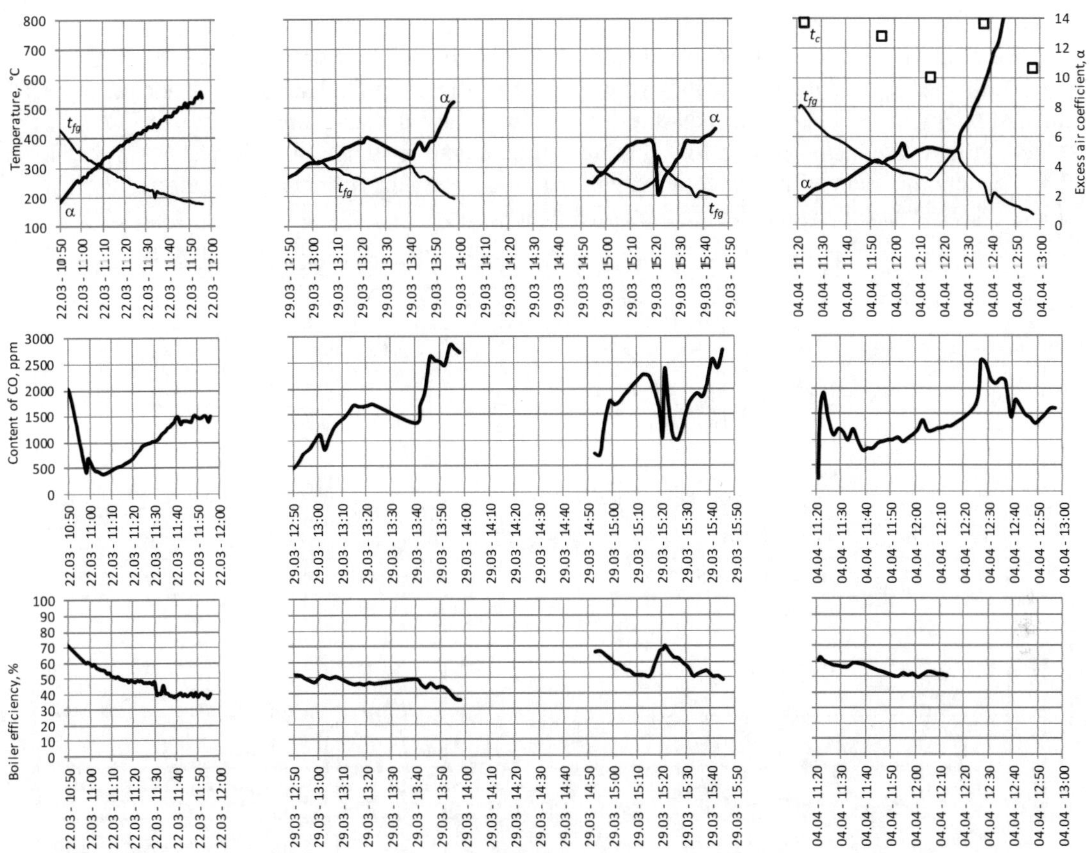

Figure 6. Results of the measurement of flue gas temperature t_{fg}, combustion temperature t_c, excess air coefficient α, CO content and thermal efficiency.

Analysis of dependencies between parameters (figure 7) has shown that there is a dependence between flue gas temperature t_{fg} and excess air coefficient α, correlation coefficient – 0,75. There is no dependency between flue gas temperature t_{fg} and CO content (correlation coefficient – 0,39) as well as between CO content and excess air coefficient α (correlation coefficient – 0,29). This suggests that burning time is not sufficient and heat extraction from gas decreases waste incineration effectiveness.

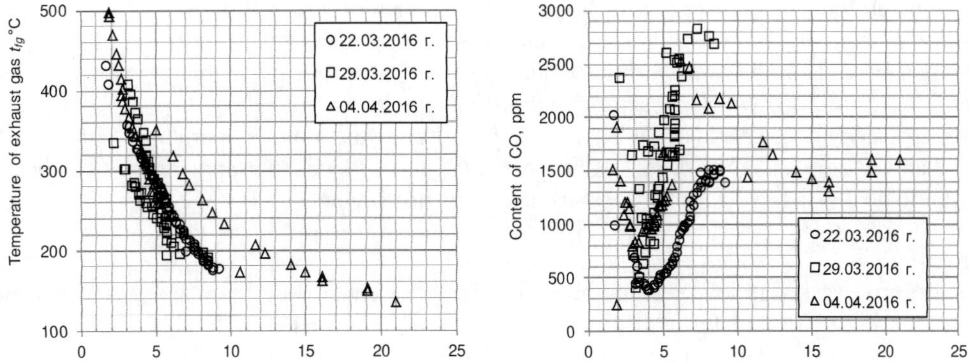

Figure 7. Dependence between parameters of combustion

Combustion temperature, measured on the third day of the test, was at 600...785°C, which is lower than theoretical values [27] and requirements in the Directive of the European Parliament and of the Council 2000/76/EC "On the incineration of waste". According to this document, plant can meet environmental requirements only if flue gas stays in it no less than 2 sec at a temperature of 850°C.

One can draw a conclusion that low combustion temperature is due to high excess air coefficient and process optimization of the incineration will help to meet the environmental requirements.

Figure 8 presents calculation data of the boiler heat power.

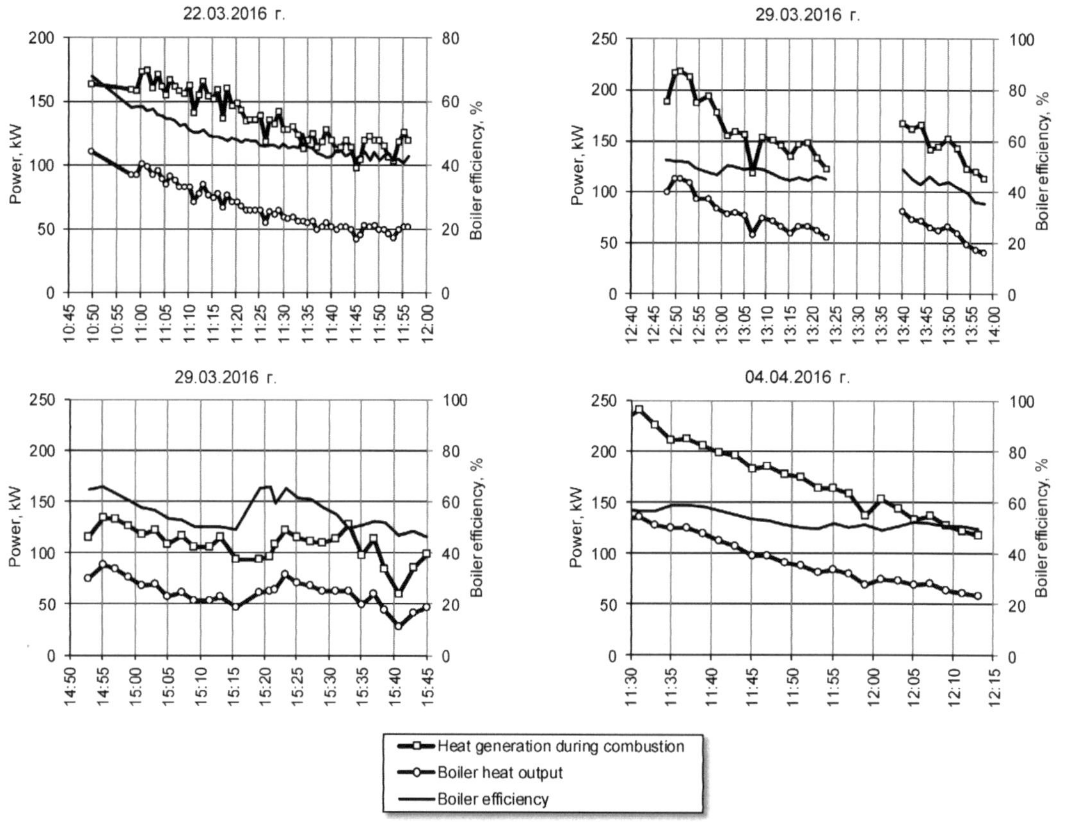

Figure 8. Boiler rated power

Conclusions for RDF briquettes incineration in the boiler with grate stoker

Combustion process in the boiler is stable and has direct dependence on burn-out time of fuel.

Mean hourly capacity for wood briquettes combustion is 50...110 kW, which is similar to the parameters for firewood and wood briquettes combustion.

Boiler efficiency doesn't exceed 70%, mean efficiency in stable mode operation – 50-60%, which is lower than certified data (84%). High CO content (up to 0,2% or 2000 ppm) in flue gas, when excess air coefficient is high, suggests that burn-out time is insufficient.

4. Conclusion

Combustion process in the low power boilers is characterized by low thermal efficiency, value of which varies within 60...70% for pyrolysis boiler and 50...70% for RDF briquettes combustion. Characteristics of RDF combustion matches up characteristics of wood and wood briquettes combustion. Combustion process was stable and plant thermotechnical parameters were in direct dependence on degree of fuel combustion. Fuel briquettes incineration had higher effectiveness in comparison with solid wood.

Flue gas, during tests, were examined for hazardous chemicals content (this paper doesn't mention about it). Findings have shown that for all tested materials estimated ground level concentration of hazardous chemicals didn't exceed maximum allowable concentration, and in most of the cases was significantly lower.

RDF briquettes incineration analysis have shown high content of CO (up to 0,2% or 2000 ppm) in flue gas at high value of excess air coefficient, which is the evidence of insufficient burn-out time.

In summary, waste incineration in the low-duty plants is feasible, but to secure complete combustion and necessary temperatures combustion process and heat extraction should be separated in terms of design.

Acknowledgments
The article was prepared within development program of the Flagship Regional University on the basis of Belgorod State Technological University named after V.G. Shukhov.

Reference

[1] Galvarino J, Balcazar C, Dias R A and Balestieri J A P 2013 Analysis of hybrid waste-to-energy for medium-sized cities *Energy* **55** 728-41.
[2] Miranda M L and Hale B 1997 Waste not, want not: the private and social costs of waste-to-energy production *Energy Policy* **25(6)** 587-600
[3] Malinauskaite an all 2017 Municipal solid waste management and waste-to-energy in the context of a circular economy and energy recycling in Europe *Energy* **141** 2013-44
[4] Trubaev P A, Verevkin O V, Grishko B M, Tarasyuk P N, Shchekin I I, Suslov D Yu and Ramazanov R S 2018 Investigation of Landfill Gas Output from Municipal Solid Waste at the Polygon *Journal of Physics: Conference Series* **1066(1)** 012015
[5] Haley C A C 1990 Energy recovery from burning municipal solid wastes: a review *Resources, Conservation and Recycling* **4(1–2)** 77-103
[6] Pavlas M, Touš M, Bébar L, Stehlík P 2010 Waste to energy – An evaluation of the environmental impact *Applied Thermal Engineering* **30(16)** 2326-32
[7] Morcos V H 1989 Energy recovery from municipal solid waste incineration – A review *Heat Recovery Systems and CHP* **9(2)** 115-26
[8] Bessi C, Lombardi L, Meoni R, Canovai A and Corti A Solid recovered fuel: An experiment on classification and potential applications *Waste Management* **47(B)** 184-94
[9] Smith R L, Sengupta D, Takkellapati S and Lee C C 2015 An industrial ecology approach to municipal solid waste management: II. Case studies for recovering energy from the organic fraction of MSW *Resources, Conservation and Recycling* **104(A)** 317-26
[10] Panepinto D, Blengini G A and Genon G 2015 Economic and environmental comparison between two scenarios of waste management: MBT vs thermal treatment *Resources, Conservation and Recycling* **97** 16-23
[11] Lombardi L, Carnevale E and Corti A 2015 A review of technologies and performances of thermal treatment systems for energy recovery from waste *Waste Management* **37** 26-44
[12] Zotto L D, Tallini A, Simone G D, Molinari G and Cedola L 2015 Energy Enhancement of Solid Recovered Fuel within Systems of Conventional Thermal Power Generation *Energy Procedia* **81** 319-38
[13] Smith R L, Sengupta D, Takkellapati S and Lee C C 2015 An industrial ecology approach to municipal solid waste management: I. Methodology *Resources, Conservation and Recycling* **104(A)** 311-6
[14] Caputo A C, Palumbo M and Scacchia F 2004 Perspectives of RDF use in decentralized areas: comparing power and co-generation solutions *Applied Thermal Engineering* **24(14–15)** 2171-87

[15] Shapiro H N and Kuehn T H 1980 Second law analysis of the Ames solid waste recovery system **Energy 5(8–9)** 985-91

[16] Li C, Gillum C, Toupin K and Donaldson B 2015 Biomass boiler energy conversion system analysis with the aid of exergy-based methods *Energy Conversion and Management* **103** 665-73

[17] Bosmans A, Vanderreydt I, Geysen D and Helsen L 2013 The crucial role of Waste-to-Energy technologies in enhanced landfill mining: a technology review *Journal of Cleaner Production* **55** 10-23

[18] Sever Akdağ A, Atımtay A and Sanin F D 2016 Comparison of fuel value and combustion characteristics of two different RDF samples *Waste Management* **47(B)** 217-24

[19] Dong C et al 2002 Experimental study on the co-firing of municipal refuse and coal in a circulating fluidization bed *Dongnan Daxue Xuebao, Ziran Kexueban* **32(1)** 95–9 [In Chinese]

[20] Boumanchar I, Chhiti Y, Alaoui F E M, Ouinani A E, Sahibed-Dine A, Bentiss F, Jama C, Bensitel M 2017 Effect of materials mixture on the higher heating value: Case of biomass, biochar and municipal solid waste *Waste Management* **61** 78-86

[21] Michaels T 2014 The 2014 ERC directory of waste-to-energy facilities Energy Recovery Council URL: http://energyrecoverycouncil.org/wp-content/uploads/2016/01/ERC_2014_Directory.pdf

[22] Murphy M L 2004 Waste-to-Energy Technology *Encyclopedia of Energy* ed C J Cleveland (Amsterdam; Boston: Elsevier Academic Press) pp 373-83

[23] Fernández-González J M, Grindlay A L, Serrano-Bernardo F, Rodríguez-Rojas M I and Zamorano M 2017 Economic and environmental review of Waste-to-Energy systems for municipal solid waste management in medium and small municipalities *Waste Management* **67** 360-74

[24] Pavlas M, Touš M, Bébar L, and Stehlík P 2010 Waste to energy – An evaluation of the environmental impact *Applied Thermal Engineering* **30(16)** 2326-32

[25] Leme M M V, Rocha M H, Lora E E S, Venturini O J, Lopes B M and Ferreira C H 2014 Techno-economic analysis and environmental impact assessment of energy recovery from Municipal Solid Waste (MSW) in Brazil *Resources, Conservation and Recycling* **87** 8-20

[26] Giugliano M, Grosso M and Rigamonti L Energy recovery from municipal waste: A case study for a middle-sized Italian district *Waste Management* **28(1)** 39-50

[27] Kornilova N V and Trubaev P A 2018 Analysis of MSW Combustion temperature in a hot water boiler with the low-capacity *Journal of Physics: Conference Series* **1066(1)** 012003

IOP Conf. Series: Materials Science and Engineering **552** (2019) 012031 doi:10.1088/1757-899X/552/1/012031

Development and research of low pressure injection burner for biogas combustion

D Y Suslov, R S Ramazanov, D O Temnikov and I V Lobanov

Department of Heat Supply and Ventilation, Belgorod State Technological University of V.G. Shukhov, Kostyukov St., 46, Belgorod, 308012, Russia

E-mail: Suslov1687@mail.ru

Abstract. The article is dedicated to the problem of using an alternative energy source – biogas in gas supply systems. A low pressure injection burner with a conical shaped heat diffuser has been designed. A diffuser placed in the burner body provides preheating of the gas-air mixture and allows to increase a flame propagation velocity. A computer simulation of the gas-air mixing process in a body has been made for 3 types of burners: without a diffuser, with a diffuser of 11 mm and 25.5 mm long. As follows from the simulation, 11mm diffuser has no adverse effect on the gas-air mixing process. While an increasing length of the diffuser up to 25.5mm leads to the decreasing cross-section area of a mixing chamber, therefore increasing air-gas mixture rate and methane concentration in the flame holes. Experimental tests have been performed in relation to biogas combustion process for those 3 types of burners. Experimental results proved that placement of 11mm diffuser in a burner body reduces the time for heating water in the tank up to 100°C by 6% (45.41 s). 25.5 mm diffuser reduces the heating time by 1.5% more (11.37 s). Thus, we can make a conclusion that the presence of a diffuser in the burner body increases a flame propagation velocity and a heat rate of the burner.

1. Introduction

One of the promising areas for the development of alternative energy, which is widely used in the European Union (EU) countries, is biogas [1-3]. The difference of biogas from the natural one consists in low methane (40-70%) content and the presence of impurities such as: carbon dioxide (30-40%), hydrogen sulfide (0-2%) and other gases (0-5%). Biogas is produced by biogas units in the process of anaerobic fermentation of organic substances. Energy plants and agricultural wastes are used as an initial substrate for biogas production [4-6]. The amount of produced biogas, as well as methane content depends on the composition of the organic part of the waste containing fats, proteins and carbohydrates [7, 8].

Most of biogas stations apply the produced biogas in cogeneration units or plants to generate electric and thermal energy [8, 9]. One of the promising fields for the use of biogas is its combustion in gas-burning units. A lot of works of our native and foreign scientists are devoted to the study of combustion processes of biogas and other LHV/LCV gases [10-17]. However, given the high content of carbon dioxide in the composition of biogas, the burning of biogas in traditional burners of natural gas will not be effective [16, 17]. Therefore, the design and testing of gas-burning equipment for the combustion of biogas with a high content of carbon dioxide is very relevant today.

2. Materials and Methods

At the present time, gas-burning units and plants are developed in such directions as [18-22]: the use of new materials; the use of devices improving gas-air mixing and the use of elements that provide pre-heating of the gas-air mixture in the burner body. Also biogas is burned with other fuels [23].

Here is the injection low pressure burner for biogas combustion that has been developed recently (Figure 1). The main design elements of the proposed gas burner are: a nozzle, a mixing chamber, exit holes for the gas-air mixture emission, a heat diffuser and primary air regulator.

The novelty of this design lies in the use of a cone-shaped heat diffuser, while the diffuser is connected with the burner cover at a smooth angle with a bend radius equal to the length of this diffuser. This helps to stabilize the gas-air mixture stream flowing inside the gas burner and decreases pressure losses. The application of a diffuser of the original form provides mixture preheating due to the process of heat transfer from the flame through the diffuser to the gas-air mixture itself. Mixture preheating allows to increase the speed of flame propagation and the temperature of combustion. The use of a primary air regulator allows to use biogas of different composition with a high efficiency factor.

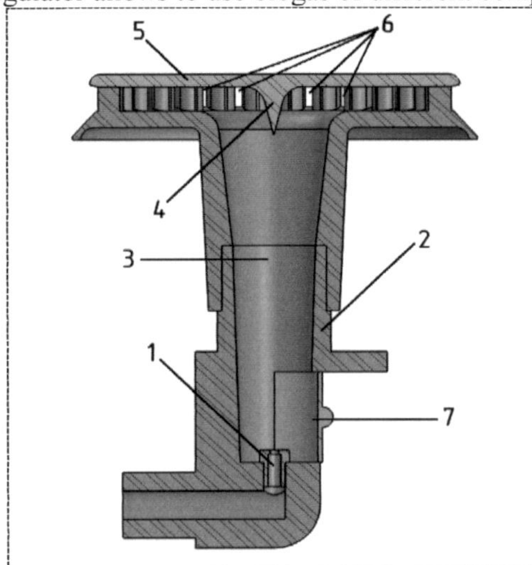

Figure 1. Gas burner: 1 - nozzle, 2 - burner body, 3 - mixing chamber, 4 - preheating diffuser, 5 - cover, 6 - holes, 7 - primary air regulator.

2.1. Theoretical research

To describe the gas-air mixture flow in the burner body, we'll use the continuity and Navier –Stokes time-averaged equations (Reynolds equation):

$$\frac{du_j}{dx_j} = 0 \tag{1}$$

$$\rho\frac{\partial u_i}{\partial t} + \rho u_j\frac{\partial u_i}{\partial x_j} = -\frac{\partial P}{\partial x_i} + \mu\frac{\partial^2 u_i}{\partial x_j \partial x_j} + \frac{\partial}{\partial x_j}\Pi_{ij} \tag{2}$$

where i,j=1,2,3; $u_1=u_x$; $u_2=u_y$; $u_3=u_z$ – time-averaged velocity vector projections; P – averaged pressure value; Π_{ij} – Reynolds stress tensor.

Reynolds stress tensor arising in the gas-air flow is defined by the expression:

$$\Pi_{ij} = -\rho < u_i' u_j' > \tag{3}$$

where u_i', u_j' – turbulent fluctuations of air-gas velocity projections, <...> – average criterion.

Considering a turbulent combustion of gas-air mixture in a burner body and at the exit of flame holes, we take k-ε model of turbulence.

Reynolds stress tensor of k-ε model is defined by the following expression:

$$\Pi_{ij} = -\rho\left(\frac{2}{3}\kappa\delta_{ij} - \nu_t\left(\frac{\partial u_i}{\partial x_j} - \frac{\partial u_j}{\partial x_i}\right)\right) \quad (4)$$

where κ – turbulent kinetic energy, ε – turbulent energy dissipation rate, ν_t – kinetic turbulent viscosity coefficient; δ_{ij} – Kronecker delta.

Mathematical modeling of unconventional fuel has difficulty [24, 25]. To study the process of gas-air environment formation in the body of the developed design burner, we'll use the simulation program complex Solid Works Flow Simulation.

2.2. Experimental research

Experimental studies of the combustion process in the burner were performed on a household gas stove Gefest 3100-07. We used a fast-acting burner with a nominal heat rate of N_{nom} = 3.05 kW, and maximum gas consumption L_{max} = 1076 l/h.

To determine the efficiency and optimal parameters of the diffuser, experiments have been performed using 3 types of burners (figure 2):

– without a diffuser;
– with 11 mm diffuser;
– with 25.5 mm diffuser.

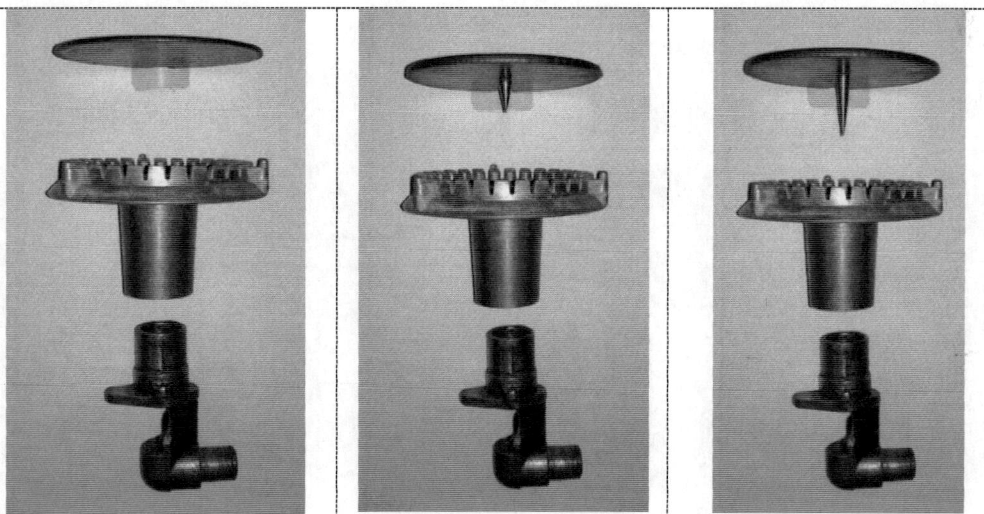

without diffuser | with diffuser L= 11 mm | with diffuser L= 25.5 mm

Figure 2. Burner designs.

Experiment procedure:

1. Measure the ambient temperature, °C;
2. Measure the temperature of cold water in a tank, °C;
3. Heat the water in a tank using the burner;
4. Measure the time of water heating in a tank 100°C.

The burner efficiency has been defined by the time spent on water heating in the tank up to 100 °C. The initial water temperature was 14°C. Finally, in order to get precise and reliable results of the experiment, we performed 3 different tests.

3. Results and Discussion

3.1. Theoretical research

The results of computer simulation provide visual pictures methane concentration and gas-air mixture velocity distribution in the burner body and at the exit of flame holes. Moreover, the concentration and gas-air mixture velocity have been measured on 7 different parts of the burner (figure 3).

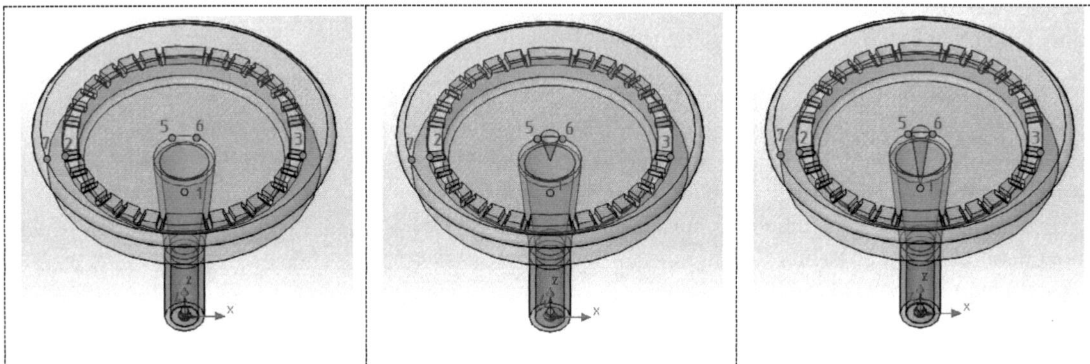

without diffuser | with diffuser L= 11 mm | with diffuser L= 25.5 mm

Figure 3. Points of measurement of concentrations and velocity of the gas-air mixture.

Simulation results of gas-air mixture velocity distribution in the body are given in figure.4.

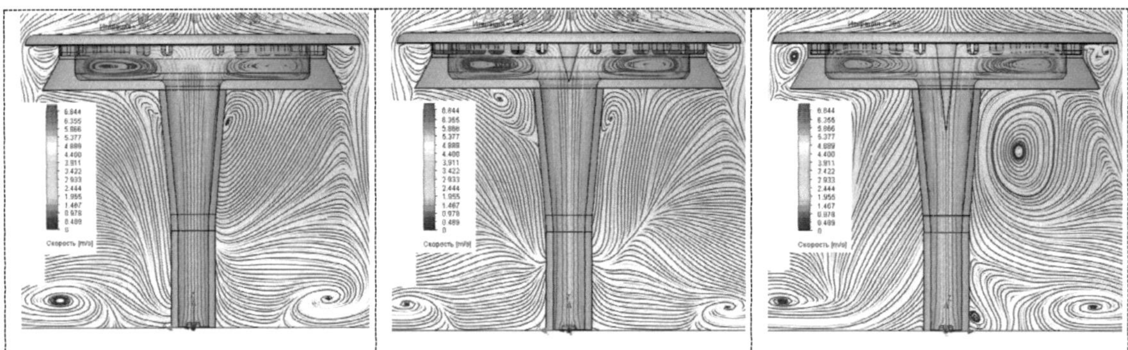

without diffuser | with diffuser L= 11 mm | with diffuser L= 25.5 mm

Figure 4. The velocity distribution of the gas-air mixture in the burner body.

Figure 4 shows that gas velocities in the central part of the burner body (point 1) without a diffuser and with a diffuser of L = 11 mm have similar values - 14.15 m/s and 14.18 m/s, respectively. And in the burner with a diffuser of L = 25.5 mm, we observe a significant speed increase (20,82 m/s).

The gas velocity at the bottom of the diffuser (points 5, 6) in the burner without the diffuser is 3.10 m/s, in the burner with the diffuser of 11 mm the speed is reduced to 2.54 m/s. And in the burner with a diffuser of 25.5 mm the speed increases to 3.45 m/s.

Simulation results of gas-air mixture velocity distribution in flame holes are given in figure 5.

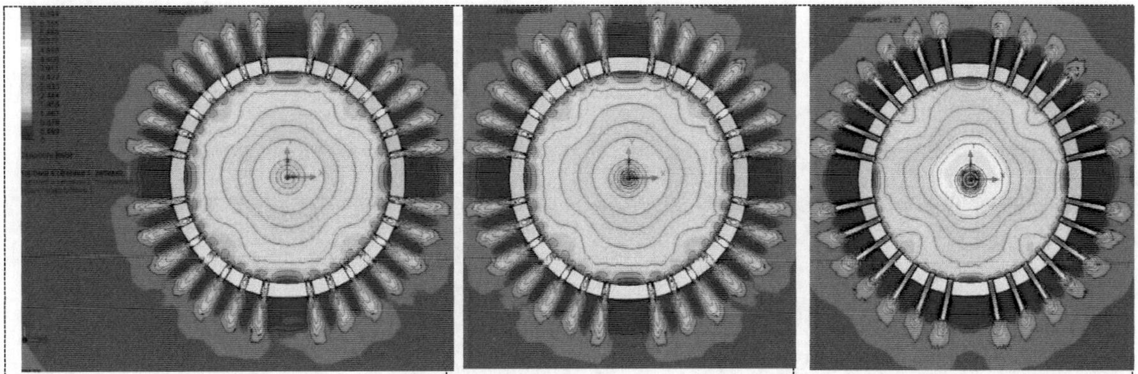

without diffuser	with diffuser L= 11 mm	with diffuser L= 25.5 mm

Figure 5. The velocity distribution of the gas-air mixture in the holes.

Figure 5 shows that gas-air mixture velocities at the exit from the flame holes (points 2, 3) in the burner without a diffuser and in a burner with a diffuser of L = 11 mm have similar values - 3.617 m/s and 3.620 m/s, respectively. Moreover, the speed of the mixture in the burner with a diffuser of L = 25.5 mm increases to 6.710 m/s.

Gas-air mixture velocities at a certain distance from the flame holes (point 7) in the burner without a diffuser and in the burner with a diffuser of L = 11 mm have similar values – 2.828 m/s and 2.834 m/s, respectively. The mixture speed in the burner with the diffuser of L = 25.5 mm increases as well (5.580 m/s).

The increased gas-air mixture velocity in the burner body with the diffuser of L = 25.5 mm is explained by the fact that the top of the diffuser is located in the mixing chamber leading to the reduction of cross-sectional area of the mixing chamber. Thus, the 11 mm diffuser does not affect the speed of the gas-air mixture in the burner body and at the flame holes. Reducing the cross-sectional area at a stable consumption rate increases the mixture flow rate. Thus, the diffuser length 11 mm does not affect the speed of the gas-air mixture in the burner body and at the flame holes.

Simulation results of methane concentration distribution in the burner body are given in figure 6.

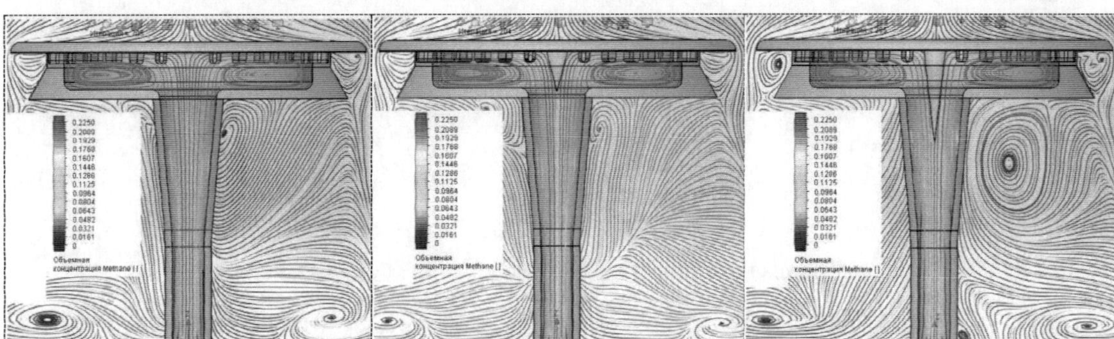

without diffuser	with diffuser L= 11 mm	with diffuser L= 25.5 mm

Figure 6. The distribution of methane concentration in the burner body.

Figure 6 shows that methane concentrations in the central part of the body (point 1) in the burner without a diffuser and in the burner with 11 mm diffusor have the same values – 26.5%, and in the burner with 25.5 mm diffuser the methane concentration increases to 31.2%.

The methane concentration at the bottom of the diffusor (points 5, 6) in the burner without the diffuser and in the burner with 11 mm diffuser have almost the same values – 21.6% and 21.7%, respectively. And in a burner with 25.5 mm diffuser, the concentration of methane slightly increases to 22.1%.

Simulation results of methane concentration distribution at the flame holes are given in figure 7.

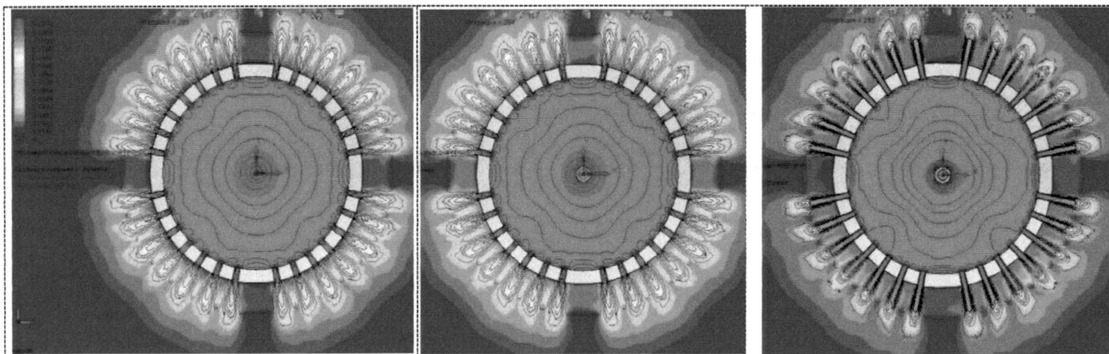

without diffuser	with diffuser L= 11 mm	with diffuser L= 25.5 mm

Figure 7. The distribution of methane concentration in the holes.

Figure 7 shows that methane concentrations at the outlet of the flame holes (points 2, 3) in the burner without a diffuser and in the burner with 11 mm diffuser have similar values - 20.5% and 20.7%, respectively. At the same time, the methane concentration in the burner with 25.5 mm diffuser slightly increases to 21%.

The methane concentration at some distance from the flame holes (point 7) in the burner without a diffuser is 17.4%. In a burner with 11mm diffuser, the concentration rises to 17.8%, and in a burner with 25.5 mm diffuser it rises to 20.2%.

The increase in methane concentration in burners with diffusers explained by the increase in the gas flow rate leading to a decrease in the amount of primary air.

According to the results of computer simulation, it can be concluded that 11 mm diffuser does not affect the distribution of methane concentration and the velocity of the gas-air mixture at the outlet of the flame holes. The 25.5 mm diffuser results in an increased gas-air mixture velocity and methane concentration.

3.2. Experimental research

The experimental results of the combustion process in the injection burner are given in table 1 and in figure 8.

Table 1. The results of experimental studies of the combustion process.

The burner design	Time, sec at °C									
	14	20	30	40	50	60	70	80	90	100
Burner without diffuser	0,00	64.94	132.30	201.75	271.67	344.04	419,88	499.89	603.62	762.97
Burner with a diffuser L=11 mm	0,00	74.43	142.10	208.73	276.03	346.27	419,45	499.58	596.03	717.56
Burner with a diffuser L=25.5 mm	0,00	69.57	135.56	196.98	265.68	332.32	403,61	477.69	571.54	706.19

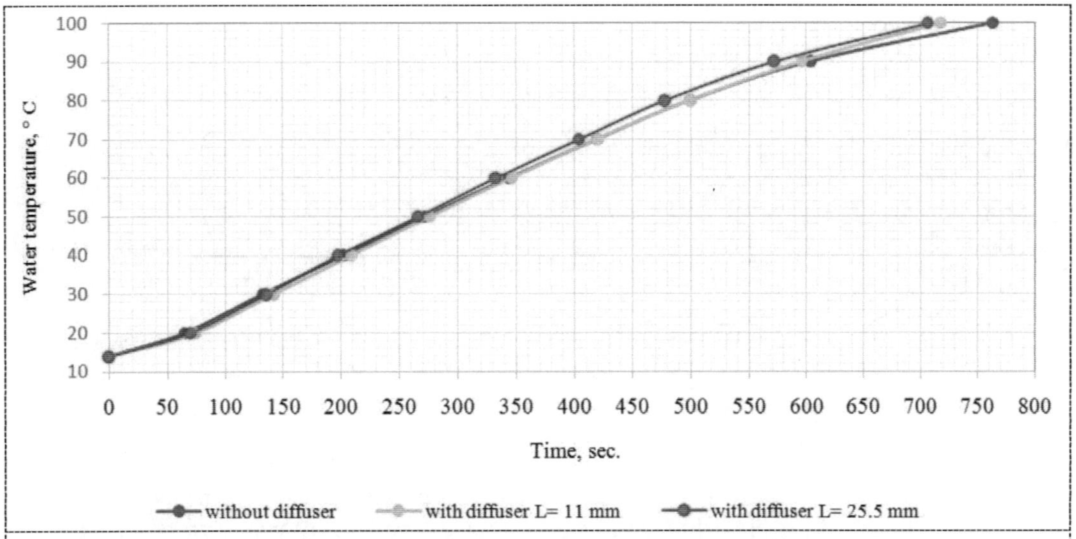

Figure 8. Graph of water temperature versus time.

At the beginning of the experiment, the water in the tank on the burner without a diffuser (132.3) is heated most quickly to a temperature of 30°C. This is explained by the fact that the part of the heat at the initial stage of the process in burners with diffusers is spent on heating this diffuser and, therefore, less heat is transferred to the water tank.

The water in the tank on a burner with the diffusor of L = 25.5 mm (403.61 s) is heated up to 70°C by 16 s faster than on a burner without a diffuser (419.88 s) and with a diffuser of L = 11 mm (419.45 s).

The water in the tank on a burner with the diffuser of L = 25.5 mm (706.19 s) is heated faster to a temperature of 100°C. The water on a burner with a diffusor L = 11 mm is heated in 717.56 s, on a burner without a diffusor – in 762.97 seconds.

Thus, 11 mm long diffuser in the burner body reduces the heating time of the water in the tank by 45.41 seconds or 6%. At the same time, an increase in the length of the diffuser to 25.5 mm shortens the heating time by 11.37 seconds or 1.5% more.

The experiment results prove that the presence of a heat diffuser in the burner allows to increase the flame propagation velocity due to a preheating of gas-air mixture in the burner body.

4. Conclusion

Here is the injection low pressure burner with a cone diffuser that has been developed recently. The theoretical studies of gas-air mixing process in the body proved by experimental tests have been performed for 3 types of burners: without a diffuser, with a diffuser of 11 mm and 25.5 mm long.

As a result of computer simulation, it has been established that the presence of a diffuser with a length of L = 11 mm has no any adverse effect on the distribution of methane and gas-air mixture velocity in the burner body and at the exit of the flame holes. An increased length L = 25.5 mm of a diffuser leads to the decreasing cross-section area of a mixing chamber, therefore increasing air-gas mixture rate (or speed) and methane concentration. Increasing the speed of the gas-air mixture can lead to disruption of the combustion process and flame lift-off.

Experimental results show that the presence of 11 mm diffuser in the burner body reduces the time for heating water in the tank to a temperature of 100°C by 6% (45.41 s). At the same time, 25.5 mm diffuser reduces the heating time by an additional 1.5% (11.37 s).

It can be concluded that the optimal solution to increase the efficiency and stability of the gas combustion process is to install an 11mm long diffuser in the burner body. This allows to increase flame propagation velocity and heat rate of the burner.

5. Acknowledgments
The reported study was funded by RFBR according to the research project № 18-38-00351.

References

[1] Wagner L 2015 Trends from the use of biogas technology in Germany *VIV Asia Biogas Conference on March 12th 2015 Bangkok* p 50
[2] Linke B 2015 *Country Report Germany IEA Bioenergy T 37*
[3] EHder B and SHul'c H 2008 *Biogazovye ustanovki* (perevod s nem. «Zorg Biogas») p 268
[4] Vedeneev A and Vedeneva T 2011 *Rukovodstvo po biogazovym texnologiyam* [Biogas Technology Guide] (DE`MI) p 84
[5] Kovalev A 2001 Effektivnost` proizvodstva biogaza na zhivotnovodcheskix fermax [Efficiency of biogas production on livestock farms] *Texnika v sel`skom xozyajstve* **3** pp 30-3 [In Russian]
[6] Kondaurov P and Marinenko E 2006 O celesoobraznosti proizvodstva i ispol`zovaniya biogaza v zamknutom cikle obrabotki otxodov sel`xozpredpriyatij s utilizaciej pobochnogo produkta anae`robnoj fermentacii [On the feasibility of the production and use of biogas in a closed cycle of processing agricultural waste using a by-product of anaerobic fermentation] *Izv. vuzov. Sev.-Kavk. region. Texn.* **10** p 105-10 [In Russian]
[7] Suslov D 2013 Diss. kand. texn. nauk Poluchenie biogaza v bioreaktore s barbotazhnym peremeshivaniem [Biogas production in a bioreactor with bubbling mixing] (Ivanovo: Ivanov. gos. xim.-tex. universitet) [In Russian]
[8] Ot polucheniya do ispol`zovaniya 2010 *Agentstvo po vozobnovlyaemy`m resursam 5-e polnost`yu pererabotannoe izdanie* (Gyul`czov Germaniya) p 214
[9] Biogas is based on renewable raw materials. *A comparative analysis of sixty-one biogas plants in Germany* (Germany: Fachagentur Nachwachsende Rohstoffe Gülzow) p 115
[10] Kushchev L A, Suslov D Y, Ramazanov R S and Shvydkaya M A 2017 Intensifikatsiya protsessa szhiganiya prirodnogo gaza v sistemakh gazosnabzheniya zhilishchno-kommunal'nogo khozyaystva [Intensification of the process of burning natural gas in gas supply systems for housing and utilities] *Vestnik BGTU im. V G Shukhova* vol 11 pp 95-9 [In Russian]
[11] Anggono W, Wardana ING, Lawes M et al. 2013 Biogas laminar burning velocity and flammability characteristics in spark ignited premix combustion *Journal of Physics: Conference Series* **423** 012015
[12] Suprianto F, Anggono W and Tanoto M 2016 Effect of Carbon Dioxide on Flame Characteristics in Biogas External Premix Combustion International Journal of Applied Engineering Research ISSN 0973-4562 vol 11 **4** pp 2240-3
[13] Lom L and Uil'yams A 1979 *Zameniteli prirodnogo gaza Proizvodstvo i svojstva* [Natural gas substitutes. Production and properties] red. N Fedorova (Moscow: Nedra) p 247 [In Russian]
[14] Anggono W, Wardana I, Pourkashanian M, Hughes K, Lawes M,Wahyudi S, Hamidi N and Hayakawa A 2012 Experimental and Numerical Simulation on Boigas Flame Propagation Characteristic in Spark Ignition Premixed Combustion Proceeding *The 3rd International Conference on Engineering and ICT ICEI2012* (Malaysia: Melaka) vol 2 pp. 290-4
[15] Yakovlev V and Komina G Optimizaciya vozduxoraspredelitel`ny`x sistem gazogorelochny`x ustrojstv s prinuditel`noj podachej vozduxa i shhelevoj kameroj smesheniya metodom matematicheskogo modelirovaniya [Optimization of air distribution systems of gas-burning installations with forced air supply and alkaline chamber for the method of mathematical modeling] *Vestnik grazhdanskix inzhenerov Sankt-Peterburgskij gosudarstvenny`j arxitekturno-stroitel`ny`j universitet* **2 (55)** pp 163-70 [In Russian]
[16] Sigal I, Marasin A, Smihula A, Sigal A and Kolchev V 2013 Experimental study of biogas burning and it use in industrial boilers *International Scientific Journal for Alternative Energy and Ecology* **17(139)** pp 84-9

[17] Ilbas M, Sahin M and Karyeyen S 2016 Combustion Behaviours of Different Biogases in an Existing Conventional Natural Gas Burner: An Experimental Study *International Journal of Renewable Energy Research* vol 6 **3** pp 1178-88

[18] Pelipenko V 2004 *Gorelki by`tovy`x gazovy`x plit Uchebnoe posobie* [Household gas stove burners. Schoolbook] (Tol`yatti: TGU) p 138 [In Russian]

[19] Bryuhanov O, ZHila V and A Pluzhnikov 2008 *Gazosnabzheniye uchebnoye posobiye dlya studentov vysshikh uchebnykh zavedeniy* [Gas supply study guide for university students] (Moscow: Izdatel'skij centr «Akademiya») p 448 [In Russian]

[20] Pelipenko V and Kruglov A 2007 Gorelki stola by`tovy`x gazovy`x plit *Santexnika, otoplenie, kondicionirovanie* [Burners table household gas stoves. Plumbing, heating, air conditioning] (Moscow: OOO "Izdatel`skij dom "MEDIA TEXNOLODZhI") **2**

[21] Ramazanov R, Suslov D and Kushhev L 2014 Razrabotka gazovoj gorelki dlya szhiganiya biogaza [Development of a gas burner for biogas combustion] *Naukoemkie texnologii i innovacii XXI nauchny`e chteniya* (Belgorod: Izd-vo BGTU) pp 195–8

[22] Ionin A, Zhila V, Arxitixovich V and Pshonik M 2012 *Gazosnabzhenie* (Izd-vo ASV) p 472

[23] Kornilova N V and Trubaev P A 2018 Analysis of MSW Combustion Temperature in a Hot Water Boiler with the Low-Capacity *J. Phys.: Conf. Ser.* **1066** 012003

[24] Kuznetsov V A and Trubaev P A 2018 Resources and Problems of the Mathematical Simulating Thermo-Technological Processes *J. Phys.: Conf. Ser.* **1066** 012024

[25] Kuznetsov V A 2017 Mathematical Model of the Radiative Heat Exchange in the Selective Gases of a Diffusion Flame *Journal of Engineering Physics and Thermophysics* **90(2)** 357-65

IOP Conf. Series: Materials Science and Engineering **552** (2019) 012032 doi:10.1088/1757-899X/552/1/012032

The study of energy performance of biogas from agricultural waste

D Y Suslov and R S Ramazanov

Department of Heat Supply and Ventilation, Belgorod State Technological University of V.G. Shukhov, Kostyukov St., 46, Belgorod, 308012, Russia

E-mail: Suslov1687@mail.ru

Abstract. The article is devoted to determination of energy indicators of the alternative source of energy - biogas derived from agricultural waste. The main energy indicators of the burning process of biogas fuel are lower heat value, Wobbe index, ignition range, flame propagation rate, as well as amount of oxygen and air required for complete combustion of biogas. The energy indicators of biogas produced from corn silage, technical fat, pig manure, cattle manure and poultry droppings are determined during the study. It has been found out that biogas obtained from technical fat and poultry droppings is characterized by the highest energy indicators, and biogas from corn silage - by the lowest indicators. Wobbe index of biogas is more than 2 times lower than index of natural gas. The biogas produced from technical fat has the highest rate of flame propagation - 23 cm/s, which is 1.5 times lower than the rate of flame of natural gas. Biogas has a wider ignition range than natural gas: the lower limit, the upper limit. The average value of the theoretic air volume required for complete combustion of biogas is 5.8 m^3, which is 1.7 times less than the value for natural gas.

1. Introduction

In many developed countries, much attention is paid to the development of alternative energy systems. One of the promising areas of alternative power generation is the production and use of biogas [1, 2]. Biogas is formed in the process of anaerobic fermentation of organic substances, which, as a rule, are waste products of various industries [3-6].

The most effective use of biogas technology is in agricultural enterprises. This is due to the proximity of the energy source of organic waste and the consumer of biogas, power plants of enterprises [7, 8]. Biogas is a gaseous fuel consisting of methane (40-70%), carbon dioxide (30-40%) and other gases (0-5%). The amount of biogas and methane produced depends on the composition of the organic part of the waste, containing fats, proteins and carbohydrates [4, 6, 9].

Currently, the biogas produced is mainly used in cogeneration plants to generate electricity and heat [4-6]. A promising direction for the use of biogas is burning in burner devices of boilers and furnace units [10]. The study of the combustion process of biogas and other low-calorie gases devoted a lot of work of domestic and foreign scientists [11-18]. However, given the high content of carbon dioxide in the composition of biogas, the burning of biogas in traditional burners of natural gas will not be effective [17, 18]. Therefore, it is relevant to determine the characteristics of the combustion process of biogas fuel with the high content of carbon dioxide and the development of burner equipment.

2. Materials and Methods

Most of the foreign and domestic biogas plants use corn silage, fats, pig manure, cattle manure and poultry droppings as an initial substrate [1-6]. The volume and composition of biogas obtained from different types of organic waste were determined as a result of experimental studies of the process of anaerobic fermentation at a temperature of 42 ° C (see table 1).

Table 1. The volume and composition of biogas obtained from different types of organic waste

Substrate type	The volume of biogas per ton of starting substrate	The volume of biogas per ton of dry matter	CH_4, %	CO_2, %	Other gases
Corn silage	213.5	710.5	51.9	45.4	2.7
Technical fat	1500.7	1509.8	67.4	29.8	2.8
Pig manure	14.7	299.6	60.8	36.5	2.7
Cattle manure	98.5	392.6	61	36.1	2.9
Poultry droppings	106.2	719.5	64.4	32.7	2.9

The main energy indicators of gaseous fuel are lower heat value, Wobbe index, flame propagation rate, lower and upper ignition limits, as well as the amount of oxygen and air required for complete combustion of biogas [19]. To determine the energy indicators of the burning process of biogas, equations based on chemical reactions of methane burning were used [20].

The lower calorific value of the gas is determined depending on the component composition of the gaseous fuel [21]. Given the composition of biogas and purification technology, in which hydrogen sulfide is removed from gas, the lower calorific value is determined by the composition of methane:

$$LHV = 357{,}97 CH_4 \tag{1}$$

357,97 – lower calorific value of the 1% methane;
CH_4 – methane content, % (by volume).

The Wobbe index is defined as the ratio of the volumetric lower calorific value to the square root of relative density of the gaseous fuel [22]:

$$W = \frac{LHV}{\sqrt{\rho_g / \rho_{air}}} \tag{2}$$

ρ_g – biogas density, kg/m^3; ρ_{air} – air density, kg/m^3.

The lower and upper limits of ignition of combustible gases is determined by the following formulas [21-24]:

$$LEL = \frac{100}{\frac{CH_4}{LEL_1}}; \tag{3}$$

$$UEL = \frac{100}{\frac{CH_4}{UEL_1}} \tag{4}$$

LFL_l, UFL_l – upper and lower limits for the ignition of methane, %.

The rate of flame propagation is an important energy parameter, which determines the combustion temperature and thermal power of gas equipment [25, 26].

The propagation rate of the flame front for combustible gas is determined by the formula [22-24]:

IOP Conf. Series: Materials Science and Engineering **552** (2019) 012032 doi:10.1088/1757-899X/552/1/012032

$$V_f = \frac{CH_4 \cdot V_{f1}}{100} \tag{5}$$

V_{f1} – normal flame propagation rate of methane, cm/s.

The main task for efficient and safe combustion of gaseous fuel is the supply to the place of combustion of the volume of oxygen required for complete combustion of 1 m^3 of gas of the given composition. In practice, gas units in the combustion area is supplied with atmospheric air, consisting of 79% nitrogen and 21% oxygen. If the amount of air is insufficient to completely burn the fuel, carbon oxide is formed in the combustion products, which is a highly toxic gas and dangerous to human health. Otherwise, when the amount of air significantly exceeds the required volume, a large amount of air remains in the combustion products, which is not involved in the combustion reaction and reduces the heat output of the burner.

The consumption of oxygen and air is to be determined on the basis of the reaction of combustion of methane in the air:

$$CH_4 + 2O_2 + 7{,}52\ N_2 = CO_2 + 7{,}52\ N_2 + 2H_2O \tag{6}$$

In accordance with the equation of combustion of methane, the volume of oxygen required for the combustion of 1 m^3 of biogas is determined by the formula:

$$V_{O2} = 0{,}01(2CH_4) \tag{7}$$

The volume of air required for complete combustion of the gas is calculated given the oxygen content:

$$V_{air} = \frac{100}{21} V_{O2} = 4{,}76 \cdot V_{O2} \tag{8}$$

3. Results and Discussion

The results of the calculation of the lower calorific value and Wobbe index are presented in figure 1.

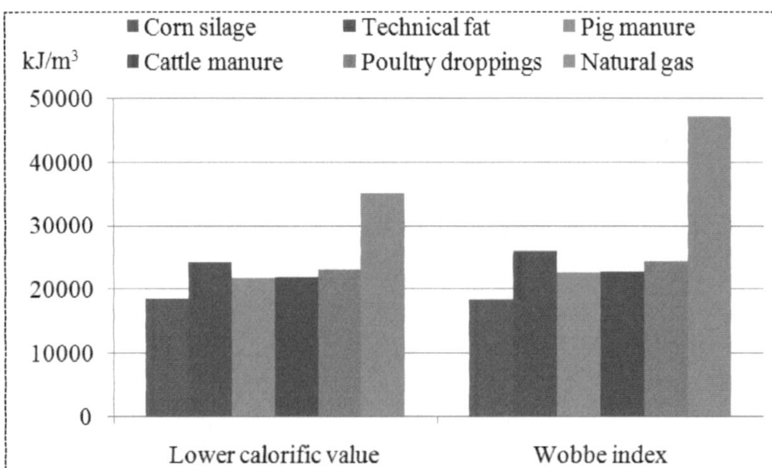

Figure 1. Lower calorific value and Wobbe index of natural gas and biogas of different composition.

Analyzing fig. 1, we can conclude that biogas obtained by processing technical fat and poultry droppings is characterized by the highest rates of lower calorific value and Wobbe index, while biogas from corn silage - by the lowest ones. It can be seen that the value of Wobbe index for biogas is more than 2 times lower than the value of natural gas.

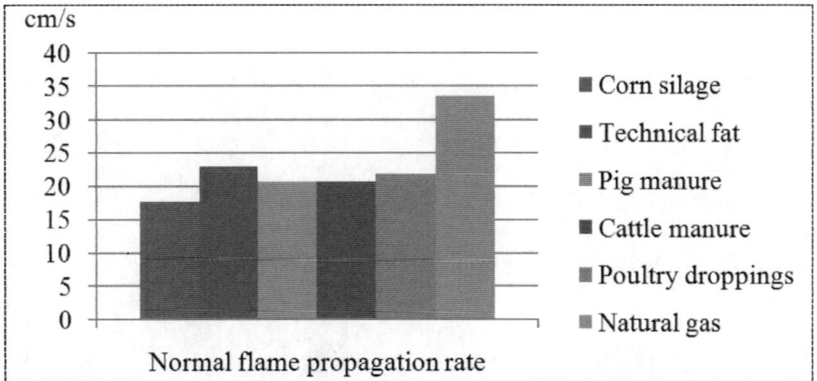

Figure 2. Normal flame propagation rate of natural gas and biogas of different composition.

From figure 2 it can be seen that the normal rate of flame propagation of biogas obtained from different types of organic waste varies in the range of 17-23 cm/s. The highest rate of flame is in biogas produced from technical fat – 23 cm/s, which is 1.5 times lower than the rate of flame of natural gas (34 cm/s).

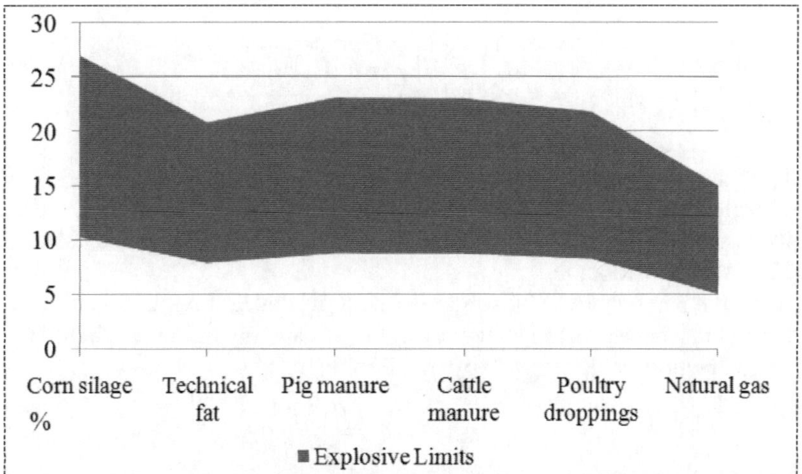

Figure 3. Explosive limits of biogas and natural gas

Taking into consideration the low methane content (figure 3), biogas has rather wide ignition range as compared to natural gas. Biogas from corn silage has the widest limits of ignition, and the narrowest ones - biogas from technical fat. So, in corn silage biogas, the lower limit is 10% and the upper limit is 27%. Due to the high lower limit of ignition, it can be concluded that biogas is a less explosive gas.

The obtained average value of the theoretic volume of air required for complete combustion of biogas is 5.8 m^3, which is 1.7 times less than the value for natural gas (figure 4).

From the comparison of the obtained energy parameters of biogas of different composition, it can be concluded that for efficient combustion of biogas of variable composition there is a need to improve the existing gas burner devices and to develop new gas burner devices.

During the operation of industrial gas units, there is a practice of transferring burners to work on a gas of a different composition due to changes in heat of combustion and gas density. To do this, one has to recalculate the design parameters of the burners [22, 28, 29].

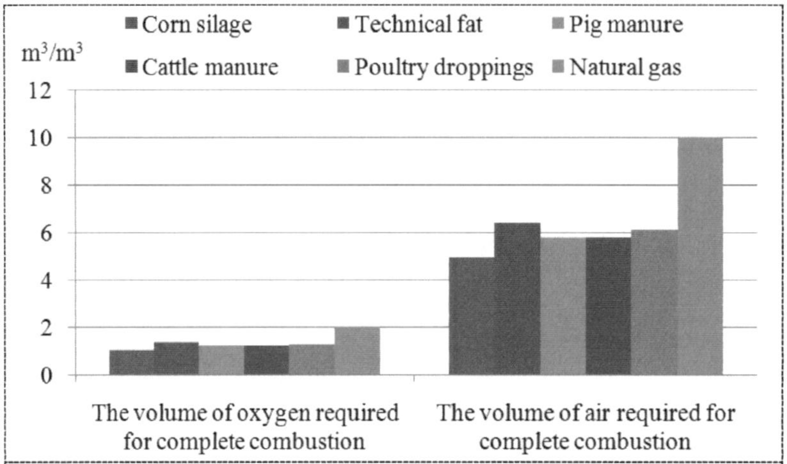

Figure 4. The volume of oxygen and air necessary for complete combustion of biogas and natural gas

To preserve the thermal power of the low-pressure gas injection burner when switching to gas fuel of the different composition, it is necessary to change the diameter of the gas nozzle, which is determined by the formula:

$$d_2 = d_1 \sqrt{\frac{LHV}{LHV'}} \sqrt{\frac{p_1 \cdot \rho_g'}{p_2 \cdot \rho_g}} \tag{9}$$

d_1 – the diameter of the gas nozzle opening when working on the gas of the initial composition, mm; LHV and LHV′ – calculated and actual combustion energy of the gas, kJ/m^3; $\rho_г$ and ρ'_g – calculated and actual density of the gas, kg/m^3; p_1 - calculated gas pressure, Pa; p_2 - gas pressure when working on gas of the different composition, Pa.

When calculating the parameters of blast burners with improved mixing, it is necessary to recalculate the area of the gas discharge openings, however, the gas and air velocities should be constant. The area of the discharge openings is determined by the formula, m^2:

$$F_1 = F\frac{LHV}{LHV'} \tag{10}$$

F – the area of the discharge openings during gas combustion of the initial composition, m^2.

Also, to preserve the heat output of the burner, it is possible to change the gas pressure before the burner while maintaining its structural dimensions, Pa:

$$p_2 = p_1\left(\frac{\rho_g'}{\rho_g}\right)\cdot\left(\frac{LHV}{LHV'}\right)^2 \tag{11}$$

However, in this case it is necessary to make a verification calculation so that the range of stable operation of the burner is not less than the set value.

Taking into consideration the variable composition of biogas, it can be concluded that for efficient combustion of biogas produced at various agricultural enterprises, it would be advisable to use a burner equipped with a device to control the flow of air into the burner body.

The design of a gas stove burner for biogas combustion has been developed. The main structural elements of the proposed gas burner are: a nozzle, a mixing chamber, gas-air mixture outlet openings, a preheating device and a regulator air primary (figure 5).

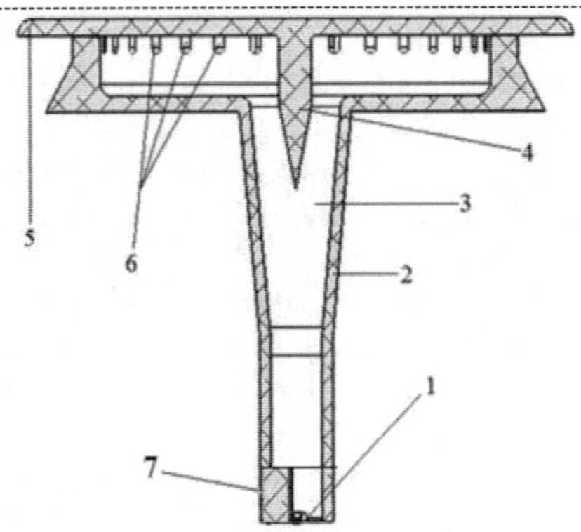

Figure 5. Gas burner: 1 - nozzle, 2 - burner body, 3 - mixing chamber, 4 - preheating device, 5 - cover, 6 - holes, 7 - primary air regulator

The use of pre-heating device provides heating of the gas-air mixture in the burner body. This allows you to increase the rate of flame propagation and the temperature of combustion. The use of a primary air regulator allows the use of biogas of different composition with high efficiency.

4. Conclusion
During the study, the energy indicators of biogas produced from various types of organic waste were identified: lower calorific value, Wobbe index, ignition range, flame propagation rate, volume of oxygen and air required for complete combustion of biogas. It has been studied out that biogas obtained from technical fat and poultry droppings is characterized by the highest energy indicators, and biogas from corn silage has the lowest indicators. At the same time, energy indicators of biogas are on average 1.5 times lower than those of natural gas. Biogas has a wider ignition range than natural gas, but the high value of lower ignition limit makes biogas less explosive. The design of a gas stove burner has been developed for efficient combustion of biogas of various compositions.

5. Acknowledgments
The reported study was funded by RFBR according to the research project № 18-38-00351.

References
[1] Wagner L 2015 Trends from the use of biogas technology in Germany *VIV Asia Biogas Conference on March Bangkok* p 50
[2] Linke B 2015 *Country Report Germany IEA Bioenergy T 37*
[3] Vedeneev A and Vedeneva T 2011 *Rukovodstvo po biogazovym texnologiyam* [Biogas Technology Guide] (DE`MI) p 84
[4] EHder B, SHul'c H 2008 *Biogazovye ustanovki* (perevod s nem. «Zorg Biogas») p 268
[5] Biogas is based on renewable raw materials. A comparative analysis of sixty-one biogas plants in Germany (Germany: Fachagentur Nachwachsende Rohstoffe Gülzow) p 115
[6] Ot polucheniya do ispol`zovaniya 2010 *Agentstvo po vozobnovlyaemy`m resursam 5-e polnost`yu pererabotannoe izdanie* (Gyul`czov Germaniya) p 214
[7] Kovalev A 2001 Effektivnost` proizvodstva biogaza na zhivotnovodcheskix fermax [Efficiency of biogas production on livestock farms] *Texnika v sel`skom xozyajstve* **3** pp 30-3 [In Russian]
[8] Kondaurov P and Marinenko E 2006 O celesoobraznosti proizvodstva i ispol`zovaniya biogaza v zamknutom cikle obrabotki otxodov sel`xozpredpriyatij s utilizaciej pobochnogo produkta anae`robnoj fermentacii [On the feasibility of the production and use of biogas in a closed cycle of processing agricultural waste using a by-product of anaerobic fermentation] *Izv. vuzov. Sev.-Kavk. region. Texn.* **10** p 105-10 [In Russian]

[9] Suslov D 2013 Diss. kand. texn. nauk Poluchenie biogaza v bioreaktore s barbotazhnym peremeshivaniem [Biogas production in a bioreactor with bubbling mixing] (Ivanovo: Ivanov. gos. xim.-tex. universitet) [In Russian]

[10] Kornilova N V and Trubaev P A 2018 Analysis of MSW Combustion Temperature in a Hot Water Boiler with the Low-Capacity *J. Phys.: Conf. Ser.* **1066** 012003

[11] Kushchev L A, Suslov D Y, Ramazanov R S and Shvydkaya M A 2017 Intensifikatsiya protsessa szhiganiya prirodnogo gaza v sistemakh gazosnabzheniya zhilishchno-kommunal'nogo khozyaystva [Intensification of the process of burning natural gas in gas supply systems for housing and utilities] *Vestnik BGTU im. V G Shukhova* vol 11 pp 95-9 [In Russian]

[12] Anggono W et al. 2013 Biogas laminar burning velocity and flammability characteristics in spark ignited premix combustion *Journal of Physics: Conference Series* **423** 012015

[13] Suprianto F, Anggono W and Tanoto M Effect of Carbon Dioxide on Flame Characteristics in Biogas External Premix Combustion International *Journal of Applied Engineering Research ISSN 0973-4562* vol 11 **4** pp 2240-3

[14] Lom L and Uil'yams A 1979 *Zameniteli prirodnogo gaza Proizvodstvo i svojstva* [Natural gas substitutes. Production and properties] red. N Fedorova (Moscow: Nedra) p 247 [In Russian]

[15] Anggono W, Wardana I, Pourkashanian M, Hughes K, Lawes M,Wahyudi S, Hamidi N and Hayakawa A 2012 Experimental and Numerical Simulation on Boigas Flame Propagation Characteristic in Spark Ignition Premixed Combustion Proceeding *The 3rd International Conference on Engineering and ICT ICEI2012* (Malaysia: Melaka) vol 2 pp. 290-4

[16] Yakovlev V and Komina G 2016 Optimizaciya vozduxoraspredelitel`ny`x sistem gazogorelochny`x ustrojstv s prinuditel`noj podachej vozduxa i shhelevoj kameroj smesheniya metodom matematicheskogo modelirovaniya [Optimization of air distribution systems of gas-burning installations with forced air supply and alkaline chamber for the method of mathematical modeling] *Vestnik grazhdanskix inzhenerov Sankt-Peterburgskij gosudarstvenny`j arxitekturno-stroitel`ny`j universitet* **2 (55)** pp 163-70 [In Russian]

[17] Kuznetsov V A and Trubaev P A 2018 Resources and Problems of the Mathematical Simulating Thermo-Technological Processes *J. Phys.: Conf. Ser.* **1066** 012024

[18] Sigal I, Marasin A, Smihula A, Sigal A and Kolchev V 2013 Experimental study of biogas burning and it use in industrial boilers *International Scientific Journal for Alternative Energy and Ecology* **17(139)** pp 84-9

[19] Ilbas M, Sahin M and Karyeyen S 2016 Combustion Behaviours of Different Biogases in an Existing Conventional Natural Gas Burner: An Experimental Study *International Journal of Renewable Energy Research* vol 6 **3** pp 1178-88

[20] Fischer M and Jiang X 2016 The chemical effects of CO_2 addition to methane on aromatic chemistry *Fuel the science and technology of Fuel and Energy* vol 183 pp 386-95

[21] Gus'kov B and Kryazhev B 1982 *Gazifikaciya promyshlennyh predpriyatij. Uchebnik dlya tekhnikumov* [Gasification of industrial enterprises. Textbook for technical schools] (Moscow: Strojizdat) p 368 [In Russian]

[22] Bryuhanov O, ZHila V and A Pluzhnikov 2008 *Gazosnabzheniye uchebnoye posobiye dlya studentov vysshikh uchebnykh zavedeniy* [Gas supply study guide for university students] (Moscow: Izdatel'skij centr «Akademiya») p 448 [In Russian]

[23] Bakulin V, Breshchenko E, Dubovkin N and Favorskij O 2009 *Gazovyye topliva i ikh komponenty. Svoystva, polucheniye, primeneniye, ekologiya. Spravochnik* [Gas fuels and their components. Properties, production, use, ecology. Directory] (Moscow: Izdatel'skij dom MEHI) p 214 [In Russian]

[24] Korol'chenko A 2007 *Processy goreniya i vzryva* [Combustion and explosion processes] (Moscow: Pozhnauka) p 266 [In Russian]

[25] Bradley D, Hicks R, Lawes M, Sheppard C and Wooley R 1998 The measurement of laminar burning velocities and Markstein numbers for isooctane-air and isooctane-n-heptane-air

mixtures at elevated temperatures and pressures in an explosion bomb *Combustion and Flame* **115** pp 126-44

[26] Gu X, Haq M, Lawes M and Wooley R 2000 Laminar burning velocity and Markstein lengths of methane-air mixtures *Combustion and Flame* **121** pp 41-58

[27] Serrano C, Hernandez. J, Mandilas C, Sheppard C and Woolley R 2008 Laminar burning behaviour of biomass gasification-derived producer gas *International Journal of Hydrogen Energy* **33** pp 851–62

[28] Isserlin A 1966 *Gazovye gorelki Izdanie 2-e pererabotannoe i dopolnennoe* [Gas burners Edition 2nd revised and enlarged] (Leningrad Moscow: Nedra) p 234 [In Russian]

[29] Blinov E 2007 *Toplivo i teoriya goreniya Razdel – podgotovka i szhiganie topliva* Ucheb.-metod. kompleks [Fuel and the theory of combustion. Section - preparation and burning of fuel. The method of teaching complex] (St. Petersburg: Izd-vo SZTU) p 119 [In Russian]

IOP Conf. Series: Materials Science and Engineering **552** (2019) 012033 doi:10.1088/1757-899X/552/1/012033

Analysis of the environmental factors influence on the efficiency of photovoltaic systems

J N Zatsarinnaya, D I Amirov and LV Zemskova

Department of Electric Stations, Kazan State Power Engineering University, Krasnoselskaya St., 51, Kazan, 420066, Russia

E-mail: waysubbota@gmail.com

Abstract. The development of civilization is associated with the creation of various methods of energy conversion. Humanity has come a long way from the water wheel to modern power plants. Though earlier people did not think about the damage to the environment from the work of a number of power plants, today this cannot be ignored. Renewable energy sources (RES) and their using by modern technologies claim to play a significant role in the world energy sector. Support for the development of RES in Russia is currently based on compensation of costs under contracts for the sale of power in the wholesale electricity and power market. According to the results of the competition of the Trading System Administrator, projects under the contracts for renewable energy power supply selected. Analysis of bids of the Trading System Administrator (TSA) showed that until 2023 will be introduced 1.8 GW of solar power plants, so the actual goal will be the correct operation of them in the weather conditions of Russia. Pollution of various types, which can be on the panels can affect the operation of solar panels (SP) and prevent the ingress of solar radiation on them. The article investigates the influence of various types of pollutants on the power generated by solar panels and promising places for the construction of solar plants in the climatic and geographical conditions of Russia, taking into account possible polluting factors, which can lead to total economic damage of up to 15 million rubles per year.

Keywords: solar panels, renewable energy sources, operation of solar modules, pollution of the solar panel, types of pollution, causes of deterioration of the solar module, the efficiency of the solar battery, the power of solar panels

1. Introduction

The development of civilization is associated with the creation of various methods of energy conversion. Humanity has come a long way from the water wheel to modern power plants. Though earlier people did not think about the damage to the environment from the work of a number of power plants, today this cannot be ignored. Nowadays, it is clear to most people that the generation of energy at thermal power plants leads to pollution of the components of the hydrosphere, atmosphere and lithosphere. Now humanity has to solve issues related not only to the creation of new energy sources, but also to their impact on the environment. For this reason, the key trend in the development of the electric power industry worldwide is the transition to renewable energy sources, which now seems to be the most successful in terms of environmental impact.

1.1. Actual state of renewable sources in the world

RES and their use of modern technologies claim to play a significant role in the global energy sector, as evidenced by the continuous growth of investments in the global energy sector, which in 2017 amounted to 279.8 billion US dollars. The past 2017 was a record year in the growth of renewable energy capacities for the entire period of their development, a similar trend continues in 2018. Since the beginning of 2000, investments in RES show stable growth or a very slight decline, which can not boast of investments in other sectors of the economy, which experienced a significant decline due to the global crisis of 2008. In 2017, investments in green energy increased by 2% compared to 2016, but this is 13% lower than the record high in 2015, which can be explained by a decrease in the specific capital costs of RES equipment. The rapid decline in the cost of equipment led to the fact that with slightly large investments in 2017, the commissioning of power using from RES was 90 GW, and a year earlier only 85 GW.

The leading countries for investment in renewable energy are the United States, Latin America, China, Germany, Italy and India. Moreover, a large share of investments (63%) is accounted for by countries with developing economies, of which 45% are only Chinese investments, which is 10% more than in 2016, the US – 14%, Asia and Oceania (excluding China and India) – 11%, North and South America (excluding Brazil and the US) – 5%, India – 4%, the middle East and Africa – 4%, Brazil – 2% [6, 12-18].

1.2. Development of RES in Russia

In Europe, the development of renewable energy began in the 2000s, in Russia, the first steps for the development of renewable energy were taken by the state in 2009. An order of the Government of the Russian Federation was adopted, in 2013 the state program "energy Efficiency and energy development" [2] was adopted, according to which 6.2 GW of power plants on renewable sources should be introduced in the country by 2020. At the beginning of 2018, the preliminary results of the program implementation were considered, which indicate a low level of budget expenditures, in particular for RES.

Support for the development of RES in Russia is currently based on compensation of costs under contracts for the sale of power in the wholesale electricity and power market. This incentive mechanism has been widely used in various sectors of the Russian economy and was able to be used in the development of pure energy through the adoption of Government Resolution No. 449 on 28 may 2013 [3].

The resolution applies to green energy generating facilities with a power of 5 to 25 MW. At the same time, the volume, structure and price of electricity are determined by the Government, based on the guaranteed return on investment for 15 years. Regulatory levels of capital and operating costs have been introduced, allowing total costs to be controlled and tariff growth to be limited. The selection of projects on RES contracts for the sale of power is carried out on the basis of the competition of the administrator of the trading system (TSA). The planned and actual results of the selection of bids from 2013-2018 (for the year of delivery) are shown in Figure 1-3, the criteria of which take into account the claimed specific capital and operating costs. There is also a diagram (Figure 4) illustrating the share of newly introduced wind farms (WF), solar power plants (SPP) and small hydropower plants (SHPP) expected in 2013 by 2020, as a result of the selection of bids until July 2018 [4, 23-24].

It can be stated that by 2020 it was planned to introduce 3.6 GW of wind farms, but only 3.4 GW were selected by the results of the competition; out of the planned 1.5 GW for the share of solar plants, 1.8 GW were selected; while only 0.14 GW of the planned 0.75 GW of small hydropower plants will be introduced by 2023. According to the government decree, by 2020 it was planned to maintain about 6 GW of generated capacities for RES, but in 2018 the total power`s value of the selected applications for the date of commissioning in 2023 in the amount of 5.4 GW, which with some lag corresponds to the government plan, but will not prevent from being implemented in full.

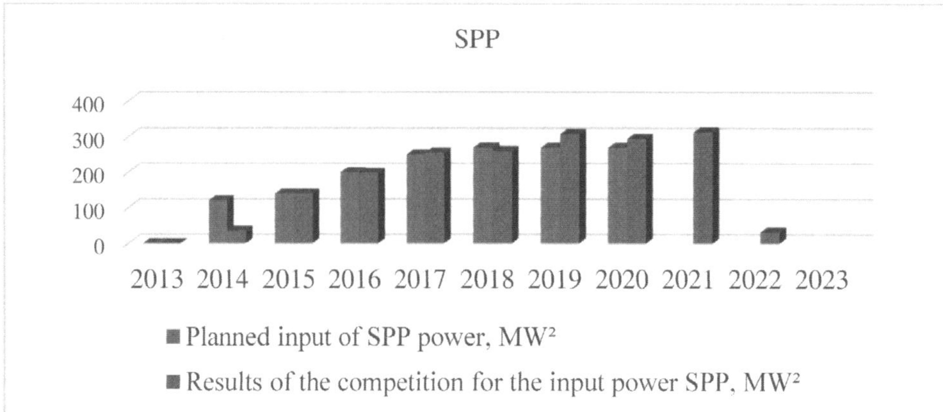

Figure 1. Comparative analysis of planned and commissioned of solar energy power.

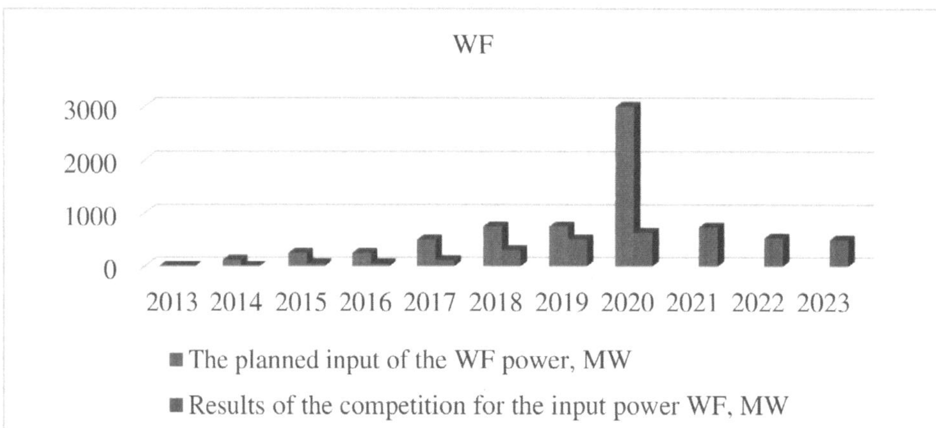

Figure 2. Comparative analysis of planned and commissioned of wind energy power.

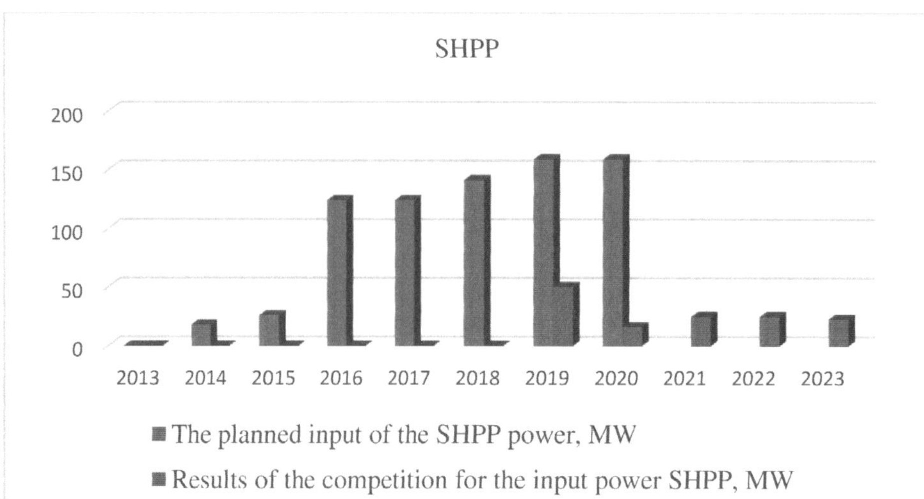

Figure 3. Comparative analysis of planned and commissioned of SHPP power.

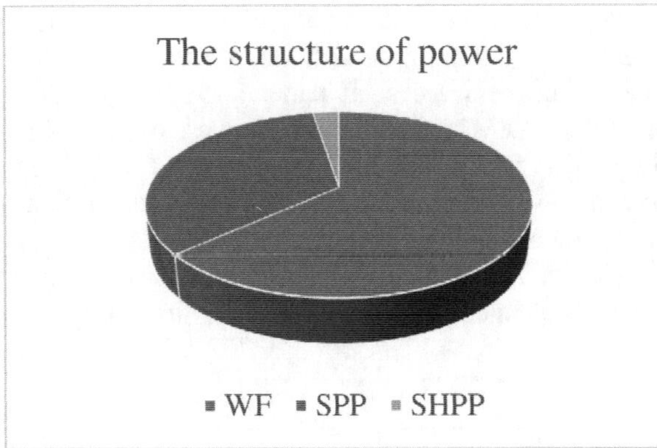

Figure 4. Shares of various types of RES that will be commissioned in Russia by 2023.

From the analysis it can be seen that at the second place in popularity of renewable energy in Russia will be solar photovoltaic installations and their operation in terms of the climate characteristics of Russia will be a great right need.

1.3. Features of the sun as an energy source

Solar radiation, which comes from space is about 7.5 kWh/year. Such power generation theoretically will not give all other renewable sources combined, which indicates the prospects of using the power of the Sun [5].

However, not all radiation that spreads towards the Earth can be used to energy generation. There are a number of factors that change the trajectory of the rays, because of what they do not reach the surface of the solar cells, disperse or absorb them. Factors can be divided into space and earth. The first group includes non-serviceable artificial objects located in near-earth space. So far, this is not a significant reason, which may not even affect the operation of solar photovoltaic installations in space. But with the growing interest in space tourism, it may become more relevant. This category should also include the earth's atmosphere, which absorbs about 19% of solar radiation. Depending on the wavelength, the atmosphere and clouds can reflect the radiation of the Sun [7, 19-22].

2. Materials and Methods

In addition to the above factors, the operation of solar panels can be affected by various kinds of pollution on the surface of the panels and prevent them from getting solar radiation. Such pollution can include dust, which can be of cosmic origin and come to the planet from different places of the solar system and the Galaxy. Several tons of it daily settles on the Earth [9].

The Earth factors affecting the operation of photovoltaic plants (PP) include soil, deserts, volcanic eruptions, the work of factories and industrial enterprises, extraction of natural resources, construction, agriculture, as they are a source of dust.

The urban environment is a source of different types of pollutants. For example, in winter housing and communal services sprinkle slippery sections of roads with sand and salt, but a strong wind can capture the salt crystals and move to closely spaced solar panels. As a result, in a few days the surface of the PP can be quite heavily contaminated.

Photovoltaic cells feeding traffic lights are most susceptible to pollution, as they are often located in open areas of highways and road junctions. Gas contamination in such areas has an impact on the plant`s power.

If we look at the current picture of the world, in the foreground of which are located hydrocarbon production, ongoing construction and anthropogenic pressure on nature as a whole, it can be conclud-

ed that, following this direction, the amount of dust and other polluting reagents will increase annually [8]. The climate and natural conditions in which solar panels are operated will also have an impact.
To study the influence of external factors on the generated power of photovoltaic cells, an experiment was conducted. During the experiment the solar panel "1.5 W Solar Panel 81X137", sodium lamp NAV-T 70W, multimeter BORT BMM-1000N (Figure 5) were used. During the experiment, the RSP-power of the solar panel (W) was found under the conditions of the experiment. The study took into account all the factors of changes in the performance of solar installations: increase and decrease in temperature, intensity and distance from the light source to the solar panel, the angle of incidence.

The panel alternately applied a layer that prevents light (dirt, salt, dust), and then turned on the light source. After that, the panel was cleaned, and the experience was repeated again.
Setting the experiment.

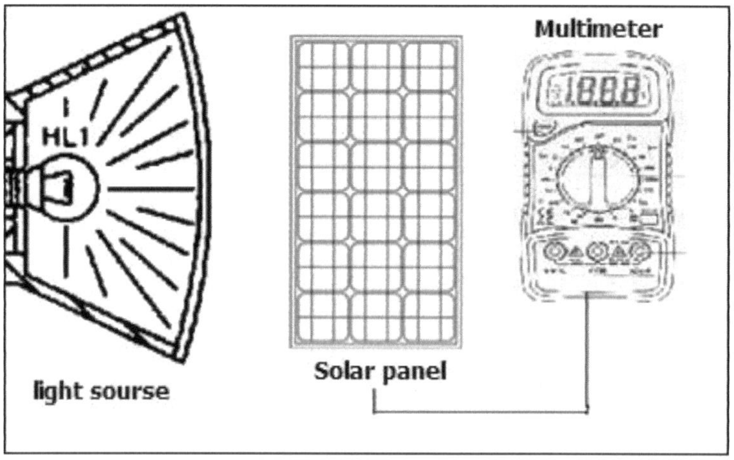

Figure 5. Laboratory setup scheme.

Each layer of the contaminant was applied with the same thickness and on a clean solar panel to equalize the conditions in all experiments. Periodically, the light source was switched off to avoid overheating of the solar panel, which could lead to a decrease in power and differences in measurements.

3. Results

The power of the solar panel was determined taking into account the pollutant factor. Current and voltage readings were taken from the multimeter connected to the solar panel. The measurement results are present in table 1.

Table 1. Measurement of solar panel operation when a contaminating factor is applied to it.

Experiment number	Pollutant factor	Voltage, U (V)	Amperage, I (A)	Power, P (W)
1	No pollutant	5,29	0,15	0,7935
2	Soil, dirt	5,2	0,1	0,52
3	Dust	5,1	0,09	0,459
4	Salt	5,25	0,11	0,5775
5	Dust, dirt, salt	5,05	0,06	0,303

The results of the experiments showed that the polluting factor can significantly reduce the productivity of the device. The values of the lost power of the solar panel from the type of pollutant are shown in figure 6.

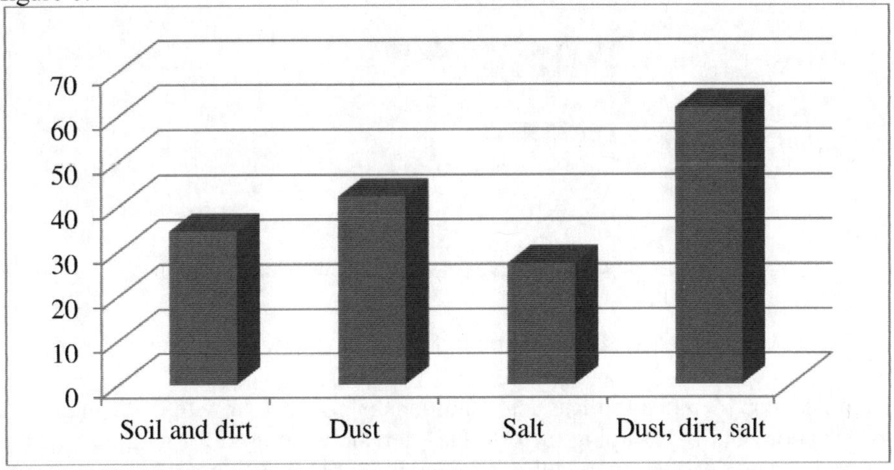

Figure 6. Dependence of lost power on the pollutant factor.

Each layer of the contaminant was applied with the same thickness and on a clean solar panel to equalize the conditions in all experiments. Periodically, the light source was switched off to avoid overheating of the solar panel, which could lead to a decrease in power and differences in measurements.

4. Discussion

The influence of all the above factors on the operation of solar panels is obvious, as they either darken the surface of the panel, or deflect the rays from the surface of the panel. Therefore, the study of the dependence of the output power of solar panels on polluting factors is very important, especially for Russia, which has a large extent of territories from East to West and has different climatic conditions. The analysis of bids of the TSA showed that 1.8 GW of solar plants would be introduced by 2023. However, according to the results of our experiment, the total power of the introduced SPP can fall by 0.5-1GW depending on the type of pollutant that can be on their surface, which can lead to a total economic damage of up to 15 million rubles per year.

Similar studies were conducted earlier by the 27 of technology for the Asian region, which included the cities of China, India and Singapore. As well as researchers from the Indian Gandinagar Institute of technology and scientists from Duke University. Scientists at the University of Massachusetts conducted experiments in 16 different locations and found that the power reduction depending on the environmental conditions ranged from 2 to 16% [9, 26].

Duke University scientists have also found that the accumulation of pollution on the SP affects energy production. In their reports, it was stated that the power produced by solar installations does not converge with what the producers indicate, it falls by 17-35% on the territory of the Indian subcontinent [10, 27]. Thus, it was concluded that the profitability of solar panels is much lower in countries with dense smoke and poor environment. Data obtained for China, where 80% of thermal power plants work on coal, and India, where emissions of harmful substances exceeded the world Health organization standards by 30 times. The analysis of data on power loss by solar panels by various scientists is presented in figure 7.

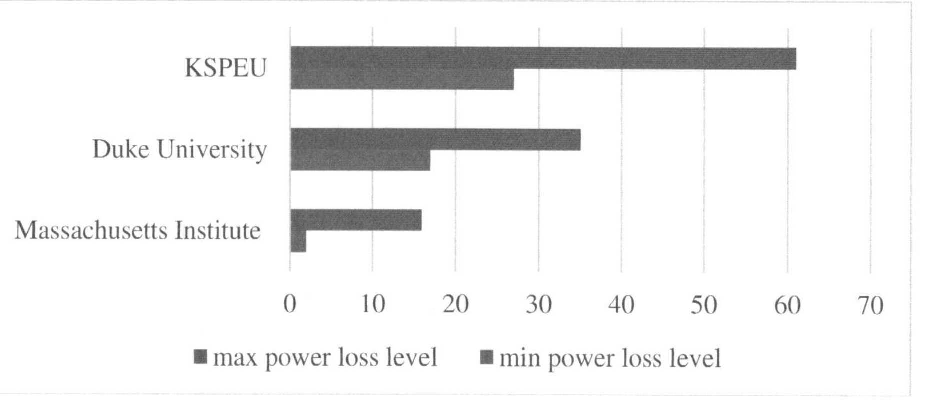

Figure 7. Dependence of the minimum and maximum levels of power losses.

These criteria cannot be applied to Russian conditions in connection with other indicators of these criteria. In our experiment, the polluting factors characteristic were used for natural conditions of our country. Dust and dirt can get on the modules installed near the road, salt on the modules designed to power the lights and traffic lights near the sea and the water that appears on panels during the rain.

5. Conclusion

The territory of the Russian Federation is 17 100 000 km2, but not everywhere the installation of SPP will be the most effective. The reason may be not only a small number of Sunny days, but also human activity. Industry has become a powerful source of dust. For this reason, in the territory of enterprises, the produced power of solar panels should be less. In cities such as Azov, Novokuznetsk, Lipetsk, Magnitogorsk, Volgograd, Norilsk, Nizhnekamsk, Chita, developed industrial sectors not only harm the health of inhabitants [10-12], but also destroy the prospects of using Solar energy in these areas. Dust is formed during coal mining. For this reason, the territory of coal mines and mines have solar installations undesirable. These areas include: Kansk- Achinsk coal basin, Kuznetsk coal basin, etc. [11, 24-25].

The most successful will be the construction of SPP in the South of the country, excluding those areas in which industry is developed and construction is underway. Promising areas include the North Caucasus, the Republic of Crimea, and the Vladivostok region. They are also called the most promising areas for the use of solar energy. In addition, the South of Western and Eastern Siberia are promising areas [28-30].

References

[1] 2018 Global status report. Renewable Energy Policy Network for the 21st Centure www.ren21.net
[2] 2018 Gosudarstvennaya programma Rossiiskoi Federacii "Energoeffektivnost i razvitie energetiki" [State Program of the Russian Federation "Energy Efficiency and Energy Development"] URL https://minenergo.gov.ru/node/323 [In Russian]
[3] 2018 Postanovlenie ot 28 maya 2013 g. N 449 O mehanizme stimulirovaniya ispolzovaniya vozobnovlyaemih istochnikov energii na optovom rinke elektricheskoi energii i moschnosti [On the mechanism of stimulating the use of renewable energy at the wholesale market of electric energy and power] URL http://www.consultant.ru/document/cons_doc_LAW_146916/ [In Russian]
[4] 2018 Departament torgovli AO «ATS», rezultati otborov proektov [ATS Trade Department, results of project selection] URL http://www.atsenergo.ru/vie/proresults [In Russian]
[5] Bystritsky G F 2012 *Osnovi energetiki* (Moscow: KNORUS) vol 3 [In Russian]
[6] Rac G I and Mordinova M A 2012 Izvestiya Baikalskogo gosudarstvennogo universiteta **82** [In

Russian]
[7] Vissarionov V I, Deryugina G V, Kuznetsova V A and Malinin N 2008 *Solnechnaya energetika* (Moscow: Izdatelskii dom MEI) [In Russian]
[8] Izmerova N F, Revicha B A, Korenberga E`I 2004 Izmeneniya klimata i zdorov`e naseleniya Rossii v XXI veke [Climate change and the health of the population of Russia in the XXI century] *Sbornik materialov mezhdunarodnogo seminara* (Moscow: Izdatel`skoe tovarishhestvo "Adamant") [In Russian]
[9] Maksimenko O 2005 *Vokrug sveta* **9** pp 108-17 [In Russian]
[10] Bezrukix P P and Degtyarev V V 2007 *Spravochnik po resursam vozobnovlyaemy`x istochnikove`nergii Rossii i mestny`m vidam topliva* (Moscow: IACz E`nergiya) [In Russian]
[11] Grun` V D et al. 2017 *Ugol` Rossii. Yubilejnoe izdanie k 70-letiyu dnya shaxtera i k 295-letiyu nachala ugledoby`chi v Rossii* (Moscow: Printleto) [In Russian]
[12] Taymarov M A, Akhmetova R V, Lavirko Y V, Sungatullin R G and ZheltuhinaYe S 2017 *KGACU news*. **1** 180-8
[13] Zatsarinnaya J N, Mingaleeva G R and Nazmeev J G 2005 *Sovremennye naukoemkie tekhnologii* vol 5 p 54 [In Russian]
[14] Kirschen D and Strbac G 2004 *Fundamentals of power system Economics*
[15] Taymarov M A 2013 Development of methods of decrease in emissions of oxides of nitrogen by coppers of thermal power plant (Kazan: KSPEU)
[16] Mingaleeva G R, Zatsarinnaya J N and Cyshevskij R V 2004 *5th International Symposium "Resource efficiency and energy saving"*
[17] [17] Zatsarinnaya J N, Rakhmatullin R R, Rizvanova G I 2013 *Bulletin of Kazan Technological University* **5** pp 278-80 [In Russian]
[18] Felix A, Farret M and Godoy S 2017 *Integration of Renewable Sources of Energy* vol 2
[19] Packer N and Al-Shemmeri T 2018 Conventional and Alternative Power Generation: Thermodynamics, Mitigation and Sustainability
[20] Boyle G 2004 *Renewable Energy* ed B Godfrey (Oxford University Press) p 456
[21] Burke M P, Chaos M, Ju Y, Dryer F L and Klippenstein S J 2012 *International Journal ofChemical Kinetics*. **44** pp 444–74
[22] Pavlasa M, Stehlíka P, Oralb J and Šikulab J 2006 *Integrating renewable sources of energy into an existing combined heat and power system* vol 31 pp 2499–511
[23] Skoplaki E and Palyvos J A 2009 On the temperature dependence of photovoltaic module electrical *performance: A review of efficiency/power correlations. Solar Energy* **83** pp 614–24
[24] Ahrenfeldt J, Thomsen T P, Henriksen U and Clausen L R 2013 *Applied Thermal Engineering* **50** pp 1407-17
[25] Sánchez A L and Williams F A 2014 *Progress in Energy and Combustion Science* **41** pp1-55
[26] 2018 Massachusetts Institute of Technology [Air pollution can put a dent in solar power] URL https://www.sciencedaily.com/releases/2018/08/180829115514.htm
[27] 2018 Duke University [Air pollution casts shadow over solar energy production] URL https://www.sciencedaily.com/releases/2017/06/170626135724.htm
[28] Kolomiecz Y G, Tarasenko A B, Tebuev V V and Sulejmanov M Z 2018 *Mezhdunarodny`j nauchny`j zhurnal Al`ternativnaya e`nergetika i e`kologiya* **4-6** pp 252-4
[29] Pristupa A Y and Ponomareva E V 2018 *Alleya nauki* **7** pp 427-32
[30] Ismagilov F R, Vavilov V E and Nurgalieva R A 2017 *Vestnik Ufimskogo gosudarstvennogo aviacionnogo texnicheskogo universiteta* **3** pp 60-5 [In Russian]

Optimization of power output for a wind turbine using methods of artificial intelligence

N V Zubova and V D Rudykh

Industrial Power Supply Systems Department, Novosibirsk State Technical University, Karl Marks Ave., 20, Novosibirsk, 630073, Russia

E-mail: zubova@corp.nstu.ru

Abstract. Percentage of wind energy in worldwide power generation increases year after year. At the present time, problems of increasing the energy efficiency of wind turbines (WT) and optimization of WT power output are of great importance due to instability of wind energy. It stimulates the development, investigation and using of intelligent systems for controlling operating regimes of wind turbines. Such systems also involve those systems that are developed using algorithms based on fuzzy logic. This paper presents the results of investigations devoted to searching for optimal membership functions for fuzzy sets of input and output variables. These variables and membership functions are used in fuzzy algorithms developed for enhancing power output of wind turbines under the given operating conditions. After analyzing the obtained results, it may be concluded that the use of symmetrical Gaussian membership functions gives the fastest convergence of the optimal power into the point.

Keywords: wind energy; energy efficiency; artificial intelligence; wind turbine; fuzzy logic; fuzzy sets; membership function; Mamdani algorithm; linguistic variables; fuzzy controller.

1. Introduction

Now, wind power engineering is being widely developed in different countries. Main designs of wind turbine nacelles, other power equipment, converters and inverters for wind turbines (WT) have been already studied. Therefore, investigations in this field are now directed to increasing the efficiency of WT operation to reduce the cost of power from wind energy, thus making wind turbines a competitive source of power related to power plants and units using the traditional fuel (i.e. coal, gas, diesel).

The most interesting way is the development of new intelligent control techniques that allow controlling WT power output depending on different operating conditions[1-3]. These control techniques can be realized by controllers based on fuzzy logic that allow controlling various WT loops, such as turning the nacelle to the wind, changing the angle of attack of the blade, changing the rotor speed, changing the blade length [4-6]. These devices have shown the better application in comparison with standard PI- and PID-controllers, because of having a regulating characteristic being close to the desired one [7-9].

A lot of publications are devoted to investigation and application of methods of artificial intelligence in wind power engineering [10-12]. Many of them contributed to this study.

This paper presents the results of the search for optimal membership functions with the help of controllers for the change of the WT rotor speed using the software package FuzzyLogicToolbox/Simulink/Matlab.

2. Materials and Methods

In any process, there are two main objectives of control – protection and optimization of operation. Being applied to wind turbines, the process of control is very important in different areas, because there is a problem of continually variable, intermittent and unpredictable nature of wind energy. Moreover, specific features of WT control systems differ depending on operating conditions (figure. 1).

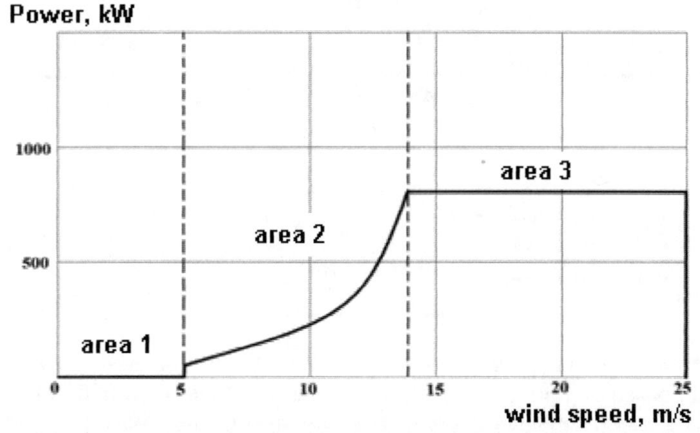

Figure 1. Energy characteristic of a wind turbine.

When a wind turbine is operated in Area 2, the change of the rotor speed is used as a control action for maximizing the useful wind energy to increase the WT energy efficiency. It is possible due to continuous acceleration or deceleration of generator rotation, thus achieving an optimal tip-speed ratio. At the nominal wind speed (Area 3), a control system limits a generator rotation speed. Therefore, generator control generally implies the optimization of wind energy conversion. In several cases, it means that generator torque is changing with the change of a wind speed that may result in additional mechanical loading on an actuator [13-15].

The present paper proposes determination of the most optimal membership functions of fuzzy sets of input and output variables that can be used in fuzzy logic algorithms of WT control loops for changing the rotor speed.

At the present time, instruments of fuzzy logic are widely used for solving the problem of increasing the energy efficiency of WT operating regimes [16-18]. Algorithms have shown good capability of maintaining the stability of network parameters and have increased the quality of energy output [19-21]. By now, several fuzzy inference systems (algorithms) has been proposed. In the present paper, investigations have been carried out using the Mamdani algorithm, which general distinctions from other algorithms are given below:

Activation is realized using the MIN-activation method;

Accumulation for combination of fuzzy sets, corresponding to terms of subconclusions related to the same output linguistic variables, is calculated as [22]:

$$\mu_{AB}(x) = \max(\mu_A(x), \mu_B(x)). \tag{1}$$

Defuzzification of output parameters is performed using the center of gravity (CoG) method or the center of area (CoA) method [22]. Defuzzification by the COG method is performed in the following way:

$$y = \frac{\sum_{i=1}^{n} x_i \mu(x_i)}{\sum_{i=1}^{n} \mu(x_i)}, \quad (2)$$

where y – result of defuzzification, x – variable corresponding to the output linguistic variable ω, μ(x) – membership function of the fuzzy set corresponding to the output variable ω after accumulation.

The structure of a fuzzy controller is illustrated in figure 2.

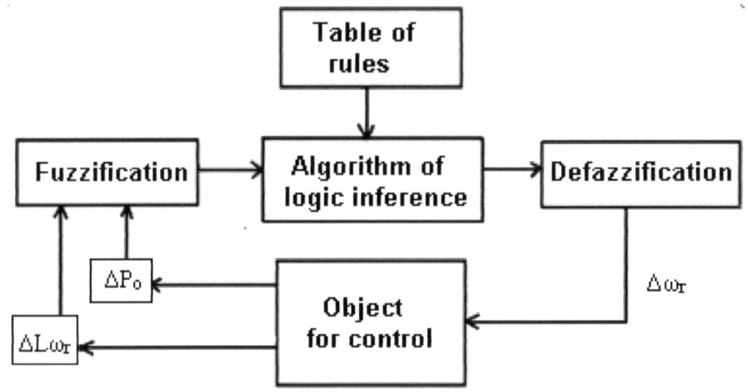

Figure 2. Structure of the algorithm for realization of generator speed control based on a fuzzy inference (ΔP_o – the change of an output power, $\Delta L\omega_r$ – a step for the change of a generator speed, $\Delta\omega_r$ – the change of an output generator speed).

Figure 3 illustrates the variety of energy characteristics of a wind turbine depending on a WT wheel rotation speed and a wind speed.

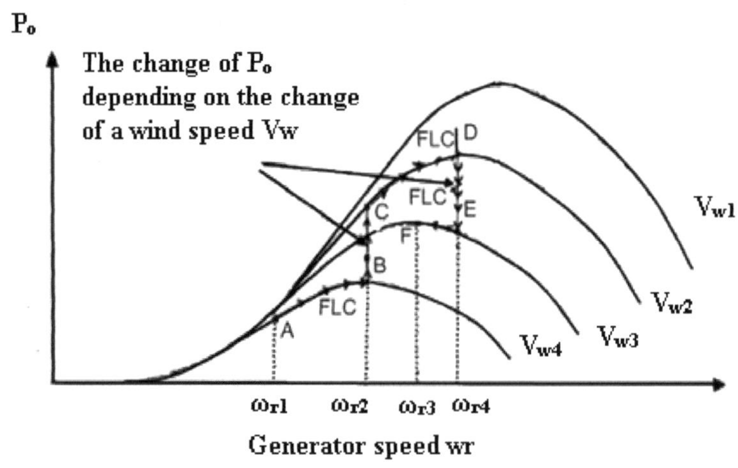

Figure 3. Dependence of output power on generator speed at various wind speeds.

It should be noted that an operating point, where the output power is maximum, has a different position for each wind speed. The main objective of control is to force a wind turbine generator to operate at this point. The change of a membership function allows achieving the maximum possible power at each wind speed.

The algorithm based on fuzzy logic for maximization of WT power generation is proposed in [23]. For certain values of a wind speed, an optimal generator speed is determined using the algorithm, until the system reaches the point of maximum power generation. For example, if generator speed is ω_{r1}, then an output power is in point A at wind speed V_{w4}. Based on a fuzzy logic inference, a generator speed is changing, until it reaches speed ω_{r2}, at which an output power has the maximum value (point B). If a wind speed increases up to V_{w2}, an output power will move to point C, then, according to the algorithm, an operating point will move to point D, thus changing a generator speed to ω_{r4}. The strategy for changing a wind speed is similar. If a wind speed decreases to V_{w3}, an output power will change and move to point E. A generator speed will decrease to the optimal value of ω_{r3} (point F), where an output power has the maximum value.

Input variables, the change of an output power ΔP_o and the following change of a generator speed $L\omega_r$, are firstly converted into corresponding fuzzy sets having human descriptive and intuitive values in the form of terms, such as "big", "medium", "small", "zero". This is realized in the "fuzzification block", where variables ΔP_o (the change of an output power), $\Delta\omega_r$ (the change of a generator speed) and $L\Delta\omega_r$ (the following change of a generator speed) are described by corresponding membership functions.

3. Results and Discussion

Under the present study, the algorithm was investigated with the use of a piecewise linear function (figure 4), a sinusoidal function (figure 5) and a Gaussian function (figure 6) [22]. Rules of a logic inference are applied using the following principle: "if the last change of an output power ΔP_o during the search for the maximum power is positive and big and the last change of the desired generator speed $L\Delta\omega_r$ is positive, then monitoring for the maximum power will be continued with the same positive direction with the big increase of $\Delta\omega_r$". Similar rules are included in the block "table of rules" and given in table 1.

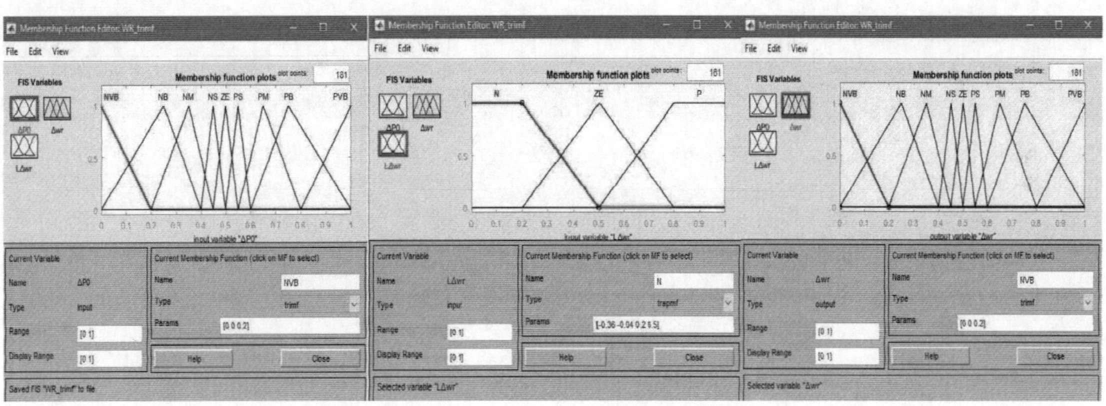

Figure 4. Piecewise linear (triangular) membership functions of input variables ΔP_0, $L\Delta w_r$ and an output variable Δw_r for the algorithm based on fuzzy logic, controlling the change of a rotor speed.

To clarify the designations presented in Table 1, their definition should be done for the change of input and output variables: PVB – Positive Very Big, PBIG – Positive Big, PMED – Positive Medium, PSMA – Positive Small, ZE – Zero (no changes), NSMA – Negative Small, NMED – Negative Medium, NBIG – Negative Big, NVB – Negative Very Big, P – Positive, ZE – Zero (no changes), N – Negative change, respectively.

As a result, output fuzzy sets demonstrating the change of a generator speed $\Delta\omega_r$ are transmitted into the "defuzzification block" where they are converted into the real values [24]. It means that output variables, such as "BIG", "MEDIUM", "SMALL", are converted into the numbers indicating measurable (normalized) values of a generator speed.

IOP Conf. Series: Materials Science and Engineering **552** (2019) 012034 doi:10.1088/1757-899X/552/1/012034

The choice of the optimal membership function is the fastest achievement of the operating point, at which the generated power has the maximum value.

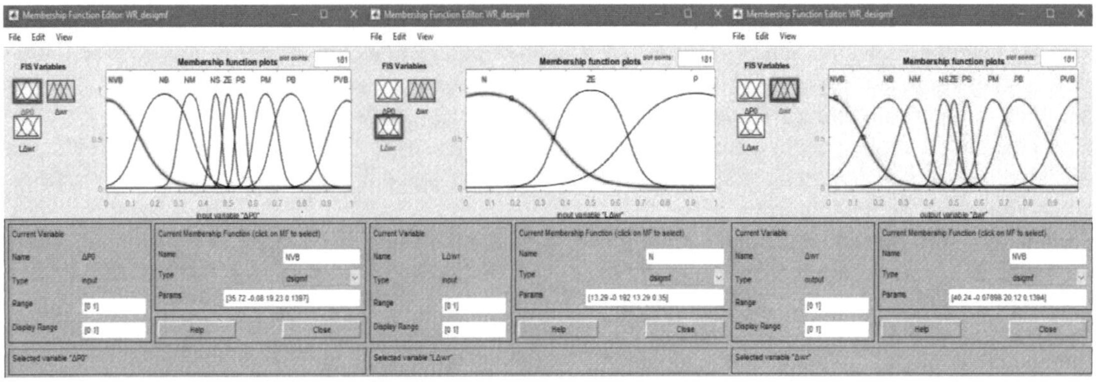

Figure 5. Sigmoid membership functions of input variables ΔP_0, $L\Delta w_r$ and an output variable Δw_r.

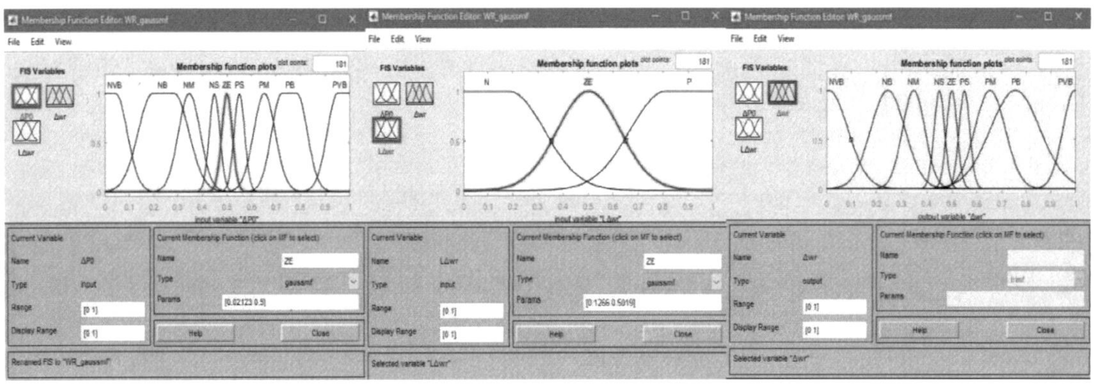

Figure 6. Gaussian membership functions of input variables ΔP_0, $L\Delta w_r$ and an output variable Δw_r.

Table 1. Rules for the algorithm based on fuzzy logic.

Δw_r \ ΔP_0	P	ZE	N
PVB	PVB	PVB	NVB
PBIG	PBIG	PVB	NBIG
PMED	PMED	PBIG	NMED
PSMA	PSMA	PMED	NSMA
ZE	ZE	ZE	ZE
NSMA	NSMA	NMED	PSMA
NMED	NMED	NBIG	PMED
NBIG	NBIG	NVB	PBIG
NVB	NVB	NVB	PVB

Experiments conducted in FuzzyLogicToolbox/Simulink/Matlab have shown that, assuming that input variables are equal, the most optimal change of an output generator speed $\Delta\omega_r$ is achieved by using a symmetrical Gaussian membership function. This result shows that, if using this function, the system can move to the point of the maximum power output in the fastest way.

The path to the fuzzy algorithms was registered from fuzzy controllers embedded in the control system of a wind turbine in a model built using Simulink/ Matlab (figure 7).This made it possible to evaluate their performance directly in the control system with specified constraints. The program uses a generator with a rated power of 1.5 MW [25].

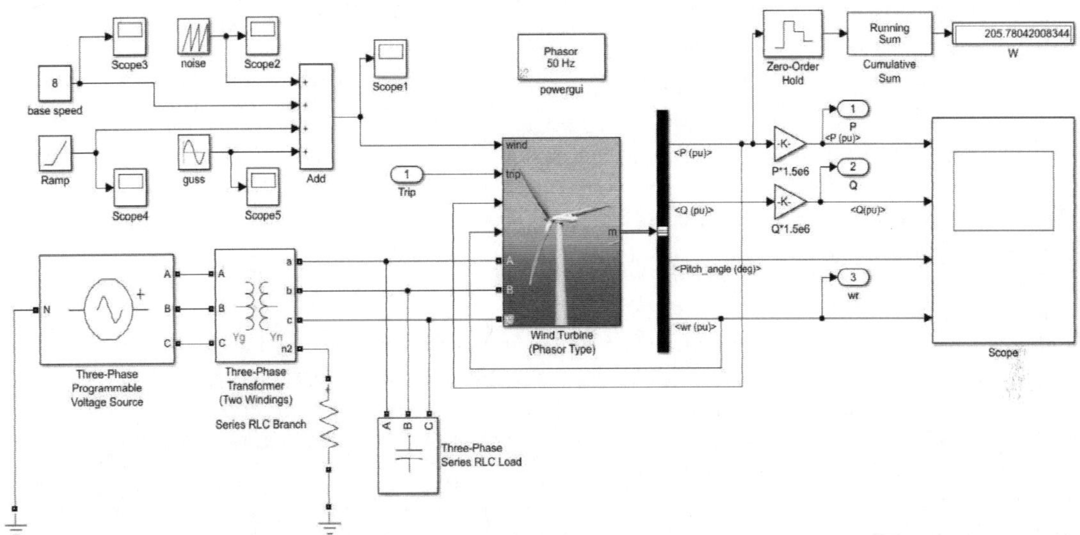

Figure 7. General view of the wind turbine model in Simulink/Matlab.

The unit for changing the rotor speed using an algorithm based on fuzzy inference is presented in figure 8. A fis-file is connected to the "Fuzzylogiccontrollerwr" block, which contains information about the corresponding control action. Input values are fed to the controller input: the last change of power and the last change of the rotor speed of the generator. In accordance with the characteristic of changes in power output of wind turbine depending on the rotor speed at various wind speeds, it is obvious that in order to reach the maximum power output: when the wind speed increases, an increase in the rotor speed should be observed and the rotor should slow down as the wind speed decreases.

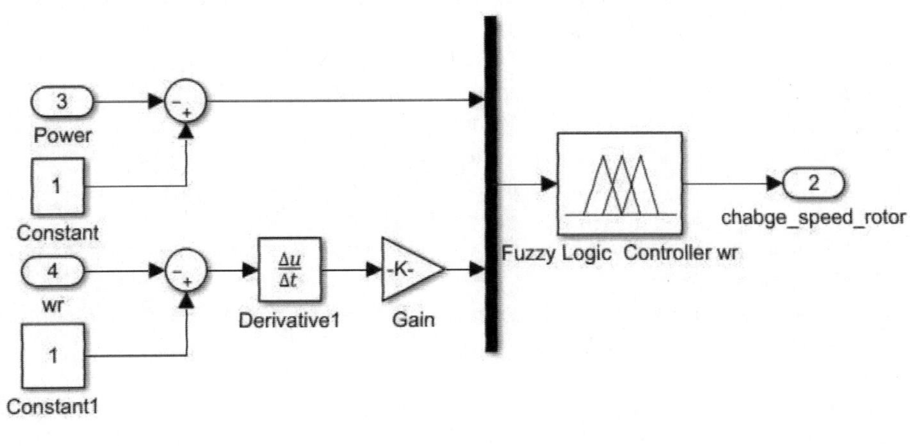

Figure 8. The control unit changes the speed of rotation of the rotor.

Table 2 shows the data of the generated energy for the same time interval under the condition of the same variability of the wind flow for different membership functions.

Table 2. Energy generated by wind turbine for different membership functions of the rotor speed controller

Membership function	W_{WT}, r.u.
Triangular	201.357
Sigmoid	202.986
Gaussian	205.781

For "Fuzzylogiccontrollerwr", there is an increase in the energy output of the wind turbine when using the Gaussian membership function with all other conditions being equal. This result shows that when using this function, the system comes to the point of maximum power generation faster than using other functions.

4. Conclusion

Advantages of fuzzy control are evident. It provides the desired step size during the search that gives fast convergence. In addition, a controller can operate with inaccurate and distorted signals. Controllers based on fuzzy logic, which membership functions of fuzzy sets of input and output variables are defined by the Gaussian distribution, can provide the maximum energy efficiency of wind turbines due to the fast tracking of the maximum power point.

Acknowledgments

The reported study was supported by Russian Foundation for Basic Research, research project No. 31 16-38-60080\15 mol_a_dk.

References

[1] Petković D 2014 Adaptive neuro-fuzzy maximal power extraction of wind turbine with continuously variable transmission *Energy* **64** pp 868 – 74

[2] Petković D, Shamshirband S, Anuar N B, Naji S, Miss L. Mat Kiah and Gani A 2015 Adaptive neuro-fuzzy evaluation of wind farm power production as function of wind speed and direction *Stochastic Environmental Research and Risk Assesment* **29(3)** pp 793 – 802

[3] Bing H, Lawu Z., Yang F and Xiang Z 2016 Individual Pitch Controller Based on Fuzzy Logic Control for Wind Turbine Load Mitigation *IET Renewable Power Generation* **10 (5)** pp 687-93

[4] Bharani R and Jayasankar K C 2015 Yaw Control of Wind Turbine Using Fuzzy Logic Controller *Power Electronics and Renewable Energy Systems* **326** pp 997-1006

[5] Macêdo A V and Mota W S 2012 Wind turbine pitch angle control using Fuzzy Logic *Transmission and Distribution: Latin America Conference and Exposition* **4**

[6] Hossain A 2013 Energy efficient wind turbine system based on fuzzy control approach *5th BSME International Conference on Thermal Engineering* **56** pp 637 – 42

[7] Udalov S N 2016 *Vozobnovlyayemaya energetika* [Renewable power engineering] (Novosibirsk: NGTU) [In Russian]

[8] Aissaouia A G, Tahourb A, Abida M, Essounboulic N and Nollet F 2013 Power control of wind turbine based on Fuzzy controllers *Energy Procedia* **42** pp 163-72

[9] Suganthi L, Iniyan S and Samuel A 2015 Applications of fuzzy logic in renewable energy systems *Renewable and Sustainable Energy Reviews* **48** pp 585-607

[10] Badran O, Abdulhadi E and El-Tous Y 2011 Fuzzy logic controller for predicting wind turbine power generation *International Journal of Mechanical and Materials Engineering (IJMME)*

6(1) pp 51-66

[11] Farh H M and Eltamaly A M 2013 Fuzzy logic control of wind energy conversion system *Journal of Renewable and Sustainable Energy* **5(2)**

[12] Shamshirband S, Petković D, Hashim R and Motamedi S 2014 Adaptive neuro-fuzzy methodology for noise assesment of wind turbine *PLoS ONE* **9(7)** p 9

[13] Rocha M M, Patrocinio da Silva J and Barbosa de Sena F das Ch 2018 Simulation of a Fuzzy Control Applied to a Variable Speed Wind System Connected to the Electrical Network *IEEE Latin America Transactions* **16(2)** pp 521-6

[14] Costa P J, Carvalho A S and Martins A J 2010 Fuzzy Logic as a Method to Optimize Wind Systems Interconnected with the Grid *Wind Power Systems, Green Energy and Technology* pp 383–405

[15] Elkhadiri S, Pr. Lamiaa Elmenzhi and Pr. Abdelouahid Lyhyaoui 2018 Fuzzy logic control of DFIG-based wind turbine *International Conference on Intelligent Systems and Computer Vision (ISCV), IEEE Conferences* pp 1-5

[16] Elfergani A, Elsharif M A A, Hamd R H A, Saad M, Naser El Naily and Faisal A. Mohamed 2018 Advanced self-tuned pitch angle control based on fuzzy logic for grid connected variable-speed wind turbine system *9th International Renewable Energy Congress (IREC), IEEE Conferences* pp 1-6

[17] Shrinath K, Paramasivam S and Palanisamy K 2017 An intelligent self-tuning fuzzy logic controller for pitch angle control for a wind turbine fed induction generator *Innovations in Power and Advanced Computing Technologies (i-PACT), IEEE Conferences* pp 1 – 5

[18] Sahoo S, Subudhi B and Panda G 2016 Pitch Angle Control for Variable Speed Wind Turbine Using Fuzzy Logic *International Conference on Information Technology (ICIT), IEEE Conferences* pp 28-32

[19] Pamuji F A and Miyauchi H 2016 A new control design of maximum power point tracking by fuzzy logic controller for wind turbine connected to low voltage grid *International Seminar on Intelligent Technology and Its Applications (ISITIA), IEEE Conferences* pp 369-72

[20] Ana Vitoria de Almeida Macêdo A V A and Wellington S M 7-10 Dec. 2014 Real time simulations of wind turbine with pitch angle control using FuzzyLogic *11th IEEE/IAS International Conference on Industry Applications* pp 1-7

[21] Van T L, Nguyen T H and Lee D C 2015 Advanced Pitch Angle Control Based on Fuzzy Logic for Variable-Speed Wind Turbine Systems *IEEE Transactions on Energy Conversion* **30(2)** pp 578-87

[22] Leonenkov A V 2005 *Nechetkoye modelirovaniye v MATLAB i fuzzyTECH* [Fuzzy modeling in MATLAB and fuzzyTECH] (St. Petersburg: BHV) p 736 (In Russian)

[23] Adzic E, Ivanovic Z, Adzic M and Katic V 2009 Maximum Power Search in Wind Turbine Based on Fuzzy Logic Control *Acta Polytechnica Hungarica* **6(1)** pp 131– 49

[24] Bimal K and Ronald J 1997 Fuzzy logic based intelligent control of a variablespeed cage machine wind generation system *IEEE Transactions on powerelectronics* **12** No.1 pp 87-95

[25] 2018 S D Shtovba "Vvedenie v teoriyu nechetkix mnozhestv i nechetkuyu logiku" [Introduction to the theory of fuzzy sets and fuzzy logic] URL http://matlab.exponenta.ru

In-service performance of hybrid geopolymer binders based class F fly ash

N I Kozhukhova[1], A Yu Teslya[1], M I Kozhukhova[2], I V Zhernovsky[1], S N Yermak[1] and Yu N Ogurtsova[1]

[1]Department of Materials and Materials Technology, Belgorod State Technological University of V.G. Shukhov, Kostyukov St., 46, Belgorod, 308012, Russia
[2]Department of Civil Engineering and the Environment, University of Wisconsin-Milwaukee, P.O. Box 784, Milwaukee, WI 53201, USA

E-mail: kozhuhovanata@yandex.ru

Abstract. Production of high-performance construction materials, meeting the requirements of construction process when development of up-to-date buildings and structures is very relevant and requires the usage of effective polycomponent binding systems, including geopolymers This work studied mechanical performance, as well as strength, water absorption and water resistance of portland cement – Class F fly ash – NaOH hybrid geopolymer binding system. The bench scale testing was conducted on different mixes, including regular portland cement-based and Class F fly ash-NaOH geopolymer binders used as reference compositions. The developed hybrid binders demonstrated, that incorporation of portland cement up to 40% as a modifying agent, reduces the formation of water soluble mineral phases such as sodium carbonates and hydroxycarbonates in low reactive class F fly ash geopolymer binders. It was found, that the presence of portland cement in the system positively affects strength development and noticeably reduces water absorption and improves water resistance of the developed binders, which is related to formation of water resistant and stable calcium hydrosilicate phases (C-S-H)

Keywords: class F fly ash, Portland cement, hybrid binders, geopolymers, water resistance, compressive strength, water resistance, water absorbtion

1. Introduction

A great interest among scientists and industries in the application of alkali-activated binders [1–6], particularly, geopolymers [7–8] can be explained by a tremendous potential and availability of Class F fly ash, a coal combustion by-product, which is as a cheap source of pozzolan and supplementary cementitious materials as well as a main component for zero-cement binding systems [9–14]. The broad spectrum of applicability of the binders driven by excellent in-service performance. However, the use of cheap off-spec materials results in production of construction composites with lower exploitation characteristic, which, therefore, limits their applicability.

Early studies [15, 16] showed that geopolymers based on low quality (i.e. low chemical reactivity) class F fly ash had lower strength and freeze-thaw performance.

Considering that geopolymers are "green" zero-cement binding systems and a very efficient approach for large-volume industrial by-products utilization it is important to look for new opportunities an effective application of off-spec materials [18, 19].

In the world practice it is common to incorporate different modifying agents and mineral admixtures into a binding system in order to upgrade the quality and in-service performance of materials [20-29]. Here, we are dealing with composite binders or "hybrid binders" in the case of geopolymers.

In the world of construction material science there is the term "hybrid material". One of the widely used area of usage of this term is paints consisting of clay and organic component, that were applied by Mayas ("Maya blue paint") [30–32].

The term "hybrid binder" is used for these systems due to its various chemical history and characteristics [33–35] as well as because of several mechanisms of structure formation involved which affect the performance of final product [36].

Along with numerous advantages of industrial by-products utilization, classical class F fly ash-based geopolymer systems have extremely low or zero water resistance, which classifies the materials to be not applicable to many outdoor applications such as geotechnical and transportation infrastructure.

This research is focused on study of basic in-service performance of hybrid geopolymer binding system "Class F fly ash – portland cement". Several mixtures were prepared and evaluated in terms of density and strength performance as we as water resistance and water absorption.

2. Materials and methods

2.1. Materials.

The Class F fly ash and sodium hydroxide NaOH pellets (with purity of 98%) were used in this research as main alumosilicate component and alkaline activator respectivelly for preparation geopolymer pastes. Portland cement CEM I 42,5N (supplied by Belgorod cement, Russia) was used as a modifying mineral admixture. The chemical composition of fly ash and portland cement presented in table 1.

2.2. Methods.

In order to evaluate the effect of portland cement admixture to density and strength pervormance as well as water absorption and water resistance of geopolymer binding system a finite element analysis with two factorial experiment was applied. The computation was performed using SigmaPlot software.

Table 1. Chemical composition of fly ash and portland cement.

№	Component	Oxides content, wt. %													
		SiO_2	Al_2O_3	TiO_2	Fe_2O_3	MnO	MgO	CaO	Na_2O	K_2O	P_2O_5	NiO	V_2O_5	ZrO_2	LOI
1	Class F fly ash	58.98	28.29	0.97	4.63	0.08	1.00	3.74	0.63	0.65	0.36	–	0.02	–	6.07
2	Portland cement	22.49	4.77	–	4.4	–	0.44	67.22	–	–	–	–	–	–	0.23

Due to strucrural and crystallographic features of vitreous phase [8] class F fly ash shows a very low chemical reactivity with alkalies based on alkali metals, particularly, with NaOH.

3. Results and discussion

In the frame of this study two sets of specimens with a total of 18 mixes of hybrid geopolymer binder pastes were prsspared and molded into the 20x20x20 mm cubes. The first set of the specimens with 9 mixes was casted and placed in oven for 24 hours, where temperature was gradually reached 70 °C and then kept constant until the end of the treatment. After, the specimens were taken out from the oven and placed to cool down in laboratory environment followed by demolding and further storage at 22±3 oC and RH 35-40% until the testing. The second set of the specimens was cured in ambient conditions at at 22±3 oC and RH 35-40%.

For the design of experiment in hybrid geopolymer binder the cement content in the range of 0-40% was chosen as a controllable imput parameter X1 and the alkaline activator content in the range of 0-10% was X2. So that the cement as a modifying agent was added to the hybrid geopolymer system in the amount of 20 and 40 % and the amount of alkaline agent was added in proportions of 5 and 10% by total weight of solid cementitious component (table 2). Compressive strength and density of hybrid composites at the age of 9 days were the output parameters of the experiment.

Table 2. Mix proportions of curing conditions

Mix ID	Solid components, %		Activator, %	*fc'*, MPa	
	Fly ash	Portland cement	NaOH	Ambient conditions	Thermal treatment
1	100	0	0	0	0
2	80	20	0	1.5	0.37
3	60	40	0	3.9	0.5
4	100	0	5	0.5	1.7
5	80	20	5	2.1	1.7
6	60	40	5	3.7	3.7
7	100	0	10	0.8	4
8	80	20	10	1.3	1.6
9	60	40	10	3.9	3.2

The nomograms obtained from the two factorial design of experiment depicted in figures 1 and demonstrated that the geopolymer binders with and without portland cement in the system cured at different conditions responded noticably differently.

For classical "fly ash – NaOH – water" geopolymer binding systems in order to achieve satisfactory strength performance thermal treatment (or thermal drying) is required. With addition of cement powder as a modifying agent into the geopolymer system allows the elimination of thermal treatment and also leads to improvement of strength performance of up to 31% of the specimens cured under ambient conditions.

Such behavior can be caused by presence of some portion of hydraulic binder in the system responsible for a higher kinetics of structure formation. It is important to mention, that the hybrid geopolymer systems subjected to thermal treatment demonstrated almost the same strength performance as reference geopolymer binders. Thermal treatment causes an intensive water evaporation from the system and so prevents an effective hydration process.

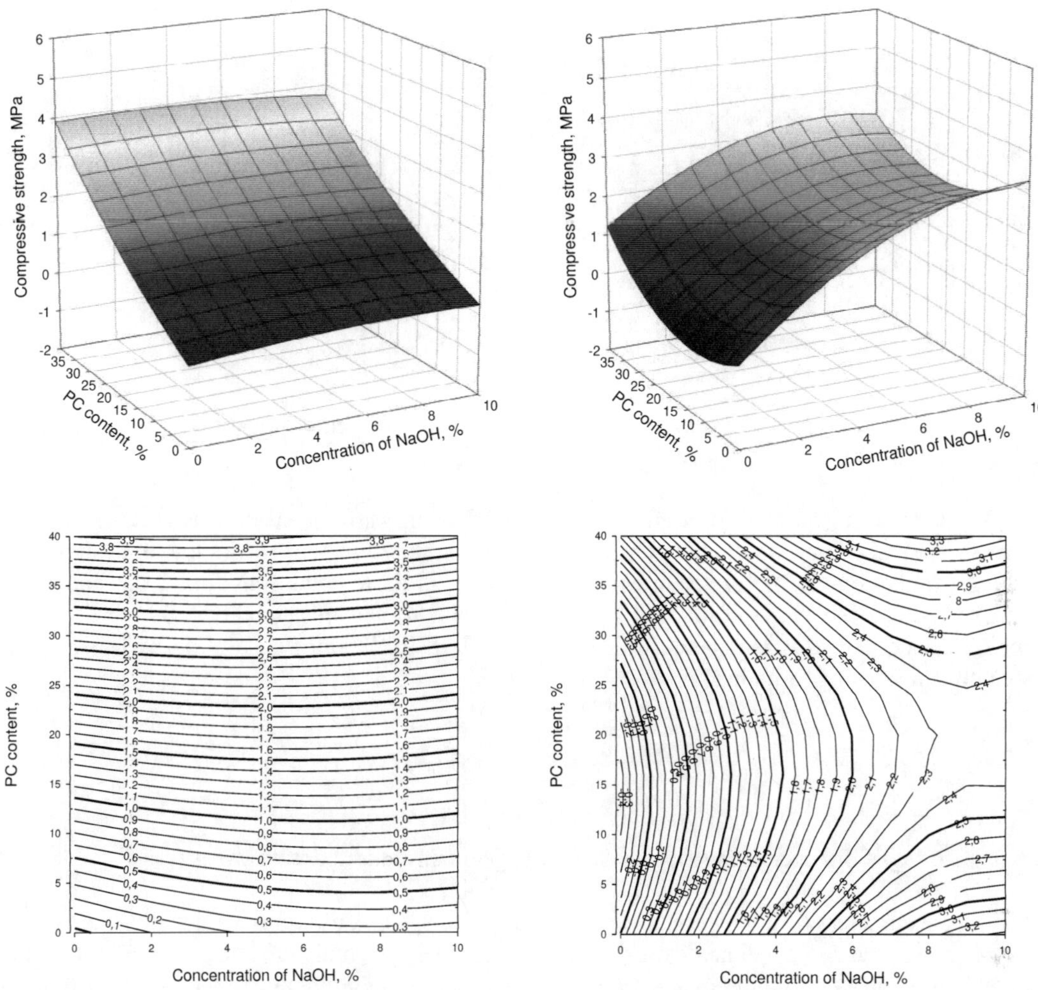

a b

Figure 1. The Effect of alkaline activator and PC content on compressive strength at a) ambient conditions; b) after thermal treatment

At the same time, geopolymerization of aluminosilicate-alkali media is intensified by temperature increase. Therefore, even an addition of up to 40% of cement into the geopolymer system which will be a subject to thermal treatment becomes detrimental in terms of strength performance.

Water resistance and water absorption of hybrid geopolymer binders

The effect of incorporation of portland cement as a modifying agent into fly ash-based geopolymer binding system was also evaluated in terms of water resistance and water absorption (figure 2, table 3). Three different mixes were prepared for this experiment which were the and a hybrid fly ash-based geopolymer binder with incorporation of 40% of portland cement (Mix 1), regular cement paste (Mix 2) fly ash-based geopolymer paste (Mix 3). After the mixes were places into molds they were stored at room temperature for 24 hours followed by thermal treatment at 70°C for another 24 hours and demolding after they cooled down to a room temperature. All specimens were testes for compressive strength at the age of 9 days.

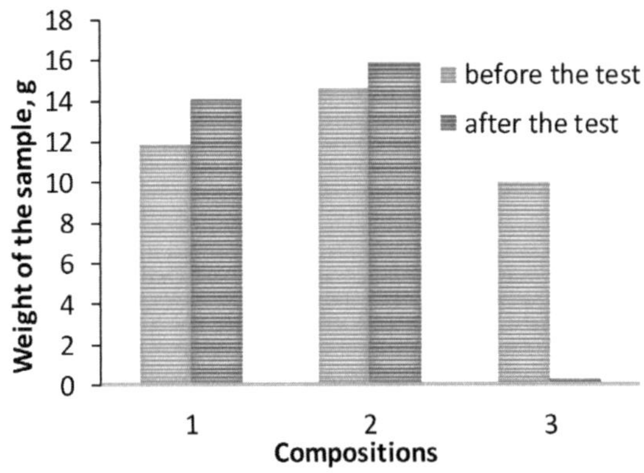

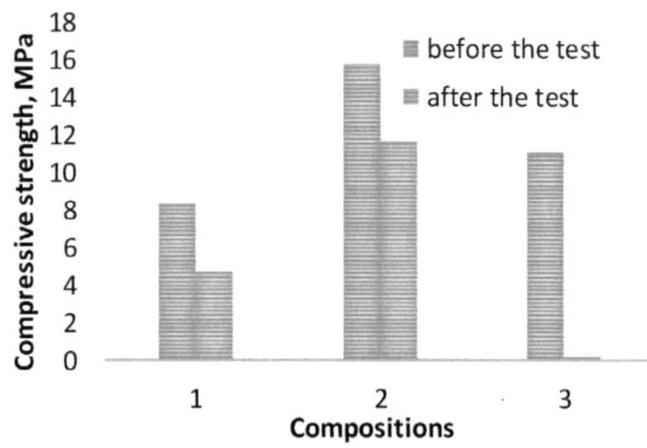

Figure 2. The effect of portland cement incorporation into geopolymer binders on a) mass change and b) compressive strength performance

In order to evaluate water absorption response, the initial mass of the specimens was measured, and then they were placed in a water tank. The specimens were stored in such a way that the water level is slightly above the top surface of specimens. After storing in these conditions for 48 days at the temperature of 23±2 °C the mass of saturated with water specimens was measured.

In accordance with Russian standard GOST 12730.3-78 [37], water absorption was measured by mass change in % using formula (1):

$$W = \frac{m_1 - m_2}{m_1} \cdot 100 \tag{1}$$

where m_1 – mass of dry specimen (g); m_2 – mass of specimen saturated with water (g).

Also, the parameters of water resistance of the prepared specimens were evaluated as a ratio of fc'_1 of dry specimen and fc'_2 specimen saturated with water using formula (2) [38]. For this experiment, specimens were placed in water tank so the water level is at least 20 mm above the top surface of specimens. The specimens were stored in such conditions for 48 hours at room temperature. For compressive strength test the specimens were taken out of the water tank, the excess of water was removed with wet cloth and then specimens were crashed to record strength data.

$$K = \frac{f_{c'2}}{f_{c'1}} \tag{2}$$,

where, fc'_1 – compressive strength of dry specimen, MPa;
fc'_2 – compressive strength of specimen saturated with water, MPa.

Materials are considered to be water resistant if the K value higher than 0.8.

As expected, Mix 2, which a portland cement paste performed very well in terms of strength before and after water saturation with a minimal mass change. However, Mix 3, a reference fly ash-based geopolymer paste completely failed after storing in water conditions, demonstrating a zero water resistance.

Table 3. Performance of developed hybrid geopolymer binders

Parameters	Mix ID		
	Mix 1	**Mix 2**	**Mix 3**
Water resistance, K	0.74	0	0.56
Water absorption, %	8.5	100	18.4
Compressive strength, MPa			
- normal conditions	8.3	15.7	11.1
- water saturated	4.6	11.7	0.2

The experimental data for Mix 1 demonstrated that the incorporation of cement powder in amount of 40 % as a modifying agent into class F fly ash-based geopolymer binding system results in reduction of water absorption of up to 82% and, also helps to improve water resistance of up to 56%.

An extremely low or zero water resistance geopolymer binders based on class F fly ash with low chemical reactivity, usually, related to presence of non-reacted fly ash in considerable amount and formation of water soluble sodium carbonates and hydroxycarbonates in the binding system as a reaction of alkaline activator and carbon dioxide, captured from atmosphere. The incorporation of portland cement into the geopolymer system causes the formation of water resistant and with higher strength calcium hydrosilicate (C-S-H) phases, typical for portland cement-based systems. This phenomenon explains an apparent improvements of strength performance as well as water resistance of the developed hybrid geopolymer binders.

4. Results and discussion

The incorporation of portland cement powder as a modifying agent into class F fly ash-based hybrid geopolymer binding system positively affects hydration reaction at ambient conditions and results in up to 30% strength improvement. Also, the developed hybrid binders demonstrated reduced water absorption and improved up to 50% water resistance vs. reference class F fly ash-based geopolymer binders, which is related to formation of water resistant and stable calcium hydrosilicate phases (C-S-H), typical for portland cement-based systems.

The incorporation of up to 40% of portland cement into geopolymer system cured at ambient conditions demonstrated a very similar strength performance vs. reference class F fly ash-based geopolymer binder cured at high temperatures.

However, while thermal treatment is favorable for classical class F fly ash-based geopolymer systems, it considerably diminishes the positive effect of portland cement in the system due to intensive evaporation of water required for hydration reaction.

Thus, the developed hybrid binding system with incorporation of up to 40% of portland cement can considerably reduce the negative effect of low chemical reactivity of class F fly ash and, so provide a considerably better in-service performance, which opens the door for wider outdoor construction application.

In addition, the elimination of thermal treatment from technological process will help to reduce ecological footprint by reduction of heat release and CO_2 emission into the atmosphere.

5. Acknowledgments

This work is realized under financial support of the Ministry of Education and Science of the Russian Federation, Federal Action Program «Research and Development in Priority Fields of Science and Technology of Russian Federation for 2014-2020», unique agreement identifier RFMEFI58317X0063, using equipment of High Technology Center at BSTU named after V.G. Shukhov/

References

[1] Fernandez-Jimenez A, Palomo A, Sobrados I and Sanz J 2006 b *Microporous Mesoporous Mater.* **91** pp 111–19.
[2] Fernandez-Jimenez A and Palomo A 2003 *Fuel* **82** pp 2259–65.
[3] Krivenko P, Sanytsky M and Kropyvnytska T 2018 *Solid State Phenomena* **276** 9-14
[4] Krivenko P V 1999 *2nd Intern. Conf. Alkaline Cements and Concretes* (Kiev) vol 5 chapter 18 pp 3-43
[5] Glukhovsky V D, Rostovskaja G S and Rumyna G V 1980 *Proceedings of the seventh international congress on the chemistry of cement* vol 3 pp 164–8.
[6] Glukhovsky V D 1981 *Slag-alkali concretes produced from fine-grained aggregate* (Kiev: Vishcha Shkolay)
[7] Davidovits J 1987 *Concr Int.* 1223–5.
[8] Davidovits J 1994 *Proceedings of the first international conference on alkaline cements and concretes* (Kiev, Ukraine) pp 131–49
[9] Kovtun M N, Kearsley E P and Shekhovtsova J A 2013 *Innovations in Concrete Construction, UKIERI Concrete* **3** 919-28.
[10] Lenka Bodnarova, Sergii Guzii, Rudolf Hela, Pavel Krivenko, Grigorii Vozniuk 2018 *Solid State Phenomena* **276** 192–7.
[11] Labrincha J et al. 2017 *Naturally Occurring Radioactive Materials in Construction* 183-252.
[12] Geddes D A, Ke X, Bernal S A, Hayes M and Provis J L 2018 *Calcined Clays for Sustainable Concrete* pp 183-8.
[13] Provis J L, Palomo A, Shi C 2015 *Cement and Concrete Research* **78** 110-25
[14] Vladimír Živica, Martin Tchingnabe Palou and Martin Križma 2014 *Building Research Journal* **61** pp 85–100.
[15] Shekhovtsova J, Zhernovsky I, Kovtun M, Kozhukhova N, Zhernovskaya I and Kearsley E 2018 *Journal of Cleaner Production* **178** pp 22–33.
[16] Kozhukhova N I, Zhernovsky I V and Sobolev K G 2018 *Bulletin of Belgorod State Technological University named after V.G. Shoukhov* **4** pp 5–12.
[17] Kovtun M, Ziolkowski M, Shekhovtsova J and Kearsley E 2016 *Journal of Cleaner Production* **133** pp 220-7.
[18] Alfimova N I, Shadskiy E E, Lesovik R V and Ageeva M S 2015 *International Journal of Applied Engineering Research* **10(24)** 45131-36.
[19] Voitovich E V, Cherevatova A V, Zhernovsky I V, Fisher H-B and Sobolev K 2014 *Materials Research Society Symposium Proceedings* **22** pp 165-70.
[20] Elistratkin M Yu, Minakova A V, Jamil A N, Kukovitsky V V and Issa Jamal Issa Eleyan 2018 *Construction Materials and Products* **1(2)** pp 37–44.
[21] Xiao Xue, Yun-Lin Liu, Jian-Guo Dai, Chi-Sun Poon, Wei-Dong Zhang and Peng Zhang 2018 *Cement and Concrete Composites* **94** pp 43-52
[22] Ping Duan, Chunjie Yan, Wenjun Luo and Wei Zhou 2016 *Materials Letters* **164** pp 172-5
[23] Tokarev Yu, Ginchitsky E, Sychugov S, Krutikov V, Yakovlev G, Buryanov A and Senkov S 2017 *Procedia Engineering* **172** pp 1161–8
[24] Salaheddine Alahrache, Frank Winnefeld, Jean-Baptiste Champenois, Frank Hesselbarth, Barbara Lothenbach 2016 *Cement and Concrete Composites* **66** pp 10-23
[25] Walid Al-Kutti, Muhammad Nasir, Megat Azmi Megat Johari· A.B.M. Saiful Islam, Abdullah A Manda and Nawaf I Blaisi 2018 *Construction and Building Materials* **159** pp 567-77
[26] Taegong Ryu, Dong-HeeLee, Jae ChunRyu, Junho Shin and Kang-Sup Chung Young Ho Kim 2018 *Hydrometallurgy*
[27] Kovtun M, Kearsley E and Shekhovtsova J 2015 *Cement and Concrete Research* **72** 1-9
[28] Flores-Vivian I, Pradoto R G K, Moini M, Kozhukhova M, Potapov V and Sobolev K 2017 *Frontiers of Structural and Civil Engineering* **1(4)** pp 436–45
[29] Kozhukhova N I and Danakin D N 2017 *Stroitel'nye materialy* **5** 33–5

[30] Van O H 1966 *Science* **154** pp 645–6
[31] Jose-Yacaman M, Rendon L, Arenas J and Puche M C S 1996 *Science* **273** pp 223–5
[32] Claudio Ferone, Giuseppina Roviello, Francesco Colangelo, Raffaele Cioffi and Oreste Tarallo 2013 *Applied Clay Science* **73** pp 42–50
[33] Kickelbick G 2014 *Hybrid Mater.* **1** 39–51.
[34] Aleman J et al. 2007 *Pure Appl. Chem.* **79** pp 1801–29
[35] Lipatov Yu S, Rosovitskii V F, Alekseyeva T T and Babkina N V 1989 *Polymer Science U.S.S.R.* **31(7)** pp 1637-41
[36] Wonsiri Punurai, Wunchock Kroehong, Adam Saptamongkol and Prinya Chindaprasirt 2018 *Construction and Building Materials* **186** pp 62–70.
[37] GOST 12730.3-78 2007 Betony. Metod opredeleniya vodopogloshcheniya [Concretes. Method of determination of water absorption] (Moscow: Standardinform)
[38] GOST 9758-86 2006 Neorganicheskiye poristyye zapolniteli dlya stroitel'nykh rabot. Metody ispytaniy [Non-organic porous aggregates for construction work. Test methods] (Moscow: Standardinform)

Research of heat exchange and efficiency of the arcs in the arc steel-melting furnaces with usual conveyor and loaded

Makarov A N, Okuneva V V and Pavlova Yu M

Departament of Electrical Supply and Electrical Engineering, Tver State Technical University, nab. Af Nikitina, 22, Tver, 170026

E-mail: tgtu_kafedra_ese@mail.ru

Abstract. The analysis of heat exchange and the efficiency of the arcs in the arc steel-melting furnaces with usual conveyor and loaded. For comparison, accepted arc furnace close in power and performance. The analysis revealed that the average efficiency of arcs furnace Consteel ASF-120co 8-9% less than particle ASF-120us. Analysis of the efficiency of the arcs of electric arc furnaces allowed to explain more 8-9% lower specific energy consumption in the furnace ASF-120co compared to the same period in the furnace ASF-120us. Ccalculation and the analysis of efficiency of various ways of input of heat in a bathtub of metal of ASF-120co and ASF-120us furnaces is performed. The most effective way is introduction of heat by means of electric arches which efficiency is 0,69-0,78. Heating of the metal charge by means of torches of gas-oxygen torches, efficiency of the method 0,45-0,55 is less effectively carried out. Heating of a metal bath by means of oxidation of coke which is continuously entered into slag for foaming of the last is carried out with the efficiency of a way of the method 0,28-0,36.

Keywords: electric arc, furnace, steel, torch, thermal radiation, efficiency

1. Introduction

The first arc steel furnace (ASF) designed by Consteel was put into operation in December 1989 at the Charlotte plant in the United States [1]. The principle of operation of the Consteel furnace with a capacity of 120 tonnes the following: the conveyor length of 30 m is supplied into the furnace charge at a speed of approximately 6 m/min. Flue gases from the furnace pass through a tunnel with conveyor, heat the mixture to 300°C, resulting in the specific energy consumption is reduced by 40-80 kWh/t. At the time of the filing of the charge in the furnace must be 20-30 tonnes of liquid metal. By 2006, the world was put into operation 22 of the furnace and in the following years their input has continued [1-3].

In 2007, OJSC "Asha metallurgical plant" (AMP) has taken the decision to replace open-hearth production at ASF using Consteel furnace [1]. Furnace ASF-120co was commissioned by the AMP in 2010. The subsequent operation of the furnace Consteel on AMP showed that one of the main advantages of these furnaces over modern furnaces of a usual design of ASF-120us – the smaller specific consumption of the electric power – in 2011-2016 years could not be achieved. Held at the furnace Consteel AMP carrying trunks showed greater 40 kWh/t specific consumption of electricity compared with the same capacity of modern ASF conventional design [4-5].

Loading of furnace charge in the furnace of a usual design is made by means of baskets from which furnace charge from above is loaded into 2 receptions on 60-70 tons. For an explanation of the raised specific expense of the electric power now in Consteel furnaces in comparison with furnaces of a usual design it is necessary to conduct a research, the including calculations of efficiency of arches and the analysis of a power consumption of ASF for 1989-2016. The results of such research and calculations are given below.

2. Calculation and the analysis of efficiency of arches in ASF of a usual design and Consteel

In [6] shows the changes in specific energy consumption in the ASF for the years 1980-2014, use them. In the late 1980's, early 1990's by the heating of the charge consumption of electricity in furnaces, Consteel was less than 80-120 kWh/t compared to furnaces of conventional design, the advantages of the Consteel furnaces compared to furnaces of conventional design was undeniable and their use in the global steel industry has increased steadily [1].

Since the mid 1990's in the production of steel in furnaces Consteel and conventional design using a variety of methods of intensification of the smelting process that led to the reduction of the difference in the index of specific energy consumption in furnaces Consteel and conventional design.

In the 1996-2003's specific energy consumption in furnaces Consteel at work on 100% scrap amounted to 380-410 kWh/t [2,3], in furnaces of conventional design 420-430 kWh/t [6]. In the early 2000's the difference in the specific electricity consumption of the furnace Consteel compared to furnaces of conventional design is reduced to 30-40 kWh/t of steel.

In the 2003-15’s continued increase in power density of up to 800-850 kVA/t in furnaces of conventional design, and up to 750 kVA/t Consteel furnaces. The increase in the power input in the furnace led to a further increase in productivity, a decrease in melting time and specific energy consumption. At present, the specific power consumption when operating at 100% scrap in Consteel furnaces is 385-395 kWh/t [5], in furnaces of conventional design 360-375 kWh/t [4]. This phenomenon requires an explanation and can be obtained by analyzing the heat exchange and the calculation and analysis of the change in the efficiency of the arc furnaces of conventional design and Consteel. Technical characteristics of the compared furnaces are given in table. 1.

Table 1. Technical characteristics of furnaces with conveyor and conventional loading.

Technical characteristics	ASF-120co	ASF-120us
Furnace capacity for production, t	120	120
The rest of the metal in the furnace, t	50	
Transformer power, MVA	90	100
The secondary voltage is linear, V	649-1000	600-1100
The voltage phase, V	375-578	347-636
Arc current, kA	50-70	50-75
Melting time, min	60-62	60
Time under current, min	50	45
Specific electricity consumption, kWh/t	416	375

The average or arithmetic mean of the energy parameters of the furnaces, was found in the results of data processing of a few dozen trunks, are robust indicators of the efficiency of furnaces. To explain increased consumption and total electric energy in the furnace with ASF-120co compared to ASF-120us about the feasible calculation of the average for the melting efficiency of the furnaces arcs. The efficiency of the arc is equal to the ratio of the useful power of the arc P_{us}, going to heat, melt the metal and slag, to the power of the arc P_a and is determined by calculating the average angular emission factor of the arc φ_{am} on the metal and metal bath [6,7,11]:

$$\eta_a = P_{us}/P_a = 0{,}08 + \varphi_{am} \quad (1)$$

According to this method, the efficiency of the arcs of Consteel furnaces and conventional design was calculated, the results of calculations coincided with the results of calculation of the efficiency of arcs described in [6] and [7]. The position of the charge in the compared furnaces is shown in fig. 1a-1c.

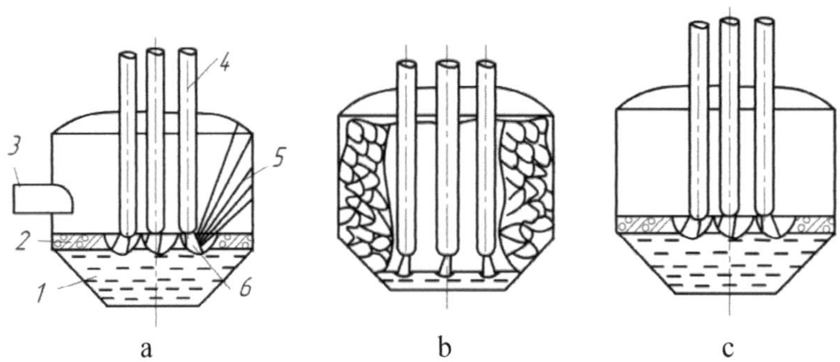

Figure 1. The working space of the electric furnace ASF-120co Consteel (1, a): 1 - metal bath; 2 - slag; 3 - the front part of the conveyor; 4 - electrodes; 5 - the emission of electric arcs on the walls and the arch; 6 - electric arc; the working space of the electric furnace ASF-120us in the process of melting the charge (1,b), η_a =0,93, and to the end of melting (1,c), η_a =0,65.

Let us use the results of calculations of the efficiency of arcs in [6,7] to plot the change in the efficiency of arcs during the operation of Consteel furnaces and conventional design under current. As shown by the results of calculations of the arc efficiency in the furnace ASF-120co [6,7] with a completely immersed in the slag arc, its limit value η_a=0,76, remaining virtually unchanged with a further increase in the height of the slag layer. (Fig. 2)

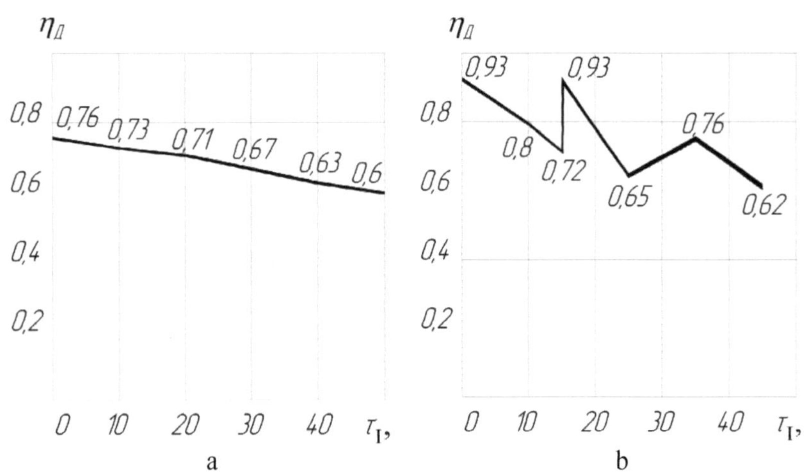

Figure 2. Change the efficiency of the arc in the furnace ASF-120co (a) and ASF-120us (b) during the operation of the furnace under current

From the analysis of the operation of the Consteel furnaces and conventional design, it is concluded that the most advanced Consteel furnaces in the early 1990's, as a result of the processes of increasing power, productivity, intensification of melting, carried out in the 1990-2000 years in furnaces of compared designs, in 2010 lost their advantages in electricity consumption and productivity compared with conventional furnaces. For comparison, efficiencies of arcs furnaces ASFco and ASFus have been found the arithmetic average of the efficiency of arcs for which data are presented on the chart. The result of calculation of efficiency of arcs to compare furnaces revealed that the average efficiency of the arcs of the furnace ASF-120co equal to 0,68, 13% less than the average efficiency of the arcs of the furnace ASF-120us, is equal to 0,78, which has a decisive influence on the specific energy consumption of furnace ASF-120co, it is 11% greater than the specific energy consumption of furnace ASF-120us.

3. Calculation and analysis of power indicators of ASF of a usual design and Consteel

Average for melting η_{as} the steady indicator characterizing energy efficiency of operation of the arc steel melting furnace (ASF). The operating mode of ASF with furnace charge loading by one, two baskets ASFus, surpasses an operating mode of ASF with furnace charge loading in efficiency of heat exchange of electric arches conveyor ASFco. The electric furnace ASF-120us of melting works practically all the time with the big efficiency of arches in comparison with the ASF-120co. More than a half of time of melting in the ASF-120us of an arch burn in a well in furnace charge and their efficiency changes from 0,93 to 0,72 depending on well height. In the furnace of a design Consteel ASF-120co of an arch all the time of melting burn not in a well, and on a liquid metal bath and their efficiency changes from 0,76 to 0,6 depending on slag layer height.

Definition of an average for melting of efficiency of arches of electric furnaces allows to calculate the following average energy performance of melting of ASF-120co and ASF120 us: the average useful specific energy entered into the furnace by arches, average specific useful energy of torches of gas-oxygen torches (GOT), average specific useful energy of oxidation of coke and electrodes. We take data for calculation from the table 2 "Energy balance" [4,8,9], will table results of calculation.

We will define the specific energy which is marked out in Q_a arches as the difference between a receipt part of specific electric energy of Q_e and electric losses of ΔQ_e of an account part:

$$Q_a = Q_e - \Delta Q_e \tag{2}$$

We will determine specific useful energy of arches by the following expression:

$$Q_{au} = Q_a \cdot \eta_{as} \tag{3}$$

Gas-oxygen torches work 12-15 minutes prior to melting, in calculations accept assimilation of heat from combustion of natural gas in GOT of 55% [8,12], the similar value of average efficiency of a torch is accepted in calculations η_{gs} =0,55. Specific useful energy of a torch of GOT Q_{ng} it is defined as the work of energy of natural Q_g gas of a receipt part of power balance on the average efficiency of a torch η_{gs}:

$$Q_{ng} = Q_g \cdot \eta_{gs} \tag{4}$$

For the ASF-120co furnace we summarize the work (4) with specific energy of the warmed scrap. Efficiency η_{FC} specific heat of Q_F from oxidation of Fe, Mn, Si is equal to unit, all heat is acquired by a metal bath [8, 14]. ΣQ_e for melting we determine consumption of electric energy by the furnace as the work of a specific expense of the electric power of Q_e on furnace capacity by production of metal G_m=120 *t*:

$$\sum Q_e = Q_e \cdot G_m \tag{5}$$

$\Sigma\Delta Q_e$ we determine losses of electric energy by for melting as performing specific electric losses of ΔQ_e on furnace capacity:

$$\sum \Delta Q_e = \Delta Q_e \cdot G_m \tag{6}$$

Power of three arches of $3P_{av}$, average for melting, is defined as private from division of the electric energy which is marked out in arches Q_a for operating time of the furnace alive τ_m:

$$3P_{av} = \frac{Q_a}{\tau_m} = \frac{\Sigma Q_e - \Sigma \Delta Q_e}{\tau_m} \tag{7}$$

We determine the useful power of arches by expression:

$$P_{au} = P_{av} \cdot \eta_{av} \tag{8}$$

For ASF-120 furnaces average current of an arch for melting makes I_a =60kA. Arc voltage U_a and length of arch l_d is determined by the following expressions:

$$U_a = (3P_{av}/3)/I_a \tag{9}$$

where U_{AC} – anode-cathode voltage drop, U_{AC}=20 V [15]; grad U_C – voltage stress in an arc column, when arcing on a liquid metal bath of grad U_C =0,8 [6,15,16].
From an account part of the table "Power Balances" [4,8,9] we define specific useful energy for melting of Q_{um} as the sum of heat content of steel of Q_{st} and Q_{sl} slag:

$$Q_{um} = Q_{st} + Q_{sl} \tag{10}$$

Efficiency η_{OC} oxidation reaction by oxygen of coke and electrodes can determine by the following expression:

$$Q_{um} = Q_{au} + Q_{ng} + Q_F + \eta_{cc} \cdot Q_c \tag{11}$$

where Qc – energy of oxidation of coke and electrodes, in calculations is used this indicator of a receipt part of "Power balance" [4,8,9].

Also the specific useful energy from oxidation of coke and electrodes of Q_{cu} coming to a metal bath is determined by the made and calculated Q_{um}, Q_{au}, Q_{ng}, Q_f, Q_c parameters η_c:

$$Q_{cu} = Q_c \cdot \eta_c \tag{12}$$

Have received all necessary expressions for calculation of energy performance of melting in arc steel-melting furnaces.
Results of calculation of average energy performance of melting ASF-120co and ASF-120us furnaces are reduced in table 2.

From results of calculation of parameters of melting stated in table 2 it is possible to draw the following conclusions.

Table 2. Average energy performance of melting.

	Parameter	Convention	Unit	ASF-120$_{co}$	ASF-120$_{us}$
1	efficiency	η	-	-	-
1.1	- arc	η_{as}	-	0,69	0,78
1.2	- torch GOB	η_{gs}	-	0,55	0,55
1.3	- oxidizing reaction Fe, Mn, Si	η_{FC}	-	1,0	1,0
1.4	- reactions of oxidation of coke and electrodes	η_c	kW·h/t	0,28	0,36
2	Specific indicators	Q	kW·h/t	-	-
2.1	- electric energy in arches	Q_a	kW·h/t	399	355
2.2	- useful energy of arches	Q_{au}	kW·h/t	275	277
2.3	- useful energy of a torch GOB	Q_{ng}	kW·h/t	59	36
2.4	- useful energy of scrap	Q_s	kW·h/t	45	-
2.5	- useful energy from oxidation Fe	Q_F	kW·h/t	79	80
2.6	- useful energy from oxidation of coke and electrodes	Q_{cu}	kW·h/t	60	47
2.7	- useful energy for melting	Q_{um}	kW·h/t	473	440
3	Consumption of electric energy for melting	ΣQ_b	kW·h/t	49920	45000
4	Losses electric for melting	$\Sigma\Delta Q_b$	kW·h/t	2040	2400
5	Average power of three arches	$3P_{av}$	mW	57,7	56,8
6	Useful power of arches	P_{au}	mW	39,8	44,3
7	Average power of one arch	P_{av}	mW	19,2	18,9
8	Welding arc voltage	U_a	V	320	315
9	length of arc	l_a	mm	375	368

4. Results of the analysis of power indicators of ASF of a usual design and Consteel

From results of calculation of parameters of melting stated in table 2 that at close electric and geometrical parameters of arches (arc current, arc voltage, arc length) specific useful energy of arches is also approximately identical and makes for the ASF-120co of Q_{au} =275 kW · h/t, for the ASF-120us of Q_{au}=277 kW · h/t. However, because the efficiency of arches of the ASF-120co is η_{av} =0,69, consumption by the furnace of specific electric energy Q_a =399 kW h/t and the general electric energy ΣQ=49920 kW · h that there are more similar indicators of the ASF-120us, components η_{av} =0,78, Q_a=355 kW · h/t, ΣQ=45000 kW·h respectively. As specific losses with the cooling water in furnaces

IOP Conf. Series: Materials Science and Engineering **552** (2019) 012036 doi:10.1088/1757-899X/552/1/012036

are approximately identical and make 48-50 kW · h/t, the difference of a specific expense of the electric power in the ASF-120co of ΔQ_a=399-355=44 kW · h/t make losses of a heat flow of radiation of arches in free space of the ASF-120co and are spent for heating of waste gases.

The higher consumption of specific useful energy on 33 kW·h/t in the ASF-120co is connected with use for 30-35% of greater mass of lime, a large slag formation and on 33 kW·h/t greater heat content of slag in comparison with the ASF-us. Reaction of oxidation with a lower efficiency at 22% of coke and electrodes in the ASF-120co it is connected with inefficient use of radiation of electrodes in these furnaces. In a receipt part of power balances of melting of furnaces ASF-120co, ASF-120us the radiation of electrodes is considered in a component of balances "Energy of coke and electrodes". In ASF-120 furnaces the power of radiation of each of electrodes is 4 mW of [6] or 20% of the power of arches. In ASF-120co during the whole time of melting electrodes radiate in the free space of the furnace blank by furnace charge. An overwhelming part of thermal radiation of electrodes in furnaces ASF-120co is absorbed by flue gases and the cooling water of panels. Energy of radiation of electrodes in furnaces ASF-120co it is useless it is spent, increasing losses with flue gases and the cooling water.

In the early 1990-ies the specific expense of the electric power in Consteel furnaces was on 80-120 kW · h/t steel less in comparison with furnace of a usual design. In the early 2000-ies because of the introduction of processes of an intensification of melting steel a difference in a specific expense of the electric power in Consteel furnaces and a usual design it is reduced to 30-45 kW·h/t. In the 2010's of the furnace of a usual design begin to be in the lead on an indicator of a specific expense of the electric power with value of this indicator on 15-35 kW·h/t less in comparison with Consteel furnaces that is explained by the higher efficiency of arches of ASF-120us in comparison with ASF-120co.

Power balances of melting and calculation of efficiency of arches of arc steel-smelting furnaces allow to define efficiency of various ways of input of energy in the furnace and designs of furnaces, to predict the total power consumption and consumption of electric energy the furnace.

5. Conclusion.

Parameters of arches (power, tension, current, length) of furnaces ASF-120co and ASF-120us are same or similar in value. However electric arches of the ASF-120us are used more effectively with the higher efficiency and bigger useful power, less the consumption of electric energy, common for melting. Heating of furnace charge in the conveyor flue gases in ASF-120co hasn't compensated decrease in efficiency of arches with η_a=0,93 at their burning on a liquid metal bath inside the well to η_a=0,76 at their burning on the metal bath covered with a thick layer of slag.

From all ways of introduction of heat to a metal bath the most effective is oxidation reaction by oxygen of iron and other metals, the efficiency of a way is equal to unit. However energy cost of 1 kW·h of oxidation of iron is 2,37 times more than cost electricity of 1 kW·h [8] because wide use of this way is economically inexpedient. Electric arches which efficiency 0,69-0,78 are in the second place by efficiency of use of heat in furnace. The third place by efficiency of use of heat for furnace charge metal heating in furnace is taken by gas-oxygen torches with the efficiency of a torch 0,45-0,55 [8,14-16].

The fourth place by efficiency of use of heat for heating of metal bath is taken by reaction of oxidation of coke and electrodes with efficiency 0,28-0,36. In furnace ASF-120us efficiency of reaction of oxidation of coke and electrodes 22% more, than in ASF-120co furnaces that is connected with useful use in furnaces ASF-120us of radiation of electrodes on solid charge. Coke is blown into a metal bath during the whole time of melting for maintenance of foamy slag. At combustion of coke in slag heat is generated, temperature of slag is slightly higher than metal bath temperature, slag transfers heat to a metal bath by thermal conductivity. At the small difference of temperatures of slag and metal the efficiency of heat exchange is small and the efficiency of heat of coke is small, η_c =0,28-0,36. At combustion of coke there is an intensive formation of high-temperature gas, gas carries away 64-72% of heat which is formed at coke combustion. As into the ASF-120co furnace enter 65% more coke, than into the ASF-120us furnace, and amount of flue gases and loss of heat with gases in ASF-120co 73% more than in the ASF-120us. Losses with gases include losses of heat radiated by electrodes.

References

[1] Borovinskih S V 2007 V Rossiyu prihodit innovacionnyj elektrostaleplavilnyj process [Russia comes innovative steelmaking process] *Elektrometallurgiya* **9** pp 13-6 [In Russian]

[2] Argenta P and Corbella M 2006 EAF integration info the blast furnace route at Wheeling Pittsburgh *MPI International* vol 2 pp 42-7

[3] Samways N 2005 Revitalized with a new continuous steelmaking process *Iron and steel technology* vol 6 pp 29-45

[4] Evstratov V G, Kiselev A D, Zinurov I Y, Shahir Z H, Mamenko J F, Shumakov A N and Gindullin M G 2012 Osobennosti teplovoj raboty electropechi DSP-120 Consteel na metallurgicheskom zavode [Features of thermal operation of an electric furnace DSP-120 Consteel on the AMW] *Elektrometallurgiya* **8** pp 2-6 [In Russian]

[5] Tulevski Yu N, Zinurov I Y and Shver V G 2011 Novye vozmozhnosti pechey Consteel [New opportunities Consteel furnaces] *Elektrometallurgiya* **6** pp 22-7 [In Russian]

[6] Makarov A N 2014 Teploobmen v elektrodugovykh i fakelnykh pechah i energeticheskih ustanovkah [The heat exchange in electric arc and torch furnaces and metallurgical, and power plants] (St. Petersburg: Lan`) [In Russian]

[7] Makarov A N 2017 Izmenenie KPD dug v processe plavki v dugovykh staleplavilnykh pechah [The change in efficiency of the arcs during the melting process in electric arc furnaces] *Metallurg* **4** pp 55-8 [In Russian]

[8] Gudim Yu A, Zinurov I Yu, Kiselev A D and Shumakov A M 2005 Ratsionalnye sposoby intensifikatsii plavki v sovremennykh dugovykh staleplavilnykh pechah [Rational methods of intensification of melting in modern arc steelmaking furnaces] *Elektrometallurgiya* **9** pp 2-6 [In Russian]

[9] Shakirov Z, Shumakov A M, Uryupin G P, Zinurov I Y and Zapolsky A A 2013 Osobennosti tekhnologii vyplavki stali v elektropechi DSP-120 Consteel i puti ee sovershenstvovaniya [Peculiarities of the technology of smelting steel in an electric furnace DSP-120 Consteel and ways of its improvement] *Elektrometallurgiya* **9** pp 2-7 [In Russian]

[10] Evstratov V G, Zinurov I Y, Shakirov Z, Shumakov A M, Sharov N P, Feldman V Z and Gindullin M T 2012 Modernizaciya staleplavilnogo proizvodstva Ashinskogo metallurgicheskogo zavoda [Modernization of steelmaking Asha metallurgical plant] *Metallurg* **9** pp. 46-50 [In Russian]

[11] Makarov A N, Rybakova V V and Galicheva M K 2014 Electromagnetic and the Arc Efficiency of Electric Arc Steel Melting Furnaces *Journal of Electromagnetic Analysis and Applications* vol 6 pp 184-92

[12] Rayle V 2015 Puti povysheniya effektivnosti metalloloma v elektrostaleplavilnykh agregatah [Ways of increase in efficiency of heating of scrap metal in electrosteel-melting units] *Elektrometallurgiya* **5** pp 26-34 [In Russian]

[13] Makarov A N 2016 Calculations of heat transfer in torch furnaces by gas volume radiation haws *Word Journal of Engineering and Technology* vol 4 pp 488-503

[14] Kozhukhov A A, Fedina V V and Merker E E 2012 Issledovanie processa vspenivanya staleplavilnogo shlaka i ego vliyaniya na teplovuiy rabotu dugovoj staleplavilnoj pechi [Research of process of foaming of steel-melting slag and its influence on thermal operation of the arc steel-melting furnace] *Metallurg* **3** pp 42-4 [In Russian]

[15] Makarov A N and Sokolov A Yu 2009 Elektricheskie geometricheskie teplovye parametry dug goryashikh v parah metallov [Electric, geometrical, thermal parameters of the arches burning in vapors of metals] *Elektrometallurgiya* **11** pp 19-24 [In Russian]

[16] Makarov A N 2017 Fundamental laws of physics and calculation of heat transfer in combustion chambers of gas-turbine plants *World Journal of Engineering and Technology* vol 5 pp 358-

75
[17] Nikolsky L E, Smolyarenko V D and Kuznetsov L N 1981 *Teplovaya rabota dugovykh staleplavilnykh pechej* [Thermal operation of arc steel-melting furnaces] (Moscow: Metallurgiya) [In Russian]
[18] Okorokov N V 1971 *Dugovye staleplavilnye pechi* [Arc steel-melting furnaces] (Moscow: Metallurgiya) [In Russian]
[19] Egorov A V 1985 *Elektroplavilnye pechi chernoj metallurgii* [Electric melting furnaces of ferrous metallurgy] (Moscow: Metallurgiya) [In Russian]
[20] Soyfer V M 2009 *Vyplavka stali v kislykh elektropechah* [Melt of steel in sour electric furnaces] (Moscow: Mashinostroyeniye) [In Russian]
[21] Makarov A N, Kuznetsov A V, Rybakova V V and Galicheva M K 2014 O vliyanii formy elektroda I parametrov dugi na ee zaglublenie v vannu matalla v dugovyh staleplavilnykh pechah [About influence of a form of an electrode and parameters of an arch on its deepening in a metal bath in arc steel-melting furnaces] *Industrial power* vol 10 pp 10-5 [In Russian]
[22] Makarov A N 2014 Theory of radiative heat exchange in furnaces, fire boxes, combustion chamber is replenished by four new laws *Science Discovery* vol 2 pp 34–42
[23] Kuznetsov M S et al 2010 Vliyanie massy metalloshihty I tolshiny shlaka na tekhnologiyu vyplavki stali v dugovoj pechi [Influence of mass of a metal charge and thickness of slag on technology of melt of steel in the arc furnace] *Electrometallurgy* **2** pp 2-6 [In Russian]
[24] Gudim Yu A, Zinurov I Yu and Kiselyov A D 2008 Sushestvuet li realnaya alternativa lomu v elektrostaleplavilnom proizvodstve [Whether there is a real alternative to scrap in electric steel production] *Electrometallurgy* **5** pp 27-30 [In Russian]
[25] Bourne K and Granderat R 2013 Potentsialnye vozmozhnosti I problemy utilizatsii tepla v staleplavilnykh pechah [Potential opportunities and problems of utilization of heat in steel-melting furnaces] *Metallurgical production and technology* **2** pp 8-19 [In Russian]

Theoretical study of the flow of cement raw material sludge through pipes

D N Perelygin[1], I P Boichuk[2], A V Grinek[2], V K Kozlov[1]

[1]Department of Technological complexes, machines and mechanisms, Belgorod State Technological University of V.G. Shukhov, Kostyukov St., 46, Belgorod, 308012, Russia.
[2]Department of Ship's Electric and Automatic Devices, Admiral Ushakov Maritime State University, Lenin's avenue, 93, Novorossiysk Krasnodar Krai, 353918

E-mail: mehanikus@mail.ru

Abstract. The flow of fluids through pipes directly depends on the physicomechanical characteristics of the material. When designing technological equipment, it is assumed that the equipment operates with the limiting physical characteristics of the liquid. The study of the flow of non-Newtonian fluids is important for the development of energy-saving methods for the cement industry. In this paper, we study the flow of cement raw sludge through rigid pipes. It is shown that when moving in pipes near the pipe walls, the sludge moves like a normal Newtonian liquid, and near the pipe it moves like a solid. This indicates that when moving through a pipe with a prolonged non-stationary effect, the sludge behaves like a normal Bingham body.The article found the flow rate, the drag coefficient, the Reynolds number, and the hydraulic resistance when the sludge moves through the pipe. In this paper, the parametric law of resistance of sludge movement in pipes is defined, which can be considered as a convenient way to calculate pressure drop. It follows from the law that, when flowing through pipes, sludge now acquires the properties of a plastic body and behaves like Bingham plastic.

Key words: cement sludge, flow through pipe, Newtonian fluids, hydraulic resistance, Reynolds, Bingham plastic.

1. Introduction

In order to set in motion the thixotropic liquid, which had been at rest for a long time, the transfer pump must develop more power [1-7]. At the moment of the beginning of the movement, the structure of the liquid is destroyed. Therefore, further the pump load is reduced. The fluid, as a result of long-term destruction under the action of shear loads, acquires properties independent of time. The stationary movement of thixotropic and other liquids should differ little from each other.

2. Flow of cement slurries

In actual production, a sludge buffer storage device is involved in the sludge dosing process, in which, as a rule, an excess amount of sludge is transported from 15 to 50 percent of the total amount, while the excess sludge is returned to the sludge pool [8]. At the same time, pre-sludge from the sludge pool is fed to the storage tanks, and then distributed to the furnaces, and from the storage tank excess sludge is returned to the sludge pool. In this regard, there is a problem of increasing the efficiency of sludge supply in cement kilns. Uneven feed of sludge into the rotary kiln with a wet production method can lead to a deterioration in the quality of the clinker firing, and sometimes to emergency operation of the unit [9].

3. Flow of the sludge through pipes

It is assumed that the equipment operates with the limiting physical characteristics of the liquid for designing technological equipment [10]. For stationary movements, the relationship between flow rate and pressure drop is set by the equation [11, 12]

$$Q = \frac{\pi R^3}{\tau_w^3} \int_0^{\tau_w} \sigma^2 f(\sigma) d\sigma ,$$

where R – pipe radius, τ_w – shear stress on the pipe wall.

In the non-stationary case, the flow of fluid through the cross section according to [13-15]

$$Q(t) = 2\pi \int_0^R r v_z(t, r) dr , \quad (1)$$

where $v_z(t,r)$- flow velocity along the pipe axis. We suppose that the dependence of speed v_z on r same as when Bingham's body is moving [11]

$$v_z = C_1\left(R^2(t,z) - r^2\right) + C_2\left(R(t,z) - r\right). \quad (2)$$

To find the constants we consider the balance of forces acting on the cylindrical element of the fluid (figure 1.)

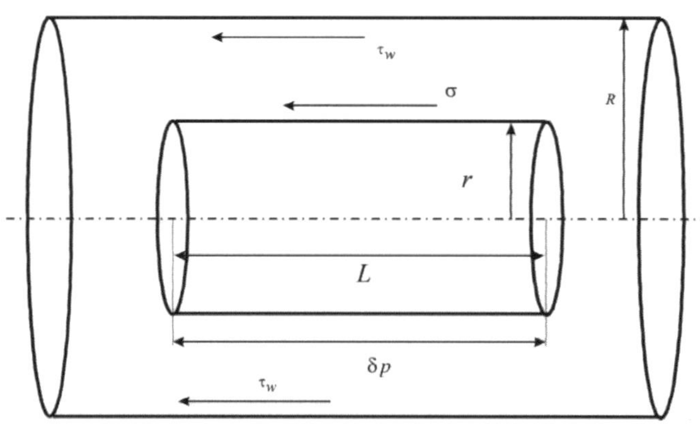

Figure 1. Scheme of flow in pipe.

$$\pi r^2 \delta p = 2\pi r L \sigma, \quad (3)$$

From where

$$\sigma = \frac{r\delta p}{2L}. \quad (4)$$

As mentioned in [16-20], the rheological relationship describing the relationship between stress and strain is

$$\varepsilon = \frac{\sigma}{E_1} + \frac{\sigma}{E_2}\left(1 - \exp\left(-\frac{E_2}{\eta_2}t\right)\right) + \frac{\sigma - \sigma_k}{\eta_1}t. \tag{5}$$

where σ - stress tensor, E - modulus of elasticity, ε - strain tensor, η - viscosity, σ_k - yield strength.

For flow in a cylindrical vessel

$$-\frac{dv_z}{dr} = \frac{\sigma}{\eta_2}\exp\left(-\frac{E_2}{\eta_2}t\right) + \frac{\sigma - \sigma_k}{\eta_1}.$$

Consequently,

$$-\int_{v_z}^{0} dv_z = \int_{r}^{R}\left(\frac{\sigma}{\eta_2}\exp\left(-\frac{E_2}{\eta_2}t\right) + \frac{\sigma - \sigma_k}{\eta_1}\right)dr. \tag{6}$$

From equation (6), on condition that $v_z = 0$ for $r = R$ we get after simple transformations

$$v_z = \frac{\delta p}{4\eta_2 L}\left(1 + \exp\left(-\frac{E_2}{\eta_2}t\right)\right)\left(R^2(t,z) - r^2\right) - \frac{\sigma_k}{\eta_1}\left(R(t,z) - r\right). \tag{7}$$

Near the axis of the pipe, where $\sigma < \sigma_k$, the sludge will move like a solid rod with a radius

$$r_0 = \frac{2\sigma_k L}{\delta p}. \tag{8}$$

Substituting r_0 from (8) into (7), we obtain the speed of quasi-solid motion

$$v_z = \frac{\delta p}{4\eta_2 L}\left((R - r_0)^2 + \left(R^2 - r_0^2\right)\exp\left(-\frac{E_2}{\eta_2}t\right)\right). \tag{9}$$

The velocity diagram will consist of a paraboloid of rotation (from the wall to a cylindrical surface of radius r_0) and a flat platform perpendicular to the axis of the pipe (figure 2).

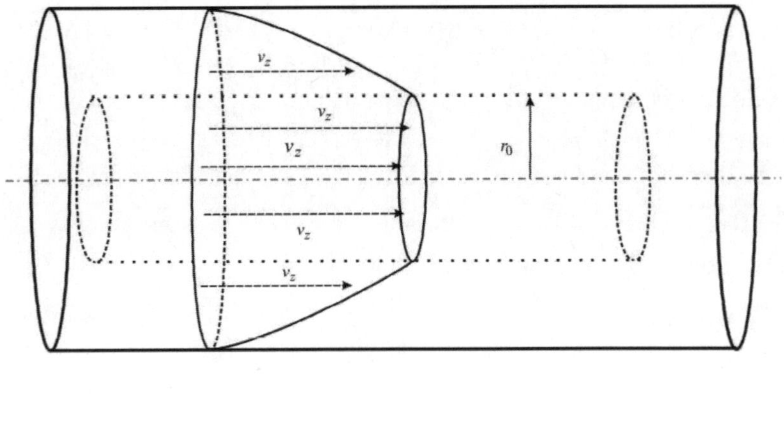

Figure 2. Velocity diagram.

The velocity profile at different points in time in the plane of symmetry of the pipe is shown in figure 3.

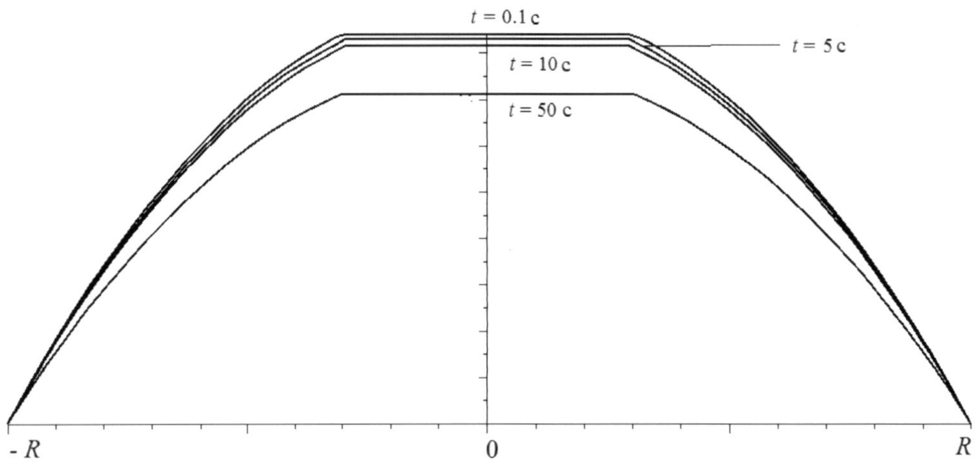

Figure 3. The dimensionless velocity profile at different points of time.

Figure 3 shows that the sludge moves like a normal Newtonian fluid when moving in pipes near the pipe walls, and near the pipe it moves like a solid. Such a movement is characteristic of Bingham liquids. Moreover, if at the initial moment of movement, the velocity of the sludge is higher than the speed of movement of the Bingham body, then over time the velocity profile tends to the profile obtained during the flow of a visco-plastic fluid. Those with prolonged non-stationary exposure, the sludge behaves like a normal Bingham body.

4. Hydraulic resistance when moving through pipes

The hydraulic resistance for the pipe is determined by the expression

$$\delta p = \lambda \frac{L}{d} \frac{\rho v_{cp}^2}{2} \tag{10}$$

where λ - coefficient of resistance, d - pipe diameter, v_{cp} - average flow rate. The drag coefficient λ depends on the physical properties of the medium — density and viscosity, as well as the average velocity — and is a function of the Reynolds number Re

$$\lambda = \lambda(\mathrm{Re}). \tag{11}$$

The average velocity of the sludge through the pipe will be

$$v_{cp} = \frac{Q}{\pi R^2}. \tag{12}$$

We calculate the flow rate Q by substituting (7) into (1)

$$Q = \frac{\pi R^2}{24\eta_2}\left(3\delta p\left(1+e^{-\frac{E_2}{\eta_2}t}\right)-8\sigma_k\right). \tag{13}$$

Turning from an independent variable r to a variable σ by the formula (8), we obtain

$$Q = \frac{\pi R^3}{\eta_2 \tau_w^3}\int_{\sigma_k}^{\tau_w}\sigma^2\left(\sigma e^{-\frac{E_2}{\eta_2}t}+(\sigma-\sigma_k)\right)d\sigma = \\ = \frac{\pi R^3 \tau_w}{4\eta_2}\left[1+\left(1-\frac{\sigma_k^4}{\tau_w^4}\right)e^{-\frac{E_2}{\eta_2}t}+\frac{1}{3}\frac{\sigma_k^4}{\tau_w^4}-\frac{4}{3}\frac{\sigma_k}{\tau_w}\right], \tag{14}$$

or

$$Q = \frac{\pi R^3 \tau_w}{4\eta_2}\cdot f\left(t,\frac{\sigma_k}{\tau_w}\right), \tag{15}$$

where

$$f\left(t,\frac{\sigma_k}{\tau_w}\right) = \left[1+\left(1-\frac{\sigma_k^4}{\tau_w^4}\right)e^{-\frac{E_2}{\eta_2}t}+\frac{1}{3}\frac{\sigma_k^4}{\tau_w^4}-\frac{4}{3}\frac{\sigma_k}{\tau_w}\right] \tag{16}$$

The dependence of the function $f(t,\sigma_k/\tau_w)$ on time in coordinates $(\lg t, \lg f)$ is presented in figure.

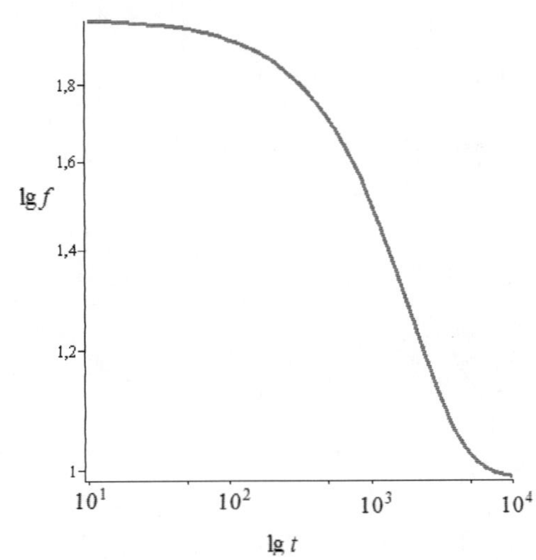

Figure 4. The dependence of the function f on time.

From Figure 4 shows that over time, the function tends to a stationary value. This means that, rushing to the limit from formulas (14) - (16), we obtain the formula for the plastic body flow rate (the Buckingham formula).

Using (16) the average speed v_{cp}, according to (12) and (14) will be equal to

$$v_{cp} = \frac{R\tau_w}{4\eta_2}\left[\left[1+\left(1-\frac{\sigma_k^4}{\tau_w^4}\right)e^{-\frac{E_2}{\eta_2}t}+\frac{1}{3}\frac{\sigma_k^4}{\tau_w^4}-\frac{4}{3}\frac{\sigma_k}{\tau_w}\right]\right]. \tag{17}$$

Because the shear stress on the pipe wall τ_w, according to equation (8), will be equal to

$$\tau_w = \frac{R\delta p}{2L},$$

formula (17) takes the form

$$v_{cp} = \frac{R^2\delta p}{8\eta_2 L}\left[1+\left(1-\left(\frac{2L\sigma_k}{R\delta p}\right)^4\right)e^{-\frac{E_2}{\eta_2}t}+\frac{1}{3}\left(\frac{2L\sigma_k}{R\delta p}\right)^4-\frac{4}{3}\frac{2L\sigma_k}{R\delta p}\right]=$$

$$=\frac{R^2\delta p}{8\eta_2 L}\cdot f\left(t,\frac{\sigma_k}{\tau_w}\right)=\frac{R^2\delta p}{8\eta_2 L}\cdot f\left(t,\frac{2L\sigma_k}{R\delta p}\right). \tag{18}$$

We rewrite equation (18)

$$\delta p = \frac{8\eta_2 L v_{cp}}{R^2} f^{-1}\left(t,\frac{2L\sigma_k}{R\delta p}\right). \tag{19}$$

After simple transformations we get

$$\delta p = \frac{64\eta_2}{\rho d v_{cp}}\frac{\rho v_{cp}^2}{2}\frac{L}{d} f^{-1}\left(t,\frac{2L\sigma_k}{R\delta p}\right), \tag{20}$$

from where

$$\lambda = \frac{64}{\mathrm{Re}} f^{-1}\left(t,\frac{2L\sigma_k}{R\delta p}\right),\ \mathrm{Re} = \frac{\rho v_{cp} d}{\eta_2}. \tag{21}$$

Find from the equation (18) the Reynolds number is

$$\mathrm{Re} = \frac{\rho d^3\delta p}{32\eta_2^2 L} f^{-1}\left(t,\frac{4L\sigma_k}{d\delta p}\right) = \frac{\rho d^2\sigma_k}{8\eta_2^2} f_1^{-1}\left(t,\frac{4L\sigma_k}{d\delta p}\right), \tag{22}$$

$$f_1\left(t,\frac{4L\sigma_k}{d\delta p}\right) = f\left(t,\frac{4L\sigma_k}{d\delta p}\right)\Big/\left(\frac{4L\sigma_k}{d\delta p}\right). \tag{23}$$

Number

$$\rho d^2\sigma_k/\eta_2^2 \tag{24}$$

is dimensionless and characterizes the visco-plastic properties of the liquid [21, 22]. If in equality

$$\lambda = \frac{64}{\text{Re}} f^{-1}\left(t, \frac{2L\sigma_k}{R\delta p}\right), \ \text{Re} = \frac{1}{8}\frac{\rho d^2\sigma_k}{\eta_2^2} f_1\left(t, \frac{4L\sigma_k}{d\delta p}\right) \quad (25)$$

Is considered $4L\sigma_k / d\delta p$ as a parameter, these equations express the parametric law of resistance of sludge movement in pipes. Figure 5 shows the graphs of the Re number for the sludge and plastic body. It can be seen from the figure that the Reynolds slurry number tends to the Re number of the plastic body over time.

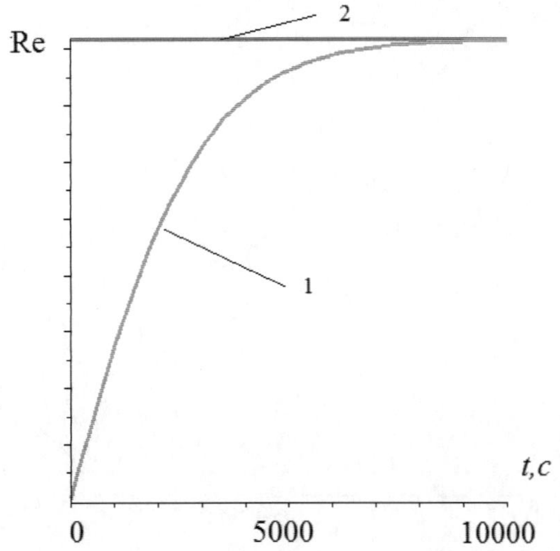

Figure 5. Dependence of the Reynolds sludge number on time: 1 – Re of sludge, 2 – Re of plastic body

Figures 6 and 7 shows the law of resistance for sludge and plastic bodies on a logarithmic scale with the same mechanical characteristics.

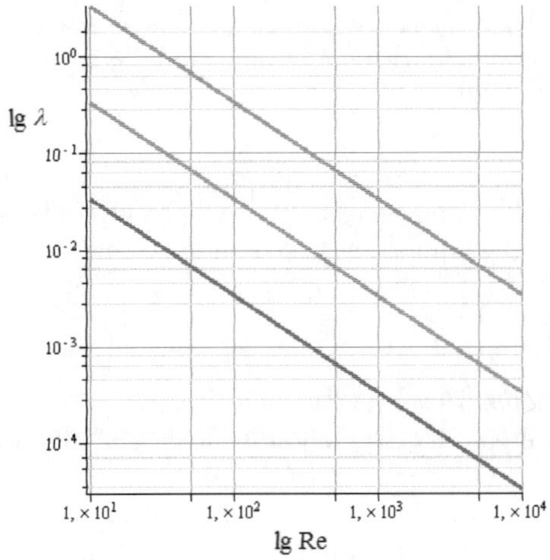

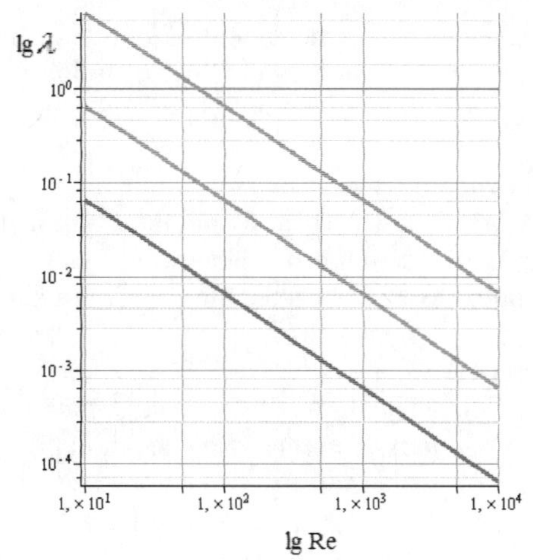

Figure 6. The law of resistance $\lambda(\mathrm{Re})$ for sludge

Figure 7. The law of resistance $\lambda(\mathrm{Re})$ for the plastic body

Figure 8 shows the family of curves of the dependence of the resistance coefficient on the number Re. The parameter is a number $\rho d^2 \sigma_k / \eta_2^2$, called the Hedstrem criterion [23-28]. This figure can be considered as a convenient way to calculate the pressure drop.

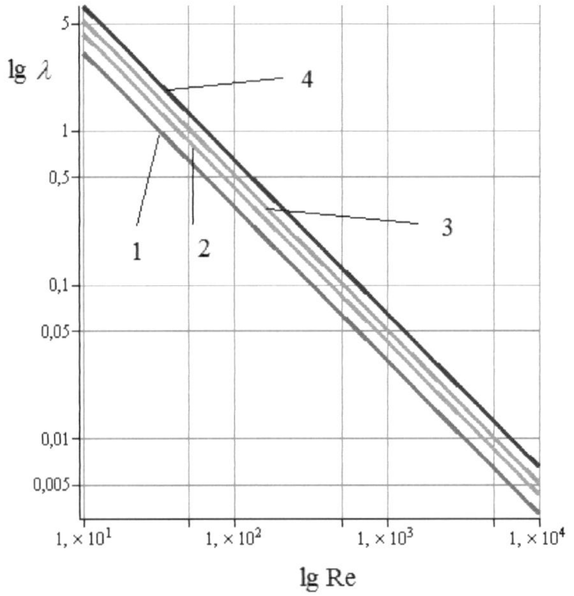

Figure 8. shows that the resistance coefficient during the flow of sludge through the pipe tends to the coefficient of resistance of the plastic body over time.

5. Conclusion

Hydraulic resistance when moving sludge through the pipes is considered in the work. Parametric law of resistance to the movement of sludge in the pipes is found. This law can be used as a convenient way to calculate the pressure drop. It follows from the law that sludge now acquires the properties of a plastic body flowing through pipes and behaves like Bingham plastic.

Acknowledgments

The work was performed within the framework of the implementation of the Development University program based on the BSTU named after V.G. Shukhov using equipment based on the Center of High Technologies BSTU named after V.G. Shukhov.

References

[1] Bernshtejn L A 1965 *Novoe v tehnologii pererabotki i transportirovanija syr'ja v cementnoj promyshlennosti* [New technologies in processing and transportation of raw materials in the cement industry](Moskow: Izdatel'stvo literatury po stroitel'stvu) [In Russian]

[2] Isaev E A 1991 Teoreticheskaya model phormirovaniya okatyshey iz sypuchih materialov v processe ih okomkovaniya [The theoretical model of the formation of pellets from bulk materials in the process of their palletization] *Nososibirsk. Izvestiya Vy`sshix uchebny`x zavedenij* **2** pp 17-21 [In Russian]

[3] Aref'ev N N 2009 Issledovaniye techeniya vyazkoplasticheskoy zhidkosti [Investigation of the flow of a viscoplastic fluid] *Vestnik AGTU* **1** pp 25–33[In Russian]
[4] Shadrina N H 2011 Matematicheskaya model techeniya zidkosty v htzistivnom sosude [Mathematical model of fluid flow in a resistive vessel] *Vestnik Nizhegorodskogo Universiteta im. N.I. Lobachevskogo* **4(5)** pp 2604–6 [In Russian]
[5] Vershinina I P 2009 Matematicheskaya model nen'yutonovskoy zidkosty vo vraschayushemsya cilindricheskom kanale [A mathematical model of the dynamics of a non-Newtonian fluid in a rotating cylindrical channel] *Izvestiya VolgGTU* **11(71)** pp 5–8 [In Russian]
[6] Paulo R de Souza Mendes 2009 Modeling the thixotropic behavior of structured fluids *Journal of Non-Newtonian Fluid Mechanics* pp 66 – 75
[7] Alexandros Syrakos 2014 Thixotropic flow past a cylinder *Journal of Non-Newtonian Fluid Mechanics* **8**
[8] Sevostyanov M V, Sevostyanov V S and Martakov I G 2018 Analysis of Design Features of Circulating Action Mixers *J. Phys.: Conf. Ser.* **1066** 012002
[9] Sivachenko L A, Sevostyanov V S and Il'ina T N 2018 Problem Tasks in the Field of Resource and Energy Saving Technologies *J. Phys.: Conf. Ser.* **1066** 012021
[10] Glagolev S N, Sevostyanov V S, Ilyina T N and Uralskii V I 2011 Production modules for combined reprocessing of technogenic materials *Chemical and Petroleum Engineering* **46(9)** 556-9
[11] Ur`ev N B 1980 *Vysokokoncentrirovannie dispersnie systemi* [Highly concentrated disperse systems] (Moscow: Himiya) [In Russian]
[12] Zakgejm A Ju 1982 Vvedenie v modelirovanie himiko-tehnologicheskih processov [Introduction to chemical Process modeling] (Moscow: Himiya) [In Russian]
[13] Boichuk I P and Tarapov I E 2002 Ptrimenenie metoda integralnich sootnosheniy dlya izucheniya dvizeniya vyazkoy zidkosti po dephormiruemym trubam [Application of the method of integral relations for studying the motion of a viscous fluid along deformable pipes] *Vestnik Har`kovskogo nacional`nogo universiteta im. V.N.Karazina. Ceriya «Matematika, prikladnaya matematika i mehanika»* **542** 13-8 [In Russian]
[14] Boychuk I P 2005 Peristal'ticheskiy transport vyazkoy zhidkosti v cilindricheskih trubah [Peristaltic transport of viscous fluid in cylindrical pipes] *Vestnik Har`kovskogo Nacional`nogo Avtomobil`no-Dorozhnogo Universiteta* **29** pp 142-3 [In Russian]
[15] Boichuk I P and Tarapov I E 2002 Application of a momentum integral model for study of unsteady viscous liquid motion in deformable tubes *Russian Journal of Biomechanics, Perm* vol 5 **4** pp 97-102
[16] Sharapov R R, Besedin P V, Boichuk I P, Kozlov V K and Andrushchak S V 2016 Mathematicheskoe modelirovanie teheniya cementnyh syr'evych shlamov pri nestacionarnich vozdeystviyzh [Mathematical modeling of the flow of cement raw sludge under non-stationary effects] *Vestnik BGTU im. V.G.Shukhova* **6** pp 134-8 [In Russian]
[17] Shayahova L D, Rahimbaev Sh M and Denisova A B 2014 Vliyanie razzhizhitelej na reologicheskie svojstva shlamov [Influence of diluents on the rheological properties of sludge] *Sement i ego primenenie* **3** pp 61-5 [In Russian]
[18] Lomachenko D V and Shapovalov N A 2013 Regulation of the rheological properties of cement slimes with the use of ore-dressing waste *Vestnik BGTU im. V.G.Shukhova* **2** pp 141-3 [In Russian]
[19] Trubaev P A and Besedin P V 2005 Criteria for the thermodynamic efficiency of cement clinker production from natural raw material *Theoretical Foundations of Chemical Engineering* **39(6)** 628-34
[20] Trubaev P A 2006 Exergy analysis of thermal processes in the building materials industry *Theoretical Foundations of Chemical Engineering* **40(2)** 175-82
[21] Eremin N F 1986 *Processy i apparaty v technologii stroitelnych materialov* [Processes and devices in the technology of building materials] (Moscow: Vy`sshaya shkola) [In Russian].

[22] Loitsyansky L G 2003 *Mechanika zhidkosty i gaza* [Fluid and gas mechanics] (Moscow: Drofa) [In Russian]
[23] Wilkinson W L 1964 *Nen'yutonovskie zhidkosty* [Non-Newtonian fluids] (Moscow: Mir) [In Russian]
[24] Ferraris C F and F de Larrard 1998 *Testing and Modelling of Fresh concrete rheology* (No. NIST Interagency/Internal Report (NISTIR)-6094)
[25] Krivilev A 2005 *Osnovy komp'juternoj matematiki s ispol'zovaniem sistemy MATLAB* [Fundamentals of computer mathematics using the system MATLAB] (Moscow: Leks-Kniga) [In Russian]
[26] Kasatkin A G 1971 *Osnovnye processy i apparaty himicheskoj tehnologii* [Basic processes and apparatuses of chemical technology] (Moscow: Himizdat) [In Russian]
[27] Hedstrem O A 1952 Flow of plastics materials in pipens *Ind. Eng. Chem.* **44** p 651
[28] Hanks R W 2004 The laminar-turbulent transition for fluids with a yields stress *AIChE Journal* **9(3)** p 306–9

Effective ways of molding man-made materials

M V Sevostyanov[1],T N Il`ina[2] and I G Martakov[1]

[1]Department of Technological Complexes Machines and Mechanisms, Belgorod State Technological University of V.G. Shukhov, Kostyukov St., 46, Belgorod, 308012, Russia
[2]Department of Heat Supply and Ventilation, Belgorod State Technological University of V.G. Shukhov, Kostyukov St., 46, Belgorod, 308012, Russia

E-mail: msev31@mail.ru

Abstract. The article is devoted to the development of multifunctional technological complexes for obtaining highly dispersed activated materials, including fast-hardening microfibrated composite mixtures. Special attention is paid to the development of conditions for their effective homogenization and the molding technology of various charges. Presented experimental design, aimed at improving the process of forming man-made mixtures. A low-tonnage technological complex for the utilization of technogenic materials with various physical-mechanical properties was developed. The urgency of the development of techniques and technology for the utilization of man-made materials by means of compaction when creating new high-tech industries is shown.

Keywords: technological complex, microfibration, composite mixtures, extrusion, granulation, additives.

1. Introduction

At the present stage of development of science and technology, one should hardly expect any "breakthroughs" in any industry of the industry using traditional technologies and physical methods of influence on the processed material. New discoveries can occur only with non-standard approaches to the object under study, often at the junction of various areas of knowledge. All of this, first of all, concerns such large-scale industries, which determine the economic and production potential of the country, such as the building materials industry, the mining industry and the metallurgical industry, the chemical, fuel, etc.

In this regard, a very promising direction of research is the study of the mechanism of interaction of particles at the submicroscopic level, which also determines new properties of composite materials [1–4].

2. Development of a multifunctional technological complex

Production of highly dispersed activated materials (including fast-hardening microfibrated composite mixtures), the development of conditions for their effective homogenization, the technology of molding various charges is a complex multi-faceted task that can be solved by creating multifunctional technological complexes (MTC) [5–11].

Taking into account the above factors, we developed and created MTC for obtaining highly activated materials with subsequent (if necessary) extrusion into granules of a given shape and size (figure 1,a,b).

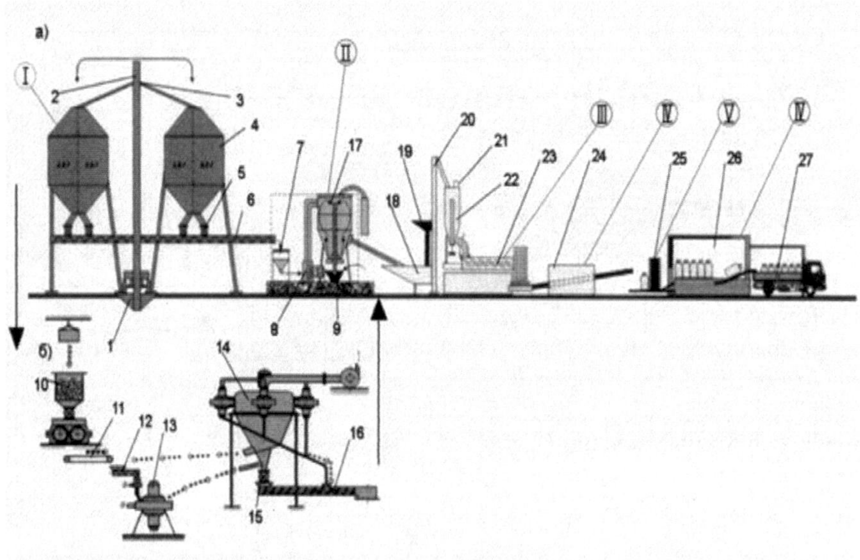

Figure 1. Multifunctional technological complex for the extrusion of mechanically activated materials: a – with the disintegrating activator; b – with the mechanical activation of additives in the vortex-acoustic dispersant.

The technological complex (figure 1, a, b) consists of the following equipment:
1-receiving hopper of raw materials components; 2-elevator; 3-way gate; 4-hopper components; 5-cell feeders; 6-screw conveyor; 7-receiving hopper of grinding-mixing complex; 8-disintegrator activator; 9- supplement receiving bin; 10-press roller shredder; 11-belt conveyor; 12-screw feeder; 13-vortex acoustic disperser; 14-separator dust collector; 15-cell feeder; 16-screw conveyor; 17-bin mixer; 18-receiving hopper of mixing-extrusion complex; 19-additives hopper; 20-vertical screw; 21-binder dispenser; 22-fluidized bed mixer; 23-press roller extruder; 24-drying unit; 25-weight dispenser; 26-warehouse of finished products; 27-motor vehicles.

The technological complex (figure 1, a) includes the following redistribution:
I – receiving and dosing department; II – crushing and mixing unit; III – block for the homogenization and extrusion of the moistened charge; IV – drying department of granules with their classification; V – unit of weight dosing and packing of granules; VI – department of warehousing and shipment of finished products.

In addition, the technological complex contains aggregates for the production of highly activated additives (figure 1, b): press roller crusher 10, vortex-acoustic dispersant 13, separator 14.

The technological complex can function both in a straightforward scheme "a" and using highly activated additives - scheme "a – b".

The most crucial unit of the MTC is the technological complex for the extrusion of materials (figures 2, 3). The complex works as follows:

Prepared charge for the molding, consisting of finely crushed materials, is fed into the receiving hopper 1 of mixture.

Highly activated additives are loaded into a receiving bin 2, equipped with a metering device. The original mixture with additives vertical screw feeder 3 is fed to the mixer 4 fluidized bed, equipped with a metering device 5 binder.

The moistened and homogenized mixture is uniformly fed into the press-roller extruder (PRE) 6 [12–16], consisting of a screw prehardening device 7 and the roller granulator 8.

Each of the units of the technological complex (screw feeder, mixer, press-roller extruder) has an individual drive and the following technical characteristics (table 1).

The PRE contains a screw prehardening device with a conical shaft (D_{avr} = 85 · 10^{-3}m), a taper of K = 0.01 and an adjustable preloading unit, as well as a roll granulator with the following characteristics:

- The outer diameter of the drum (press matrix) –D_{dr} = 450 · 10^{-3}m;
- Drum width – B_{dr} = 120 · 10^{-3}m;
- The drum wall thickness – b = 15 · 10^{-3}m;
- The number of holes in the drum – Z_{hl} = 1080 pcs;
- The number of pressing rolls –Z_r = 2 pcs;
- Roll diameter – D_r = 140 · 10^{-3}m;
- Roll width – B_r = 75 · 10^{-3}m;
- Roll rotational speed – n_r = 90 rpm.

Table 1. Technical characteristics of technological complex units.

№pos	Technical characteristics of technological complex units	Values of technical characteristics of units of the technological complex		
		Verticalscrew	Mixer	Press-rollerextruder
1	Installed power of the electric drive, kW	1,5	4	11,5
2	The frequency of rotation of the working body, rpm	–	400	90
3	Screw pitch 10^{-3},m	60	–	35 prehardener
4	The angle of the blades of the mixer shaft, deg.	–	35	–

The humidity of the formation of the charge is determined by its physical-mechanical characteristics (dispersion, plasticity, etc.) and mineralogical composition [17–20].

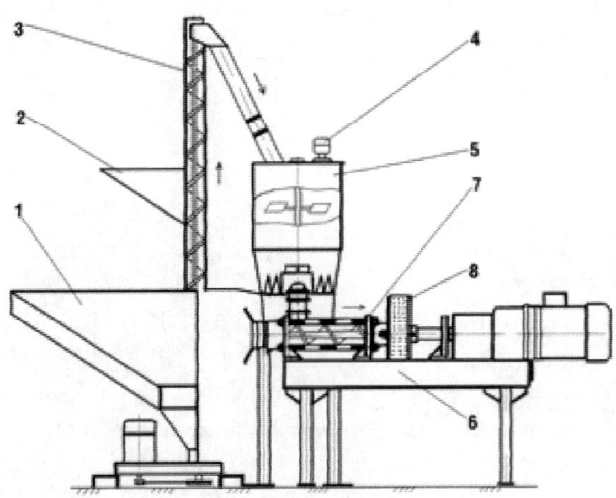

Figure 2. Technological complex for extruding materials:1-receiving hopper of the initial mixture; 2-receiving hopper of finely ground additives; 3-vertical screw feeder; 4-fluidized bed mixer; 5-metering device binder; 6-press roller extruder; 7-screw prehardening device; 8-roll granulator.

The technological complex developed by us (figure 1) has a multifunctional purpose and can be used to perform various tasks:

- Produce highly dispersed materials in the manufacture of dry building mixtures;

- Getting quick-hardening composite mixes using microfibers;
- Production of highly activated additives for various technological purposes;
- Getting highly activated cement-sand mortars with additives for the production of heat-insulating materials (foam blocks, shaped products, etc.)
- Production of extruded materials of various technological purposes (fillers for the reinforcement of foam blocks, Na-bentonite for waterproofing underground structures, calcium-containing organic fertilizers, granulated animal feed, etc.).

Figure 3. Press-roller extruder.

For compacting materials by the pressing method, a press-roll unit of stepwise compaction was developed, equipped with a device for prehardening of the charge (figure 4 a, b). This invention provides: at the 1st stage pre-compaction with elastic rollers to remove the gaseous phase from loose, low-flow materials and squeezing excess moisture from moisture saturated pressing rollers with giving the compressed bodies a given geometric shape and size. [21–25].

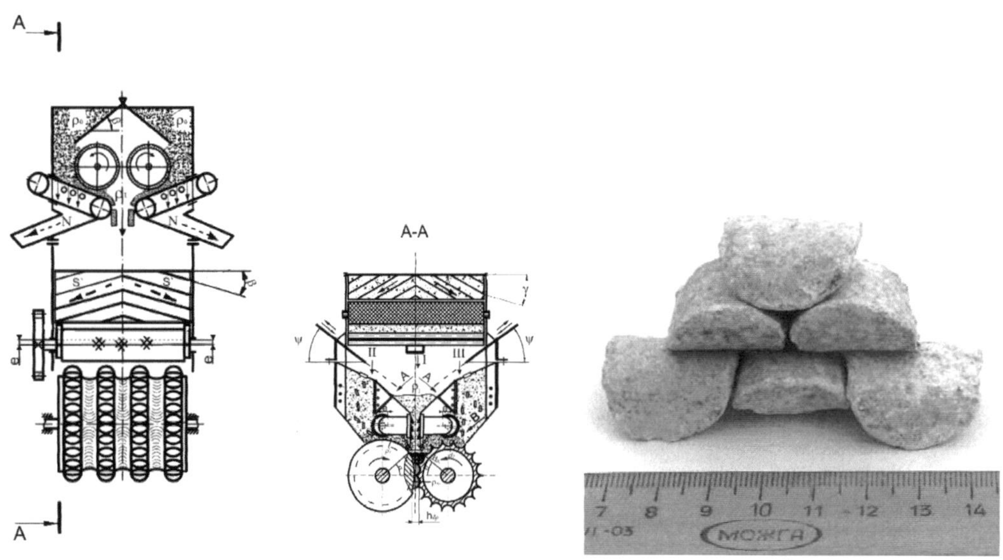

a b

Figure 4. Press roller unit of stepwise consolidation and briquettes produced on it. the scheme of the press-roller unit, b) the resulting briquettes

IOP Conf. Series: Materials Science and Engineering **552** (2019) 012038 doi:10.1088/1757-899X/552/1/012038

We have developed effective methods of forming man-made materials and aggregates for their implementation, which can be used in the disposal of a wide range of powdered, viscoplastic and moisture saturated materials used in various industries.

Our complex design, technological, theoretical and experimental studies of the units developed at the level of inventions confirmed their efficiency and operational reliability.

The units can be used both in standalone mode and as part of energy-saving technological complexes for solving various high-tech technological problems.

Taking into account the results of experimental studies and the developed technological regulations for the process of utilization of man-made raw materials, we have designed and manufactured a technological complex for the production of various building materials and products based on them.

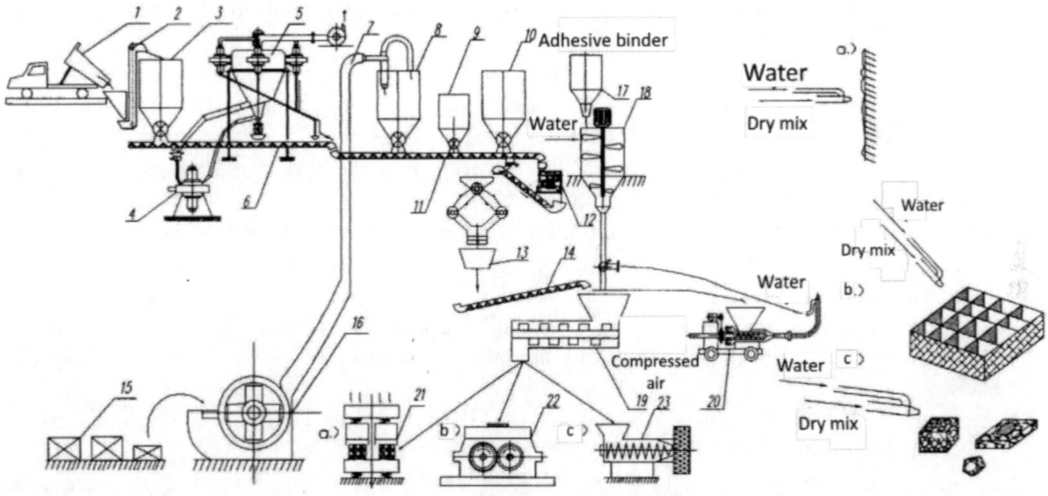

Figure 5. Technological complex for the production of composite building materials: 1 - vehicles with powdered binder; 2 - elevator; 3 - hopper with powdered binder; 4 - vortex dispersant; 5 - a combined action separator; 6, 11, 14 - screw feeders; 7 - pneumatic transport;8 - hopper with fiber filling; 9 - hopper with additives; 10 - hopper with material; 12 - centrifugal grinding and mixing unit; 13 - rotary circulation mixer; 15 - warehouse of man-made fibrous materials; 16 - percussion unit; 17 - hopper with a liquid binder; 18 - mixer-homogenizer; 19 - paddle mixer; 20 - ejection spray installation; 21 - vibropress; 22 - press roller unit; 23 - press roller extruder.

The developed technological complex has a multifunctional purpose and allows to solve issues related both to the utilization of man-made materials and the receipt of new types of building materials based on the developed composite mixtures.

Disposal methods include both the production of various types of molded products from man-made materials (figure 5, position 21, 22, 23) and the implementation of other technological operations: the production of thermal insulation coatings, etc. (figure 3, position 20 a, b, c).

3. Conclusion

In this way, our research and design and technological developments allowed us to create technological complexes that ensure the utilization of powdered and viscous-plastic man-made materials with different physical and mechanical properties, and the production of new types of building materials and products. The developed equipment and technologies for utilization of man-made materials can be used both at existing enterprises of the building materials industry and when creating new high-tech industries.

Multifunctional technological complexes, which include the main stages of various production of building materials industry (dosing, grinding, mixing, molding, drying, etc.) are a good material and technical base of practice for students and graduate students, conducting research and development, the integration of creative thought university scientists from various departments (technologists, mechanics, specialists in electric drives and automated control systems, ecologists, economists, etc.), as well as real-life production Competitive products business facility.

Acknowledgments
The work was performed within the framework of the implementation of the Development University program based on the BSTU named after. V.G. Shukhov using equipment based on the Center of High Technologies BSTU n.a. V.G. Shukhov.

References
[1] Rybiev I A 2002 *Stroitel`noe materialovedenie* [Construction Materials] (Moscow:Vy`sshaya shkola) [In Russian]
[2] Poluxina L M and Barambojm N K 1993 Mexanoximicheskaya aktivaciya polimerov pri dispergirovanii. Texnologicheskie problemy` izmel`chenii i mexanoaktivacii [Mechanochemical activation of polymers during dispersion. Technological problems of grinding and mechanical activation] *Materialy` nauchno-texnicheskogo seminara stran sodruzhestva. Mogilev* pp 171-78 [In Russian]
[3] Sevostyanov V S, Shestakov V L, Dubinin N N and Sevostyanov M V 2010 *Processy` v proizvodstve stroitel`ny`x materialov* [Processes in the production of building materials] (Belgorod: BGTU im. V.G.Shukhova) [In Russian]
[4] Glagolev S N, Sevostyanov V S, Ilyina T N and Uralskii V I 2011 Production modules for combined reprocessing of technogenic materials *Chemical and Petroleum Engineering* **46(9)** 556-9
[5] Sivachenko L A, Sevostyanov V S and Il´ina T N 2018 Problem Tasks in the Field of Resource and Energy Saving Technologies *J. Phys.: Conf. Ser.* **1066** 012021
[6] Laskorin B I, Barsky L A and Persits B Z 1984 *Bezotxodnaya texnologiya pererabotki mineral`nogo sy`r`ya. Sistemny`j analiz* [Waste-free technology of mineral processing. System analysis] (Moscow: Nedra) [In Russian]
[7] Lvov V V 2013 *Novy`e texnologii v tonkom i sverxtonkom izmel`chenii mineral`nogo sy`r`ya. Zapiski Gornogo instituta* [New technologies in fine and ultra-fine grinding of mineral raw materials. Notes of the Mining Institute.] vol 202 pp 115-7 [In Russian]
[8] Gridchin A M, Sevostyanov V S, Lesovik V S, GlaGolev S N, Sevostyanov M V and Funikov I M 2007 Texnologicheskie kompleksy` dlya proizvodstva porizovanny`x zapolnitelej iz texnogenny`x materialov [Technological complexes for the production of porous aggregates from technogenic materials] *Izvestiya vy`sshix uchebny`x zavedenij. Stroitel`stvo* **7** pp 22-8 [In Russian]
[9] Glagolev S N, Sevostyanov V S, Gridchin A M, Uralsky V I, Sevostyanov M V and Yadykina V V 2013 Resurso-e`nergosberegayushhie moduli dlya kompleksnoj utilizacii texnogenny`x materialov [Resource-energy-saving modules for the integrated recycling of man-made materials] *Vestnik BGTU im. V.G.Shukhova* **6** pp 102-6 [In Russian]
[10] Sevostyanov M V, Il`ina T N, Kuznetsova I A, Osokin A V and Martakov I G 2016 Resursosberegayushhij texnologicheskij kompleks dlya proizvodstva granulirovanny`x stabiliziruyushhix dobavok shhebenochno-mastichnogo asfal`tobetona [Resource-saving technological complex for the production of granular stabilizing additives of crushed stone-mastic asphalt concrete] *Vestnik Tambovskogo gosudarstvennogo texnicheskogo universiteta* vol 22 pp 272 – 9 [In Russian]
[11] Sevostyanov V S, Shatalov A V, Sevostyanov M V, Krutov V S and Shatalov V A 2018 Technics and Technology of Processing Basalt Waste *J. Phys.: Conf. Ser.* **1066** 012027
[12] Gridchin A M, Sevostyanov V S, Lesovik V S and Sevostyanov M V 2002 *Press-rulonnyy ekstruder* [Press-roll extruder] Patent RF No 2207247 [In Russian]

[13] Sevostyanov M V, Dubinin N N 2003 Konstruktivno-texnologicheskoe sovershenstvovanie press-valkovy`x e`kstruderov [Design and technological improvement of press-roll extruders] *Sovremenny`e texnologii v promy`shlennosti stroitel`ny`x materialov i industrii: Sb. stud. dokl. Mezhdunarodnogo kongressa* (Belgorod: BGTU im. V.G.Shukhova) **3** pp 31-5 [In Russian]

[14] Sevostyanov V S, Goryagin P Yu, Obolonsky V V and Ermilov R A 2018 Resource-Saving Technological Complex for Polymer Waste Processing *J. Phys.: Conf. Ser.* **1066** 012025

[15] Sevostyanov M V, Il`ina T N, Uvarov V A and Shinkarev L I 2015 Sposoby` kompaktirovaniya texnogenny`x materialov i texnicheskie sredstva dlya ix realizacii [Ways of compacting man-made materials and technical means for their implementation] *Vestnik BGTU im. V.G.Shuxova* 2 pp 107-11 [In Russian]

[16] Gridchin A M, Sevostianov V S, Lesovik V S and Sevostyanov M V 2003 *Mnogotselevoy granulyator* [Multipurpose granulator] Svidetel'stvo RF na poleznuyu model' No 30244 [In Russian]

[17] Sevostyanov M V, Sevostyanov V S and Martakov I G 2018 Analysis of Design Features of Circulating Action Mixers *J. Phys.: Conf. Ser.* **1066** 012002

[18] Gridchin A M, Sevostyanov V S, Lesovik V S, Gorlov A S, Perelygin D N, Romanovich A A and Kolesnikov A V 2006 E`nergo-resursosberegayushhie kompleksy` tonkogo i sverxtonkogo izmel`cheniya materialov [Energy-saving complexes of fine and ultrafine grinding of materials] *Izvestiya vy`sshix uchebny`x zavedenij. Stroitel`stvo* **11-12** pp 60-8 [In Russian]

[19] Kravchenko V P 2010 Aktivaciya domenny`x shlakov [Activation of blast furnace slag] *Vestnik Priazovskogo derzhavnogo texnicheskogo univer-siteta. Texnicheskie nauki* **2** pp 17-20 [In Russian]

[20] Il`ina T N 2009 *Processy` aglomeracii v texnologiyax pererabotki dispersny`x materialov* [Agglomeration processes in the processing technologies of dispersed materials] (Belgorod: BSTU V G Shukhova) [In Russian]

[21] Sevostyanov V S, Goryagin P Y, Sevostyanov M V and Ermilov R A 2018 Constructive and Technological Features of Units for Milling Polymer Material *J. Phys.: Conf. Ser.* **1066** 012006

[22] Sevostyanov V S, Sverguzova S V, Sevostyanov M V, Spirin M N, Shinkarev L I and Fetisov D D 2011 Press-valkovy`j agregat dlya formovaniya texnogenny`x materialov s maloj nasy`pnoj massoj [Press-roll unit for the formation of man-made materials with a low bulk weight] *Sb. dokl. IV Mezhdunar. nauch.-prakt. konf. «E`kologiya – obrazovanie, nauka, promy`shlennost` i zdorov`e»* (Belgorod: BGTU im. V.G.Shukhova) pp 125-30 [In Russian]

[23] Ilyina T N, Sevostyanov V S, Uralskii V I, Sevostyanov M V and Shkarpetkin E A 2010 Research, design, calculations, and operating experience. Processes and equipment for chemical and oil-gas production: Mechanism of stagewise polydispersed material granulation *Chemical and Petroleum Engineering* **46(3)** 193-200

[24] Sevostyanov V S, Sverguzova S V, Sevostyanov M V, Spirin M N, Shinkarev L I and Fetisov D D 2012 *Texnika i texnologiya briketirovaniya poroshkoobrazny`x i vyazkoplastichny`x texnogenny`x materialov* [Technique and technology of briquetting powdery and viscous-plastic man-made materials] (Vestnik BGTU im. V.G.Shuxova) **3** pp 87-90 [In Russian]

[25] Sevostyanov M V, Sevostyanov V S, Shinkarev L I and Malkov A V 2012 Izuchenie processa dvizheniya skorostny`x potokov shixty` iz pitayushhix ustrojstv val`cevogo pressa [Studying the process of movement of high-speed flows of the charge from the feeding devices of a roller press] *Mezhvuz. sb. nauch. statej ITOM BGTU im. V.G. Shuxova «E`nergosberegayushhie texnologicheskie kompleksy` i oborudovanie dlya proizvodstva stroitel`ny`x materialov»* (Belgorod: BGTU im. V.G.Shukhova) pp 322-8 [In Russian]

Process of charge prevention with low bulk density

M V Sevostyanov[1], T N Il'ina[2] and I G Martakov[1]

[1]Department of Technological Complexes Machines and Mechanisms, Belgorod State Technological University of V.G. Shukhov, Kostyukov St., 46, Belgorod, 308012, Russia

[2]Department of Heat Supply and Ventilation, Belgorod State Technological University of V.G. Shukhov, Kostyukov St., 46, Belgorod, 308012, Russia

E-mail: msev31@mail. ru

Abstract. The processes of agglomeration of dispersed materials, the relevance of their use in the utilization of man-made powdered materials. Shows the need for prehardening of the granulated charge when they are balling with balling. This is especially true for man-made materials with low bulk density, which is an obligatory stage for the formation of micro-granules, the subsequent growth and compaction of which occurs under the action of centrifugal forces. Analytical expressions for calculation of rotation frequency of the forming rolls providing necessary conditions for realization of processes of filtration of gaseous and liquid phases of a composite mix, and also for reception of microgranules of the set density at a vibro-roll method of preliminary compaction are presented. The obtained dependences are the basis for the development of a vibration-centrifugal granulator for the step-by-step process of forming technogenic dispersed materials with low bulk density.

Keywords: Granulation, technogenic materials, compacting rolls, filtration of liquid and gaseous phases, vibration centrifugal granulator.

1. Introduction

In the processing technologies of dispersed materials in many industries to obtain molded products of a given quality, as a rule, use agglomeration processes carried out by extrusion, briquetting, pelleting, etc. The choice of molding method depends on the physicochemical properties of powdered materials, their dispersion, requirements for physical-mechanical parameters of the finished product and other factors [1]. The ever-increasing volume of powdered and dust-like wastes in the production of building materials and products necessitates the utilization of them by compaction [2–5]. However, some man-made materials have low bulk density (less than 200 kg / m^3), high dispersion. To obtain a granulated product from such charges, a stepwise molding process is necessary with the obligatory stage of preliminary compaction and the formation of granules centers [6–8]. Such materials include powdered waste from the production of expanded perlite, vermiculite, lime, etc. The study of the characteristic features of the molding process of man-made materials, as well as obtaining analytical dependencies of the filtration processes of the gaseous and liquid phases with the development of recommendations for choosing a method and apparatus for prehardening of a three-phase disperse system is the purpose of this work.

2. Materials and Methods

There are various technological methods of removing the gaseous phase during the formation of powder mixtures: vibrocompaction, vacuum, mechanical action using centrifugal forces, rolling - rolled, screw prehardening, etc. Each of these technological methods has its advantages and disadvantages [9-13].

In our opinion, the most effective way to remove the gaseous phase (deaeration) during compaction of powdered materials is a combination of vibration and mechanical action. This can provide not only a rational method of preparation of the mixture being compacted for granulation with minimal energy consumption, but also allows combining two technological methods: removal of the gaseous phase and obtaining micrograins (microgranulate), which are the centers of granulation.

The technical solution of this technological method can be the vibro-roller method of prehardening of the charge (figure 1).

The process of prehardening of the charge is greatly influenced by its physical-mechanical parameters: the flow ability of the material, depending on the particle size distribution, moisture content of the material and the shape of its particles; coefficients of internal and external friction; the rate of removal of the gaseous phase and others [14].

The main factors determining the flow ability of powdered materials are friction and adhesion of particles between themselves, making it difficult for them to move.

When the charge layer moves from top to bottom and is mechanically compacted in the rolls, the gaseous phase (air) is filtered through a layer of material between the cylindrical surfaces of the rolls and the sidewalls of the loading device. The air coming out through the layer of material rises up, worsening the material flow ability and the conditions for its supply to the roll deformation zone.

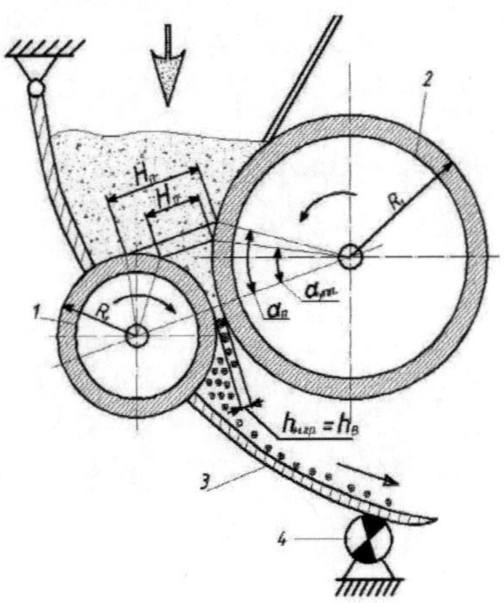

Figure 1. Scheme for the calculation of the prehardening process 1,2 - sealing rolls; 3 - vibro-grooves; 4 - vibrator.

The filtration rate of the gaseous phase passing through the cross-sectional area of H_0 thickness and width B can be determined by the formula:

$$\vartheta_{gas} = \frac{Q_{gas} \cdot \rho_{gas}}{H_0 \cdot B(\rho_{gr} - \rho_0)}. \quad (1)$$

where Q_{gas} – gas phase consumption, m^3 / s;ρ_{gr} – density of formed microgranules, kg / m^3;ρ_0 – initialdensity (bulk density) of the material, kg / m^3.

The stage of preliminary compaction of the charge and its microgranulation is followed by the stage of final molding of the granules with the application of dynamic action. In this regard, it is advisable to know the range of variation of the compaction rate of the material in the study of the granulation of materials with different physical and mechanical characteristics: from materials with high bulk density (iron ore production waste ρ_0>4 · 10^3 kg / m^3), carbonate-clay materials (ρ=(2…4)· 10^3 kg / m^3); as well as man-made materials with a bulk mass of ρ_0=800 - 2000 kg / m^3 and materials with a low initial bulk mass, ρ_0<800 kg / m^3 and less (exfoliated vermiculite, perlite, captured dust of rotary kilns of cement, lime and other industries).

The minimum pre-compaction rate of the charge is determined on the basis of conditions that exclude the free passage of powdered materials through the roll gap, which determines the diameter d_{gr} of the microgranulate particles or the height of the microgranules h_{gr} (due to their irregular geometric shape). In this case, the flow ability of compacted material q should be taken into account, which determines the free passage of material through the roll gap h_b (figure 1).

The maximum speed of microgranulation ϑ_{gr} is limited by the rate of filtration of the gaseous phase, moving towards the compacted layer of powdered material and making it difficult to deliver to the sealing zone between the rollers.

The critical filtration rate of the gaseous phase through a layer of powdered material limits the value ϑ_{gr} and is determined by the value of the Reynolds criterion [15]:

$$Re = \frac{Ar^{0.68}}{600}\left[\frac{H_{ch}}{h_{gr} + (R_1 + R_2)(1 - \cos\alpha_{hr})}\right]. \quad (2)$$

whereH_{ch} – thickness of compacted charge in the area of its seal, m; R_1, R_2 – forming rollers radius, m; h_{gr} – size of microgranules, m (figure 1).

Assuming that the volumetric flow rate of the gaseous phase through the upper section of the thickening zone$H_0 = h_{gr} + (R_1 + R_2)(1 - \cos\alpha_{hr})$ is $Q_{gas} = BH_0\vartheta_{gas}$ determine the critical filtration rate of the gaseous phase:

$$\vartheta_{gas} = \vartheta_{max}\left(\frac{\rho_{gr}}{\rho_0} - 1\right)\frac{h_{gr}}{h_{gr} + (R_1 + R_2)(1 - \cos\alpha_{hr})}. \quad (3)$$

According to $Re = \frac{\vartheta_{cr} \cdot d_{sz}}{\vartheta}$,

where d_{sz} – suspended size of forming charge particles, m;ϑ – kinematic viscosity of the gaseous phase, m^2/s.

As well as expressions (2) and (3), we obtain the maximum pre-compaction rate of the charge:

$$\vartheta_{max} = \frac{Ar^{0.68} \cdot H_0 \cdot \vartheta \cdot \rho_0}{600 \cdot h_{gr} \cdot (\rho_{gr} - \rho_0) \cdot d_{sz}}. \quad (4)$$

Thus, in the development of design-technological parameters of the roller-type device for the implementation of the process of microgranulation of powder particles, the condition:

$$\vartheta_{min} \le \vartheta_{for} \le \vartheta_{max}:$$

$$\pi D_1 n_v = \frac{Ar^{0.68} \cdot H_0 \cdot \vartheta \cdot \rho_0}{600 \cdot h_{gr} \cdot (\rho_{gr} - \rho_0) \cdot d_{sz}}. \quad (5)$$

So the rotational speed of the forming roll:

$$n_v = \frac{Ar^{0.68} \cdot H_0 \cdot \vartheta \cdot \rho_0}{1884 \cdot D_1 \cdot (\rho_{gr} - \rho_0) \cdot d_{sz}}. \quad (6)$$

If the process of prehardening of a moldable powder mixture is not accompanied by significant energy consumption, then the process of deformation of a compacted moisture-saturated mixture is accompanied by both structural and mechanical changes of the formed samples (reorientation and packaging of polydisperse particles, their adhesion-autohesion interaction, etc.), and the migration of the liquid phase from the zones greatest stress.

The liquid phase, which fills the pores between the particles, due to its incompressibility, provides, together with the solid phase of the structural system, elastic resistance to external stresses, and when the yield strength of the material is exceeded, it moves to reduced pressure zones.

When studying the process of migration of the liquid phase in a powdery mixture and using Darcy's law [16], an analytical expression was obtained to determine the necessary time for filtering a liquid through a granular layer of a moldable mixture:

$$\tau_f = 0.5\eta_g \cdot r_0 \cdot X_0 \frac{\vartheta_g^2}{S^2 \cdot \Delta\bar{P}}, \quad (7)$$

where ϑ_g – volume of filtered liquid, m^3;τ_f – time of filtration, s; S – surface of filtration, m^2; $\Delta\bar{P}$– pressure difference, N·m^{-2};η_g–viscosity of the filtered liquid, Pa·s; R_{res}–resistance of the granular medium, m^{-1};

$$R_{res} = r_0 \cdot X_0 \frac{\vartheta_g}{S}$$

where $r_0 X_0$ – resistivity and reduced thickness of the granular medium, respectively, m^{-1}.

The obtained analytical expression is of great importance for the practice of forming moisture-saturated powder mixtures and obtaining high-quality granules from them.

Regardless of the method of force (pressing, centrifugal molding, vibration, etc.) in the molding of granules and the design of the unit for granulating materials, it is necessary to observe the filtration conditions of the gaseous and liquid phase, defined by expressions (6) and (7), respectively.

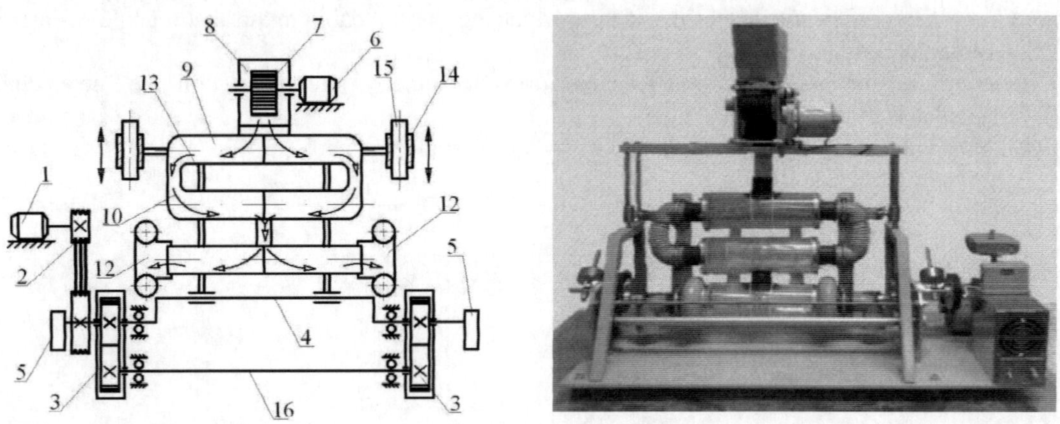

a b

Figure 2. Kinematic scheme (a), general view of the vibration-centrifugal granulator (b): 1 - main drive motor; 2 - V-belt transmission; 3 - gear pair; 4 - eccentric shaft; 5 - counterweights; 6 - electric device for warning and microgranulation of the material; 7 - elastic microgranulation compaction roller; 8 - vibration gutter; 9,10,11 - material granulation drums; 12 - toroidal chambers; 13 - movable frame; 14 - crawlers; 15 - rack guide; 16 - intermediate shaft

3. Results

Our studies of the filtration conditions of the gaseous and liquid phases during the compaction of powdered materials, the obtained analytical expressions for calculating the time of air filtration and the migration of the liquid phase in the moldable mixture at the stage of its preliminary compaction, were used in the design and creation of highly efficient process equipment for the step-by-step granulation process - vibration - centrifugal granulator [17, 18] (figure 2).

Studies [19] have shown that to ensure an effective process of microgranulation of moisture-saturated mixtures and filtration of the liquid phase, it is necessary to observe the condition:

$$n_f \leq \frac{2\sqrt{F(1-\mu^2)} \cdot S^2 \cdot \Delta\bar{P}}{0{,}5\pi \cdot \eta_g \cdot r_0 \cdot X_0 \cdot \vartheta_g^2 \cdot \sqrt{\pi \cdot E_f \cdot R_f}}. \tag{8}$$

where μ– Poisson's ratio, for an elastic material, such as rubber μ=0,4…0,5;E_f – modulus of elasticity of the forming surface of the roll, for rubber E_f=(10…100)·10^5 N/m^2; F – compressive force, N / m.

To ensure the specified performance of the granulator, in addition, it is necessary to take into account its throughput in obtaining quality products, which should also take into account the obtained valuesn_f.

The granulation (formation of compacted embryos), which further stabilizes the quality characteristics of the granules (their geometrical dimensions and shape, density distribution by volume, etc.), is important for the structure formation of granules during their stepwise formation. Prehardening obtained the following expression:

$$n_f \leq \frac{1}{2P}\sqrt{\frac{[K_{mn} - \sigma_r(1+K_\varphi)\zeta]}{[K_{hr} \cdot \rho_0 \cdot r_{0gr}(r_0 + r_{0gr})](1+K_\varphi)}}. \tag{9}$$

where K_{mn} – coefficient taking into account inter-contact interaction of particles during their compaction;K_φ– coefficient taking into account the internal friction of the particles.

ValuesK_{mn} andK_φ related by ratios, $K_{mn} = 2K_{sc}\cos\varphi$, $K_\varphi = \sin\varphi$,
where K_{sc}– the coefficient of adhesion of the particles to the moldable mixture;φ – angle of internal friction of particles in the mixture.

The coefficient of internal friction f is related to the angle of internal friction φ by dependence $f = \tan\varphi$ or $\varphi = \arctan f$.

Using the values $K_{mn} = 2K_{sc}\cos\varphi$, and $K_\varphi = \sin\varphi$, finally we get

$$n_f \leq \frac{1}{2P}\sqrt{\frac{2K_{sc}\cos\varphi - \sigma_r(1+K_\varphi)\zeta}{K_{hr} \cdot \rho_0 \cdot r_{0gr}(r_0 + r_{0gr})(1+\sin\varphi)}}, \tag{10}$$

where $$\zeta = \frac{2K_\varphi}{1+K_\varphi} = \frac{2\sin\varphi}{1+\sin\varphi}.$$

4. Discussion

The analytical expressions obtained by us characterizing the general conditions of the process of microgranulation of particles by the method of their volume compression allow us to establish not only the speed parameter of the dynamic effect on compacted particles for a given technical solution (sealing rolls), but also allow to determine the effect of other physicomechanical characteristics and technological parameters on microgranulation process. This, in turn, allows the use of various physical and chemical methods developed by us to improve the process of granulometry.

Analytical expressions (7, 9, 11) were used to calculate the frequency of rotation of the sealing rolls of the pilot-industrial vibratory-centrifugal granulator developed by us (figure 2). For materials with low bulk density, for example, pearlite-containing mixtures, ρ_{bulk} = 150 kg/m^3, ρ_{gr} = 1000 kg/m^3 with the parameters of the sealing device l = 0,041 m, Δl = 0,002 m, (l, Δl – length and width the gap between the rolls, respectively), D_B = 0,25 m, the rotational speed of the forming roll should not exceed values of $n_f \leq 2{,}7$ s^{-1}.

To ensure the conditions for the removal of the gaseous phase and the compaction of the charge, the rotational speed of the forming rolls (7) is $n_f \leq 0{,}04$ s^{-1}, i.e. the removal of the gaseous phase occurs at low speeds of the rolls.

According to the experimentally obtained data for perlite mixtures, we take: K_{sc} = 0,93·106 Pa, σ_r= 2,0 · 106 Pa, r_0 = 0,05·10^{-3} m, r_{0gr} = 2 · 10^{-3} m. The frequency of rotation of the sealing rolls, calculated by equation 10, is $n_f \leq 2{,}8$ s^{-1}.

For conditions E_f = 1 · 106 Pa, μ = 0,5 (for rubber), l = 0,041 m, F = 0,22 · 106 n / m (for pearl-containing mixtures), the rotational speed of the forming rolls calculated by equation (9) is n_{Φ} .v ≤ 2,0 s^{-1}.

Thus, to ensure the conditions for removal of the gaseous phase, filtration of the liquid, compaction of the charge with the formation of microgranules for pearlite-containing mixtures at the VCG productivity (Q = 50-100 kg/h), the rotational speed of the forming roll should be n_f = 1,1 – 2,0 s^{-1}.

Comprehensive studies of dispersed materials with different physicochemical properties allowed us to develop a classification of dispersed materials with recommendations on the organization of the processes of their agglomeration [20].

5. Conclusion

The obtained analytical expressions describing the processes of removing the gaseous phase, the movement of the liquid phase in the moldable mixture have found their application in the development of the roller device for pre-compaction of the mixture of the vibratory-centrifugal granulator.

This device allows to realize the stage of preliminary compaction of a moldable mixture with a bulk density of less than 200 kg/m^3 with the formation of microgranules, the subsequent growth and compaction of which is carried out in a forming unit consisting of three drums.

Acknowledgments

The work was performed within the framework of the implementation of the Development University program based on the BSTU named after V.G. Shukhov using equipment based on the Center of High Technologies BSTU named after V.G. Shukhov.

References

[1] Il'ina T N 2009 *Processy` aglomeracii v texnologiyax pererabotki dispersny`x materialov* [The agglomeration processes in the processing technologies of dispersed materials] (Belgorod: BGTU im. V.G.Shukhova) [In Russian]

[2] Glagolev S N, Sevostyanov V S, Il'ina T N and Uralsky V I 2010 Texnologicheskie moduli dlya kompleksnoj pererabotki texnogenny`x materialov [Technological modules for complex processing of technogenic materials] *Himicheskoe i neftegazovoe mashinostroenie* **9** pp 43-5 [In Russian]

[3] Sevostyanov M V, Il'ina T N, Uvarov, V A and Shinkarev L I 2015 Sposoby` kompaktirovaniya texnogenny`x materialov i texnicheskie sredstva dlya ix realizacii [Ways of compacting man-made materials and technical means for their implementation] *Vestnik BGTU im. V.G.Shuxova* **2** pp 107-11 [In Russian]

[4] Sevostyanov M V and Il'ina T N 2014 Klassifikacionny`j analiz texnicheskix sredstv dlya kompaktirovaniya texnogenny`x materialov [Classification analysis of technical means for

the compaction of man-made materials] *Naukoemkie texnologii i innovacii: sb. dokladov Yubilejnoj Mezhdunar. nauch.-prakt. konf., posvyashhennoj 60-letiyu BGTU im. V.G. Shuxova* (Belgorod: BGTU im. V.G.Shukhova) **4** pp 121-6 [In Russian]

[5] Sevostyanov M V, Il'ina T N, Shkarpetkin E A, Koshchukov A V and Emelyanov D A 2013 Utilizaciya poroshkoobrazny`x otxodov promy`shlenny`x predpriyatij [Utilization of powdered industrial waste] *Sb. nauchn. tr. Mezhd. Nauchno-prakticheskoj konf. «Aktual`ny`e problemy` razvitiya nauki i obrazovaniya»* (Moscow: «AR-Konsalt») pp 12-3 [In Russian]

[6] Il'ina T N, Sevostyanov V S, Uralsky V I, Sevostyanov M V and Shkarpetkin E A 2010 Mexanizm postadijnogo granuloobrazovaniya polidispersny`x materialov [The mechanism of stepwise granulation of polydisperse materials] *Ximicheskoe i neftegazovoe mashinostroenie* **5** pp 11-4 [In Russian]

[7] Sevostyanov M V and Dubinin N N 2003 Konstruktivno-texnologicheskoe sovershenstvovanie press-valkovy`x e`kstruderov [Design and technological improvement of press-roll extruders] *Sovremenny`e texnologii v promy`shlennosti stroitel`ny`x materialov i industrii: Sb.stud.dokl. Mezhdunarodnogo kongressa* (Belgorod: BGTU im. V.G.Shukhova) **3** pp 31-5 [In Russian]

[8] Sevostyanov M V 2004 Issledovanie uslovij formovaniya materialov v press-valkovom e`kstrudere [Study of materials forming conditions in a press-roll extruder] *Mezhvuzovskij sbornik statej* (Belgorod: BGTU im. V.G.Shukhova) pp115-20 [In Russian]

[9] Rieschel H 1971 Uber den Verdichtungs vorgang beim Briketieren Aufbereitungs *Technik* **11** pp 691-8

[10] Ravych B M 1975 *Briquetting in non-ferrous and ferrous metallurgy* [Briketirovanie v czvetnoj i chernoj metallurgii] (Moscow: Ugletexizdat) pp 142-76 [In Russian]

[11] Gridchin A M et al. 2006 E`nergosberegayushhaya texnika i texnologiya dlya kompleksnoj pererabotki prirodny`x i texnogenny`x materialov [Energy-saving equipment and technology for complex processing of natural and man-made materials] *Steklo mira* **6** [In Russian]

[12] Min'ko N I, Laz'ko E A and Doroganov E A 2009 Effect of finely disperse cullet on glass batch briquetting *Glass and Ceramics, Springer US* **1** pp 305–9

[13] Mulevanov S V, Min'ko N I, Kemenov S A, Osipov A A and Bykov V N 2009 Vibrational spectroscopy analysis of the structure of multicomponent phosphorus-containing silicate glasses *Glass and Ceramics, Springer US* vol 66 **3-4** pp 117-9

[14] Klassen P V, Grishaev I G and Shomin I P 1991 *Granulirovanie* [Granulation] (Moscow: Khimia) [In Russian]

[15] Il'ina T N 2009 Structural and mechanical properties of pelletized fine materials *Chemical and Petroleum Engineering* vol 45 **3-4** pp 115-8

[16] Zhuzhikov V A 1971 *Teoriya i praktika razdeleniya suspenzij* [Theory and practice of separation of suspensions] (Moscow: Khimia) pp 26-7 [In Russian]

[17] Il'ina T N, Sevostyanov M V, Shkarpetkin E A and Uralsky V I 2011 *Vibracionno-centrobezhny`j granulyator* [Vibration-centrifugal granulator] Patent RF No 2412753 [In Russian]

[18] Il'ina T N, Sevostyanov M V and Shkarpetkin E A 2010 Konstruktivno-texnologicheskoe sovershenstvovanie agregatov dlya granulirovaniya poroshkoobrazny`x materialov [Constructive-technological improvement of aggregates for granulating powdered materials] *Vestnik BGTU im. V.G.Shukhova* **2** pp 100-2 [In Russian]

[19] Sevostyanov V S, Il'ina T N, Sevostyanov M V and Shkarpetkin E A 2011 Issledovanie uslovij processa mikrogranulirovaniya v dispersny`x sistemax [Investigation of the conditions of the process of microgranulation in disperse systems] *Vestnik BGTU im. V.G.Shukhova* **1** pp 81-6 [In Russian]

[20] Il'ina T N 2013 Klassifikaciya dispersny`x materialov i rekomendacii po processam ix aglomeracii [Classification of dispersed materials and recommendations on the processes of their agglomeration] *Ximicheskoe i neftegazovoe mashinostroenie* **4** pp 17-9 [In Russian]

Analysis of separator operating conditions in a technological module with vibration-centrifugal closed-cycle grinding unit for obtaining high-dispersity materials

V S Sevostyanov[1], V I Uralsky[1], I P Boichuk[2], R A Ermilov[1] and E A Lukyanova[1]

[1]Department of Technological Complexes Machines and Mechanisms, Belgorod State Technological University of V.G. Shukhov, Kostyukov St., 46, Belgorod, 308012, Russia
[2]Department Operation of Ship Electrical Equipment and Automation, Admiral Ushakov Maritime State University, Lenin St., 93, Novorossisk, 353918, Russia

E-mail: ermilov.rodion@gmail.com

Abstract. The demand for up-to-date building materials and components for their production is constantly growing. The purchasing of foreign analogues is often unreasonable and costly, and the domestic products not always meet the high requirements specified for them. In light of this, it is necessary to develop up-to-date knowledge-intensive technologies of producing building materials of various purpose, as well as materials and products of ceramic, varnish-and-paint, glass and other industries, the basis of which is the application of high-dispersive powders. The extensive use of fine-ground and ultrafine-ground materials resulted in designing a wide range of grinders of various types (ball, vibration, planetary, centrifugal, jet grinders etc.). In each certain unit the special conditions for grinding are created, which allows using them for processing materials with various physical and mechanical characteristics. In this work the mathematical model of closed grinding cycle with combined-action air separator is suggested. In this model the integral functions of partition in air separator have been obtained. The calculated and experimental values of the integral functions of partition in air separator have been compared.

Key words. Technological complex, finely-dispersed product, centrifugal grinding-mixing unit, combined-action air separator, closed grinding cycle.

1. Introduction.

One of the most promising trends of improving the technology of fine and ultrafine grinding of materials is designing the energy-saving technologies of the comprehensive processing and recycling of technogenic materials with the use of selective closed-cycle grinding units [1-3].

In this regard, the implementation of stage-by-stage grinding of materials, providing their micro-defect structure at the first stage, i.e. in press-rolling grinder, and the subsequent selective grinding (of heavy-impact, impact-abrasive or abrasive mechanism of action) in vibration-centrifugal grinding units, is of special interest for science and practice. These units can be of open-cycle and closed-cycle types of grinding [4-8].

2. The main part.

For obtaining high-dispersity materials we have carried out scientific and technological research and a number of studies in creating a special process module. The patent-protected design of aggregates: centrifugal grinding-mixing units (CGMU), have been designed, implementing a set of successive dynamic loadings, which contributes to the efficient obtaining of high-dispersed powders, with the possibility of varying their values or type of action (from heavy-impact to abrasive) depending on the grindability of the ground material [14,15].

2.1. Centrifugal grinding-mixing unit.

A distinctive feature of CGMU (fig. 2) is implementing in one production machine the stages of coarse, fine and ultrafine grinding, which is achieved by various motion trajectories of grinding chambers for providing the corresponding working modes of the grinding media: for coarse grinding – intensive impact load; for fine grinding – impact-abrasive action; and for ultrafine grinding – intensive abrasion [16]. The operating principle of the unit is as follows. The source material is fed through the loading spout and the restraint bar to the upper grinding chamber, making vertical reciprocating movement. This results in passing high energy to grinding bodies, which contributes to their intensive impact action on the material [18,10].

The lengthwise movement of material within the grinding chamber is provided by the natural propping-up with the loaded material. As a result the ground material through the restraint bar, which keeps the grinding bodies inside the grinding chamber, passes to the discharge sleeve, and then to the middle grinding chamber, which also contains grinding bodies. In the middle grinding chamber the material, together with grinding bodies, moves by the ellipsoidal trajectory. The material is ground and through the special restraint bar gets to the lower grinding chamber, where the material under the action of grinding bodies and at the circular motion of the chamber undergoes the intensive abrasion. The dispersity of the end product is mostly determined by the abrasive action in the lower grinding chamber [11,12].

2.2. Combined-action air separator.

In order to improve the efficiency of grinding process in CGMU and to expand the technological opportunities of the module we have designed a combined-action air separator-granulator (patent application № 2018102629), which was tested in high-dispersity materials production line [13].

The structure of separator includes two steps. The first step is intended for separating the coarse product, and the second – for separating fine product.

The separator-granulator operates in the following way. The gas-material flow, consisting of the source material and the energy carrier, passes tangentially to the lower blunted cone through the loading spout. Ascending spirally, the largest particles of the mix are pressed by the centrifugal force to the inner surface of blunted cones, and, losing their speed due to the increasing classifier's section, fall to the grooved channel of the blunted cone and are discharged through the discharge sleeve. The degree of mix classification can be regulated by changing the energy carrier's pressure. To provide the free flow of particles through the discharge sleeve its tilt angle to the vertical axis should be more than natural slope of the material. The nonobservance of this condition makes the process of self-induced discharge of particles more difficult.

By changing the technological configuration of the bellmouth and slide gates, the aerodynamic characteristics of the unit can be varied (the speed of gas-material flow, hydraulic resistance) [9,12].

3. Theoretical research

The designed units are used in the grinding processing system (GPS) (figure.1), consisting of CGMU and combined-action air separator.

In figure 1 the closed-cycle GPS is shown. The ground material from hopper 5, preliminarily mixing with the return flow from separator 2, is fed to CGMU. In CGMU the material is ground. The ground material is fed with pneumatic conveyor to separator 2, which divides it to the fine product (finished powder) and coarse product (return product). The end product is separated from gas in the dust-catching system 3, and the return product goes to the entrance of CGMU for regrinding. The grain-size composition $R_1(\delta)$ and stock flow B_1 of CGMU product depend on the grain-size composition of the raw material $R_C(\delta)$ and its stock flow B_C, as well as on the grain-size composition $R_2(\delta)$ and the stock flow B_2 of the return product.

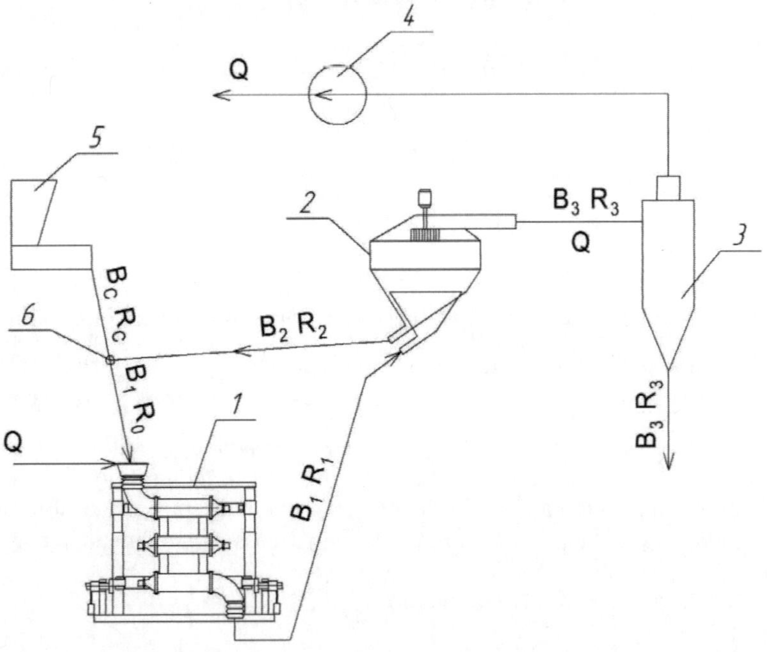

Figure 1. The scheme of closed grinding cycle: 1 – CGMU, 2 – separator, 3- finished powder catching system, 4 – fan, 5 – hopper, 6 – mixing unit

Let us suppose that the process in GPS is steady-state and Q=const. Let the mathematical model of transformation of the material grain-size composition in main units of the scheme be also known. Then the mathematical model of the closed grinding cycle consists of the following set of equations [17,19]

$$R_1(\delta)=F_M\left(R_0(\delta),B_1,Q,P_i\right), \tag{1}$$

$$\varphi_\delta=\varphi_\delta\left(\delta,B_1,Q,q_i\right), \tag{2}$$

$$\bar{\varphi}=\int_0^1\varphi_\delta(\delta)dR_1, \tag{3}$$

$$R_3(\delta)=\frac{1}{\bar{\varphi}}\int_0^{R_1}\varphi_\delta(\delta)dR_1, \tag{4}$$

$$R_2(\delta)=\frac{1}{1-\bar{\varphi}}\int_0^{R_1}\left(1-\varphi_\delta(\delta)\right)dR_1, \tag{5}$$

$$B_3 = B_1 \bar{\varphi}, \quad (6)$$

$$B_2 = B_1 (1 - \bar{\varphi}), \quad (7)$$

$$R_0(\delta) = R_c(\delta)\bar{\varphi} + R_2(\delta)(1 - \bar{\varphi}), \quad (8)$$

$$B_1 = B_c + B_2, \quad (9)$$

$$B_c = B_3, \quad (10)$$

$$\Sigma \Delta p_i (B_i, Q) = H_B(Q), \quad (11)$$

where φ_δ - separator's partition curve, F_M — symbolic notation of a certain grinding kinetics equation; P_i — vector of parameters, controlling the grinding process; q_i — vector of parameters, controlling the separation process; Δp_i — pressure loss at various parts of GPS; H_B - fan head. In this system the equation (1) describes the transformation of grain-size composition in GPS. The equations (2)-(7) describe the alteration of grain-size composition of the material at the partition in the separation. The alteration of grain-size composition at mixing the raw material and return material flows are described by the equations (8)-(10). The equation (11) presents the process of pneumatic conveying of material in GPS.

The function of CGMU separation is presented in figure 2.

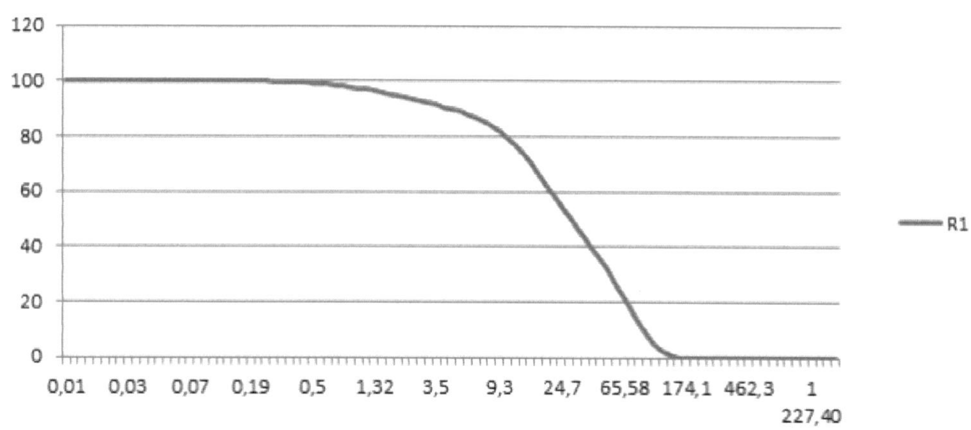

Figure 2. Integral partition curve of CGMU

To describe a separator's partition curve let us use one-parameter dependence [20]

$$\varphi_\delta \left(\frac{\delta}{\delta_{гр}} \right) = \left(1 + \exp\left(s \left\{ \left(\frac{\delta}{\delta_{гр}} \right)^2 - 1 \right\} \right) \right)^{-1}, \quad (12)$$

where s – free parameter [21-25]. As parameter s doesn't exceed 0.12-0.14, the partition curve for s=0.12 is presented in figure 3.

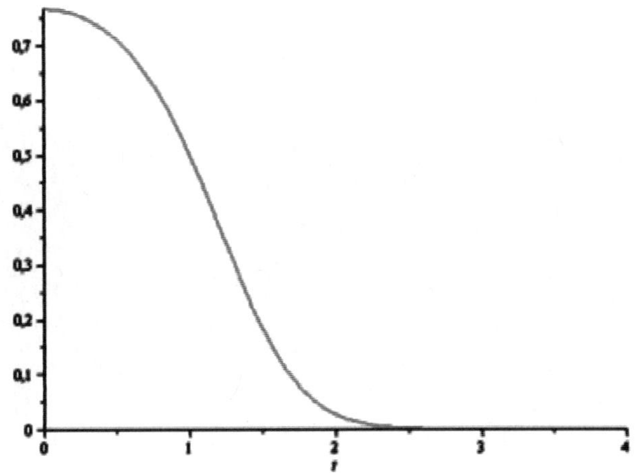

Figure 3. Partition curve of the separator

As it is not possible to get the analytical solution of the problem (1)-(11), combined with condition (12), the given equations were solved by means of successive-approximation method. For the comparison with experimental values the grain-size composition of coarse and fine product after the separator treatment was taken. In figure 4 the integral functions of separation R_1,R_2,R_3 are presented.

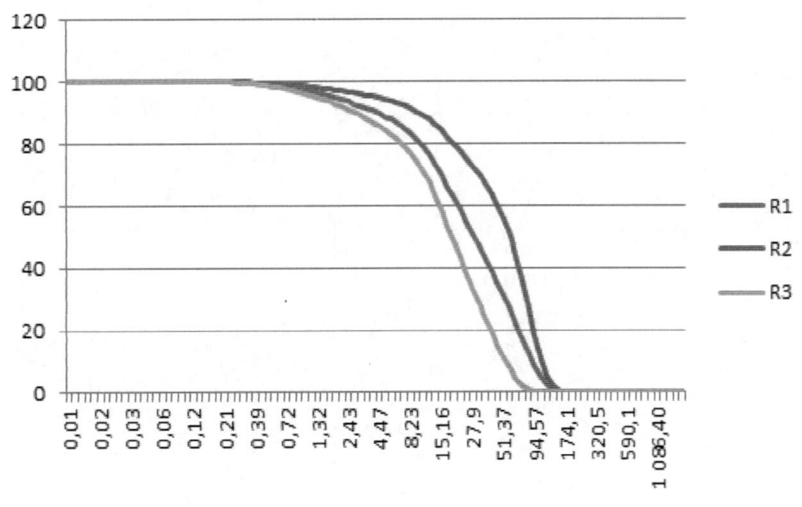

Figure 4. Integral functions of partition R_1, R_2, R_3.

4. Experimental research.

For experimental research we used ferruginous quartzites refinement tailings (FQRT) with grain-size composition, presented in table 1.

Table 1. Grain-size composition of FQRT samples.

	Sieve apertures, mm					
Sieve residue, %	3	2	1	05	025	008
	10.2	35.1	38.9	7.1	3.6	0.6

IOP Conf. Series: Materials Science and Engineering **552** (2019) 012040 doi:10.1088/1757-899X/552/1/012040

After the experimental research of the production module, the grain-size analysis of the collected samples was carried out, the findings of which are presented in figure 5 and figure 6.

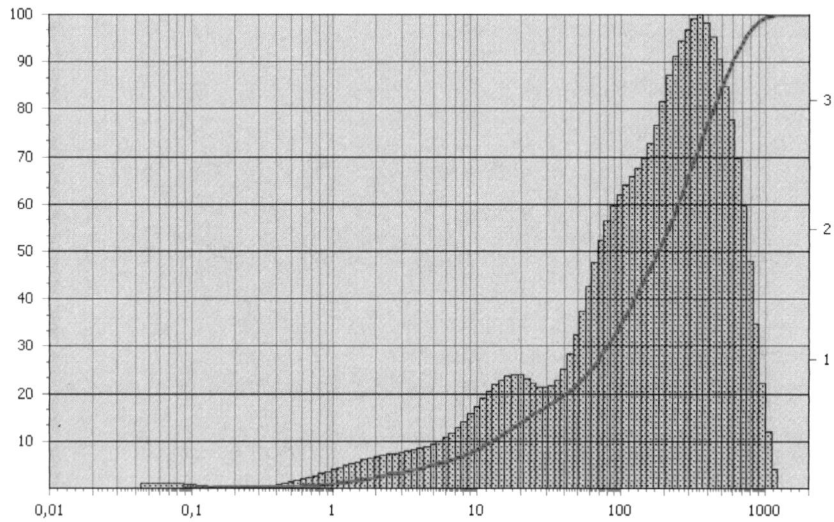

Figure 5. Grain-size analysis of grit.

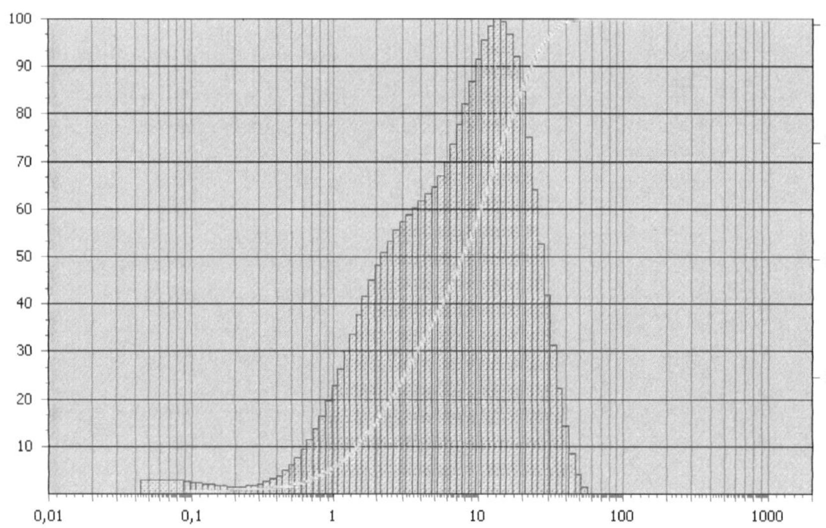

Figure 6. Grain-size analysis of end product.

Comparison of calculated and experimental values of R_3 is given in figure 7. The difference between experimental and calculated data is explained by the fact that for calculation not real partition curve, but the theoretical partition curve was taken (12).

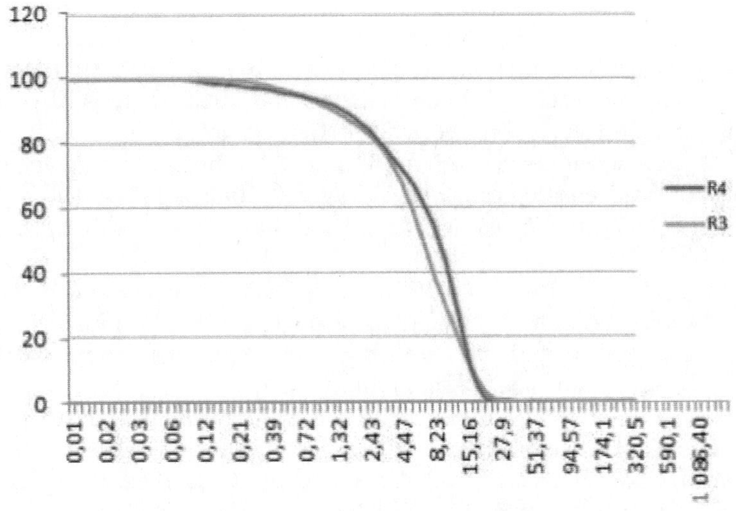

Figure 7. Comparison of the partition integral functions in the separator by the end product. R_3 –calculated, R_4 – experimental.

5. Conclusions.

The carried-out design-engineering developments of patent-protected aggregates and their tests have allowed designing a technological complex for fine grinding of materials with various physical and mechanical properties. A mathematic model of the closed grinding cycle with combined-action air separator was suggested. Within this model the integral functions of partition in air separator have been obtained. The grain-size compositions of fine-dispersed particles, obtained by means of the designed units, combined in a technological complex for high-dispersity materials production, have been researched. The comparative analysis of calculated and experimental values of integral functions of partition in air separator allows making a conclusion about the high enough accuracy degree of the approximate analytical solution.

Acknowledgments
The article was prepared within development program of the Flagship Regional University on the basis of Belgorod State Technological University named after V.G. Shukhov.

References

[1] Klimenko V G, Balakhonov A V and Elistratkin M Yu 2010 Aktivaciya osnovnogo domennogo shlaka produktami termoobrabotki prirodnogo gipsa [The activation of blast-furnace slag with rock gypsum heat processing products] *Izv. vuzov. Ximiya i ximicheskaya texnologiya* **53** pp 7-10 [In Russian]

[2] Glagolev S N, Sevostyanov V S, Ilyina T N and Uralskii V I 2011 Production modules for combined reprocessing of technogenic materials *Chemical and Petroleum Engineering* **46(9)** 556-9

[3] Sevostyanov V S, Goryagin P Y, Sevostyanov M V and Ermilov R A 2018 Constructive and Technological Features of Units for Milling Polymer Material *J. Phys.: Conf. Ser.* **1066** 012006

[4] Mischenko S V and Tkachev A G 2008 *Uglerodny`e nanomaterialy`. Proizvodstvo, svojstva, primenenie* [Carbon nanomaterials. Production, properties, usage] (Moscow: Machine

building) [In Russian]
[5] Gridchin A M, Lesovik V S, Sevostyanov V S, Uralsky V I and Sinitsa E V 2006 Pomol`no-smesitel`ny`j agregat [Grinding-mixing unit] Patent RF no 2277973 [In Russian]
[6] Sevostyanov M V, Sevostyanov V S and Martakov I G 2018 Analysis of Design Features of Circulating Action Mixers *J. Phys.: Conf. Ser.* **1066** 012002
[7] Sivachenko L A, Sevostyanov V S and Il´ina T N 2018 Problem Tasks in the Field of Resource and Energy Saving Technologies *J. Phys.: Conf. Ser.* **1066** 012021
[8] Pirumov A I 1974 *Obespy`livanie vozduxa* [Dedusting of air] (Moscow: Stroyizdat) [In Russian]
[9] Mizonov V E and Ushakov S G 1989 *Ae`rodinamicheskaya klassifikaciya poroshkov* [Aerodynamic classification of powders] (Moscow: Chemistry) [In Russian]
[10] Idelchik I E 1975 Spravochnik *po gidravlicheskim soprotivleniyam* [Hydraulic resistance guidebook] (Moscow: Machine building) [In Russian]
[11] Ushakov S G and Zverev N I 1974 *Inercionnaya separaciya py`li* [Inertial separation of dust] (Moscow: Energia) [In Russian]
[12] Sobolev A A, Melnikov N A and Tyutyunnik L O 2011 Dvizhenie chasticz v vozdushnom potoke [The particles movement in air flow] *Nauchnoe izdanie «Vektor nauki TGU»: Nauchno-teoreticheskij zhurnal* **3** pp 82-6 [In Russian]
[13] Sevostyanov V S, Goryagin P Yu, Obolonsky V V and Ermilov R A 2018 Resource-Saving Technological Complex for Polymer Waste Processing *J. Phys.: Conf. Ser.* **1066** 012025
[14] Sevostyanov V S, Goryagin P Y, Sevostyanov M V and Ermilov R A 2018 Constructive and Technological Features of Units for Milling Polymer Material *J. Phys.: Conf. Ser.* **1066** 012006
[15] Sinitsa E V, Uralsky V I, Lesovik V S, Sevostyanov V S, Gridchin A M and Uralsky A V 2009 E`nergosberegayushhie pomol`ny`e kompleksy` dlya polucheniya mexanoaktivirovanny`x kompozicionny`x smesej [Energy-saving grinding complexes for mechano-activated composite mixtures production] *Izv. Vuzov. Stroitel`stvo.* **5** pp 68–79 [In Russian]
[16] Sevostyanov V S, Shatalov A V, Sevostyanov M V, Krutov V S and Shatalov V A 2018 Technics and Technology of Processing Basalt Waste *J. Phys.: Conf. Ser.* **1066** 012027
[17] Uvarov V A, Sharapov R R and Kharlamov E V 2016 Obosnovanie krivoj razdeleniya vozdushnogo separatora [Substantiation of partition curve of air separator] *Mexanizaciya stroitel`stva* **9** pp 13-5 [In Russian]
[18] Shishkin S F and Tekhov S M 1991 Raschet processa izmel`cheniya v zamknutom cikle [Calculation of grinding process in the closed cycle] *Ximiya i ximicheskaya texnologiya* **34** pp 117-9 [In Russian]
[19] Kharlamov E V, Sharapov R R and Stepanov M A 2014 Metodika rascheta apparata dlya razdeleniya zhelezosoderzhashhego mineral`nogo sy`r`ya [Calculation methodology of the apparatus for ferruginous mineral raw materials separation] *Mexanizaciya stroitel`stva* **3** pp 28-35 [In Russian]
[20] Molerus O 1967 Stochastisches Modellder Gleichgewichtsichtung *ChemieIng. Technik* vol 13 pp 792-6
[21] Saint Etienne V and Cordonnier A 1997 The current state of development of the TVS high-efficiency dynamic classifier *Zement Kalk Gips International* **50** pp 540-55
[22] Flament G 1999 LV technology small change, large gain *Asian Cement Magasine* **6** pp 29-31.
[23] Kershow M and Yardi I 1989 Analysis of O-SEPA separators new at Blue Circle Australia *World Cement* **11** pp 400-5
[24] Onuma E and Furukawa T 1982 O-SEPA a new high-perfomance air-classifier *World Cement* **10** pp 13-24
[25] Walter H Duda 1985 *Cement Data Book*

Investigation of cavitation and vibration processes in the technology of treatment of the contaminated liquids

V S Sevostyanov, M M Belykh and K I Sevostyanova

Department of Technological Complexes Machines and Mechanisms, Belgorod State Technological University of V.G. Shukhov, Kostyukov St., 46, Belgorod, 308012, Russia

E-mail: insegir@yandex.ru

Abstract. This article considers the problems of treatment of contaminated liquids. Scientific works of home and foreign authors were analysed and according to this analysis it was made the conclusion. Experimental studies of the process of treatment of the contaminated liquids, in which the cavitation process in the technology of treatment of the contaminated media were investigated, as well as the process of treatment of the contaminated media in a vibration and centrifugal unit . As a result of the research, it was discovered that the cavitation method can be effectively used to treat water from heavy mechanical impurities, which makes it possible to carry out further filtration in fine filters. A vibration-rotary unit, developed in Belgorod State Technological University n.a. V G Shukhov, (Russia) at the Department of Technological Complexes, Machines and Mechanisms, was taken as a technical object, where experimental investigation of treatment of the contaminated liquids was carried out. It was also carried out the research of the stepwise process of treatment of the contaminated liquids in the developed unit. On the basis of these results the authors propose to use a stepwise method of treatment of liquid media: cavitation effect - for the preliminary phase of treatment and vibration -rotary effect – for the final stage of treatment.

Keywords. liquid, contaminated liquids, vibration and centrifugal unit, rotational speed, filtration chamber, cavitation, vibration, sieve, amplitude of oscillations, treatment.

1. The relevance of treatment of contaminated liquid media

The problem of industrial wastewater treatment and water treatment for technical and household purposes is becoming more and more important. The complexity of treatment is connected with an extraordinary diversity of impurities in the effluent, the amount and composition of which is constantly changing due to the appearance of new industries and changes in the existing technology [1].

The rapid growth of cities, the active development of industry and agriculture - all these factors make the situation of pollution of rivers and lakes with untreated plums more and more complicated every year [2]. Most of the effluents of enterprises fall directly into the reservoirs, which leads to a gradual deterioration of the ecological situation. The principle of wastewater treatment is a complex, multi-stage process [3]. The level of purity of liquids used in manufacturing [4], clinical practice [5], agriculture [6], the service sector and in everyday life [7], as well as the degree of subsequent regeneration of valuable components and disposal of harmful impurities determine the industrial potential, the ecological situation, and the standard of living of the population in general [8].

A new direction in the treatment of industrial wastewater is the creation of non-waste technologies. To a greater extent, it can be achieved by returning the treated water back to manufacturing. Taking into account the growing cost of water consumed, the creation of closed production cycles is becoming increasingly important. In particular cases, enterprises manage to switch over to completely waste-free technology. And sometimes enterprises can profit from the sale of waste, if they are pretreated and meet the necessary standards [9].

About 50% of the river water is exposed to man-made impact every year, including the result of the discharge of 425 • 109 m3 of wastewater. The main causes of pollution of water bodies are ineffective technologies for water treatment, as well as the system of regeneration of used water [10]. Due to these reasons, the requirements of SanR&S 2.1.4.1175-02, which determine the physical - chemical and microbiological indicators of water, are violated [11].

Liquid environment is not only a valuable resource that is used daily by man, but also is the main resource for many industrial enterprises. It includes water, engine oils, technical fluids, and others [12]. Water at enterprises is used for technological needs, for example, for cooling and washing equipment. Usually water is used in a cold and hot state, as well as in the form of steam. Moreover, water has got an economic potential. First, it is purchased for a certain price, and secondly, it is paid for when it is discharged into the environment.

The used water is not the water that the company acquired, it is the water with specific characteristics and properties. This is a complex cocktail of a mixture of various contaminants. Such water cannot be simply dumped back into the river or into the well; it must be cleaned [13]. If the company does not treat it up after using, liquid waste will adversely affect the livelihoods of cities and towns, and the consequences for the environment will be most disastrous. Dirty water getting into the reservoirs will disrupt the ecosystem for many kilometers around [14].

Every enterprise that uses liquid media must be equipped with sewage treatment plants, but such facilities are not cheap: both their installation and operation. Therefore, there is a need to develop and upgrade the equipment that will provide the necessary quality of liquid media treatment, in accordance with the established standards [15].

Water quality is more dependent on the concentration of pollutants and compounds in it. If the concentration of pollutants is exceeded, it has a detrimental effect not only on humans, but also on the biological situation in the region. In this regard, it is necessary to exercise strict control over the treatment of water from harmful substances, as well as to ensure the maximum permissible concentration of MAC of wastewater [16].

It order to solve the problems successfully it is necessary to have a methodology that generalizes a complex of theoretical, research, and technological issues of the development and application of complex systems for the liquid media treatment [17].

The purpose of this work is to study the influence of the processes of preliminary (cavitation) and final (vibration-rotary) treatment on the contaminated liquids.

2. Experimental investigation of the process of treatment of the contaminated liquids

Our analysis of home and foreign literature [16–22] suggests that the problems of treatment of the contaminated liquid media, which are not always subjected to treatment at the industrial enterprises, have not been fully studied so far. In this regard, the most important task is working out measures and constructive decision in order to reduce the negative impact of the contaminated liquids on the environment. Earlier, some experimental studies of the technical means and methods of regenerating of the contaminated liquids [19] were carried out, as well as the development and study of the technical agents for cleaning engine oils [20]. It has been established that vibration-centrifugal treatment of the contaminated liquids allows to speed up the filtration process, as well as to improve its quality.

2.1. Study of the cavitation process in the technology of treatment of the contaminated liquids

Figure 1 shows an experimental plant using the cavitation process. Technical characteristics of the plant: the volume of the treatment chamber $V = 200$ ml; rated voltage $U = 12$ V; power consumption

$N_{el.en}$ = 5 W; consumed current I = 2,4 A; nominal frequency of rotation of the motor shaft with blades $n_{el.en}$ = 2500 ÷ 3000 rev / min.

Figure 1. Experimental plant for treatment of the liquids: 1 – electrical engine; 2 – filtration cell; 3 – blades; 4 – switch ; 5 – bed; 6 – power unit

In each experiment, it was used 200 ml of the liquid with 30, 25 and 15% contamination of the total volume. Engine oil, soil, slag and other impurities were the contaminants. Figure 2 shows the contaminated liquid.

30% 25% 15%

Figure 2. Contaminated liquid.

A number of experiments with treatment of the contaminated liquid in a cavitation chamber in which special blades were carried out.

According to the results of the experimental investigation, it was discovered that the cavitation effect in the liquid being treated was created owing to the cavitation blades - it occurred a break of discontinuity in the liquid with the formation of cavitation micro bubbles filled with a mixture of contaminants, the effect of "cold boiling". As the blades rotated, the microbubbles collapsed and separated heavy and light impurities in the liquid. Experimental studies were conducted using an experimental plant at the frequency of rotation of the shaft of the electrical engine: 2500 ÷ 3000 rpm the operation time varied within 10-12 minutes.

When conducting experimental studies, it was discovered that by the fourth minute a gradual separation of light and heavy contaminants began to occur. By the seventh minute, there was a pronounced separation, and by the eleventh - twelfth minute, heavy contaminants precipitated. With 30% liquid contamination, heavy substances precipitated within some 12 minutes. At 25% contamination of the liquid, it took 11.5 minutes to be clean, and at 15%, heavy contaminants precipitated within 10.2 minutes.

After switching of the plant the liquid settled for 10 minutes. As a result of the treatment, due to the use of the cavitation method, the contaminated liquid was clarified and purified from heavy impurities. At 30% contamination in the liquid being filtered there were 5 ÷ 7% of impurities, at 25% - 4 ÷ 6% on average, and at 15% - about 3 ÷ 5%. Figure 3 shows contaminated and treated liquids.

a b

Figure 3. The state of the liquids: a – before treatment; b – after treatment.

According to the results of the experimental studies of the cavitation method, it was constructed graphical dependences of the volume of the liquid being cleaned (*V*, ml.) on time (*t*, min.) with various amounts of contaminants. Figure 4 shows these graphical dependencies.

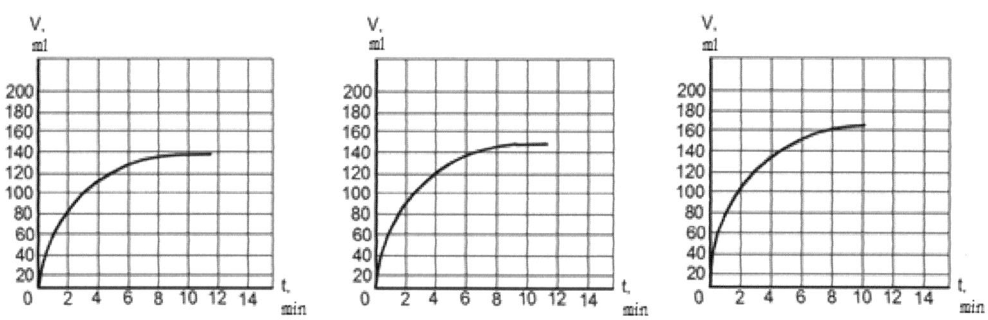

30% 25% 15%

Figure 4. Dependences of the volume of the liquid being cleaned (*V*, ml.) on time (*t*, min.) with various amounts of contaminants.

As a result of the research, it was discovered that the cavitation method can be effectively used to treat water from heavy mechanical impurities, which makes it possible to carry out further filtration in fine filters.

2.2. Study of the process of treatment of the contaminated liquid media in a vibratory centrifugal unit

A vibration-centrifugal unit [21] developed in the Belgorod State Technological University named after V G Shukhov (Russia) by the staff of the department of technological complexes, machines and mechanisms [22], which has a number of technological advantages: can be used as a technical object for the realization of the established regularities:

- it provides the possibility of varying the number of parameters (frequency of vibratory and centrifugal effects, the period of staying of the medium being cleaned in the unit, the value of the hydrodynamic effect, etc.);
- it allows you to change the nature of external influences (the value and duration of inertial effects, the sequence of operations, etc.) and the cyclical nature of the liquid to be cleaned;
- it provides the possibility to implement both internal and external recycling of the streams being treated in the unit;
- it creates the conditions for interference of additional positively influencing factors: temperature, ultrasound exposure, the introduction of chemical reagents, etc. on the filtering process;
- it has a simple design, not complex degree of installation and dismantling of the unit itself, its components, etc.
- it allows to realize the operational multifunctionality of the developed unit: to combine physical and mechanical, physical and chemical processes, to organize the implementation of staged technological processes, etc. [23].

At the first stage the analysis of filtering elements for water treatment was carried out, which showed that the most effective filtering elements are activated carbon and sand [24].

They assembled the filtration chambers where some gauze, activated carbon and sand were used. The particle size of the activated carbon was 1–2 mm. The particle size of sand is 0.5 ÷ 0.25 mm. The thickness of the filtering coal layer o - 40 mm, sand - 20 mm [25].

The filtration chamber was installed on an experimental plant of a vibrating and centrifugal unit, which is shown in Figure 5.

Technical characteristics of the experimental plant: the rotational speed of the eccentric shaft - 10 ÷ 330 rpm; vibration amplitude - 40 mm; the volume of the filtration chamber is 200 ml.

Figure 5. An experimental plant of a vibration and centrifugal unit with a filtration chamber vertically installed.

Experimental research used water that was contaminated with slag, engine oil and soil. In each experiment, 500 ml of the liquid was filtered. In a series of experiments the liquid was contaminated at 10, 20, and 30% of the total liquid volume [26].

It was carried out a series of experimental studies of filtration of the contaminated liquids in an experimental plant of a vibration- centrifugal unit at different frequencies of rotation of the eccentric shaft on which the filtration chamber was fixed. The rotational speed of the eccentric shaft was determined by the speed sensor. A series of the experiments at 10, 20 and 30% contamination showed that 500 ml of the liquid at a frequency of rotation of the eccentric shaft 110 ÷ 140 rpm. is treated from dirt for 21.5 minutes on average. After treatment of the initial 500 ml some 470 ÷ 480 ml of the liquid treated left. The remaining impurities in the filtered liquid averaged 2–5%. With the frequency of rotation of the eccentric shaft 220 ÷ 240 rpm, the filtration time was reduced to an average of 13 minutes. Only some 485 ml of the liquid treated left and the amount of the contaminants averaged 5 ÷ 8%. When the rotational speed of the eccentric shaft was increased to 300 ÷ 320 rpm, the filtration time averaged 10 minutes. Average 480 ÷ 490 ml of the liquid treated left and the amount of the contaminants increased to 8 ÷ 15% on average.

Experimental studies have shown that the amount of contaminants increases with increasing rotation frequency of the eccentric shaft. Therefore, to ensure the highest degree of purification, it was decided to carry out a series of experiments with several filtration chambers connected in series (Figure 6).

Figure 6. An experimental vibration and centrifugal plant for treatment of liquids.

To improve the filtration conditions of the liquids, it was decided to install each filtration chamber at a certain angle [27]. The first chamber was installed at an angle of 13 ÷ 14°, the second - 8 ÷ 10°, and the third chamber, in view of the need to increase treatment time, was installed horizontally.

Each experiment was carried out with a different frequency of rotation of the eccentric shaft. In the first experiment, 500 ml of the contaminated liquid was filtered in 25.5 minutes on average, at an oscillation frequency of n = 110 ÷ 140 rpm. While 440 ml of the filtrate left. In the second experiment, 500 ml of the contaminated liquid was filtered within 14.5 minutes at n = 220 ÷ 240 rpm. Average 460 ml of the water treated left. In the third experiment, 500 ml of the liquid was filtered within 11 minutes, at n = 300 ÷ 320 rpm. Average 475 ml of the water treated left, but at the same time, the amount of contaminants increased by 8 ÷ 10% on average.

While carrying out a series of experimental studies, about 6% of the total amount of contaminants was observed in the liquid being treated, of which there were only minor mechanical impurities and slag. There was no oil in the liquid. The contamination at the subsequent stages of the experimental studies decreased to approximately 2–4%.

As a result of the experimental studies, it was discovered that to implement the process of filtering contaminated liquids in a vibration and centrifugal unit, several filtration chambers should be sequentially installed at tilt angles of 0 ÷ 14° (with decreasing tilt angles of chambers with their sequential installation) and eccentric shaft rotation speed n = 220 ÷ 240 rpm This allows to speed up the filtration time to 14.5 minutes and reduce the amount of pollutants in the liquid being treated to 2 ÷ 4%.

2.3. Investigation of the stepwise process of treatment of the contaminated liquids

When conducting experimental analysis, a cavitation impact chamber was used, which provided preliminary treatment of the liquid from the coarse contaminants.

In a series of experimental investigation it was used 200 ml. of the liquid contaminated by 20% of the total volume. Slag, engine oil and other inclusions were contaminants in the liquids [28].

A series of experiments showed that the liquid is treated from large impurities in 11.5 minutes on average. After the cavitation process, the liquid precipitated for 10 minutes. As a result of the treatment, the liquid was clarified and cleaned from heavy impurities.

The next step was the filtration of the liquid at the experimental plant of the vibration and centrifugal unit.

Sand and activated carbon were the filter elements in the chambers. The thickness of the filter layer was 80mm [29].

A series of the experimental investigation were conducted at different frequencies of rotation of the eccentric shaft of the unit. In the first series of the experimental investigation, the liquid, cleared from large impurities, was moved into the unit, where it was cleaned from small impurities. The average treatment time was 25.5 minutes, with a shaft rotational speed of n = 110 ÷ 140 rpm. In the second series of experimental investigation, the fluid was cleaned for 14.5 minutes on average at a rotational frequency of n = 220 ÷ 240 shaft revolutions per minute. In the third series of experiments, the fluid was cleaned for 11 minutes on average at a shaft rotation frequency n = 300 ÷ 320 rpm.

According to the results of experimental investigation, the graphic dependences of the volume of the filtered liquid (*V*, ml.) on the filtration time (*t*, min.) were constructed. Figure 7 presents these dependencies.

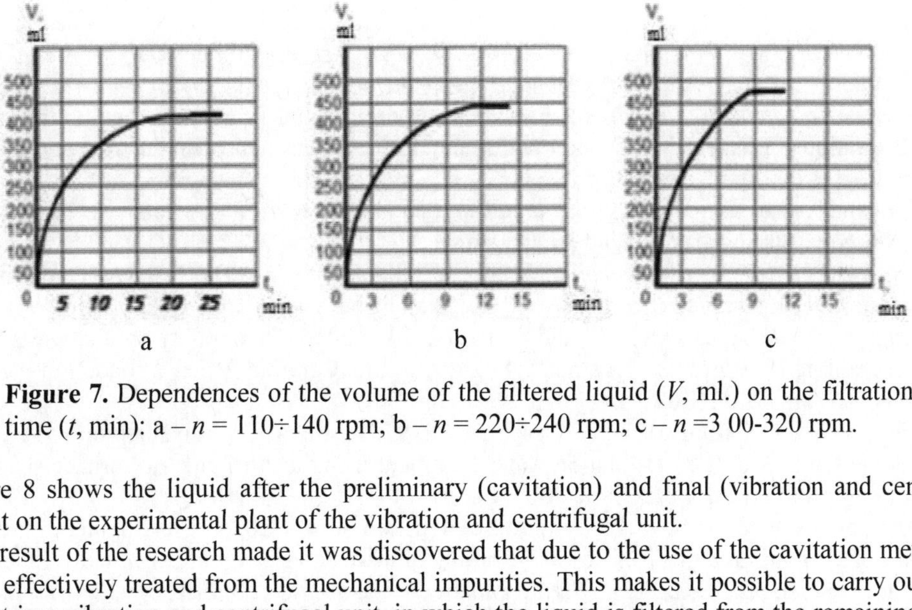

Figure 7. Dependences of the volume of the filtered liquid (*V*, ml.) on the filtration time (*t*, min): a – *n* = 110÷140 rpm; b – *n* = 220÷240 rpm; c – *n* =3 00-320 rpm.

Figure 8 shows the liquid after the preliminary (cavitation) and final (vibration and centrifugal) treatment on the experimental plant of the vibration and centrifugal unit.

As a result of the research made it was discovered that due to the use of the cavitation method, the liquid is effectively treated from the mechanical impurities. This makes it possible to carry out further treatment in a vibration and centrifugal unit, in which the liquid is filtered from the remaining impurities. The time spent on the preliminary (cavitation) treatment was 21.5 minutes. The final treatment time in a vibration and centrifugal unit at *n* = 110 ÷ 140 rpm was 25 minutes, the remaining contaminants were 2 ÷ 4%., when *n* = 220 ÷ 240 rpm - 14.5 minutes. The contaminants were about 2 ÷ 4%. when *n* = 300-320 rpm - 11 minutes. The contaminants were - 8 ÷ 10%.

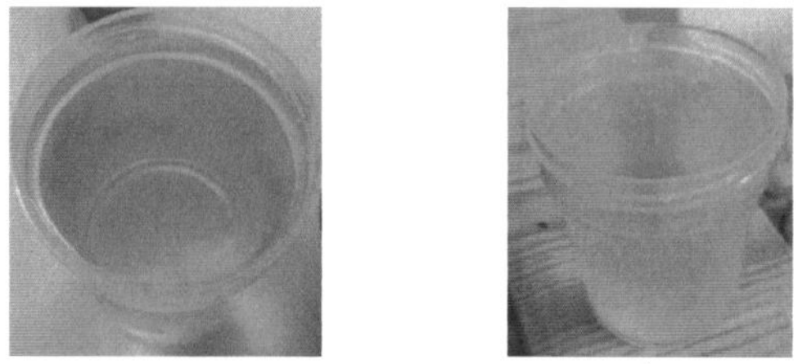

a b

Figure 8. The liquid after some stages of treatment:
a – n = 300÷320 rpm, b – n = 110÷240 rpm.

Thus, our experimental investigation in treatment of the contaminated liquids in a vibration and centrifugal unit with the provision of a preliminary cavitation effect on the medium being treated showed that the best indicators are achieved when using several treatment stages - preliminary (cavitation) and final (vibration and centrifugal one).

Acknowledgments
The article was prepared within development program of the Flagship Regional University on the basis of Belgorod State Technological University named after V G Shukhov.

References

[1] Royev G A 1993 *Ochistny`e sooruzheniya:Oxrana okruzhayushhej sredy` : Uchebnik dlya vuzov po special`nosti "Proektirovanie,sooruzhenie i e`kspluataciya gazonefteproizvodstv i gazoneftexranilishh"* [Treatment Facilities: Environmental Protection: Textbook for Higher Institutions "Design, construction and maintenance of gas and oil industries and oil storage tanks"] (Moscow: Nedra) [In Russian]

[2] Lukinykh N A et al 1978 *Metody` doochistki stochny`x vod* [Wastewater treatment method] (Moscow: Stroiizdat) pp 150-4 [In Russian]

[3] 2018 *Aktual`nost` ochistki promy`shlenny`x stochny`x vod* [The relevance of industrial wastewater treatment] URL http://www.mediana-filter.ru/ochistka_stokov2.html

[4] Sverguzova S V, Sapronova Z A and Starostina I V 2016 Disposal of Synthetic Surfactants-Containing Wastewater Treatment Sludge in the Ceramic Brick Production *Procedia Engineering* **150** 1610-6

[5] Sukhanov E V, Sapronova Zh A, Sverguzova S V, Denisova L V and Sapronov D V 2017 Some features of the coagulation water treatment using dust of electric furnace steelmaking *Ecology and Industry of Russia* **21(1)** 24-9

[6] Sverguzova S V, Sapronova Zh A and Svtatchenko A V 2018 Electrokinetic phenomena accompanying treatment of soy milk production wastewater by agent based on electric arc furnace dust *IOP Conference Series: Materials Science and Engineering* 327(4) 042096

[7] Sapronova Zh, Sverguzova S, Sulim K, Svyatchenko A and Chebotaeva E 2018 Sewage treatment in megacities by modified chestnut tree waste *IOP Conference Series: Materials Science and Engineering* **365(2)** 022058

[8] Gorbachev E A *Proektirovanie ochistny`x sooruzhenij vodoprovoda iz poverxnostny`x istochnikov* [Design of sewage treatment plants from surface sources] (Moscow: Publishing House of Association of Building Higher Institutions) [In Russian]

[9] Lopanov A N 2009 *Monitoring i e`kspertiza bezopasnosti zhiznedeyatel`nosti* [Monitoring and examination of life safety] (Belgorod: BSTU VG Shukhova) [In Russian]

[10] Novikov Yu V 2005 *E`kologiya, okruzhayushhaya sreda i chelovek* [Ecology, environment and human] (Moscow: FAIR-PRESS) **3** [In Russian]

[11] Onishhenko G G 2002 *Sanitarno-e`pidemiologicheskie pravila i normativy` SanPiN 2.1.4.1175-02* [Sanitary and epidemiological regulations and standards SanPiN 2.1.4.1175-02] **13**

(Moscow: Minzdrav Rossii) [In Russian]

[12] Shevtsov M N 2010 *Vodosnabzhenie promy`shlenny`x predpriyatij* [Water supply of industrial enterprises] (Khabarovsk: TOGU) [In Russian]

[13] Sverguzova S V, Sakalova G V and Malyovanyi M S 2014 E`ffektivnost` ochistki stochny`x vod gal`vanicheskogo proizvodstva adsorbcionny`m metodom [Efficiency of sewage treatment of galvanic production by the adsorption method] *Vestnik BGTU im V G Shukhova* **4** pp 153-6 [In Russian]

[14] 2018 Dlya chego nuzhny` ochistny`e sooruzheniya? [Why do we need sewage treatment plants?] URL http://новости-регионов.рф/detail1247.html [In Russian]

[15] Kalygin A V 2004 *Promy`shlennaya e`kologiya: uchebnoe posobie dlya studentov vy`sshix uchebny`x zavedenij* [Industrial ecology training materials for students of Higher Institutions] (Moscow: Izdatel`skij centr "Akademiya") [In Russian]

[16] 2018 *Pavlov D V* Metody` ochistki stochny`x vod [Pavlov D V Wastewater treatment methods] URL http://enviropark.ru/course/view.php?id=79

[17] Orlov N S 2000 *Metodologiya razrabotki kompleksny`x sistem ochistki zhidkix texnologicheskix sred na osnove baromembranny`x processov* [Methodology of development of complex systems for treatment of liquid technological medium based on baromembrane processes] thesis... Dr Tech Sc (Moscow: RGB OD) [In Russian]

[18] Karmanov A P and Polina I N 2015 *Texnologiya ochistki stochny`x vod* [Wastewater treatment technology] (Syktyv: SLI) [In Russian]

[19] Belykh M M et al 2016 *Texnicheskie sredstva i sposoby` regeneracii zagryaznenny`x zhidkix sred* [Technical tools and methods for the regeneration of contaminated liquid medium] (Belgorod: BSTU VG Shukhova) pp 1532-5 [In Russian]

[20] Belykh M M, Shinkarev L I and Sevostyanov V S 2017 *Issledovanie i razrabotka texnicheskix sredstv dlya ochistki mashinny`x masel* [Research and development of technical tools for engine oils treatment] (Belgorod: BSTU) pp 542-7 [In Russian]

[21] Gridchin A M, Lesovik V S, Sevostyanov V S, Uralsky V I and Sinitsa E V 2006 Pomol`no-smesitel`ny`j agregat [Grinding and mixing unit] Pat. RF no 2277973 [In Russian]

[22] 2018 Kafedra texnologicheskix kompleksov, mashin i mexanizmov [Department of technological systems, machines and mechanisms] URL http://tkmm.bstu.ru

[23] 2018 Kommercheskaya deyatel`nost` [Commercial activity] URL: http://tkmm.bstu.ru/businesses

[24] 2018 Vse o fil`tracii vody` 2 - Fil`traciya cherez aktivirovanny`j ugol` [Everything about water filtration 2 – Filtration through activated carbon] URL http://aquatex.com.ua/blog/vse_o_filtratsii_vody_2s_filtratsiya_cherez_aktivirovannyj_ugol

[25] Ayukaev R I and Melzer V Z 1985 *Proizvodstvo i primenenie fil`truyushhix materialov dlya ochistki vody`* [Producing and using of filter materials for water treatment] (Leningrad: Stroiizdat) [In Russian]

[26] Knyazev V P 2005 *E`kologiya. Osnovy` restavracii* [Ecology. Basics of restoration] (Moscow: Architecture) [In Russian]

[27] Pamfilova K D et al 1985 *Posobie po proektirovaniyu sooruzhenij dlya ochistki i podgotovki vody`: SNiP 2.04.02-84. Texnologicheskie sxemy` i konstruktivny`e osobennosti tonkoslojny`x otstojny`x sooruzhenij* [Handbook for the design of facilities for water treatment and preparation of water: Building Regulations 2.04.02-84. Technological schemes and design features of thin-layer settling facilities] (Moscow: CITP Gosstroya SSSR) [In Russian] **24**

[28] Ostroumov S A 2005 *Zagryaznenie, samoochishhenie i vosstanovlenie vodny`x e`kosistem* [Pollution, self-treatment and restoration of aquatic ecosystems] (Moscow: MAKS Press) [In Russian]

[29] Gudkov A G 2003 *Mexanicheskaya ochistka stochny`x vod* [Mechanical wastewater treatment] (Vologda: VoGTU) [In Russian]

Resource and energy-saving technologies of complex processing and utilization of technogenic materials

N T Shein[1], V S Sevostyanov[2], V V Obolonsky[1], M V Sevostyanov[2], P Yu Goryagin[2] and V A Babukov[2]

[1] LLC TC «Ecotrans», Serafimovicha St., 72, Belgorod, 308006, Russia.
[2] Department of Technological complexes, machines and mechanisms, Belgorod State Technological University of V.G. Shukhov, Kostyukov St., 46, Belgorod, 308012, Russia.

E-mail: goryagin.pawel@yandex.ru

Abstract. At present, a very topical environmental problem is the increase in the formation of industrial and solid municipal waste (ISMW). This leads to a deterioration of the ecological situation, and, consequently, threatens human life and health, has a detrimental effect on the biosphere as a whole. Therefore, a special place in the organization of rational use of natural resources should take a comprehensive system of measures for the processing and utilization of various man-made materials that are the product of human production. The purpose of research and development projects is the creation of resource-saving technology, technological systems and special equipment for integrated processing and recycling of various man-made materials. One of the conditions for the implementation of resource-saving technology is the use of internal energy and material resources of the processing plant.

Keywords. Complex processing, utilization, technogenic materials, resource saving, energy saving, industrial and solid municipal waste.

1. Introduction

At present, a very urgent problem is the increase in the scale of solid municipal waste (MSW) and the urgent need for their integrated processing. Partial sorting of MSW with the subsequent disposal of non-utilizable waste at landfills [1] is a highly inefficient method of disposal [2, 3], and the existing technologies for the combustion of MSW are an additional source of global pollution by emissions [4]. In addition, any improvements in combustion technologies do not eliminate the formation of dioxins and furans, which, in turn, necessitates the subsequent disposal of significant volumes of harmful substances after cleaning the flue gases. Existing technologies are economically unprofitable and are realized only under conditions of state subsidies [5].

2. Technologies of complex processing and utilization of technogenic materials

In recent years, due to the steadily increasing environmental tensions in large cities, metropolitan areas, at the highest state level, more and more attention is paid to the development of resource-efficient technologies for complex processing of MSW [6-8].

Each technological direction of complex processing of MSW exists as a set of separate processes and is economically inefficient due to high cost (due to the need to acquire external energy and material resources).

In connection with the above, the purpose of research and experimental development is the creation and mastering of the technology of complex processing of MSW.

Research and production groups of LLC TC "Ecotrans" and BSTU named after V.G. Shukhov are carried out the development of resource-saving technologies and special equipment for complex processing of various technogenic materials [9, 10]: waste wood processing industry, organic waste (polymers, plastics, etc.), basalt waste, cullet, etc.

In this technology of complex processing of MSW, the implementation of various technological directions (flows), connected with each other and ensuring the use of internal energy and material resources is carried out. At the same time, there is no need to purchase these resources from outside. Combining the areas of complex processing with technological interconnections makes it possible to obtain a new technology that ensures low cost due to the use of domestic energy and material resources. The variety of the products obtained, which can be produced at low cost using the principle of technological interconnections, is quite large and is determined by the specific conditions of production (the nomenclature of waste, the need for any product, etc.). Figure 1 presents the block diagram of technological processes for the specific conditions of LLC TC "Ecotrans".

2.1. Areas of integrated processing of MSW

Taking into account the accumulated experience and the requirements of maximum economic efficiency for LLC TC "Ekotrans", the most appropriate are the following areas of integrated processing of MSW:

I. Reprocessing MSW to obtain RDF-fuel.

II. Production of briquetted products for various purposes from wood waste, with their subsequent processing to produce thermal or electrical energy in heat and power generators.

III. Production of hydrocarbon fuels from raw materials, resulting from thermolysis. This technology of processing of technogenic materials is implemented in LLC "Research and production enterprise Thermolysis".

IV. Recycling of polymeric wastes from MSW and production of silica-polymer-containing products from them.

V. Comprehensive processing and recycling of basalt fiber multifunctional wastes for the purpose of producing dry building mixes with fibro-fillers.

The main objective of the proposed technology for complex processing of MSW is to ensure a minimum amount of waste to be transferred to a landfill for disposal. This is possible with very careful separation of the mass of waste into components and the full use of the organic part of MSW. This requirement is implemented in the preliminary preparation of raw materials for subsequent processing. In the absence of sales / consumption of the products produced in any of the directions, the prepared raw materials are sold in other resource-saving technologies.

Technological operations of preparing MSW for further processing include: sorting of waste, grinding them to the required size and drying to 8-10% moisture. The prepared raw material is further processed in accordance with the requirements imposed on the final product. The final products of complex processing of solid chemical waste are:

- According to item I:

a) loose, non-caking mass with a particle size of not more than 25x25 mm - RDF-fuel for cement, lime and expanded-clay furnaces;

b) loose briquettes produced by a vibroimpulsive press - fuel for cement, lime and burned furnaces, supplied with a burned material.

- According to item II:

a) high-quality briquettes from crushed wood without inclusions of high-polymer materials

b) fuel dust and lower quality briquettes for own consumption from polluted wood with inclusions of various materials.

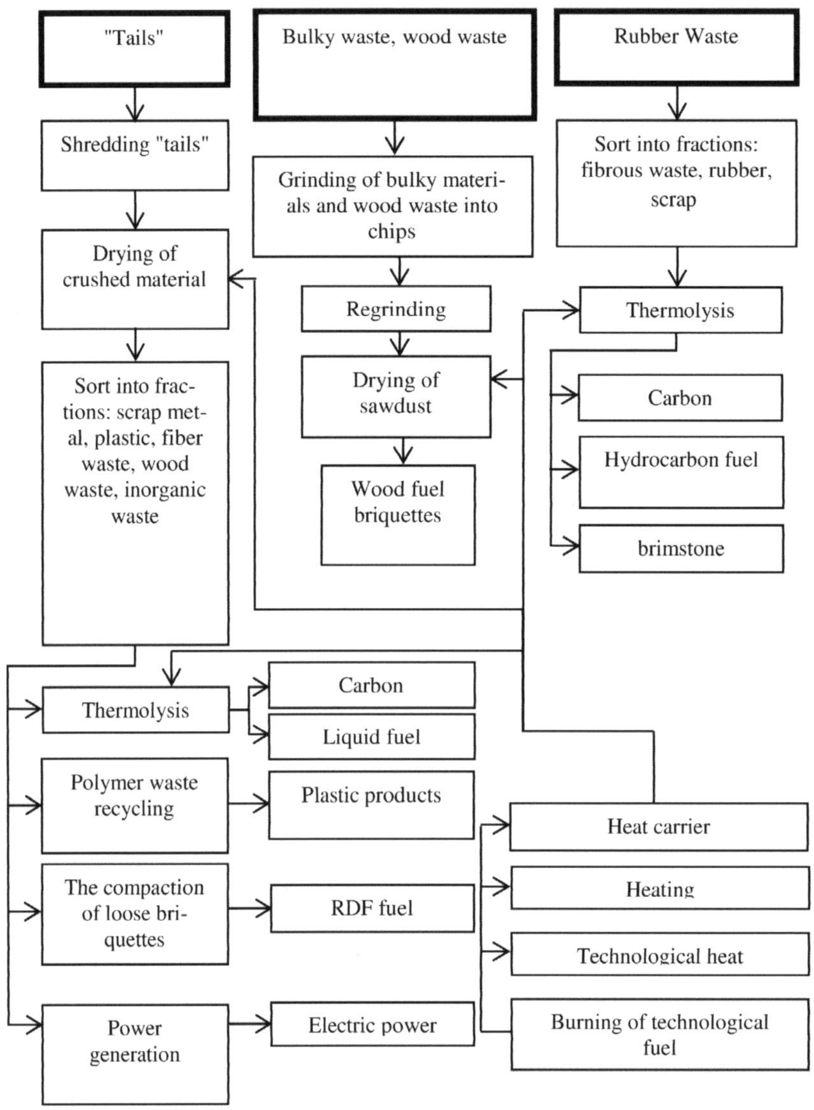

Figure 1. Scheme of separate process streams with internal technological connections.

- According to item III: electric power produced by a 200 kW diesel generator and light heating oil for diesel engines of excavators, bulldozers, tractors, cars, etc.
- According to item IV: plastic pipes, piece plastic products with various fillers, etc.
- According item V:

a) piece plastic products with a filler - micro powder from basalt waste, ready-mixed concrete and products from it with the addition of crushed basalt fibers - fiber fillers;

b) a heavy ground for pouring out the MSW on the landfill, obtained by styling of disposal basalt fiber with soil impurities.

Each of the technological operations of this technology has a technical, economic and environmental focus and can be used in real production. However, the implementation of the above technologies requires specialized equipment that takes into account the various physical and mechanical characteristics of man-made materials: various densities, geometrical sizes and forms, low flow ability, high humidity, etc.

2.2. Processing of technogenic fibrous materials

In the developed technology, innovative processes are implemented for the complex processing of MSW, such as high-speed grinding of wet raw materials in the stream, contact drying of crushed raw materials, thermolysis, etc. One of the promising areas in the field of resource conservation is the processing of anisotropic and complex in composition man-made fibrous materials, in particular basalt fiber waste [11]. By origin, they can be divided into the following groups: 1 - waste is generated during the production of insulating materials; 2 - waste from the construction industry; 3 - greenhouse waste, which are basalt fiber mats with rhizomes of plants, etc.

As a result of processing, basalt fiber can serve as a fiber filler of various composite mixtures and products. The peculiarity of basalt fiber processing is that the waste is initially a single conglomerate containing various inclusions (figure 2) [12, 13].

Figure 2. Waste of basalt fiber.

As a result of the conducted research to study the specifics of processing and utilization of basalt fibers, a technological complex was developed for the processing of technogenic fibrous materials (figure 3).

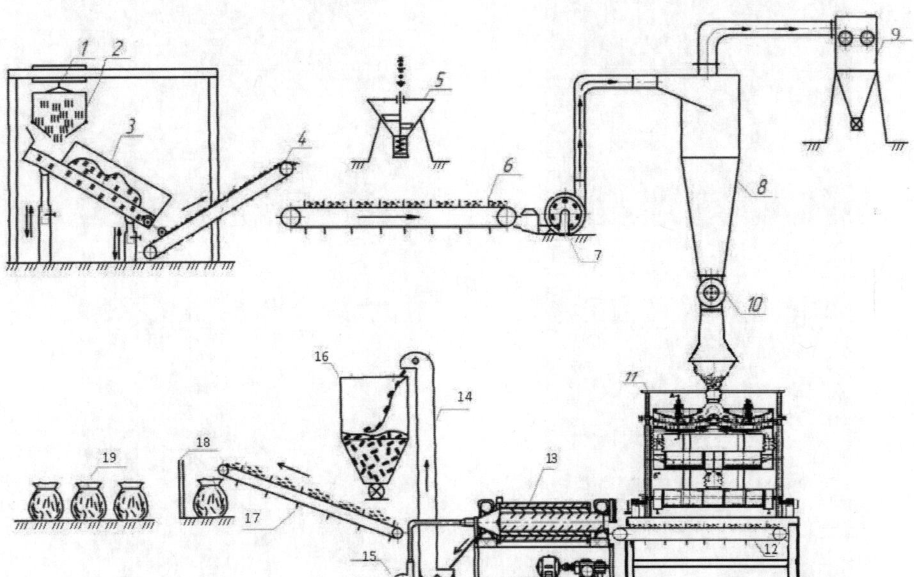

Figure 3. Technological complex for processing of technogenic fibrous materials: 1 - hoist; 2 - loading container; 3 - receiving tray; 4, 12, 17 - plate conveyor; 5 – additive dispenser; 6 - volumetric dosing; 7 - hammer crusher; 8 - cyclone; 9 - sleeve filter; 10 - cell feeder; 11 - vibration centrifugal unit; 13 – drum-screw unit-classifier; 14 - elevator; 15 – fan; 16 - bunker; 18 - scales; 19 –big bag.

IOP Conf. Series: Materials Science and Engineering **552** (2019) 012042 doi:10.1088/1757-899X/552/1/012042

The advantages of the technological complex developed by us are its versatility - the possibility of obtaining a wide range of products from various technogenic fibrous materials.

The vibration-centrifugal unit developed by us (figure 3, pos. 11) [14-16] provides complex processing of man-made fibrous materials and the production of various types of products for use in innovative technologies: DE agglomerated basalt waste (fiber length $-l = (5 \div 15) \cdot 10^{-3}$ м) dry construction mixtures, thermal insulation coatings, concrete products, etc; fine fiber filler ($l \leq (5 \div 10) \cdot 10^{-6}$м) -fillers of nanostructured composite mixtures (figure 4), mechanically activated of silica-containing composite mixtures with air and hydraulic binders; highly concentrated microfiber fillers in granular form, etc.

a)

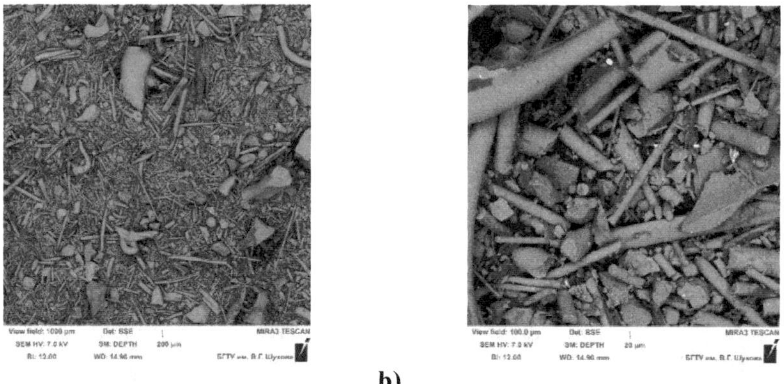

b)

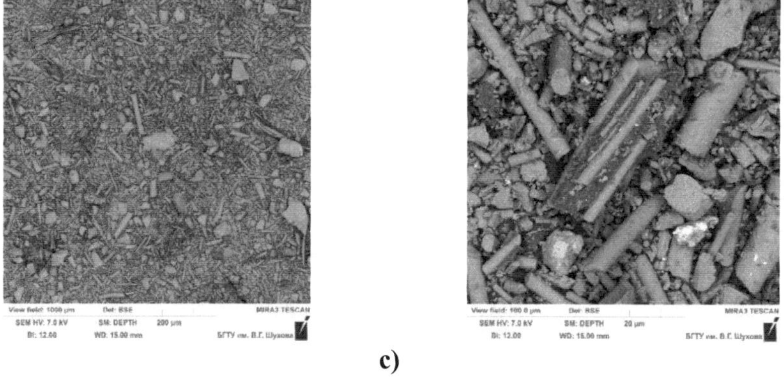

c)

Figure 4. Microfiber fillers from finely dispersed basalt fibers with varying degrees of magnification; grinding time: a) 5 min; b) 10 min; c) 20 min.

The resulting types of products can be used both in traditional building technologies and in promising innovative technologies, for example, in the production of architectural and building products using 3D printing.

One of the areas of use of basalt fiber as a fiber filler is the production of polymeric products for various purposes. Raw materials for the production of these products can be recycled technogenic polymeric materials. For the production by extrusion of compacted polymer products based on technogenic polymeric materials, a technological complex was developed (figure 5).

Technological complex works as follows: the source material using hoist 1 is fed to the receiving tray 2, in the lower part of which a belt conveyor 3 is installed. Then, using a plate conveyor, the raw material enters the rotor-centrifugal unit of combined action 4. In the coarse grinding chamber the material is crushed by disk mills and under the action of the air flow, moves to the fine grinding chamber, where it is ground by needle-milling working parts to a powder state. After the rotary centrifugal unit of the combined action, the raw material enters the separator 5 and is divided into fractions, large of which are returned to the grinding sludge. The material crushed to the required size enters the cyclone 6. When precipitated, it is dosed into the screw mixer 7, where it is mixed with additives supplied from the bunker 8. The composite mixture enters the receiving funnel 9 of the elevator 10, which raises the mixture into the receiving bunker 11. Using a cell feeder 12 in a certain dosage, the material enters the auger thermal heater 13, where, under the influence of high temperature, it turns into a homogeneous mass. Then it enters the extruder with an annular matrix 14. The extruded mixture through the plate conveyor 15 enters the drum-screw unit 16, where the material is cooled. The cooled granules are weighed on scales 17, packed in big bags 18 and shipped to the warehouse with the help of a hoist 19.

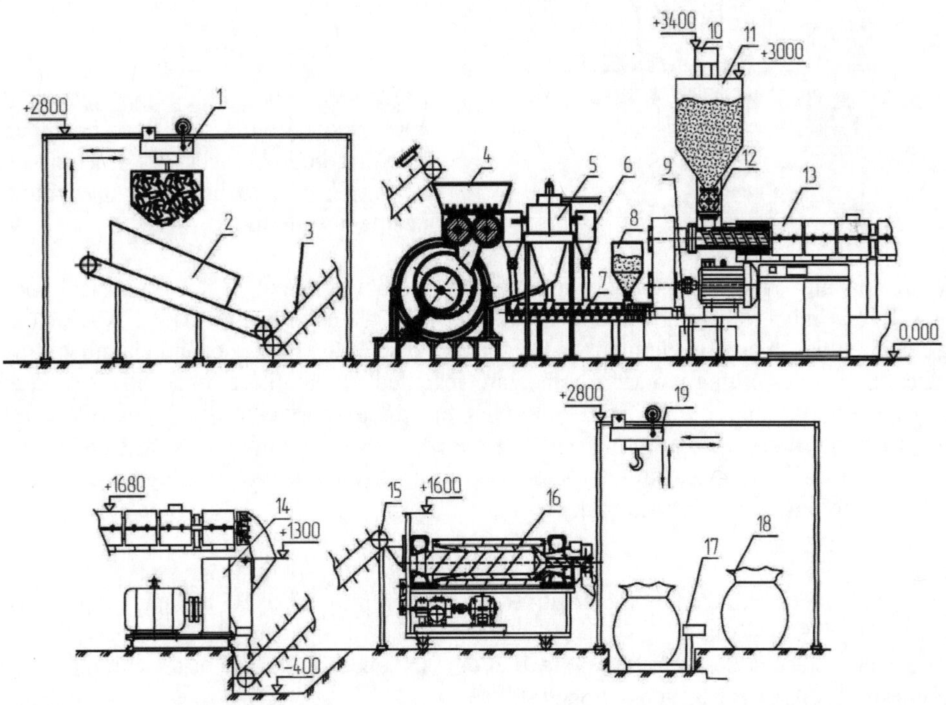

Figure 5. Technological complex for the production of extruded products from man-made polymeric materials: 1, 19 – hoist; 2 — receiving tray; 3, 15 – lamellar conveyor; 4 – rotary centrifugal unit of combined action; 5 – separator; 6 – cyclone; 7 – screw mixer; 8 – bunker of additional components; 9- receiving hopper; 10 – elevator; 11 – receiving hopper; 12 – cell feeder; 13 – screw thermal heater; 14 – extruder with ring die; 16 – drum-screw unit; 17 – scales; 18 – big bag.

In the processing of technogenic materials, one of the energy-intensive stages is the grinding process [17, 18]. In this regard, our research is aimed at the development and technological improvement of multifunctional grinding equipment. In the development of energy-efficient equipment, we paid special attention to the physical and mechanical characteristics of man-made polymeric materials such as: elasticity, plasticity, melting temperature, etc.

The combined-action rotor-centrifugal unit developed by us (figure 6) takes into account various technological conditions for processing technogenic polymeric materials, their physical and mechanical characteristics: the initial state of materials (shape, size, flow ability, etc.) and their strength characteristics [19-21]. The design of the unit allows realizing the possibility of varying the tearing and abrasive force, the degree of thermal heating of the material and its cooling. In the rotor-centrifugal unit of combined action, high durability of the most wear-out assemblies and parts of the unit is provided, the possibility of introducing additional components into the unit to obtain composite mixtures, the stepwise grinding process of man-made polymeric materials, the increased energy intensity of the second stage and the use of working bodies with a developed working surface - a needle cutter [22-24] and others. The rotor-centrifugal unit contains a body consisting of two crushing chambers 1 horizontally placed one behind the other and fine grinding 2 [25, 26].

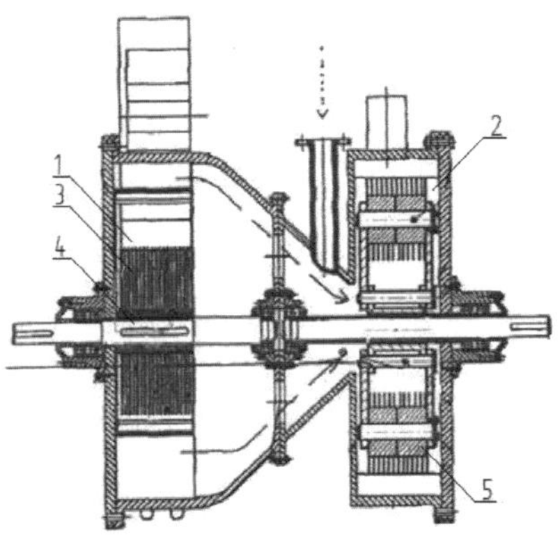

Figure 6. The rotor-centrifugal unit of combined action: 1 - crushing chamber; 2 - fine grinding chamber; 3 - crushing chamber shaft; 4 - crushing chamber rotor; 5 - needle milling tools.

Inside the crushing chamber 1, there is an eccentric set about the axis of the cylindrical body of the bandage, which is lined with removable elements (profiled plates). On the shaft 3 is the rotor of the crushing chamber 4 with cutting elements in the form of disc mills. In the second chamber (fine grinding chamber) 2, needle-milling working bodies are installed 5, which are made of a set of core elements assembled in bags and rigidly fixed on the holder. In the rotary centrifugal unit of the combined action, the principle of stepwise grinding with the realization of the combined effect on the processed material is implemented. This ensures the reduction of energy-intensive grinding process, as well as expands the technological capabilities of the unit.

3. Results

As a result of research conducted by the creative group of employees of BSTU named after V.G. Shukhov and LLC TC "Ecotrans":

1) an innovative resource-energy-saving technology and special equipment for complex processing of municipal solid waste have been developed;

2) promising areas of the developed technology have been identified;

3) the specificity of processing basalt wastes and technogenic polymeric materials was studied;

4) technological complexes for processing technogenic fibrous materials and producing a wide range of extruded products from technogenic polymeric materials have been developed;

5) special equipment has been developed that takes into account the specifics of processing anisotropic and technogenic materials that are complex in composition.

Acknowledgments
The article was prepared as part of the state assignment number 9.11523.2018/11.12.

References

[1] Trubaev P A, Verevkin O V, Grishko B M, Tarasyuk P N, Shchekin I I, Suslov D Yu and Ramazanov R S 2018 Investigation of Landfill Gas Output from Municipal Solid Waste at the Polygon *J. Phys.: Conf. Ser.* **1066** 012015
[2] Sverguzova S V 2008 *Promy`shlennaya e`kologiya kak nepremennoe uslovie stabil`nogo razvitiya* [Industrial ecology as an indispensable condition for stable development] (Belgorod: BSTU VG Shukhova) [In Russian]
[3] Orletskaya L V 2006 Cenny`e vtorichny`e resursy` [Valuable secondary resources] *Specializirovanny`j informacionno-analiticheskij zhurnal «Recikling otxodov»* **6** pp 3-4 [In Russian]
[4] Kornilova N V and Trubaev P A 2018 Analysis of MSW Combustion Temperature in a Hot Water Boiler with the Low-Capacity *J. Phys.: Conf. Ser.* **1066** 012003
[5] Koshlich Y, Trubaev P and Belousov A 2018 An approach of energy resources control system design *International Multidisciplinary Scientific GeoConference Surveying Geology and Mining Ecology Management SGEM* **Vol 18(2.1)** 63-9
[6] Sevostyanov M V 2016 Resursosberegayushhee oborudovanie dlya kompleksnoj pererabotki texnogenny`x materialov [Resource-saving equipment for complex processing of man-made materials] *Vestnik BGTU im. V G Shukhova* **4** pp 140-5 [In Russian]
[7] Sevostyanov V S, Shinkarev L I, Sevostyanov M V, Makridin A A and Solopov N V 2011 *Texnicheskie osnovy` pererabotki i utilizacii texnogenny`x materialov* [Technical basis for the processing and disposal of man-made materials] (Belgorod: BSTU VG Shukhova) [In Russian]
[8] Sevostyanov V S, Mikhailichenko S A, Sevostyanov M V, Spirin M N, Kolesnikov A V and Ural A V 2007 E`nergosberegayushhie texnologicheskie kompleksy` i agregaty` dlya utilizacii texnogenny`x materialov [Energy-saving technological complexes and units for the disposal of man-made materials] *Vestnik BGTU im. V G Shukhova* 1 pp 85-90
[9] Sevostyanov V S, Uralsky V I, Sevostyanov M V and Nosov O A 2015 *Tekhnologicheskie kompleksy i oborudovanie dlya pererabotki i utilizatsii tekhnogennykh materialov* [Technological complexes and equipment for processing and utilization of technogenic materials] (Belgorod: BSTU VG Shukhova) [In Russian]
[10] Sevostyanov V S, Ural'skiy V I, Sevostyanov M V, Babukov V A and Martakov I G 2018 *Malotonnazhnye tekhnologicheskie kompleksy i oborudovanie (osnovy nauchnykh issledovaniy – prakticheskoe rukovodstvo)* [Small-tonnage technological complexes and equipment (the basis of scientific research - practical guidance)] (Belgorod: BSTU VG Shukhova) [In Russian]
[11] 2018 *Bazal`tovy`e texnologii* [Basalt technologies] URL http://www. basaltech.org.
[12] Sivachenko L A, Rovsky A M and Reutsky I A 2016 Cepny`e texnologicheskie agregaty` mnogocelevogo naznacheniya i ix razvitie [Chain technological units of multi-purpose use and their development] *Vestnik Belorussko-Rossiyskogo universiteta* **1** pp 78-86 [In Russian]
[13] Sevostyanov V S, Sivachenko T L, Ilina T N, Sivachenko L A, Sevostyanov M V and Mykhailichenko S A 2015 Igol`chaty`j izmel`chitel` materialov [Needle-type material grinder] Patent RF no 151633 [In Russian]
[14] Sinitsa E V, Sevostyanov V S, Uralskiy V I and Uralskiy A V 2013 *Teoreticheskie osnovy` sozdaniya centrobezhny`x izmel`chitelej s selektivny`m dinamicheskim vozdejstviem na material* [Theoretical basis for the creation of centrifugal grinders with selective dynamic effects on the material] (Belgorod: BSTU VG Shukhova) [In Russian]
[15] Sevostyanov M V, Poluektova V A, Sevostyanov V S, Sirota V V, Uralskiy V I, Martakov I G and Babukov V A 2018 Ustrojstvo i sposob pererabotki voloknisty`x texnogenny`x

materialov dlya polucheniya fibronapolnitelej (varianty`) [A device and method for processing fibrous technogenic materials to produce fibro-fillers (versions)] *Application for a Patent RF* [In Russian]

[16] Babukov V A, Belyaev D P and Sevostyanov M V 2018 Vibro-centrobezhny`j agregat kombinirovannogo dejstviya dlya pererabotki bazal`tovy`x otxodov [Vibro-centrifugal unit of combined action for the processing of basalt waste] *X Mezhdunarodny`j molodezhny`j forum «Obrazovanie. Nauka. Proizvodstvo»* (Belgorod: BSTU VG Shukhova) [In Russian]

[17] Ovchinnikov P F 1993 Sravnenie razlichny`x vidov izmel`chitelej [Comparison of different types of shredders] *Mater. pervoj konf. po sravneniyu razlichny`x vidov izmel`chitelej* (Odessa: OSMA) Part 1 [In Russian]

[18] Sivachenko L A, Sharoykina E A, Sevostyanov V S and Sivachenko T L 2010 Voprosy` razvitiya texniki i texnologii izmel`cheniya materialov [Questions of the development of technology and technology of grinding materials] *Energosberegayushchie tekhnologicheskie kompleksy i oborudovanie dlya proizvodstva stroitel'nykh materialov: mezhvuz. sb. st.* (Belgorod: BSTU VG Shukhova) pp 315-21 [In Russian]

[19] Gogol E V, Mingazetdinov I Kh, Gumerova G I, Yegorova O S, Maltsev S A, Grigorieva I G and Tunakova Yu A 2013 Analiz sushhestvuyushhix sposobov utilizacii i pererabotki otxodov polimerov [Analysis of existing methods of disposal and recycling of polymers waste] *Vestnik Kazanskogo texnologicheskogo universiteta* **10** pp 163-8 [In Russian]

[20] Sevostyanov V S, Nosov O A, Berezhnoy O L and Vardanyan G R 2015 Texnika i texnologiya izmel`cheniya vtorichnogo polimernogo sy`r`ya [Technique and technology of grinding secondary polymeric raw materials] *E`nergo- i resursosberegayushhie e`kologicheski chisty`e ximiko-texnologicheskie processy` zashhity` okruzhayushhej sredy`: Sb. dokl. mezhdunarodnoj nauchno-prakticheskoj konferencii* (Belgorod: BSTU VG Shukhova) pp 50-4 [In Russian]

[21] Sevostyanov V S, Nosov O A, Vardanyan G R and Berezhnoy O L 2015 Pererabotka ispol`zovannoj polie`tilenovoj tary` [Recycling of used plastic packaging] *Sbornik dokladov V Mezhdunarodnoj nauchno-prakticheskoj konferencii «E`kologiya: obrazovanie, nauka, promy`shlennost` i zdorov`e»* (Belgorod: BSTU VG Shukhova) pp 152-6 [In Russian]

[22] Sevostyanov V S, Sivachenko T L and Mikhailichenko S A 2015 Texnologicheskie apparaty` s iglofrezerny`mi rabochimi organami dlya kompleksnoj pererabotki kompozicionny`x materialov [Technological devices with needle-milling workers for complex processing of composite materials] *Vestnik BGTU im. V G Shukhova* **2** pp 50-6 [In Russian]

[23] Sevostyanov V S and Sivachenko T L 2016 Iglofrezerny`e izmel`chiteli, ix texnologicheskie vozmozhnosti i puti razvitiya [Needle milling grinders, their technological capabilities and ways of development] *Vestnik Belorussko-Rossiyskogo universiteta* **1** pp 69-77 [In Russian]

[24] Sevostyanov V S, Sivachenko T L, Sevostyanov M V, Goryagin P Yu and Babukov V A 2018 Nauchno–prakticheskie osnovy` sozdaniya iglofrezerny`x izmel`chitelej mnogocelevogo naznacheniya [Scientific and practical bases for creating multi-purpose needle-shredded shredders] *Vestnik BGTU im. V G Shukhova* **3** pp107-16 [In Russian]

[25] Goryagin P Yu, Ermilov R A 2017 Texnika i texnologii dlya pererabotki texnogenny`x polimerny`x materialov [Techniques and technologies for processing technogenic polymeric materials] *Molodezh` i nauchno-texnicheskij progress: Sb. dokl. X mezhdunarodnoj nauchno-prakticheskoj konferencii studentov, aspirantov i molody`x ucheny`x. v 4 t.* (Gubkin: LLC Assistant Plus) pp 83-8 [In Russian]

[26] Glagolev S N, Gridchin A M, Sevostyanov V S, Mikhaylichenko S A and Makridin A A 2012 Ustanovka dlya izmel`cheniya voloknisty`x materialov [Installation for grinding fibrous materials] Patent RF no 2446015 [In Russian]